de Gruyter Studies in Mathematics 12

Andrei Borisovich Shidlovskii

Transcendental Numbers

With a foreword by W. Dale Brownawell

Translated from the Russian by Neal Koblitz

W DE G Walter de Gruyter
Berlin · New York 1989

Author
Andrei Borisovich Shidlovskii
Faculty of Mechanics and Mathematics
Department of Number Theory
Moscow State University
119899 Moscow, V-234, USSR

Translator
Neal Koblitz
Department of Mathematics
University of Washington
Seattle, WA 98195
USA

Title of the Russian original edition: *Transtsendentnye chisla.* Publisher: Nauka, Moscow 1987

1980 Mathematics Subject Classification (1985 Revision):
Primary: 11-02; 11J81. Secondary: 11J82; 11J85.

Library of Congress Cataloging-in-Publication Data

Shidlovskiĭ, A. B.
[Transtsendentnye chisla. English]
Transcendental numbers / Andrei Borisovich Shidlovskiĭ ; translated from the Russian by Neal Koblitz.
p. cm. – (De Gruyter studies in mathematics ; 12)
Translation of: Transtsendentnye chisla.
Includes bibliographical references.
ISBN 0-89925-437-3 (U.S. : alk. paper)
1. Numbers, Transcendental. I. Title. II. Series.
QA247.5.S4813 1989
512'.73–dc20 89-17116
CIP

Deutsche Bibliothek Cataloging-in-Publication Data

Šidlovskij, Andrej Borisovič:
Transcendental numbers / Andrei Borisovich Shidlovskii. Transl. from the Russian by Neal Koblitz. – Berlin ; New York : de Gruyter, 1989
(De Gruyter studies in mathematics ; 12)
Einheitssacht.: Transcendentnye čisla <engl.>
ISBN 3-11-011568-9
NE: GT

⊚ Printed on acid-free paper.

Typesetting with TEX: Neal Koblitz, Seattle, and Danny Lee Lewis, Berlin. Printing: Gerike GmbH, Berlin. Binding: Dieter Mikolai, Berlin. Cover design: Rudolf Hübler, Berlin.

Foreword

In a series of fundamental papers beginning in 1954, A. B. Shidlovskii greatly extended the theoretical scope of Siegel's original method for algebraic independence and proved the algebraic independence of large classes of explicitly given numbers. The development of the resulting Siegel–Shidlovskii theory has been carried out almost exclusively by Shidlovskii and his students.

Three possible interrelated reasons may have contributed to the relative lack of Western participation (Lang, Mahler and Väänänen are notable exceptions) in this development. First, since the fundamental classical results require many ingenious technical considerations not met elsewhere, the newcomer may have more difficulty than usual in understanding the basic principles of the method and in isolating potential areas for original research. Secondly, interesting connections to other branches of mathematics had not been very apparent. Finally, a set of central problems had not been collected and publicized to attract newcomers.

Professor Shidlovskii's present book pays careful attention to the first point. His many successful students certify that he knows how to lead novices step-by-step into the heart of the method. Even in the development of quite standard results, the book oftentimes brings simplifications and improvements of the original presentations. Moreover, the historical comments at the ends of the chapters serve as a guide to the original literature.

As an example of how completely this literature is covered, we note that, in the final chapter, one of the classics of the field becomes available to the Western reader for the first time. The elegant notion of an irreducible set of functions was one of Shidlovskii's earliest contributions, and it has been revived recently to obtain effective bounds on measures of algebraic independence. However, not only was the basic paper on irreducibility never translated; it was published in a journal which is but rarely available in Western libraries.

The Siegel–Shidlovskii theory was developed for its own intrinsic interest and with its own generally elementary, although ingenious, techniques. However, interaction with other major areas of mathematics increasingly characterizes the various newer developments. It may be useful to examine this phenomenon a bit more closely.

For example, the proofs given here of several of Salikhov's previously unpublished results invoke basic ideas about algebraic functions and differential operators on polynomial rings. The notions of linear and algebraic irreducibility given by Oleinikov and Salikhov are interesting also from the point of view of differential equations. Their applications in the Siegel–Shidlovskii theory involve asymptotic expansions or formal solutions of differential equations. Ramis' study of the coefficient growth of formal solutions (Gevrey classes) has shown that $\mathbf{E} = \mathbf{E}^*$ (cf. p. 407).

In contrast to the generally complete coverage of the classical Siegel–Shidlovskii theory, the text omits some important work of Nesterenko because of its extensive prerequisites. Nesterenko builds on commutative algebra and classical algebraic geometry to bound the order of zero of a polynomial in z and the coordinates of an $n \times n$ system (A) of linear homogeneous first order differential equations $Y' = AY$ over $\mathbf{C}(z)$. His work has made a deep impact on other branches of transcendental number theory. It has also proved useful in establishing bounds for the computational complexity of certain algorithms in commutative algebra, and it has even furnished the basis for the solution of problems of classical commutative algebra via the effective versions of Hilbert's Nullstellensatz. As part of the argument for the bound on the order of zero, Nesterenko establishes a natural correspondence between certain polynomial ideals and G-invariant varieties of solutions of (A) satisfying them, where G is the differential Galois group of (A).

The notion of the differential Galois group G is a powerful concept which should continue to play a major role in future developments. It may be defined in the following way: Let F denote the field generated over $\mathbf{C}(z)$ by a fundamental solution of (A). Then F is closed under differentiation, and G is defined to be the group of all $\mathbf{C}(z)$-automorphisms of F which commute with differentiation and which fix $\mathbf{C}(z)$. Identifying G with its action on the space V of solutions of (A), Kolchin showed that G is an algebraic subgroup of $GL(n)$ whose dimension equals the transcendence degree of F over $\mathbf{C}(z)$. He also established a Galois correspondence analogous to the ordinary one; so knowledge about the structure of the groups G provides a tool to investigate F and (A).

For example, the differential Galois group was the key to understanding Siegel's concept of normality for the system (A) above. Although Siegel introduced this notion in 1949, it was awkward to state, and it found no real application even within the Siegel–Shidlovskii theory since it appeared virtually impossible to verify. However Beukers, Brownawell and Heckman recently showed that the essential part of Siegel's original condition can be expressed very naturally in terms of G, namely that the action of G on V be a direct sum of inequivalent irreducible representations. Consequently, the relevant condition for algebraic independence is that (A) decompose into a direct sum of irreducible, pair-wise unrelated systems, where the differential Galois group of a $k \times k$ system contains either $SL(k)$ or $Sp(k)$. This turns out to be a condition which can then be verified for large classes of generalized hypergeometric equations.

This indication of the newer developments falling outside the scope of the book is meant to bring the readers of this translation more up-to-date and, at the same time, to illustrate and perhaps encourage the growing involvement of other areas in the Siegel–Shidlovskii theory. To help focus attention for future investigation, let us collect here a few of the outstanding complexes of open problems of this and related areas:

1) Prove results on algebraicity, irrationality or transcendence of values assumed near the border of convergence by generalized hypergeometric functions with rational parameters, which are not E-functions, but only G-functions. Wolfart has begun, and Beukers and Wolfart have continued, this investigation for the classical Gaussian hypergeometric functions. Also in this connection see the recent book of André on G-functions.

2) Contrary to the general theme of the Shidlovskii program, the values of given algebraically independent coordinates of an E-function solution for a system (A) may fail to generate a field of maximal transcendence degree at an exceptional set of algebraic numbers (pp. 149, 150). Determine this exceptional set in the general case. Compare the results of Wolfart mentioned above.

3) Investigate Siegel's conjecture on the generation by hypergeometric functions of the algebra of E-functions satisfying linear differential equations over $\mathbf{C}(z)$. Start with the special case of first order non-homogeneous equations (p. 184). Investigate the situation for G-functions, which are conjectured to "come from geometry."

4) Reformulate Shidlovskii's criterion of irreducibility (p. 406) in terms of the differential Galois group. Specifically, characterize Shidlovskii irreducibility for hypergeometric equations in terms of their parameters.

5) Determine the differential Galois groups of differential fields generated by solutions of (possibly more than) one generalized hypergeometric differential equation. In particular, determine the differential Galois group of reducible hypergeometric equations. For a single equation, see the recent work of Gabber, Katz, Beukers, Heckman, Ramis, Boussel and others. Determine the transcendence degree over $\mathbf{C}(z)$ of differential fields generated by (several) generalized hypergeometric functions.

Professor Shidlovskii's book will be recognized as a classic of the Siegel–Shidlovskii theory. Its accessible introduction to and authoritative survey of the field provide a solid foundation for continuing progress. The challenging questions which remain should provide stimulation for further generations.

University Park, August 1989 *W. D. Brownawell*

Preface to the English edition

The English translation is essentially identical in content with the Russian original published by Nauka in 1987. Some misprints and inaccuracies have been removed, and some interesting contributions made since the publication of the original version are summarized in an appendix at the end of the book. Most important here are the papers [8] by W. D. Brownawell and [104] by Yu. V. Nesterenko on effective estimates of algebraic independence measures of the values of E-functions; the papers [107: 1–3] by V. Kh. Salikhov on hypergeometric functions, and the paper [91] by F. Beukers, W. D. Brownawell, and G. Heckman.

I would like to express my deep gratitude to Professor Neal Koblitz for his excellent translation. I am also grateful to Walter de Gruyter & Co. for making the book available to those mathematicians who do not know Russian.

Moscow, June 1989 *A. B. Shidlovskii*

Preface

This book is devoted to one of the directions of research in the theory of transcendental numbers. It includes an exposition of the fundamental results concerning the arithmetic properties of the values of E-functions which satisfy linear differential equations with coefficients in the field of rational functions.

The notion of an E-function was introduced in 1929 by Siegel, who created a method of proving transcendence and algebraic independence of the values of such functions. An E-function is an entire function whose Taylor series coefficients with respect to z are algebraic numbers with certain arithmetic properties. The simplest example of a transcendental E-function is the exponential function e^z. In some sense Siegel's method is a generalization of the classical Hermite–Lindemann method for proving the transcendence of e and π and obtaining some other results about arithmetic properties of values of the exponential function at algebraic points.

In the course of the past 30 years, Siegel's method has been further developed and generalized. Many papers have appeared with general theorems on transcendence and algebraic independence of values of E-functions; estimates have been obtained for measures of linear independence, transcendence and algebraic independence of such values; and the general theorems have been applied to various classes of concrete E-functions. The need naturally arose for a monograph bringing together the most fundamental of these results. The present book is an attempt to meet this need.

The first chapter contains the simplest information about approximation of real and algebraic numbers which relates to the theme of the book. The second chapter presents the results of Hermite and Lindemann. We give several different methods of proof.

In the third chapter, we develop the method which serves as the corner stone for this branch of transcendence theory. We prove several general theorems about transcendence and algebraic independence of the values of E-functions in the case when these functions are not connected by algebraic equations over the field of rational functions. In the fourth chapter, the method in the third chapter is generalized to the case when the E-functions are connected by algebraic equations over the field of rational functions.

Chapters 5–10 describe various methods of proving that some functions are algebraically independent over the field of rational functions. This enables us to apply the general theorems in Chapters 3–4 to concrete classes of E-functions.

Chapters 11–13 are devoted to estimates for measures of linear independence, transcendence and algebraic independence of the values of E-functions in different cases.

The concluding remarks give a survey of some results obtained by various authors which are somewhat close to the contents of the earlier chapters. The book ends with an extensive bibliography. In order for the book to be usable by a wide circle of readers, the proofs are given in considerable detail. Some supplementary facts are stated without proof. Most of the chapters conclude with some remarks about related research.

The lemmas, theorems and formulas are numbered independently in each chapter. When referring to formulas of other chapters, the chapter number is given before the formula number. For example, (3.62) refers to formula (62) of Chapter 3. In references to the bibliography, two numbers are given: the author number and the article number. For example, [2:3] refers to the third article under A. Baker (who is author number 2 in the bibliography).

The author dedicates this book to the memory of his parents Boris A. Shidlovskii (1884-1942) and Aleksandra V. Shidlovskaia (1887-1976).

The author wishes to thank V. Kh. Salikhov and V. V. Kazakov for allowing him to use their as yet unpublished work on linear irreducibility of differential equations (see Chapter 10); V. A. Oleinikov for her suggestions and help; and O. N. Vasilenko, A. I. Galochkin, Yu. N. Makarov, V. Kh. Salikhov and V. G. Chirskii for reading chapters of the manuscript and making corrections.

The author owes a special debt of gratitude to the editor, Yu. V. Nesterenko, for his meticulous editing, for several important remarks and improvements in the exposition, and for giving the author his proofs of Siegel's Theorem in Chapter 9 and Theorem 8 in Chapter 10.

A. Shidlovskii

Contents

Notation

$\{a\}$	—the set consisting of the element a
$\{a_1, \ldots, a_n\}$	—the set consisting of $a_1, \ldots, a_n$
$\emptyset$	—the empty set
$A \setminus B$	—the set of elements in A not in B
$A \times B$	—the cartesian product of the sets A and B
$\{x\}$	—the fractional part of x
$[x]$	—the integral part of x
$\mathbf{N}$	—the set of natural numbers
$\mathbf{Z}$	—the ring of rational integers
$\mathbf{Z}^+$	—the set of nonnegative rational integers
$\mathbf{Q}$	—the field of rational numbers
$\mathbf{R}$	—the field of real numbers
$\mathbf{R}^m$	—m-dimensional real Euclidean space
$\mathbf{C}$	—the field of complex numbers
$\mathbf{A}$	—the field of all algebraic numbers
$\mathbf{K}$	—a fixed algebraic number field over $\mathbf{Q}$
$[\mathbf{K} : \mathbf{Q}]$	—the degree of the algebraic extension $\mathbf{K}$ over $\mathbf{Q}$
$\mathcal{N}(\alpha)$	—the norm of an element α of $\mathbf{K}$
$\mathbf{Z}_\mathbf{A}$	—the ring of all algebraic integers
$\mathbf{Z}_\mathbf{K}$	—the ring of algebraic integers in the field $\mathbf{K}$
$\mathbf{I}$	—an imaginary quadratic field
$\mathbf{V}$	—an arbitrary field or ring
$\mathbf{V}[x_1, \ldots, x_m]$	—the ring of polynomials in $x_1, \ldots, x_m$ over the field (ring) $\mathbf{V}$
$\mathbf{V}(x_1, \ldots, x_m)$	—the field of rational functions in $x_1, \ldots, x_m$ over the field $\mathbf{V}$
$\deg f(x)$	—the degree of the polynomial $f(x)$
$\deg_\mathbf{x} f(x_1, \ldots, x_m)$	—the total degree of $f(x_1, \ldots, x_m)$ in the variables $x_1, \ldots, x_m$
$\deg_{x_k} f(x_1, \ldots, x_m)$	—the degree of $f(x_1, \ldots, x_m)$ in x_k
H_f	—the height of the polynomial $f = f(x_1, \ldots, x_m) \in \mathbf{C}[x_1, \ldots, x_m]$ (the maximum modulus of its coefficients)
$\overline{\|f\|}$	—the size of a form or polynomial f with coefficients in $\mathbf{K}$ (the maximum modulus of the coefficients and all of their conjugates in $\mathbf{K}$)
$\deg \alpha$	—the degree of the algebraic number α
H_α	—the height of the algebraic number α
$\overline{\|\alpha\|}$	—the size of the algebraic number α (the maximum modulus of the conjugates of α in $\mathbf{K}$)

$\deg \operatorname{tr}^0_{\mathbf{V}} \mathbf{U}$	—the homogeneous transcendence degree of the set **U** over the field **V**
$\deg \operatorname{tr}_{\mathbf{V}} \mathbf{U}$	—the transcendence degree of the set **U** over the field **V**
$\deg \operatorname{tr}^0 \mathbf{U}$	—the homogeneous transcendence degree of **U** over **Q**
$\deg \operatorname{tr} \mathbf{U}$	—the transcendence degree of the set **U** over **Q**
$\ker \varphi$	—the kernel of the vector space homomorphism φ
$\operatorname{im} \varphi$	—the image of the mapping of sets φ
$\dim L$	—the dimension of the vector space L
$\dim \mathcal{I}$	—the dimension of the ideal $\mathcal{I}$
$\operatorname{ord} f(z)$	—the order of zero of the analytic function $f(z)$ at $z = 0$
E	—the class of E-functions which satisfy linear differential equations with coefficients in $\mathbf{C}(z)$
$\mathbf{E}^*$	—the analogous class of E-functions with the second definition of E-function
$\Vert a_{i,k} \Vert_{i,k}$	—the matrix with entries $a_{i,k}$
$\begin{pmatrix} a_{1,1} & \dots & a_{1,n} \\ \dots & \dots & \dots \\ a_{m,1} & \dots & a_{m,n} \end{pmatrix}$	—the matrix with entries $a_{1,1}, \dots, a_{m,n}$
$\vert a_{i,k} \vert_{i,k}$	—the determinant of the matrix with entries $a_{i,k}$
$L_{\mathbf{K}}(\alpha_1, \dots, \alpha_m; H)$	—the linear independence measure of the numbers $\alpha_1, \dots, \alpha_m$ relative to the field **K**
$\Phi_{\mathbf{K}}(\alpha; s; H)$	—the transcendence measure of α relative to **K**
$\Phi^0_{\mathbf{K}}(\alpha_1, \dots, \alpha_m; s; H)$	—the homogeneous algebraic independence measure of $\alpha_1, \dots, \alpha_m$ relative to **K**
$\Phi_{\mathbf{K}}(\alpha_1, \dots, \alpha_m; s; H)$	—the algebraic independence measure of $\alpha_1, \dots, \alpha_m$ relative to **K**

If $\mathbf{K} = \mathbf{Q}$, then the subscript **K** is omitted in the above notation.

In this book, square brackets are not used for grouping algebraic expressions. They are used for references to the bibliography, and also for certain special symbols, of which the most frequently used is the symbol for the integral part of a real number.

Introduction

§ 1. Approximation of algebraic numbers

We start with a brief survey of the basic directions of research in the theory of transcendental numbers. This book will be concerned with the results of one of these directions.

A number α is said to be *algebraic* if it is a root of a polynomial

$$f(x) = a_n x^n + \cdots + a_1 x + a_0, \qquad f(x) \not\equiv 0,$$

with rational coefficients. It is easy to prove that a root of a polynomial $f(x)$ with algebraic coefficients, $f(x) \not\equiv 0$, is an algebraic number.

A complex number α (which may be a real number) is said to be *transcendental* if it is not algebraic. Thus, a transcendental number is not a root of any polynomial with algebraic coefficients which is not identically zero.

Irrational numbers were studied even by the ancient Greeks. For example, Euclid's *Elements* contains a proof that $\sqrt{2}$ is irrational.

More generally, problems involving the arithmetic properties of various numbers have been studied for a very long time. An example is "squaring the circle," which is a famous problem of antiquity. This is the problem of constructing with ruler and compass a square having the same area as a given circle. If we set the radius of the circle equal to 1 unit, then this problem reduces to determining whether it is possible to construct with ruler and compass a segment of length $\sqrt{\pi}$. Over a period of centuries, attempts to square the circle led to increased interest in the properties of the number π.

In 1766, J. H. Lambert [36:1] obtained the first proof that π is irrational. He used the continued fraction expansion of $(e-1)/2$ that had been found by Euler in order to obtain the continued fraction expansions of the functions $\tan(x)$ and $(e^x - e^{-x})/(e^x + e^{-x})$, with which he was then able to show that the numbers $\tan(x)$ and e^x are irrational for any rational nonzero x. Since $\tan(\pi/4) = 1$, this implies, in particular, that π is irrational. We note that Lambert's proof was not completely rigorous. It was Legendre [38:1] who later made the proof rigorous, at the same time proving irrationality of a whole class of continued fractions which includes the ones considered by Lambert.

Many famous mathematicians made conjectures about the existence of transcendental numbers. For example, in 1744 in his book [16:1] Euler stated without proof that a number of the form $\log_a b$, where a and b are rational numbers with b not equal to a rational power of a, must be a transcendental number.

The first rigorous proof of the existence of transcendental numbers was due to Liouville in 1844 [41:1,2]. He proved the following theorem: *if α is a root of an irreducible polynomial with rational coefficients of degree n, $n \geq 1$, then there exists a constant $c(\alpha) > 0$ such that the following inequality holds for any rational integers p and q, with $q > 0$ and $p/q \neq \alpha$:*

$$|\alpha - \frac{p}{q}| > \frac{c(\alpha)}{q^n}.$$

This theorem tells us that an algebraic number cannot be approximated "too closely" by rational numbers. Since it is easy to give examples of irrational numbers having "arbitrarily close" approximations by rational numbers, Liouville's theorem made it possible for the first time to construct examples of transcendental numbers.

In 1874, Cantor [9:1] gave another proof of the existence of transcendental numbers, using set theory. He showed that the set of all algebraic numbers is countable, while the set of real numbers is uncountable, from which it follows that transcendental numbers exist. Moreover, almost all numbers (in the sense of Lebesgue measure) are transcendental.

In 1909, Thue [78:1,2] developed a method that could be used to significantly strengthen Liouville's theorem. He was able to prove a *finiteness theorem for the number of solutions of a Diophantine equation of the form*

$$a_n x^n + a_{n-1} x^{n-1} y + \cdots + a_1 x y^{n-1} + a_0 y^n = m, \qquad n \geq 3, \tag{1}$$

in which the left hand side is an irreducible binary form with integer coefficients and m is an integer. Several mathematicians have refined and generalized Thue's theorem.

A number of methods for proving transcendence are based on the study of approximations of algebraic numbers. For example, Mahler [42:3] proved that *if $P(x)$ is a polynomial which takes natural number values for $x = 1, 2, \ldots$, then the following number is transcendental for any q:*

$$\alpha = 0.q_1 q_2 \ldots q_n \ldots,$$

where q_n is the base-q expansion of the integer $P(n)$. In particular, taking $P(x) = x$ and $q = 10$, we see that

$$\alpha = 0.12345678910111213\ldots$$

is a transcendental number.

Liouville's theorem set the stage for a fundamental direction of research – the theory of Diophantine approximation, which concerns approximation of algebraic numbers and is closely connected to the study of Diophantine equations and transcendental numbers. The theory of approximations of algebraic numbers is usually regarded as part of transcendental number theory.

This book includes only some of the simplest facts about the approximation of real and algebraic numbers by rationals and by algebraic numbers, namely, the facts which are most closely connected with the questions we shall be considering.

§ 2. The classical method of Hermite–Lindemann

The examples of transcendental numbers that were constructed using Liouville's theorem were all in a rather narrow class of numbers. One usually encountered significant difficulties proving transcendence of the numbers that arise naturally in pure and applied mathematics, numbers such as e, π, e^{π}; in those cases it was possible to prove transcendence only by using analytic methods. In transcendental number theory in general, most of the fundamental results about the arithmetic properties of numbers are proved by analytic methods.

The first such method for the exponential function e^z was created by Hermite in 1873 [27:1]. He used this method to prove that e is transcendental. Developing Hermite's method further, Lindemann in 1882 [40:1,2] proved the transcendence of π. Since it was well known that segments that can be constructed by ruler and compass have lengths in a certain set of algebraic numbers (see Wantzel [86:1]), Lindemann thereby proved the impossibility of squaring the circle.

Lindemann also proved that *if ξ is a nonzero algebraic number, then e^{ξ} is transcendental.* From this it followed that, *if ξ is an algebraic number not equal to 1, then* $\ln \xi$ *is transcendental.*

All of these results of Hermite and Lindemann are contained in a general theorem proved by Lindemann and known by his name:

If $\alpha_1, \dots, \alpha_n$ are distinct algebraic numbers, and if $c_1, \dots, c_n$ are algebraic numbers not all zero, then

$$c_1 e^{\alpha_1} + \cdots + c_n e^{\alpha_n} \neq 0.$$

Complex numbers $\alpha_1, \dots, \alpha_m$ are said to be *algebraically independent* if $P(\alpha_1, \dots, \alpha_m) \neq 0$ for any polynomial $P = P(x_1, \dots, x_m)$ with algebraic coefficients which is not identically zero. If this is not the case, then the numbers $\alpha_1, \dots, \alpha_m$ are said to be *algebraically dependent.* The concept of algebraic independence is a generalization of the notion of transcendence. If a set of numbers is algebraically independent, then each of them is transcendental.

Lindemann's theorem is equivalent to the following:

If $\xi_1, \dots, \xi_m$ are algebraic numbers which are linearly independent over the field of rational numbers, then the numbers $e^{\xi_1}, \dots, e^{\xi_m}$ are algebraically independent.

Thus, the Hermite–Lindemann method was used to give a complete answer to the question of transcendence and algebraic independence of the values of the exponential function at algebraic points.

After the work of Hermite and Lindemann, attempts were naturally made to generalize their methods and results to other functions. But nothing succeeded for almost half a century. In the meantime, many well-known mathematicians introduced refinements and simplifications in the proofs of Hermite and Lindemann, without changing their basic ideas (see the Remarks at the end of Chapter 2). In this way they did not obtain any essentially new results. But several theorems were published establishing transcendence of the values of certain functions at algebraic points. These theorems were deduced from Lindemann's theorem using various transformations.

Weierstrass, who simplified the method of Hermite–Lindemann in [87:1], had a high opinion of Lindemann's work, calling his theorem on transcendence of the values of the exponential and logarithmic functions at algebraic points "one of the most beautiful theorems of arithmetic."

Hilbert, who published a variant of the proofs of Hermite and Lindemann in [28:1], said in his report at the Second International Congress of Mathematicians in Paris in 1900 [28:2]: "Hermite's theorems on the arithmetic of the exponential function and their further development by Lindemann will undoubtedly remain an inspiration for mathematicians of future generations."

The next important progress in transcendental number theory occurred at the end of the 1920's in the papers of A. O. Gel'fond and C. L. Siegel, and in the mid-1930's in work by Gel'fond and T. Schneider. These mathematicians initiated two of the most important analytic directions of research in transcendental number theory; they are still playing a fundamental role in this branch of number theory.

§ 3. Methods arising from the solution of Hilbert's Seventh Problem, and their subsequent development

In his address to the Second International Congress of Mathematicians on August 8, 1900, Hilbert proposed 23 problems "the study of which is likely to stimulate the further development of our science." In the seventh of these problems he conjectured that *a number of the form α^β, where α is an algebraic number not 0 or 1 and β is an algebraic irrational number, is transcendental.* In particular, he mentioned the numbers $2^{\sqrt{2}}$ and $e^\pi = i^{-2i}$.

This statement is a generalization of the conjecture of Euler mentioned above concerning transcendence of numbers of the form $\log_a b$. Hilbert replaced rationality by algebraicity, and put the resulting assertion in a somewhat different but equivalent form.

The first step toward a solution of Hilbert's seventh problem was taken only in 1929 by Gel'fond [24:1,2]. He developed a new analytic method, based on a type of interpolation, which enabled him to prove Hilbert's conjecture in the special case when the exponent β is an imaginary quadratic irrationality. In particular, Gel'fond's theorem implied that $e^{\pi} = i^{-2i}$ is transcendental.

In 1930, Kuz'min [35:1] extended Gel'fond's method to the case when β is a real quadratic irrationality. He thus proved transcendence of $2^{\sqrt{2}}$. Gel'fond's original method was used by various authors to obtain several new results about the arithmetic properties of numbers.

Then, in 1934, Gel'fond [24:3,4] developed another analytic method for proving transcendence, and was able to use it to give a complete solution to Hilbert's seventh problem. In the same year, Schneider [68:1] independently obtained a proof of Hilbert's conjecture by a different method. In 1934–41, Schneider [68:2–5] proved a series of deep theorems about transcendental numbers connected with elliptic functions, modular functions and abelian integrals.

Gel'fond's method of 1934 was used to obtain many other results.

In 1949 in [24:7], Gel'fond was able to generalize his second method, and he used the resulting technique to prove several theorems about algebraic independence of numbers. One such theorem states that *if α is an algebraic number not 0 or 1, and if β is an algebraic number of the third degree, then the numbers α^{β} and α^{β^2} are algebraically independent.*

It is very difficult to prove algebraic independence of algebraic powers of algebraic numbers. For example, it has not yet been possible to extend Lindemann's theorem to values of an exponential function with algebraic base.

Let α be an algebraic number not 0 or 1, and let β be an algebraic number of degree n. We consider the numbers

$$\alpha^{\beta}, \ \alpha^{\beta^2}, \ \ldots, \ \alpha^{\beta^{n-1}}, \qquad n \geq 3. \tag{2}$$

Gel'fond conjectured that *the numbers (2) are algebraically independent.* In 1949 in [24:7], he proved that the numbers (2) include at least two which are algebraically independent. Following that, a series of papers appeared in which Gel'fond's method was used to prove that, starting from some n, the numbers (2) include three algebraically independent numbers, and similarly for four algebraically independent numbers. For example, in 1972 Shmelev [71:2] proved that there are three algebraically independent numbers if $n \geq 19$, and in 1974 Waldschmidt [84:2] proved that there are four if $n \geq 31$.

In 1982, Nesterenko [52:6–8] generalized Gel'fond's method and showed that *the numbers (2) include at least* $[\log_2(n+1)]$ *algebraically independent numbers.* This was the first result which showed that the number of algebraically independent numbers in (2) grows with n.

Nesterenko's method later made it possible to significantly improve this result. Philippon [59:2] and Nesterenko [52:9,10] proved that *there are at least* $[n/2]$ *algebraically independent numbers among the numbers in (2).* Here Nesterenko also obtained a quantitative version of this statement.

Gel'fond's third method was also used by many mathematicians to obtain new results.

The methods of Hermite–Lindemann and Gel'fond also made it possible to obtain quantitative information about the arithmetic properties of numbers in the form of estimates for linear forms or polynomials in these numbers with integer coefficients. Gel'fond's methods were used by him and by other authors to establish a series of such estimates.

We return to the problem of approximating algebraic numbers. In Liouville's theorem the constant $c(\alpha)$ is effective, i.e., it can be computed for any given α. However, the generalization of Liouville's theorem due to Thue and other theorems proved by Thue's method are not effective. Thue's method does not provide a way to compute the constants in the inequalities which occur. This circumstance explains why Thue's theorem on Diophantine equations is also not effective: it states that there are finitely many solutions x and y to the equation (1), but does not give bounds for all possible solutions x and y. For many years none of the attempts to make Thue's method effective met with success.

The problem of making Thue's method effective is connected with the problem of obtaining effective lower bounds for linear forms with integer coefficients in the logarithms of a set of algebraic numbers (see Gel'fond [24:8]). In 1966, Baker [2:2] was able to strengthen Gel'fond's 1934 method in an essential way, and establish effective estimates for linear forms in the logarithms of algebraic numbers. This enabled him to prove stronger effective versions of Liouville's theorem, to make Thue's theorem on Diophantine equations effective, and also to obtain several other results in number theory (see [2:2–4,6]).

Several authors used Baker's methods to refine and generalize his results. Of this work one should note Fel'dman's theorems [17:5,6] generalizing the estimate for a linear form in logarithms, lowering the power in the estimate in the effective Liouville theorem, and making Thue's theorem on Diophantine equations effective.

In recent years both in the USSR and other countries there has been increasing interest in aspects of the theory of transcendental numbers connected with the methods that arose as a result of Hilbert's seventh problem.

This book does not treat results coming from the methods of Gel'fond and Schneider and their generalizations. The reader can become familiar with those results from the monographs by Gel'fond [24:8], Schneider [68:6], Siegel [72:4], LeVeque [39:1], Lang [37:1], Baker [2:5], Waldschmidt [84:1], Cijsow [14:1], Fel'dman [17:7,8], and also from the collected works of Gel'fond [24:10]. A survey of the results obtained by these methods as of 1967 can be found in an article by Fel'dman and Shidlovskii [18:1].

§ 4. Siegel's method and its further development

The method of Hermite–Lindemann is based on two properties of the exponential function: 1) e^z satisfies the functional equation $f(x)f(y) = f(x+y)$, and 2) e^z is the solution of the simplest linear differential equation $y' = y$.

In the methods of Gel'fond and Shneider, which were used to solve Hilbert's seventh problem, one considers an exponential function α^z to an algebraic base α. This function also satisfies the functional equation 1), and the differential equation it satisfies has the form $y' = (\ln\alpha)y$, where $\ln\alpha$ is a transcendental number. The fact that the power series expansion of α^z has transcendental coefficients, as does the differential equation this function satisfies, is a source of the difficulties which arise when proving algebraic independence results about its values.

After the creation of the Hermite–Lindemann method, the problem naturally arose of extending Hermite's and Lindemann's results to other functions, satisfying more general linear differential equations. Already Legendre [38:1] had considered the functions

$$f_\alpha(x) = \sum_{n=0}^{\infty} \frac{x^n}{n!\alpha(\alpha+1)\cdots(\alpha+n-1)}, \qquad \alpha \neq 0, -1, -2, \ldots, \tag{3}$$

which are solutions of the linear differential equation

$$xy'' + \alpha y' = y,$$

and had proved that *the numbers $f_\alpha(x)/f'_\alpha(x)$ are irrational if α and x are rational, $x \neq 0$.*

In 1910, Stridsberg [76:1] proved that each of the numbers $f_\alpha(x)$ and $f'_\alpha(x)$ is irrational under the same assumptions on α and x, and in 1927 in [43:1], Maier showed that they are not quadratic irrationalities. But for a long time no one was able to prove transcendence of these numbers or the values of other functions which satisfy linear differential equations.

In 1929 in the first part of [72:3], Siegel gave a new method for proving transcendence and algebraic independence of the values at algebraic points of a class of entire functions. This method is a direct generalization of the classical method of Hermite and Lindemann. Siegel did not require that the functions satisfy a functional equation, but he did need them to be solutions of a linear differential equation with coefficients in the field of rational functions.

A stimulus for developing his method, as Siegel himself notes, was Maier's paper [43:1] devoted to the study of the irrationality of the values of certain functions. It was Maier's work that suggested to Siegel the idea of constructing linear approximating forms from the functions under consideration. Another essential ingredient in Siegel's method was a generalization of an idea of Thue in [78:2] concerning approximation of algebraic numbers.

The basic result which Siegel obtained in [72:3] relates to the functions

$$K_\lambda(z) = \sum_{n=0}^{\infty} \frac{(-1)^n}{n!(\lambda+1)\cdots(\lambda+n)} \left(\frac{z}{2}\right)^{2n}, \qquad \lambda \neq -1, -2, \ldots,$$

which satisfy the linear differential equation

$$y'' + \frac{2\lambda+1}{z} y' + y = 0$$

and also the relation $K_\lambda(z) = f_{\lambda+1}(-z^2/4)$, where $f_\lambda(z)$ is given by (3). The function $K_\lambda(z)$ differs only by the factor $\Gamma(\lambda+1)^{-1}(z/2)^{2\lambda}$ from the Bessel function $J_\lambda(z)$ (where $\Gamma(z)$ is Euler's gamma-function), and $K_0(z) = J_0(z)$.

Siegel proved that *if λ is a rational number which is not half of an odd integer, and if ξ is a nonzero algebraic number, then the numbers $K_\lambda(\xi)$ and $K'_\lambda(\xi)$ are algebraically independent.* He also proved an analog of Lindemann's theorem for $K_\lambda(z)$ and $K'_\lambda(z)$: *the set of numbers $\{K_{\lambda_j}(\xi_i), K'_{\lambda_j}(\xi_i)\}$, $i = 1, \ldots, m$, $j = 1, \ldots, n$, are algebraically independent for rational $\lambda_1, \ldots, \lambda_n$ satisfying certain natural restrictions and for nonzero algebraic $\xi_1, \ldots, \xi_m$ having distinct squares.* In addition, he obtained a lower bound for the modulus of a polynomial with integral coefficients in $J_0(\xi)$ and $J'_0(\xi)$ for nonzero algebraic ξ.

Siegel's method can be applied to a certain class of entire functions with algebraic Taylor series coëfficients satisfying some arithmetic conditions. Siegel called these entire functions E-functions. These E-functions must also be solutions of linear differential equations with coefficients in the field of rational functions.

In 1949 in [72:4], Siegel presented his method in the form of a general theorem about algebraic independence of the values at algebraic points of a set of E-functions which satisfy a system of first order homogeneous linear differential equations with coefficients in the field of rational functions. This monograph did not contain any concrete new results not in [72:3].

Siegel's general theorem reduces the proof of an arithmetic property to the verification of a certain analytic normality condition for sets of products of powers of the functions under consideration. Siegel was able to verify this normality condition and apply his general theorem only in the case of E-functions, each of which satisfies a first or second order linear differential equation. It has not yet been possible to verify the normality condition for solutions of differential equations of order greater than two. Thus, despite the apparent generality of Siegel's theorem, it has few applications to concrete functions.

Until the mid-1950's, except for the two works of Siegel cited above, not a single paper appeared on Siegel's method. Then, in 1954 in [70:1,9], a theorem was published that was similar to Siegel's but with less restrictive assumptions. In this theorem Siegel's normality condition for products of powers of the functions was replaced by a certain irreducibility condition for the same functions. The latter condition is also only a sufficient condition, but it is simpler to state and imposes fewer restrictions on the functions. It is still complicated to verify this condition, but it can be done in several cases where Siegel's normality condition has not been verified. This theorem made it possible to apply Siegel's method more extensively to sets of E-functions which are solutions of a nonhomogeneous system of linear differential equations. In particular, it gives transcendence and algebraic independence of the values of certain E-functions which satisfy 3-rd and 4-th order linear differential equations.

In 1955 in [70:2] a theorem appeared with a necessary and sufficient condition for algebraic independence of the values of E-functions. This condition was a rather natural one: algebraic independence of the functions over the field of rational functions. A detailed proof of the theorem is given in [70:8], and a refinement of the method of proof is described in [70:19]. This theorem made it possible to prove transcendence and algebraic independence of the values at algebraic points of many concrete E-functions which are solutions of linear differential equations of arbitrary order.

In 1955–1962, several theorems were proved about the arithmetic properties of values of E-functions which are connected by algebraic equations over the field of rational functions [70:3–6,11,14]. One of them concerned algebraic independence of the values at algebraic points of a subset of the E-functions in the case when the basic set of E-functions under consideration is algebraically dependent over the field of rational functions. Thus, the possibilities of applying the method to prove transcendence and algebraic independence of the values of E-functions at algebraic points were carried to their natural limits.

Over the past 30 years many authors have conducted research on generalizing Siegel's method. This has led to a large number of new results. The 1929 work of Siegel [72:3] was the point of departure for one of the most important approaches to proving transcendence and algebraic independence of the values of analytic functions.

Many applications of the general method to concrete E-functions have appeared. For this purpose it was necessary to develop and generalize methods for proving that functions are algebraically independent over the field of rational functions. Quantitative versions of the theorems on the arithmetic properties of the values of E-functions were established in the form of inequalities which give lower bounds for the modulus of linear forms and polynomials with integer coefficients in the values of the functions under consideration. The method was also applied to another class of analytic functions which have a finite radius of convergence; Siegel called them G-functions. This research found application to the theory of Diophantine equations.

This book contains an exposition of the fundamental results on the arithmetic properties of the values of E-functions. Unfortunately, limitations of space prevent us from including some interesting research. Part of the results will only be stated without proof. Since the exponential function e^z is the simplest transcendental E-function, we shall examine in detail the classical results of Hermite and Lindemann. Proofs will be given based on different types of ideas.

We note that this book does not treat the metric theory of transcendental numbers. The reader can become acquainted with that theory from the books by Khinchin [30:1,2] and Sprindzhuk [73:1,3]. Nor does this book contain transcendence results in p-adic fields.

Chapter 1
Approximation of real and algebraic numbers

§ 1. Approximation of real numbers by algebraic numbers

The following notation will be used throughout the book: $\mathbf{N}$ is the set of natural numbers, $\mathbf{Z}$ is the ring of rational integers, $\mathbf{Z}^+$ is the set of nonnegative rational integers, $\mathbf{Q}$ is the field of rational numbers, $\mathbf{R}$ the field of real numbers, and $\mathbf{C}$ the field of complex numbers, and $\mathbf{R}^m$ denotes m-dimensional real Euclidean space.

Let $\alpha \in \mathbf{R}$. For various $p \in \mathbf{Z}$ and $q \in \mathbf{N}$ we consider the modulus of the difference

$$|\alpha - \frac{p}{q}|. \tag{1}$$

Since the set of rational numbers is everywhere dense in the set of real numbers, it follows that for a suitable choice of p and q the magnitude in (1) can be made arbitrarily small. Thus, it makes sense to consider its *relative* smallness, i.e., how small we can make it if q — the denominator of the fractional approximation — is not allowed to exceed some prescribed number.

Let $\varphi(q)$ be a function which is positive for all $q \in \mathbf{N}$. We consider the inequality

$$0 < |\alpha - \frac{p}{q}| < \varphi(q).$$

For a given α it would be interesting to know: for which functions $\varphi(q)$ does this inequality have an infinite set of solutions $(p, q) \in \mathbf{Z} \times \mathbf{N}$, and for which functions $\varphi(q)$ does the inequality have no solution as soon as q exceeds some bound?

We say that a real number α *has rational approximations p/q of order $\varphi(q)$* if there exists a constant $c > 0$ depending only on α and the function $\varphi(q)$ such that the inequality

$$0 < |\alpha - \frac{p}{q}| < c\,\varphi(q)$$

has infinitely many solutions $(p, q) \in \mathbf{Z} \times \mathbf{N}$.

The most common choice of function is a power function

$$\varphi(q) = \frac{1}{q^\nu}, \qquad \nu > 0.$$

In that case one is asking about the set of solutions of the inequality

$$0 < |\alpha - \frac{p}{q}| < \frac{c}{q^\nu}$$

for different positive values of ν and c.

Suppose that α is a rational number: $\alpha = a/b$, $(a, b) \in \mathbf{Z} \times \mathbf{N}$, $(a, b) = 1$. For any $q \in \mathbf{N}$ there exists $p \in \mathbf{Z}$ such that

$$\frac{p}{q} < \frac{a}{b} \leq \frac{p+1}{q}.$$

Then

$$0 < |\alpha - \frac{p}{q}| \leq \frac{1}{q}. \tag{2}$$

If we let q take on different values, we find that there exists an infinite set of fractions p/q, $p/q \neq \alpha$, which satisfy (2). Thus, α has rational approximations p/q of order $1/q$.

On the other hand, for any fraction p/q, $p/q \neq a/b$, we have

$$|\alpha - \frac{p}{q}| = |\frac{a}{b} - \frac{p}{q}| = \frac{|aq - bp|}{bq} \geq \frac{1}{bq}. \tag{3}$$

From the inequality in (3) it follows that, for any constant c with $0 < c \leq 1/b$ the inequality

$$0 < |\alpha - \frac{p}{q}| < \frac{c}{q}$$

has no solutions p/q.

This gives us the following criterion for irrationality.

Theorem 1. *If $\varphi(q) > 0$ for all $q \in \mathbf{N}$, the inequality*

$$0 < |\alpha - \frac{p}{q}| < \varphi(q)$$

has infinitely many solutions $(p, q) \in \mathbf{Z} \times \mathbf{N}$, and

$$\lim_{q \to +\infty} q\,\varphi(q) = 0,$$

then α is an irrational number.

The order of approximation by rational numbers is different for different real numbers. We shall soon show that any irrational real number has rational approximations p/q of order $1/q^2$, and also that there exist real numbers with "arbitrarily high" order of approximation.

Dirichlet's Theorem. *If* $\alpha \in \mathbf{R}$, $t \in \mathbf{R}$, *and* $t \geq 1$, *then there exist* $p \in \mathbf{Z}$ *and* $q \in \mathbf{N}$ *such that*

$$|\alpha - \frac{p}{q}| < \frac{1}{qt}, \qquad 0 < q \leq t. \tag{4}$$

Proof. Let $\{a\}$ and $[a]$ denote the fractional part and the integer part of a, respectively. We set $T = [t] + 1$, and consider the $T + 1$ numbers:

$$0 = \{\alpha \cdot 0\},\ \{\alpha \cdot 1\}, \ldots,\ \{\alpha(T-1)\},\ 1, \tag{5}$$

where

$$\{\alpha x\} = \alpha x - [\alpha x], \qquad 0 \leq \{\alpha x\} < 1, \qquad x = 0, 1, \ldots, T-1.$$

All of the points in (5) belong to the interval $0 \leq y \leq 1$. We divide this interval into T equal parts of length $1/T$:

$$\frac{k}{T} \leq y < \frac{k+1}{T}, \qquad k = 0, 1, \ldots, T-2, \qquad \frac{T-1}{T} \leq y \leq 1. \tag{6}$$

Each of the points in (5) lies in one of the intervals (6). Since the number of points in (5) is greater than the number of intervals in (6), it follows that there must be an interval which contains two of the points in (5). We now consider such an interval. There are two possible cases.

1. The interval is not the one at the extreme right. Suppose it contains the points $\{\alpha x_1\}$ and $\{\alpha x_2\}$, where $x_2 > x_1$. Then we have

$$|\{\alpha x_2\} - \{\alpha x_1\}| = |\alpha(x_2 - x_1) - ([\alpha x_2] - [\alpha x_1])| < \frac{1}{T} < \frac{1}{t}.$$

We set $x_2 - x_1 = q$ and $[\alpha x_2] - [\alpha x_1] = p$. Obviously, $0 < q \leq T - 1 \leq t$. We then have the inequalities

$$|\alpha q - p| < \frac{1}{t}, \qquad 0 < q \leq t, \tag{7}$$

from which (4) follows.

2. Our interval is the extreme right interval in (6). Suppose it contains the point $\{\alpha x_1\}$ as well as 1, where $x_1 \neq 0$. Then

$$|\{\alpha x_1\} - 1| = |\alpha x_1 - ([\alpha x_1] + 1)| \leq \frac{1}{T} < \frac{1}{t}.$$

We set $x_1 = q$ and $[\alpha x_1] + 1 = p$. We then have $0 < q \leq T - 1 \leq t$, and again we obtain (7), from which (4) follows.

We note that if $t \in \mathbf{N}$ in the hypothesis of the theorem, then the proof is a little simpler: we replace T by t, replace 1 by $\{\alpha t\}$ in (5), and consider only the first case.

The method used to prove Dirichlet's Theorem is known as the *Dirichlet pigeon-hole principle*. It is based on the simple idea that *if m objects are placed in n boxes with $m > n$, then at least one box must contain two or more objects.*

Suppose that $\alpha \in \mathbf{Q}$, $\alpha = a/b$, $(a, b) \in \mathbf{Z} \times \mathbf{N}$, and $(a, b) = 1$. In this case we showed that the inequality (3) holds for any fraction p/q not equal to α. Hence, if $t \geq b$, then (4) has only the trivial solution $p/q = a/b$. If $t < b$, then by Dirichlet's Theorem the inequality (4) has a solution with denominator q satisfying $q \leq t < b$. Hence, the denominators of all nontrivial solutions of (4) for all possible values of t are bounded, and so (4) has only a finite number of solutions (p, q). Consequently, for $\alpha \in \mathbf{Q}$ the theorem gives us some information about approximations of a rational number by rational numbers with smaller denominator.

If α is irrational, then $|\alpha q - p| > 0$ for any solution (p, q) of (4). There are infinitely many solutions (p, q) of (4) corresponding to different t. In fact, if there were only finitely many solutions, then we would have the inequality $|\alpha q - p| \geq \beta > 0$ for some β for all solutions (p, q) of (4). But $1/t \to 0$ as $t \to \infty$, and so this would contradict (4) for t sufficiently large.

Theorem 2. *For any irrational number $\alpha \in \mathbf{R}$, the inequality*

$$|\alpha - \frac{p}{q}| < \frac{1}{q^2} \tag{8}$$

has infinitely many solutions $(p, q) \in \mathbf{Z} \times \mathbf{N}$, and hence the set of denominators of the solutions to (8) is unbounded.

Proof. By Dirichlet's Theorem, for any solution (p, q) of (4) we have

$$|\alpha - \frac{p}{q}| < \frac{1}{qt} \leq \frac{1}{q^2}.$$

But, by what was proved before, (4) has infinitely many solutions (p, q) corresponding to different t. Consequently, the inequality (8) also has an infinite set of solutions $(p, q) \in \mathbf{Z} \times \mathbf{N}$.

From this theorem it follows that any irrational real number has approximations of order $1/q^2$.

From Theorem 1 we know that an irrational number $\alpha \in \mathbf{R}$ can be represented in the form

$$\alpha = \frac{p}{q} + \frac{\theta_q}{q^2}, \qquad (p, q) \in \mathbf{Z} \times \mathbf{N}, \qquad |\theta_q| < 1, \tag{9}$$

where q is not an arbitrary natural number, but is a number which can be chosen to be arbitrarily large, i.e., there exists an infinite sequence of pairs $(p_n, q_n) \in \mathbf{Z} \times \mathbf{N}$, $q_{n+1} > q_n$, such that we can set $p = p_n$ and $q = q_n$ in (9).

The representation of an irrational real number in the form (9) is often used in number theory and in other branches of pure and applied mathematics.

It is easy to construct examples of irrational real numbers having rational number approximations of order greater than any desired power of q. Consider the number

$$\alpha = \sum_{k=1}^{\infty} \frac{1}{a^{k!}}, \qquad a \in \mathbf{N}, \qquad a \geq 2. \tag{10}$$

We denote

$$\sum_{k=1}^{n} \frac{1}{a^{k!}} = \frac{p_n}{q_n}, \qquad q_n = a^{n!}, \qquad p_n \in \mathbf{N}, \qquad n = 1, 2, \ldots .$$

Then

$$\alpha - \frac{p_n}{q_n} = r_n = \sum_{k=n+1}^{\infty} \frac{1}{a^{k!}} > 0, \qquad n = 1, 2, \ldots ,$$

and

$$r_n = \frac{1}{a^{(n+1)!}} \left(1 + \frac{1}{a^{(n+2)!-(n+1)!}} + \cdots \right) < \frac{1}{a^{(n+1)!}} \left(1 + \frac{1}{a} + \frac{1}{a^2} + \cdots \right) =$$
$$= \frac{a}{a-1} \frac{1}{a^{(n+1)!}} = \frac{a}{a-1} \frac{1}{q_n^{n+1}} \leq \frac{1}{q_n^n}.$$

Hence,

$$0 < \alpha - \frac{p_n}{q_n} < \frac{1}{q_n^n}, \qquad n = 1, 2, \ldots .$$

The last inequality implies that for any nonnegative $\nu \in \mathbf{R}$ the inequality

$$0 < |\alpha - \frac{p}{q}| < \frac{1}{q^\nu} \tag{11}$$

has a solution $(p, q) \in \mathbf{Z} \times \mathbf{N}$, $q \geq 2$. But any solution of (11) with a given value of ν is also a solution of the same inequality for any smaller value of ν. This implies than for any $\nu \in \mathbf{R}$, $\nu > 0$, the inequality (11) has infinitely many solutions.

For $a = 10$ the series (10) gives us the number with decimal expansion

$$\alpha = 0.110001000\ldots ,$$

which has the digit 1 in the $(n!)$-th place to the right of the decimal point and 0's everywhere else.

It is also quite simple to construct examples of this sort using continued fractions. Moreover, it is not hard to prove the stronger result that there exist irrational real numbers having rational approximations which are better than any given function of q.

Theorem 3. *Given any positive function $\varphi(q)$ on the natural numbers, there exists an irrational $\alpha \in \mathbf{R}$ such that the inequality*

$$|\alpha - \frac{p}{q}| < \varphi(q)$$

has infinitely many solutions $(p, q) \in \mathbf{Z} \times \mathbf{N}$.

Proof. We define a sequence $\{\gamma_n\}$, $\gamma_n \in \mathbf{R}$, $\gamma_n \geq 0$, in such a way that the sequence of natural numbers

$$m_1 = 1, \qquad m_n = \left[\log_2\left(\frac{1}{\varphi(2^{m_{n-1}})} + \gamma_{n-1}\right)\right] + 1, \qquad n = 2, 3, \ldots,$$

satisfies the condition $m_{n+1} > 2m_n$. We then set

$$\alpha = \sum_{k=1}^{\infty} \frac{(-1)^k}{2^{m_k}}, \tag{12}$$

and

$$q_n = 2^{m_n}, \qquad p_n = q_n \sum_{k=1}^{n} \frac{(-1)^k}{2^{m_k}}, \qquad n = 1, 2, \ldots.$$

Since (12) is an alternating series with decreasing terms, we have the inequalities

$$\begin{aligned} 0 < \frac{1}{2^{m_{n+1}}} - \frac{1}{2^{m_{n+2}}} < \left|\alpha - \frac{p_n}{q_n}\right| < \frac{1}{2^{m_{n+1}}} < \\ < \frac{1}{2^{\log_2(1/\varphi(2^{m_n}))}} = \varphi(2^{m_n}) = \varphi(q_n), \qquad n = 1, 2, \ldots. \end{aligned} \tag{13}$$

Since $m_{n+1} > 2m_n$, it follows from (13) that

$$0 < |\alpha - \frac{p_n}{q_n}| < \frac{1}{2^{m_{n+1}}} < \frac{1}{2^{2m_n}} = \frac{1}{q_n^2}.$$

Now α is irrational by Theorem 1, and then Theorem 3 follows from (13).

The set of real numbers having very good rational approximations is relatively small. Almost all real numbers (in the sense of Lebesgue measure) are not approximated very well by rational numbers. This follows from a theorem proved by Khinchin in 1924 [30:1]. We shall state it here without proof.

Khinchin's Theorem. *Let $f(x)$ be a positive continuous function for $x \geq c$, where $c > 0$, such that $xf(x)$ is non-increasing. Then the inequality*

$$|\alpha - \frac{p}{q}| < \frac{f(q)}{q}$$

has an infinite set of solutions $(p, q) \in \mathbf{Z} \times \mathbf{N}$ for almost all α provided that the integral

$$\int_c^{+\infty} f(x)dx$$

diverges; while if this integral converges, then for almost all α the above inequality has only finitely many solutions $(p, q) \in \mathbf{Z} \times \mathbf{N}$.

This theorem implies, in particular, that *the inequality*

$$|\alpha - \frac{p}{q}| < \frac{1}{q^2 \ln q}$$

has an infinite set of solutions for almost all α, while the inequality

$$|\alpha - \frac{p}{q}| < \frac{1}{q^2 \ln^{1+\epsilon} q}$$

for any $\epsilon > 0$ has only finitely many solutions for almost all α. Thus, we have a good estimate of the order of approximation by rational numbers of almost all real numbers.

We note that the irrational real numbers which have the worst approximation by rational numbers p/q are those whose continued fraction expansion has bounded quotients: their best approximations are of order $1/q^2$. The set of such real numbers has the cardinality of the continuum, but it has Lebesgue measure zero (see Khinchin [30:2]).

Dirichlet's Theorem can be proved in other ways, for example, using continued fractions or Farey series. But the simplest method of proof is to use the Dirichlet pigeon-hole principle. The pigeon-hole principle can be applied in many situations in the theory of Diophantine approximation when continued fractions or Farey series cannot be used, such as certain theorems on simultaneous approximation of a set of numbers, and also in many proofs in the theory of transcendental numbers. Some examples of such problems will be examined in the next section.

§ 2. Simultaneous approximation

We consider a linear form of several real variables with integer coefficients bounded by some number and not all zero. We shall discuss how small it is possible to make the modulus of this form by suitably choosing the coefficients.

Theorem 4 (Dirichlet). *If $\alpha_1, \dots, \alpha_m$ are real numbers, $m \ge 1$, and if $t \in \mathbf{R}$, $t \ge 1$, then there exist $a_1, \dots, a_m, b \in \mathbf{Z}$ such that*

$$|a_1\alpha_1 + \cdots + a_m\alpha_m - b| < \frac{1}{t^m}, \qquad 0 < \max_{1\le k\le m} |a_k| \le t. \tag{14}$$

Proof. We set $T = [t] + 1$ and consider the following $T^m + 1$ numbers: the T^m fractional parts

$$\gamma = \{a_1\alpha_1 + \cdots + a_m\alpha_m\}, \qquad 0 \le \gamma < 1, \tag{15}$$

where each of the $a_1, \dots, a_m$ independently runs through the T values $0, 1, \dots, T-1$, and also the number 1. We divide the interval $0 \le y \le 1$ into T^m parts of length $1/T^m$:

$$\frac{k}{T^m} \le y < \frac{k+1}{T^m}, \quad k = 0, 1, \dots, T^m - 2, \qquad \frac{T^m - 1}{T^m} \le y \le 1. \tag{16}$$

The points in (15) along with 1 all belong to the interval $0 \le y \le 1$. Hence, these $T^m + 1$ points each fall in one of the T^m intervals (16). Since the number of points is greater than the number of intervals, there must be an interval containing at least two points. Two cases are possible.

1. The interval containing two of our points is not the one at the extreme right. Suppose that this interval contains the points $\gamma' = \{a_1'\alpha_1 + \cdots + a_m'\alpha_m\}$ and $\gamma'' = \{a_1''\alpha_1 + \cdots + a_m''\alpha_m\}$. Then

$$\begin{aligned} &|\{a_1''\alpha_1 + \cdots + a_m''\alpha_m\} - \{a_1'\alpha_1 + \cdots + a_m'\alpha_m\}| = \\ &\quad = |a_1\alpha_1 + \cdots + a_m\alpha_m - b| < \frac{1}{T^m} = \frac{1}{([t]+1)^m} < \frac{1}{t^m}, \end{aligned} \tag{17}$$

where

$$\begin{aligned} a_k &= a_k'' - a_k', \qquad k = 1, \dots, m; \\ b &= [a_1''\alpha_1 + \cdots + a_m''\alpha_m] - [a_1'\alpha_1 + \cdots + a_m'\alpha_m], \end{aligned} \tag{18}$$

in which $0 < \max_{1\le k\le m} |a_k| \le T - 1 = [t] \le t$.

2. Our interval is the one at the extreme right. Suppose that it contains the points

$$\gamma = \{a_1\alpha_1 + \cdots + a_m\alpha_m\}, \qquad 0 < \max_{1\le k\le m} |a_k| \le T - 1 \le t, \tag{19}$$

and 1. Then

$$|\{a_1\alpha_1 + \cdots + a_m\alpha_m\} - 1| = |a_1\alpha_1 + \cdots + a_m\alpha_m - b| \le \frac{1}{T^m} < \frac{1}{t^m}, \tag{20}$$

where $b = [a_1\alpha_1 + \cdots + a_m\alpha_m] + 1$.

The inequality (14) follows from (17) and (18) in the first case, and from (19) and (20) in the second case.

This proof is completely analogous to the proof of Dirichlet's Theorem in § 1; in fact, the latter theorem follows from Theorem 4 with $m = 1$.

We now prove a theorem on simultaneous approximation of a set of numbers.

Kronecker's Theorem. *If $\alpha_1, \ldots, \alpha_m$ are arbitrary real numbers, $m \ge 1$, and if $t \in \mathbf{N}$, then there exist $p_1, \ldots, p_m \in \mathbf{Z}$ and $q \in \mathbf{N}$ such that*

$$|\alpha_k - \frac{p_k}{q}| < \frac{1}{qt}, \qquad k = 1, \ldots, m, \qquad 0 < q \le t^m. \tag{21}$$

Proof. In $\mathbf{R}^m$ we consider the points

$$M_x(\{\alpha_1 x\}, \ldots, \{\alpha_m x\}), \qquad x = 0, 1, \ldots, t^m, \tag{22}$$

where $\{\alpha_k x\} = \alpha_k x - [\alpha_k x]$, $0 \le \{\alpha_k x\} < 1$, $k = 1, \ldots, m$; and we consider the unit cube, i.e., the set of points

$$M(y_1, \ldots, y_m), \qquad 0 \le y_k < 1, \qquad k = 1, \ldots, m. \tag{23}$$

We divide the segment $0 \le y_i \le 1$ of each coordinate axis into t parts of length $1/t$, and we use the corresponding hyperplanes to divide the unit cube into t^m small cubes, each of which consists of all points $Q(y_1, \ldots, y_m)$ whose coordinates satisfy the inequalities

$$\frac{k_1}{t} \le y_1 < \frac{k_1 + 1}{t}, \quad \ldots, \quad \frac{k_m}{t} \le y_m < \frac{k_m + 1}{t}, \tag{24}$$

where the k_i, $i = 1, \ldots, m$, are an m-tuple of integers from $\{0, 1, \ldots, t - 1\}$.

All of the $t^m + 1$ points in (22) are contained in the unit cube (23). Each of these points falls in exactly one of the small cubes (24). Hence, there is some small cube which contains at least two of the points in (22), say

$$M_{x_1}(\{\alpha_1 x_1\}, \ldots, \{\alpha_m x_1\}), \qquad M_{x_2}(\{\alpha_1 x_2\}, \ldots, \{\alpha_m x_2\}), \qquad x_2 > x_1.$$

Since these two points lie in the same cube, it follows that the absolute value of the difference between any two corresponding coordinates is less than $1/t$. That is,

$$|\{\alpha_k x_2\} - \{\alpha_k x_1\}| < \frac{1}{t}, \qquad k = 1, \ldots, m;$$

in other words, if we set $q = x_2 - x_1$ and $p_k = [\alpha_k x_2] - [\alpha_k x_1]$ for $k = 1, \ldots, m$, then we have

$$|\alpha_k q - p_k| < \frac{1}{t}, \qquad k = 1, \ldots, m, \qquad 0 < q \le t^m. \tag{25}$$

These inequalities imply (21).

For $m = 1$ Kronecker's Theorem becomes Dirichlet's Theorem (with $t \in \mathbf{N}$).

Just as in the case of Dirichlet's Theorem, from Kronecker's Theorem it is easy to obtain the following

Corollary. *Under the assumptions in Kronecker's Theorem, if at least one of the $\alpha_1, \ldots, \alpha_m$ is irrational, then the inequalities*

$$|\alpha_k - \frac{p_k}{q}| < q^{-1-\frac{1}{m}}, \qquad k = 1, \ldots, m, \; .$$

have an infinite set of solutions $p_1, \ldots, p_m \in \mathbf{Z}$, $q \in \mathbf{N}$.

We next prove a theorem giving a bound for linear forms which we shall need later.

Theorem 5. *If $\alpha_1, \ldots, \alpha_m$ are arbitrary complex numbers, $m \ge 2$, and if $H \in \mathbf{N}$, then there exist numbers*

$$a_k \in \mathbf{Z}, \qquad |a_k| \le H, \qquad k = 1, \ldots, m, \qquad \max_{1 \le k \le m} |a_k| > 0, \tag{26}$$

such that the linear form

$$L = a_1\alpha_1 + \cdots + a_m\alpha_m \tag{27}$$

satisfies the inequality

$$|L| \le cH^{1-\tau m}, \tag{28}$$

where $\tau = 1$ if all of the α_k are real and $\tau = 1/2$ if one or more of them is complex, and

$$c = \begin{cases} \sum_{k=1}^m |\alpha_k|, & \text{if } \tau = 1, \\ \sqrt{2}\sum_{k=1}^m |\alpha_k|, & \text{if } \tau = 1/2. \end{cases}$$

Proof. The theorem is obvious if $H = 1$; so we suppose that $H \geq 2$.

We consider all possible linear forms of the form (27), where the a_k independently run through all integer values satisfying the inequalities

$$|a_k| \leq \left[\frac{H}{2}\right], \qquad k = 1, \ldots, m.$$

The number of such forms L is equal to

$$\left(2\left[\frac{H}{2}\right] + 1\right)^m. \tag{29}$$

Their values satisfy the inequalities

$$|L| \leq \gamma\left[\frac{H}{2}\right], \qquad \gamma = \sum_{k=1}^{m} |\alpha_k|. \tag{30}$$

We consider two cases (we exclude the obvious case $\gamma = 0$).

1. All of the $\alpha_1, \ldots, \alpha_m$ are real numbers. Then the values of all of our forms L are contained in the interval with endpoints at $\pm\gamma\left[\frac{H}{2}\right]$, which has length $2\gamma\left[\frac{H}{2}\right]$. We divide this interval into

$$\left(2\left[\frac{H}{2}\right] + 1\right)^m - 1 \tag{31}$$

equal subintervals. The number of forms L (see (29)) is greater than the number (31) of subintervals. Hence, there exists a subinterval containing the values of two different forms L. Let these two forms be

$$L' = a_1'\alpha_1 + \cdots + a_m'\alpha_m, \qquad L'' = a_1''\alpha_1 + \cdots + a_m''\alpha_m. \tag{32}$$

Then

$$|L' - L''| \leq \frac{2\gamma\left[\frac{H}{2}\right]}{\left(2\left[\frac{H}{2}\right] + 1\right)^m - 1}.$$

If H is an even number, then $2\left[\frac{H}{2}\right] = H$, and

$$|L' - L''| \leq \frac{\gamma H}{(H + 1)^m - 1} < \gamma H^{1-m}.$$

If H is an odd number, then $2\left[\frac{H}{2}\right] = H - 1$, and

$$|L' - L''| \le \frac{\gamma(H-1)}{H^m - 1} < \gamma H^{1-m}.$$

Thus,

$$|L' - L''| < \gamma H^{1-m}.$$

We set $a_k = a'_k - a''_k$, $k = 1, \dots, m$. Then the form

$$L = L' - L'' = a_1\alpha_1 + \cdots + a_m\alpha_m$$

satisfies the conditions

$$|a_k| \le 2\left[\frac{H}{2}\right] \le H, \qquad \sum_{k=1}^{m} |a_k| > 0,$$

and

$$|L| < cH^{1-m}, \qquad c = \gamma.$$

2. At least one of the numbers $a_1, \dots, a_m$ is complex. In this case it follows from (30) that the values of the forms L lie in a square centered at the origin with sides parallel to the coordinate axes of length equal to $2\gamma\left[\frac{H}{2}\right]$.

We divide the sides of this square into M equal segments, where M is the integer satisfying the inequalities

$$\left(2\left[\frac{H}{2}\right]+1\right)^{m/2} - 1 \le M < \left(2\left[\frac{H}{2}\right]+1\right)^{m/2}. \tag{33}$$

Using lines through the division points parallel to the coordinate axes, we divide our square into M^2 small squares. From (33) we have

$$M^2 < \left(2\left[\frac{H}{2}\right]+1\right)^{m}.$$

This inequality means that the number of small squares is less than the number (29) of linear forms L. Hence, there exists a small square which contains the values of two different forms L. Suppose that these two forms are the ones in (32). We then find that $|L' - L''|$ is no greater than the diagonal of a small square. This observation, together with (33), enables us to say that the following inequalities hold for any $H \geq 2$ and $m \geq 2$:

$$|L' - L''| \leq \frac{\sqrt{2}\,\gamma\, 2\left[\frac{H}{2}\right]}{M} \leq \frac{\sqrt{2}\,\gamma\, 2\left[\frac{H}{2}\right]}{\left(2\left[\frac{H}{2}\right]+1\right)^{m/2} - 1}. \tag{34}$$

If H is an even number, then $2[H/2] = H$, and, since $m \geq 2$, we conclude from (34) that

$$|L' - L''| \leq \frac{\sqrt{2}\,\gamma\, H}{(H+1)^{m/2} - 1} \leq \sqrt{2}\,\gamma\, H^{1-\frac{m}{2}}.$$

If H is an odd number, then $2[H/2] = H - 1$, and we similarly conclude from (34) that

$$|L' - L''| \leq \frac{\sqrt{2}\,\gamma\,(H-1)}{H^{m/2} - 1} = \sqrt{2}\,\gamma\, H^{1-\frac{m}{2}}\, \frac{1 - H^{-1}}{1 - H^{-m/2}} \leq \sqrt{2}\,\gamma\, H^{1-\frac{m}{2}}.$$

We then argue as in the first case in order to obtain the inequality (28) with $c = \sqrt{2}\,\gamma$.

§ 3. Approximation of algebraic numbers by rational numbers

The arithmetic operations applied to algebraic numbers give us algebraic numbers again. Hence, the set of all algebraic numbers forms a field, which we shall denote $\mathbf{A}$.

Let α be an algebraic number. There exists a unique irreducible polynomial $f(x) \in \mathbf{Q}[x]$ with leading coefficient 1 which has α as a root. This polynomial is called the *minimal polynomial* of α, and its degree is called the *degree* of α, denoted $\deg \alpha$. If α is an algebraic number of degree n, then the roots $\alpha = \alpha_1, \alpha_2, \ldots, \alpha_n$ of the minimal polynomial of α are called the *conjugates* of α. We set

$$\overline{|\alpha|} = \max_{1 \leq k \leq n} |\alpha_k|,$$

and call this number the *size* of the algebraic number α.

We now prove Liouville's Theorem on the approximation of an algebraic number by rational numbers.

Liouville's Theorem. *If α is a real algebraic number of degree n, $n \geq 1$, then there exists a constant $c = c(\alpha) > 0$ such that the following inequality holds for* *any* $p \in \mathbf{Z}$ *and* $q \in \mathbf{N}$, $p/q \neq \alpha$:

$$|\alpha - \frac{p}{q}| > \frac{c}{q^n}. \tag{35}$$

Proof. Suppose that α is a root of the irreducible polynomial

$$f(x) = a_n x^n + \cdots + a_1 x + a_0, \qquad f(x) \in \mathbf{Z}[x], \qquad a_n > 0,$$

and $\alpha = \alpha_1, \ldots, \alpha_n$ are the conjugates of α. Then

$$|f(x)| = a_n |x - \alpha| \prod_{k=2}^{n} |x - \alpha_k|. \tag{36}$$

Let p and q be arbitrary integers with $q > 0$, $p/q \neq \alpha$. We consider two cases.

1. $n > 1$. If p and q are such that $|\alpha - \frac{p}{q}| \geq 1$, then

$$|\alpha - \frac{p}{q}| \geq \frac{1}{q^n}. \tag{37}$$

On the other hand, if $|\alpha - \frac{p}{q}| < 1$, then we have

$$|\frac{p}{q}| < |\alpha| + 1 \leq \overline{|\alpha|} + 1. \tag{38}$$

In addition, $f(p/q) \neq 0$, since the irreducible polynomial $f(x)$ cannot have a rational root for $n > 1$. We now set $x = p/q$ in (36). If we note that

$$\left| f\left(\frac{p}{q}\right) \right| = \frac{|a_n p^n + \cdots + a_1 p q^{n-1} + a_0 q^n|}{q^n} \geq \frac{1}{q^n}$$

and, by (38),

$$\prod_{k=2}^{n} |\frac{p}{q} - \alpha_k| \leq \prod_{k=2}^{n} (|\frac{p}{q}| + |\alpha_k|) < (2\overline{|\alpha|} + 1)^{n-1},$$

then we obtain

$$\frac{1}{q^n} < a_n (2\overline{|\alpha|} + 1)^{n-1} |\alpha - \frac{p}{q}|. \tag{39}$$

Now the inequality (35) follows from (37) and (39), if we set

$$c = c(\alpha) = \frac{1}{a_n (2\overline{|\alpha|} + 1)^{n-1}}.$$

2. $n = 1$. We have $\alpha = -a_0/a_1$. Then

$$|\alpha - \frac{p}{q}| \geq \frac{1}{a_1 q} = \frac{c}{q}.$$

In this case the theorem follows from the last inequality, if we take a smaller value of c in order to get strict inequality.

The assumption that $p/q \neq \alpha$ in Liouville's Theorem relates only to the case $n = 1$, since α is irrational if $n > 1$. The case $n = 1$ of Liouville's Theorem was actually proved in § 1 in (3).

Corollary. *Under the conditions of Liouville's Theorem, there exists a constant* $c = c(\alpha) > 0$ *such that the inequality*

$$0 < |\alpha - \frac{p}{q}| < \frac{c}{q^n}$$

has no solutions $(p, q) \in \mathbf{Z} \times \mathbf{N}$.

This means that an algebraic number α of degree n does not have approximations by rational numbers of order better than $1/q^n$. If $n = 2$, this together with Theorem 2 implies that the best order of approximation of a real quadratic irrationality by rational numbers is $1/q^2$. Thus, quadratic irrational numbers are among the real numbers whose order of approximation by rational numbers is the worst possible. One can view Liouville's Theorem as giving a definitive answer to the question of the order of approximation of quadratic irrationalities by rational fractions.

Liouville's Theorem provides a necessary condition for a number α to be algebraic; in other words, it gives us a sufficient condition for transcendence. We state this as a theorem.

Theorem 6. *Suppose that α is a real number such that for any $\nu \in \mathbf{R}$, $\nu > 0$, the inequality*

$$0 < |\alpha - \frac{p}{q}| < \frac{1}{q^\nu} \tag{40}$$

has an infinite set of solutions $(p, q) \in \mathbf{Z} \times \mathbf{N}$. *Then α is transcendental.*

Proof. Suppose that α were an algebraic number of degree n. According to Liouville's Theorem, there would exist a constant $c(\alpha) > 0$ such that for any $(p, q) \in \mathbf{Z} \times \mathbf{N}$, $p/q \neq \alpha$, we have

$$|\alpha - \frac{p}{q}| > \frac{c(\alpha)}{q^n}. \tag{41}$$

We set $\nu = n + 1$ in (40), and choose a solution for which $1/q < c(\alpha)$. Then

$$0 < |\alpha - \frac{p}{q}| < \frac{1}{q^{n+1}} < \frac{c(\alpha)}{q^n},$$

which contradicts (41), and thereby proves the theorem.

Theorem 7. *There exist transcendental numbers.*

Proof. We need only construct an example of a transcendental number. In § 1 we considered the number

$$\alpha = \sum_{n=1}^{\infty} \frac{1}{a^{n!}}, \qquad a \in \mathbf{N}, \qquad a \geq 2,$$

and we showed that for any $\nu \in \mathbf{R}$, $\nu > 0$, the inequality (11) has infinitely many solutions $(p, q) \in \mathbf{Z} \times \mathbf{N}$. By Theorem 6, this means that α is a transcendental number.

Theorem 6 enables us to construct many other transcendental numbers. For example, a number whose continued fraction expansion has quotients which increase sufficiently rapidly will be transcendental. Transcendental numbers which satisfy the condition in Theorem 6 are called *Liouville numbers.*

We stated Liouville's Theorem for a real algebraic number α. It also holds for a complex number α. However, there is no interest in considering rational approximations to a complex number not in $\mathbf{R}$, since no rational number can be closer to such an α than the size of the imaginary part of α.

§ 4. Approximation of algebraic numbers by algebraic numbers

By the *height* $H = H_\psi$ of a polynomial $\psi = \psi(x) \in \mathbf{C}[x]$ we mean the maximum modulus of its coefficients. Now suppose that $f(x)$ is the minimal polynomial of an algebraic number α. If we multiply it by the least common denominator of its coefficients, we obtain a primitive polynomial $\varphi(x) \in \mathbf{Z}[x]$ which also has α as a root. (Recall that a polynomial in $\mathbf{Z}[x]$ is called *primitive* if there is no integer greater than 1 which divides all of its coefficients.) Then the *height* $H = H_\alpha$ of the algebraic number α is defined to be the height of the irreducible primitive polynomial $\varphi(x) \in \mathbf{Z}[x]$ which has α as a root.

An algebraic number α is called an *algebraic integer* if all of the coefficients of its minimal polynomial $f(x)$ are rational integers. One can prove that *if* α *is a root of a polynomial* $\varphi(x) \in \mathbf{Z}[x]$ *with leading coefficient* 1 *(not necessarily irreducible), then* α *is an algebraic integer.*

The sum, difference, or product of two algebraic integers is an algebraic integer. Consequently, the set of all algebraic integers forms a ring, which we shall denote $\mathbf{Z_A}$. If $\alpha \in \mathbf{A}$, then there exists an $r \in \mathbf{N}$ such that $r\alpha \in \mathbf{Z_A}$. If α is a root of $\psi(x) \in \mathbf{Z}[x]$, $\psi(x) \not\equiv 0$, then we can choose r to be the modulus of the leading coefficient of $\psi(x)$.

By an *algebraic number field* we mean the extension field of $\mathbf{Q}$ obtained by adjoining an algebraic number θ, i.e., the set of numbers $Q(\theta)$, where $Q(x)$ runs through all rational functions in $\mathbf{Q}(x)$ whose denominator does not vanish at θ. Such a field will be denoted $\mathbf{Q}(\theta)$. Here θ is called a *generating element* of the algebraic number field $\mathbf{Q}(\theta)$. Different elements of this field can be chosen to serve as a generating element. All of them have the same degree h. Generating elements are also called *primitive elements* of the algebraic number field. By the *degree of the algebraic number field* $\mathbf{K} = \mathbf{Q}(\theta)$ we mean the degree of any generating element; this degree is denoted $[\mathbf{K} : \mathbf{Q}] = [\mathbf{Q}(\theta) : \mathbf{Q}]$. The degree h of an algebraic number field $\mathbf{Q}(\theta)$ is equal to the maximum number of linearly independent elements over $\mathbf{Q}$.

Suppose that $\mathbf{K} = \mathbf{Q}(\theta)$ is an algebraic number field, $[\mathbf{K} : \mathbf{Q}] = h$, and $\theta = \theta_1, \ldots, \theta_h$ are the conjugates of θ. Every element $\alpha \in \mathbf{K}$ can be represented uniquely in the form

$$\alpha = r(\theta) = c_0 + c_1\theta + \cdots + c_{h-1}\theta^{h-1}, \qquad c_i \in \mathbf{Q}, \qquad i = 0, 1, \ldots, h-1.$$

The numbers

$$\alpha_i = r(\theta_i), \qquad i = 1, \ldots, h, \tag{42}$$

are called the *conjugates of* α *in the field* $\mathbf{K}$. They are the conjugates of α (in the sense defined before, i.e., the roots of the minimal polynomial), but each conjugate is repeated $h/\deg\alpha$ times. Here $\alpha_2, \ldots, \alpha_h$ do not have to belong to the field $\mathbf{K}$.

When we are considering an algebraic number field $\mathbf{K} = \mathbf{Q}(\theta)$, we shall always suppose that the numbering $\theta = \theta_1, \ldots, \theta_h$ is fixed. According to (42), we thereby also fix the numbering of the conjugates of any $\alpha \in \mathbf{K}$.

It is easy to show that, if $\alpha_1, \ldots, \alpha_h$ are the conjugates of α in $\mathbf{K}$, and if $Q(x) \in \mathbf{Q}(x)$ is such that $Q(\alpha)$ is defined, then the elements $Q(\alpha_1), \ldots, Q(\alpha_h)$ are the conjugates of $Q(\alpha)$ in $\mathbf{K}$.

Given any finite set of algebraic numbers $\alpha_1, \ldots, \alpha_n$, there exists an algebraic number field $\mathbf{K} = \mathbf{Q}(\theta)$ which contains these numbers.

If α is an element of $\mathbf{K}$, we define its size $\overline{|\alpha|}$ by the equality

$$\overline{|\alpha|} = \max_{1 \le i \le h} |r(\theta_i)|, \qquad \alpha = r(\theta).$$

This definition agrees with the meaning of $\overline{|\alpha|}$ in § 3. We obviously have

$$\overline{|\alpha \pm \beta|} \le \overline{|\alpha|} + \overline{|\beta|}, \qquad \overline{|\alpha\beta|} \le \overline{|\alpha|}\,\overline{|\beta|}.$$

The set of algebraic integers contained in the field **K** forms a ring, which we shall denote $\mathbf{Z_K}$. We always have $\mathbf{Z} \subset \mathbf{Z_K}$.

The *norm* of an element $\alpha \in \mathbf{K}$ in the field **K** is defined to be the product of all of the conjugates of α in **K**. We denote the norm by $\mathcal{N}(\alpha)$:

$$\mathcal{N}(\alpha) = r(\theta_1)\cdots r(\theta_h), \qquad \alpha = r(\theta).$$

We list some properties of the norm:

1. $\mathcal{N}(\alpha) \in \mathbf{Q}$ for all $\alpha \in \mathbf{K}$;
2. $\mathcal{N}(\alpha) \in \mathbf{Z}$ for all $\alpha \in \mathbf{Z_K}$;
3. $\mathcal{N}(\alpha) = 0$ if and only if $\alpha = 0$;
4. $\mathcal{N}(\alpha) = \alpha^h$ if $\alpha \in \mathbf{Q}$;
5. $\mathcal{N}(\alpha\beta) = \mathcal{N}(\alpha)\mathcal{N}(\beta)$ for all $\alpha, \beta \in \mathbf{K}$;
6. $\mathcal{N}(a\alpha) = a^h\mathcal{N}(\alpha)$ for $a \in \mathbf{Q}$ and $\alpha \in \mathbf{K}$.

In what follows, when we are considering only a single $\alpha \in \mathbf{A}$, we shall always use $\mathcal{N}(\alpha)$ to denote its norm in the algebraic number field $\mathbf{Q}(\alpha)$ generated by α.

If α is a root of a nonzero polynomial $f(x)$ over K, i.e., if $f(x) \in \mathbf{K}[x]$, then $\alpha \in \mathbf{A}$. The *degree of α over* **K**, denoted $\deg_{\mathbf{K}} \alpha$, is the degree of the monic irreducible polynomial in $\mathbf{K}[x]$ which has α as a root.

We now prove a lemma which is based on the symmetric polynomial theorem. Let **V** be a commutative ring with unit. A polynomial $F(\alpha_1, \dots, \alpha_n) \in \mathbf{V}[\alpha_1, \dots, \alpha_n]$ is called a *symmetric polynomial* in $\alpha_1, \dots, \alpha_n$ if it does not change when $\alpha_1, \dots, \alpha_n$ are subjected to any permutation. We denote

$$\sigma_1 = \alpha_1 + \cdots + \alpha_n,$$

$$\sigma_2 = \alpha_1\alpha_2 + \alpha_1\alpha_3 + \cdots + \alpha_{n-1}\alpha_n,$$

$$\cdot \quad \cdot \quad \cdot \quad \cdot \quad \cdot \quad \cdot$$

$$\sigma_n = \alpha_1 \cdots \alpha_n.$$

These are called the *elementary symmetric polynomials* in $\alpha_1, \dots, \alpha_n$; they are (up to sign) the coefficients of the polynomial $(x - \alpha_1)\cdots(x - \alpha_n)$. The symmetric polynomial theorem states that *any symmetric polynomial* $F(\alpha_1, \dots, \alpha_n) \in \mathbf{V}[\alpha_1, \dots, \alpha_n]$ *can be expressed uniquely in the form* $F(\alpha_1, \dots, \alpha_n) = H(\sigma_1, \dots, \sigma_n)$, *where* $H(\sigma_1, \dots, \sigma_n) \in \mathbf{V}[\sigma_1, \dots, \sigma_n]$.

Lemma 1. *Suppose that* $\alpha \in \mathbf{A}$, $\deg \alpha = n$, *and* $\alpha = \alpha_1, \dots, \alpha_n$ *are the conjugates of* α. *Further suppose that*

$$F = F(x_1, \dots, x_k; \alpha_1, \dots, \alpha_n) \in \mathbf{Q}[x_1, \dots, x_k; \alpha_1, \dots, \alpha_n], \qquad k \ge 0,$$

and that, as a polynomial in $\alpha_1, \ldots, \alpha_n$ *with coefficients in* $\mathbf{Q}[x_1, \ldots, x_k]$, F *is a symmetric polynomial in* $\alpha_1, \ldots, \alpha_n$. *Then*

$$F = F(x_1, \ldots, x_k) \in \mathbf{Q}[x_1, \ldots, x_k];$$

and $F \in \mathbf{Q}$ *in the case* $k = 0$. *In addition, if we also have* $\alpha \in \mathbf{Z_A}$ *and*

$$F = F(x_1, \ldots, x_k; \alpha_1, \ldots, \alpha_n) \in \mathbf{Z}[x_1, \ldots, x_k; \alpha_1, \ldots, \alpha_n], \qquad k \geq 0,$$

then it follows that

$$F = F(x_1, \ldots, x_k) \in \mathbf{Z}[x_1, \ldots, x_k];$$

and $F \in \mathbf{Z}$ *in the case* $k = 0$.

Proof. We consider F as a polynomial in $\alpha_1, \ldots, \alpha_n$ with coefficients in $\mathbf{Q}[x_1, \ldots, x_k]$. Since F is a symmetric polynomial in $\alpha_1, \ldots, \alpha_n$, it follows by the theorem on symmetric polynomials that it can be written as a polynomial in the elementary symmetric polynomials $\sigma_1, \ldots, \sigma_n$ with coefficients in $\mathbf{Q}[x_1, \ldots, x_k]$. But the elementary symmetric polynomials $\sigma_1, \ldots, \sigma_n$ are equal (up to sign) to the coefficients of the minimal polynomial of the algebraic number α, and so they are elements of $\mathbf{Q}$. Hence, $F = F(x_1, \ldots, x_k) \in \mathbf{Q}[x_1, \ldots, x_k]$, and $F \in \mathbf{Q}$ in the case $k = 0$. The proof of the second part of the lemma is similar; we further use the fact that all of the coefficients of the minimal polynomial of α lie in $\mathbf{Z}$.

One can generalize the problem of approximating real numbers by rational numbers by studying the approximation of complex numbers by certain classes of algebraic numbers. Given a complex number α, we consider the modulus of the difference

$$|\alpha - \theta| \tag{43}$$

for different θ, where θ is an algebraic number. Since the field $\mathbf{A}$ is everywhere dense in the complex numbers (not just the reals), it follows that α can be approximated arbitrarily closely by elements $\theta \in \mathbf{A}$. We might study the approximations of α by elements θ of a fixed algebraic number field $\mathbf{K}$, or by the primitive elements of that field, or by all algebraic numbers θ of some fixed degree.

It is natural to estimate the magnitude in (43) using positive functions $\varphi(H, k)$, where $H = H_\theta$ is the height of θ and $k = \deg \theta$. We consider the inequality

$$|\alpha - \theta| < \varphi(H, k),$$

and ask: for which functions $\varphi(H, k)$ does this inequality have only finitely many solutions, and for which functions does it have an infinite set of solutions in algebraic numbers θ?

The behavior of the difference (43) for different algebraic numbers θ of degree k and height H is closely related to the behavior of $|P(\alpha)|$ for a fixed complex number α and different polynomials $P(x) \in \mathbf{Z}[x]$ of degree k and height H. In fact, if $P(\theta) = 0$, then $|P(\alpha)|$ is small whenever α is close to θ, i.e., whenever the difference (43) is small. Conversely, if $|P(\alpha)|$ is small, then at least one of the differences $|\alpha - \theta_s|$, $1 \leq s \leq k$, is small, where $\theta_1, \ldots, \theta_k$ are the roots of $P(x)$. This means that, in some sense, the magnitude $|P(\alpha)|$ characterizes the order of approximation of α by algebraic numbers of height H and degree k.

Usually, for fixed $\alpha \in \mathbf{C}$ we want to find a lower bound

$$|P(\alpha)| > \varphi(H, k), \tag{44}$$

where $\varphi(H, k)$ is a function of the height H and the degree k of the polynomial $P(x)$, which holds for all polynomials $P(x) \in \mathbf{Z}[x]$, not necessarily irreducible, for which $P(\alpha) \neq 0$. This problem generalizes to the case of several complex numbers $\alpha_1, \ldots, \alpha_m$ and polynomials $P(x_1, \ldots, x_m) \in \mathbf{Z}[x_1, \ldots, x_m]$.

In this section we shall obtain the simplest type of bounds for $|\alpha - \theta|$, $|P(\alpha)|$, and $|P(\alpha_1, \ldots, \alpha_m)|$ when $\alpha, \alpha_1, \ldots, \alpha_m$ are algebraic; in Chapters 11–13 we shall do the same for certain classes of transcendental numbers.

There is an estimate of the form (44) for $\alpha \in \mathbf{A}$ which is a generalization of Liouville's Theorem. For $n \geq 1$ the inequality (35) can be rewritten in the following form:

$$|q\alpha - p| > \frac{c}{q^{n-1}}, \qquad \frac{p}{q} \neq \alpha. \tag{45}$$

We denote

$$P = P(x) = a_1 x + a_0, \qquad a_0 \in \mathbf{Z}, \qquad a_1 \in \mathbf{N}, \qquad H_P = H.$$

Now from (45) it follows that, if $P(\alpha) \neq 0$, then

$$|P(\alpha)| > \frac{c}{a_1^{n-1}} \geq \frac{c}{H^{n-1}}.$$

We shall show that a similar result holds for polynomials $P(x) \in \mathbf{Z}[x]$ of any degree k. We first prove a lemma, which is of independent interest.

Lemma 2. *If*

$$f(x) = b_n x^n + \cdots + b_1 x + b_0, \qquad f(x) \in \mathbf{C}[x], \qquad b_n b_0 \neq 0, \qquad H_{f(x)} = H,$$

and $f(\alpha) = 0$, *then the following inequalities hold:*

$$\frac{|b_0|}{H + |b_0|} < |\alpha| < \frac{H + |b_n|}{|b_n|}, \qquad |b_n \alpha| < 2H.$$

Proof. It is enough to prove the upper bound in the first inequality: the lower bound follows from the upper bound, because $1/\alpha$ is a root of the polynomial $b_0x^n + \cdots + b_{n-1}x + b_n$; and the last inequality is an immediate consequence of the upper bound.

If $|\alpha| \le 1$, then the upper bound obviously holds. Now suppose that $|\alpha| > 1$. Then

$$|b_n||\alpha|^n = |b_{n-1}\alpha^{n-1} + \cdots + b_1\alpha + b_0| \le H(|\alpha|^{n-1} + \cdots + |\alpha| + 1) =$$
$$= H\frac{|\alpha|^n - 1}{|\alpha| - 1} < H\frac{|\alpha|^n}{|\alpha| - 1},$$

from which we obtain

$$|\alpha| < \frac{H + |b_n|}{|b_n|}.$$

The lemma is proved.

Corollary. *If α is a nonzero algebraic number of height h, then its size satisfies the bounds*

$$\frac{1}{2h} < \overline{|\alpha|} < h + 1 \le 2h. \tag{46}$$

Theorem 8. *Suppose that $\alpha \in \mathbf{A}$, $\deg\alpha \le n$, $H_\alpha \le h$, and*

$$P = P(x) \in \mathbf{Z}[x], \qquad \deg P \le k, \qquad H_P \le H.$$

Then either $P(\alpha) = 0$, or else the following inequality holds:

$$|P(\alpha)| \ge \frac{c^k}{H^{n-1}}, \qquad c = \frac{1}{3^{n-1}h^n}. \tag{47}$$

Proof. Without loss of generality we may assume that $\deg\alpha = n$, $H_\alpha = h$, $\deg P = k$, $H_P = H$. Suppose that $P(\alpha) \ne 0$. Then $H \ge 1$. We let b_n denote the leading coefficient in the primitive and irreducible polynomial in $\mathbf{Z}[x]$ having α as a root, where $b_n > 0$. Then $\xi = b_n^k P(\alpha) \in \mathbf{Z_A}$.

Of $n = 1$, then

$$|\xi| = b_n^k|P(\alpha)| \ge 1, \qquad |P(\alpha)| \ge \frac{1}{b_n^k} \ge \frac{1}{h^k},$$

which proves (47) in this case.

Now suppose that $n > 1$. We have

$$|\mathcal{N}(\xi)| = b_n^k |P(\alpha)| \prod_{i=2}^{n} b_n^k |P(\alpha_i)| \geq 1. \tag{48}$$

Using Lemma 2, we obtain the bound

$$b_n^k |P(\alpha_i)| \leq b_n^k H(1 + |\alpha_i| + \cdots + |\alpha_i|^k) \leq b_n^k H(1 + |\alpha_i|)^k < (3h)^k H. \tag{49}$$

From (48) and (49) we have

$$|P(\alpha)| > \frac{1}{b_n^k (3h)^{k(n-1)} H^{n-1}} \geq \frac{1}{(3^{n-1} h^n)^k H^{n-1}},$$

which completes the proof of the theorem.

Theorem 9. *If α is an algebraic number of degree n, $n \geq 1$, then there exists a constant $c = c(\alpha) > 0$ such that the following inequality holds for any algebraic number θ of degree k, $k \geq 1$, and height H for which $\theta \neq \alpha$:*

$$|\alpha - \theta| > \frac{c^k}{H^n}. \tag{50}$$

Proof. Suppose that θ is not equal to α and is a root of the irreducible polynomial

$$P(x) = a_k x^k + \cdots + a_1 x + a_0, \qquad P(x) \in \mathbf{Z}[x], \qquad a_k > 0.$$

There are three possible cases.

1. $k = 1$. In this case the theorem follows from Liouville's Theorem.

2. $k > 1$ and $P(\alpha) = 0$. Then $k = n$, and θ is a conjugate of α. Let $\delta = \delta(\alpha)$ be any constant such that

$$0 < \delta < \min_{1 \leq i < j \leq n} |\alpha_i - \alpha_j|.$$

Then

$$|\alpha - \theta| > \delta \geq \frac{(\sqrt[n]{\delta}\,)^k}{H^n} = \frac{c^k}{H^n}. \tag{51}$$

The inequality (51) shows that the theorem holds in this case.

3. $k > 1$ and $P(\alpha) \neq 0$. If $|\alpha - \theta| \geq 1$, then

$$|\alpha - \theta| \geq \frac{1}{H^n}. \tag{52}$$

On the other hand, if $|\alpha - \theta| < 1$, then

$$|\theta| < |\alpha| + 1. \tag{53}$$

We write

$$|\alpha - \theta| = \frac{|P(\alpha)|}{|P_1(\alpha)|}, \tag{54}$$

and we bound $|P_1(\alpha)|$ from above. We have

$$\begin{aligned} P_1(x) &= \frac{P(x)}{x-\theta} = \frac{P(x)-P(\theta)}{x-\theta} = \\ &= a_k\frac{x^k-\theta^k}{x-\theta} + a_{k-1}\frac{x^{k-1}-\theta^{k-1}}{x-\theta} + \cdots + a_1\frac{x-\theta}{x-\theta} = \\ &= a_k(x^{k-1}+\theta x^{k-2}+\cdots+\theta^{k-1}) + \\ &\qquad + a_{k-1}(x^{k-2}+\theta x^{k-3}+\cdots+\theta^{k-2}) + \cdots + a_1 = \\ &= \sum_{l=0}^{k-1} g_l(\theta)x^l, \end{aligned} \tag{55}$$

where

$$g_l(\theta) = a_{l+1} + a_{l+2}\theta + \cdots + a_k\theta^{k-l-1}, \qquad l = 0, 1, \ldots, k-1. \tag{56}$$

From (56) and (53) we find that

$$\begin{aligned} |g_l(\theta)| &< H(1+|\theta|+\cdots+|\theta|^k) \le H(1+|\theta|)^k < \\ &< (|\alpha|+2)^k H, \qquad l = 0, 1, \ldots, k-1, \end{aligned} \tag{57}$$

and from (55) and (57), if we use the Corollary to Lemma 2 (see (46)), we find that

$$\begin{aligned} |P_1(\alpha)| &< (|\alpha|+2)^k H(1+|\alpha|+\cdots+|\alpha|^k) \le \\ &\le (|\alpha|+2)^k(1+|\alpha|)^k H < ((h+2)(h+3))^k H. \end{aligned} \tag{58}$$

By Theorem 8, there exists a constant $c_0 = c_0(\alpha)$, $0 < c_0 \le 1$, such that

$$|P(\alpha)| > \frac{c_0^k}{H^{n-1}}, \qquad c_0 = \frac{1}{3^{n-1}h^n}. \tag{59}$$

Substituting (58) and (59) into (54), we obtain the bound

$$|\alpha - \theta| > \frac{c^k}{H^n}, \qquad c = \frac{c_0}{(h+2)(h+3)} < 1. \tag{60}$$

Now (52) and (60) imply the desired inequality (50). The theorem is proved.

Theorem 10. *Suppose that α is a complex number such that for any $\nu \in \mathbf{R}$, $\nu > 0$, the inequality*

$$|\alpha - \theta| < \frac{1}{H_\theta^\nu}$$

has an infinite set of solutions in algebraic numbers θ of degree $\deg \theta \le k$ and height H_θ. Then α is transcendental.

The proof of Theorem 10 is just like that of Theorem 6, except that Theorem 9 is used in place of Liouville's Theorem.

We shall later have need of a theorem which generalizes Theorem 8 to the case of a polynomial in several algebraic numbers. Its proof is similar to the proof of Theorem 8, but it also uses the fact that there exists an algebraic number field $\mathbf{K} = \mathbf{Q}(\theta)$ containing the given algebraic numbers $\alpha_1, \dots, \alpha_m$.

Theorem 11. *Suppose that $\alpha_1, \dots, \alpha_m$ are algebraic numbers, and h is the degree of an algebraic number field $\mathbf{K} = \mathbf{Q}(\theta)$ which contains these numbers. Then there exists a constant $c = c(\alpha_1, \dots, \alpha_m) > 0$ such that for any polynomial*

$$P = P(z_1, \dots, z_m) \in \mathbf{Z}[z_1, \dots, z_m], \qquad \deg_z P \le k, \qquad H_P \le H,$$

either $P(\alpha_1, \dots, \alpha_m) = 0$ or else

$$|P(\alpha_1, \dots, \alpha_m)| \ge \frac{c^k}{H^{h-1}}. \tag{61}$$

Proof. Since our constant c will satisfy the condition $0 < c \le 1$, without loss of generality we may assume that $\deg_z P = k$, $H_P = H$. Suppose that $P(\alpha_1, \dots, \alpha_m) \ne 0$. Let $\alpha_i = \alpha_i^{(1)}, \dots, \alpha_i^{(h)}$ be the conjugates of α_i in the field $\mathbf{K}$, $i = 1, \dots, m$, and let $r \in \mathbf{N}$ be chosen so that $r\alpha_i \in \mathbf{Z_K}$, $i = 1, \dots, m$. Then

$$\xi = r^k P(\alpha_1, \dots, \alpha_m) \in \mathbf{Z_K}, \qquad \xi \ne 0. \tag{62}$$

Two cases are possible.

1. $h = 1$. Then, by (62), we have

$$|\xi| = r^k |P(\alpha_1, \dots, \alpha_m)| \ge 1, \qquad |P(\alpha_1, \dots, \alpha_m)| \ge \frac{1}{r^k}. \tag{63}$$

2. $h > 1$. We denote

$$A_j = P(\alpha_1^{(j)}, \dots, \alpha_m^{(j)}), \qquad j = 1, \dots, h.$$

The numbers $A_1, \ldots, A_h$ are conjugates in the field $\mathbf{K}$. Hence, by (62) and the properties of the norm,

$$|\mathcal{N}(\xi)| = |\mathcal{N}(r^k A_1)| = r^{kh}|A_1 \cdots A_h| \geq 1,$$

and so

$$|A_1| \geq \left(r^{kh} \prod_{j=2}^{h} |A_j|\right)^{-1}. \tag{64}$$

If we write

$$P(z_1, \ldots, z_m) = \sum_{0 \leq k_1 + \cdots + k_m \leq k} c_{k_1,\ldots,k_m} z_1^{k_1} \cdots z_m^{k_m}, \qquad c_{k_1,\ldots,k_m} \in \mathbf{Z},$$

then

$$|A_j| \leq H\left(1 + \sum_{i=1}^{m} \overline{|\alpha_i|}\right)^k = c_0^k H, \qquad c_0 = 1 + \sum_{i=1}^{m} \overline{|\alpha_i|}. \tag{65}$$

From (64) and (65) we obtain:

$$|P(\alpha_1, \ldots, \alpha_m)| > \frac{1}{(r^h c_0^{h-1})^k H^{h-1}} = \frac{c^k}{H^{h-1}}, \qquad c = \frac{1}{r^h c_0^{h-1}}. \tag{66}$$

Now (63) and (66) imply (61), as desired.

We note that, using the Corollary to Lemma 2, we can express the constant c in terms of the heights of the numbers $\alpha_1, \ldots, \alpha_m$.

§ 5. Further refinements and generalizations of Liouville's Theorem

Liouville's Theorem gives an upper bound for the order of approximation of an algebraic number by rational numbers. Hence, after the theorem was published, the question naturally arose: *Given an algebraic number α of degree n, $n \geq 3$, find a constant $\nu = \nu(\alpha)$ such that the inequality*

$$\left|\alpha - \frac{p}{q}\right| < \frac{1}{q^{\nu+\epsilon}} \tag{67}$$

has only finitely many integer solutions $p \in \mathbf{Z}$, $q \in \mathbf{N}$ *if* $\epsilon > 0$ *and infinitely many solutions if* $\epsilon < 0$. From Liouville's Theorem it follows that $\nu(\alpha) \leq n$ for any algebraic number α of degree n.

In 1909, Thue [78:2] published a method for obtaining results on the approximation of algebraic numbers. He used this method to lower significantly the bound for $\nu(\alpha)$. He showed that $\nu(\alpha) \leq n/2 + 1$ for any algebraic number α of degree $n \geq 2$. In 1908 in [78:1] he had obtained a similar result, but only for numbers of the form $\sqrt[n]{a/b}$, where $a, b \in \mathbf{N}$.

Thue's Theorem. *If* α *is an algebraic number of degree* n, $n \geq 2$, *and if* ϵ *is any positive number, then the inequality*

$$\left|\alpha - \frac{p}{q}\right| < \frac{1}{q^{(n/2)+1+\epsilon}} \tag{68}$$

has only finitely many solutions $p \in \mathbf{Z}$, $q \in \mathbf{N}$.

In Thue's method one constructs a polynomial $P(x, y) \in \mathbf{Z}[x, y]$ which vanishes at the point (α, α) along with several of its successive partial derivatives with respect to x, and which satisfies a "sufficiently good" upper bound on the modulus of its coefficients. In the proof of Thue's Theorem one supposes that (68) has infinitely many solutions, and one chooses two solutions p_1/q_1 and p_2/q_2 with large denominators. One then shows that $P(p_1/q_1, p_2/q_2) \neq 0$, after which Thue's Theorem can be obtained in the same manner as Liouville's Theorem before.

Thue's Theorem made it possible for Thue to prove the theorem mentioned in the Introduction on finiteness of the number of solutions of the Diophantine equation (0.1).

In 1921, Siegel [72:1] used Thue's method to lower significantly the bound for the exponent in (67). In particular, his result implies that $\nu(\alpha) < 2\sqrt{n}$ for any algebraic number α of degree $n \geq 2$. In the same paper Siegel extended Thue's method so that it can be applied to approximations of an algebraic number by algebraic numbers.

Subsequent attempts to lower the bound for $\nu(\alpha)$ through the use of two solutions of (67) did not lead to major improvements. However, Dyson in 1947 [15:1] and Gel'fond in 1948 [24:6] independently were able to show that $\nu(\alpha) < \sqrt{2n}$ for any algebraic number α of degree $n \geq 2$.

It seemed that if one could construct a polynomial in several variables with properties similar to those of Thue's polynomial, then, using the existence of many solutions of (67), one could obtain a further lowering of the bound for $\nu(\alpha)$. However, for a long time no one was able to do this, because of the difficulty of constructing the polynomial.

In 1955 K. F. Roth [63:1] overcame these difficulties, and proved that $\nu(\alpha) = 2$ for any algebraic number α of degree $n \geq 2$.

Roth's Theorem. *If α is an algebraic number of degree n, $n \geq 2$, and if ϵ is any positive number, then the inequality*

$$\left|\alpha - \frac{p}{q}\right| < \frac{1}{q^{2+\epsilon}}$$

has only finitely many solutions $p \in \mathbf{Z}$, $q \in \mathbf{N}$.

We note that when $n = 2$, Liouville's Theorem gives a more precise result than the theorems of Thue, Siegel and Roth.

Roth's result was a major advance in the theory of approximations of an algebraic number by rational numbers, but it did not resolve everything. Further research was needed in order to determine for which positive functions $\varphi(q)$ with rate of growth slower than any power function does the inequality

$$\left|\alpha - \frac{p}{q}\right| < \frac{1}{q^2\varphi(q)}, \qquad \alpha \in \mathbf{A}, \qquad \deg\alpha = n, \qquad n \geq 3,$$

have finitely many or infinitely many solutions $p \in \mathbf{Z}$, $q \in \mathbf{N}$.

Let us again consider Liouville's Theorem. The constant c in it is *effective*, i.e., it can be computed once α (and hence the degree n and height h) is given. Using the corollary of Lemma 2, if $n \geq 2$ in Liouville's Theorem we can set, for example, $c = 1/(5h)^{n-1}$.

Unlike Liouville's Theorem, the much stronger theorems of Thue, Siegel and Roth are not effective. This accounts for the fact that Thue's theorem on Diophantine equations (see the Introduction), which is proved using his theorem on approximation of algebraic numbers, is also ineffective. This theorem establishes finiteness of the number of solutions to a Diophantine equation of the type under consideration, but it does not give any bound for the solutions. Thue's method does not enable one to obtain an effective strengthening of Liouville's Theorem. For many years the attempts to solve this problem did not meet with success.

It was observed (see Gel'fond's book [24:8]) that the problem of making Thue's method effective is connected with the problem of obtaining effective bounds for linear forms in the logarithms of several algebraic numbers with coefficients in $\mathbf{Z}$. However, Gel'fond was able to obtain effective bounds only in the case of logarithms of two algebraic numbers.

In 1966, Baker [2:1] significantly strengthened Gel'fond's 1934 method, obtaining an effective bound for the modulus of a linear form in the logarithms of several algebraic numbers. This made it possible for him to prove an effective strengthening of Liouville's Theorem and an effective version of Thue's theorem on Diophantine equations, along with some other results. Several mathematicians then used Baker's method to improve and generalize his work. We shall give two theorems of N. I. Fel'dman from his 1971 paper [17:6] which give an effective power sharpening of the exponent in Liouville's Theorem and a corresponding effective version of Thue's theorem on Diophantine equations.

Theorem 12. *If* $\alpha \in \mathbf{A}$, $\deg \alpha = n$, $n \geq 3$, *then there exist effective constants* $a_1 > 0$ *and* $c_1 > 0$, *depending only on* α, *such that for any* $p \in \mathbf{Z}$ *and* $q \in \mathbf{N}$

$$\left|\alpha - \frac{p}{q}\right| > c_1 q^{a_1 - n}.$$

Theorem 13. *Let* $f = f(x, y)$ *be an irreducible form with coefficients in* $\mathbf{Z}$, *and let* $\deg_{x,y} f \geq 3$. *Then there exist effective constants* $a_2 > 0$ *and* $c_2 > 0$, *depending only on the form* f, *such that all integral solutions of the Diophantine equation*

$$f(x, y) = m, \qquad m \in \mathbf{Z},$$

satisfy the inequality

$$\max(|x|, |y|) < c_2 m^{a_2}.$$

Remarks

The reader can become acquainted with the theory of approximation of real numbers from the books by Khinchin [30:2], Cassels [10:1], Lang [37:5], Shidlovskii [70:35] and Schmidt [67:4]. Basic facts about algebraic numbers are contained in the books by Fel'dman [17:6] and Shidlovskii [70:35]. For more detailed treatments of algebraic number theory, the reader is referred to Borevich and Shafarevich [7:1], Hecke [26:1], Weyl [88:1], Lang [37:4]. A discussion of Liouville numbers can be found in Maillet's monograph [44:1], which apparently was the first book devoted to the properties of transcendental numbers.

Topics in the theory of approximation of algebraic numbers are discussed in considerable detail in the monographs by Stolarsky [75:1] and Fel'dman [17:6]. Proofs of Thue's and Roth's theorems are in [17:6], Siegel's theorem is proved in [70:35], and Roth's theorem is proved in [10:1]. Roth's own proof of his theorem is in [63:1].

Surveys of work on approximation of real and algebraic numbers can be found in Koksma's book [31:1] and in the article [18:1] by Fel'dman and Shidlovskii. The latter article gives bounds which are more precise than those in Theorems 8–10 above.

We indicate some results which generalize the Thue-Siegel-Roth Theorem.

In 1921 in [72:1], Siegel proved the following. *Let* $\alpha \in \mathbf{A}$, $\deg \alpha = n$, $n \geq 2$.

Then:

1) *if* $\deg_{\mathbf{K}} \alpha = m$, $m \geq 2$, *and if* $s \in \mathbf{N}$, $s < m$, *then for any* $\epsilon > 0$ *the inequality*

$$|\alpha - \theta| \leq H_\theta^{-\left(\frac{m}{s+1}+s+\epsilon\right)}$$

has only finitely many solutions in primitive numbers θ *of* **K**;

2) *if* $\nu, s \in \mathbf{N}$ *and if* $s < \nu$, *then for any* $\epsilon > 0$ *the inequality*

$$|\alpha - \theta| \leq H_\theta^{-\nu\left(\frac{n}{s+1}+s+\epsilon\right)}$$

has only finitely many solutions $\theta \in \mathbf{A}$, $\deg \theta \leq \nu$.

In papers by Siegel [72:2], Schneider [68:3] and Mahler [42:2], theorems of the following type are proved: if $\alpha \in \mathbf{A}$ has "very good" approximations of order $\varphi(q)$ by rational numbers, then the solutions of the corresponding inequality

$$\left|\alpha - \frac{p}{q}\right| < \varphi(q)$$

are very sparse.

Various classes of numbers have been proved to be transcendental using the theorems of Schneider and Mahler. For example, in 1937 in [42:3], Mahler proved that the number α in § 1 of the Introduction is transcendental.

In 1956 Le Veque [39:1] generalized Roth's theorem, proving that *if* $\alpha \in \mathbf{A}$, *then the inequality*

$$|\alpha - \theta| \leq H_\theta^{-k}, \qquad k > 2, \tag{69}$$

has only finitely many solutions θ *from a fixed algebraic number field* **K**.

In 1969, Wirsing [89:1] generalized Roth's theorem to the case of approximations of an algebraic number by algebraic numbers of bounded degree. He proved that *if* $\alpha \in \mathbf{A}$ *and* $t \in \mathbf{N}$, *then for any* $k \geq 2t$ *the inequality (69) can have only finitely many solutions* $\theta \in \mathbf{A}$ *of degree not exceeding* t.

Other authors have generalized Roth's theorem in various directions. The best results of Schmidt, obtained in 1970-1975, can be found in [67:1-4]. He generalized Roth's theorem to the case of approximations of an algebraic number α by elements of **A** of bounded degree and to different cases of simultaneous approximation of algebraic numbers by numbers in **Q** or **A**. We shall only give one of his results.

Let $\alpha \in \mathbf{A}$, $t \in \mathbf{N}$, *and* $\epsilon > 0$. *Then there are only finitely many numbers* $\theta \in \mathbf{A}$, $\deg \theta \leq t$, *for which*

$$|\alpha - \theta| < H_\theta^{-(t+1+\epsilon)}.$$

We conclude with the following observation. One can use Liouville's Theorem to construct examples of transcendental numbers which have good approximations by rational numbers. On the other hand, if one uses the results on approximation of an algebraic number by irrational algebraic numbers, then one can construct examples of transcendental numbers which have the smallest possible best order of approximation by rational numbers, i.e., they have bounded terms in their continued fraction expansion. In [44:1] Maillet constructed some transcendental numbers with bounded continued fraction expansion. His proof is based on the fact that his numbers have good approximations by quadratic irrationalities.

Chapter 2
Arithmetic properties of the values of the exponential function at algebraic points

§ 1. Transcendence of *e*

Theorem 1. *The number* e *is irrational.*

Proof. e is given by the series

$$e = \sum_{k=0}^{\infty} \frac{1}{k!}. \tag{1}$$

We denote

$$A_n = n! \sum_{k=0}^{n} \frac{1}{k!} \quad \in \mathbf{N}, \qquad n = 1, 2, \ldots; \tag{2}$$

$$a_n = n! \sum_{k=n+1}^{\infty} \frac{1}{k!} > 0, \qquad n = 1, 2, \ldots. \tag{3}$$

We have

$$\begin{aligned} a_n &= \frac{1}{n+1}\left(1 + \frac{1}{n+2} + \frac{1}{(n+2)(n+3)} + \cdots\right) < \\ &< \frac{1}{n+1}\left(1 + \frac{1}{2} + \frac{1}{2^2} + \cdots\right) = \frac{2}{n+1}, \qquad n = 1, 2, \ldots. \end{aligned} \tag{4}$$

Combining (3) and (4), we obtain:

$$0 < a_n < \frac{2}{n+1} \leq 1, \qquad n = 1, 2, \ldots. \tag{5}$$

Using (1)–(3) and (5), we see that

$$n!e = (A_n + a_n) \notin \mathbf{N}, \qquad n = 1, 2, \ldots. \tag{6}$$

Now suppose that e is a rational number: $e = p/q$ with $p, q \in \mathbf{N}$. Setting $n = q$, we have $ne \in \mathbf{N}$, which contradicts (6). The theorem is proved.

Theorem 2. *e is not a quadratic irrationality.*

Proof. Suppose the contrary, i.e., that

$$ae^2 + be + c = 0, \qquad a, b, c, \in \mathbf{Z},$$

where a, b and c are not all zero. By Theorem 1, we have $a \neq 0$ and $c \neq 0$. We may suppose that $a > 0$. We have

$$ae + b + ce^{-1} = 0, \qquad a > 0, \;\; c \neq 0. \tag{7}$$

The number $1/e$ is given by the series

$$\frac{1}{e} = \sum_{k=0}^{\infty} \frac{(-1)^k}{k!}. \tag{8}$$

We denote

$$B_n = n! \sum_{k=0}^{n} \frac{(-1)^k}{k!}, \qquad B_n \in \mathbf{Z}, \qquad n = 1, 2, \ldots, \tag{9}$$

and

$$\begin{aligned} b_n &= n! \sum_{k=n+1}^{\infty} \frac{(-1)^{k-n-1}}{k!} = \\ &= \frac{1}{n+1} - \frac{1}{(n+1)(n+2)} + \frac{1}{(n+1)(n+2)(n+3)} - \cdots. \end{aligned} \tag{10}$$

Note that b_n is given by an alternating series whose terms decrease monotonically in absolute value. Hence,

$$0 < \frac{1}{n+1} - \frac{1}{(n+1)(n+2)} < b_n < \frac{1}{n+1}. \tag{11}$$

From (1), (7) and (8), along with the definitions (2), (3), (9) and (10), we obtain:

$$\begin{aligned} n!(ae + b + ce^{-1}) &= (aA_n + bn! + cB_n) + (aa_n + (-1)^{n+1} cb_n) = \\ &= C_n + c_n = 0. \end{aligned} \tag{12}$$

By (2) and (9), we have

$$C_n = (aA_n + bn! + cB_n) \in \mathbf{Z}, \qquad n = 1, 2, \ldots. \tag{13}$$

We choose any n satisfying the two inequalities

$$n \geq 2a + |c|, \qquad (-1)^{n+1} c > 0. \tag{14}$$

Since $a > 0$, it now follows from (5), (11) and (14) that

$$0 < c_n = a a_n + (-1)^{n+1} c b_n < \frac{2a + |c|}{n+1} < 1. \tag{15}$$

But the condition (13) and the inequality (15) contradict the equation (12). The theorem is proved.

We note that the proofs of Theorems 1 and 2 are based on the fact that e and $1/e$ have rational approximations p/q of higher order than $1/q$.

To prove that e is transcendental, we must show that e cannot be a root of a polynomial of arbitrary degree with rational coefficients. This is not as simple as proving the much weaker assertions in Theorems 1 and 2. We shall apply Hermite's method for the proof.

We first prove two lemmas.

Lemma 1. *If $g(x) \in \mathbf{Z}[x]$, then for any $k \in \mathbf{N}$ all of the coefficients of the k-th derivative $g^{(k)}(x)$ are divisible by $k!$.*

Proof. Since differentiation is a linear operation, it suffices to prove the lemma for the polynomial x^s, $s > 0$. But the k-th derivative of x^s is zero if $k > s$, and if $1 \leq k \leq s$, then it is equal to $k!\binom{s}{k}x^{s-k}$, in which the binomial coefficient $\binom{s}{k}$ is an integer. The lemma is proved.

Lemma 2. *Let $f(x)$ be a degree ν polynomial with real coefficients, and set*

$$F(x) = f(x) + f'(x) + \cdots + f^{(\nu)}(x). \tag{16}$$

Then

$$e^x \int_0^x f(t)e^{-t}dt = F(0)e^x - F(x), \tag{17}$$

where x is a real or complex number.

Proof. Integrating by parts, we obtain the relation

$$\int_0^x f(t)e^{-t}dt = f(0) - f(x)e^{-x} + \int_0^x f'(t)e^{-t}dt. \tag{18}$$

If we repeat this process $\nu + 1$ times, we arrive at the equality

$$\int_0^x f(t)e^{-t}dt = F(0) - F(x)e^{-x},$$

from which (17) follows.

The equality (17) is called "Hermite's identity."

Theorem 3 (Hermite). *e is transcendental.*

Proof. We suppose the contrary, i.e., that e is an algebraic number of degree m. Then

$$a_m e^m + \cdots + a_1 e + a_0 = 0, \qquad a_0 \neq 0, \qquad a_k \in \mathbf{Z}, \quad k = 0, 1, \ldots, m. \tag{19}$$

If we set $x = k$ in Hermite's identity (17), where $k = 0, 1, \ldots, m$, we obtain:

$$F(0)e^k - F(k) = e^k \int_0^k f(t)e^{-t}dt, \qquad k = 0, 1, \ldots, m. \tag{20}$$

We multiply both sides of (20) by a_k, and then add the resulting equations for $k = 0, 1, \ldots, m$. Using (19), we find that

$$-\sum_{k=0}^{m} a_k F(k) = \sum_{k=0}^{m} a_k e^k \int_0^k f(t)e^{-t}dt. \tag{21}$$

The equality (21) holds for any polynomial $f(x)$ with real coefficients. We set

$$f(x) = \frac{1}{(n-1)!} x^{n-1}(x-1)^n \cdots (x-m)^n, \tag{22}$$

where n is a large natural number satisfying some conditions which will be given later.

We show that the left side of (21) will then be a nonzero rational integer, while the right side will have absolute value less than 1. This contradiction will prove the theorem.

The polynomial $f(x)$ in (22) has 0 as a root of multiplicity $n-1$ and $1, \ldots, m$ as roots of multiplicity n. Hence,

$$f^{(l)}(0) = 0, \qquad l = 0, 1, \ldots, n-2, \tag{23}$$

$$f^{(n-1)}(0) = (-1)^{mn}(m!)^n, \tag{24}$$

$$f^{(l)}(k) = 0, \qquad l = 0, 1, \ldots, n-1; \quad k = 1, \ldots, m. \tag{25}$$

By Lemma 1, the l-th derivative of $x^{n-1}(x-1)^n \cdots (x-m)^n$ has coefficients which are divisible by $l!$. This implies that for $l \geq n$ the coefficients of $f^{(l)}(x)$ are divisible by n. Hence, from (16), (23) and (24) we find that

$$F(0) = \sum_{l=n-1}^{(m+1)n-1} f^{(l)}(0) = (-1)^{mn}(m!)^n + nA, \qquad A \in \mathbf{Z}, \tag{26}$$

and from (25) we find that

$$F(k) = \sum_{l=n}^{(m+1)n-1} f^{(l)}(k) = nB_k, \qquad B_k \in \mathbf{Z}, \quad k = 1, \ldots, m. \tag{27}$$

Now let n be any integer satisfying the conditions

$$(n, m!) = 1, \qquad n > |a_0|. \tag{28}$$

It follows from (26) and (27) that all of the terms on the left in (21) are integers. Here the conditions (19) and (28) along with (26) imply that $a_0 F(0)$ is not divisible by n. But all of the other terms $a_k F(k)$ are divisible by n. Hence, the left side of (21) is a nonzero integer; in particular,

$$\left|\sum_{k=0}^{m} a_k F(k)\right| \geq 1. \tag{29}$$

We now find an upper bound for the right side of (21). On the interval $0 \leq x \leq m$ each of the factors $x - k$ in (22), $0 \leq k \leq m$, is bounded by m. Thus,

$$|f(x)| \leq \frac{m^{(m+1)n-1}}{(n-1)!}, \qquad 0 \leq x \leq m,$$

and

$$\begin{aligned} \left|\sum_{k=0}^{m} a_k \int_0^k f(t) e^{k-t} dt\right| &\leq \frac{m^{(m+1)n-1}}{(n-1)!} \sum_{k=0}^{m} |a_k| \int_0^k e^{k-t} dt < \\ &< \frac{m^{(m+1)n}}{(n-1)!} e^m \sum_{k=0}^{m} |a_k| = C_0 \frac{C^n}{(n-1)!}, \end{aligned} \tag{30}$$

where the constants C_0 and C do not depend on n.

From (21) and (29)–(30) we obtain the inequality

$$1 \leq \left|\sum_{k=0}^{m} a_k F(k)\right| < C_0 \frac{C^n}{(n-1)!},$$

which is impossible when n is large, since the right side approaches zero as $n \to \infty$. This contradiction proves the theorem.

Let us consider the equalities in (20). From (26) and (27) it follows that $F(k)$, $k = 0, 1, \ldots, m$, are integers. If we estimate the right side of (20) in the same way as we did for the right side of (21), we see that it approaches zero as $n \to \infty$. Hence, for every n the fractions

$$\frac{F(k)}{F(0)}, \qquad k = 1, \ldots, m,$$

are simultaneous approximations to the values e^k, $k = 1, \ldots, m$. Thus, our proof that e is transcendental is based on a construction (using Hermite's identity) of a sequence of simultaneous approximations of powers of e.

§ 2. Transcendence of π

Theorem 4. *π is irrational.*

Proof. Let $f(x)$ be any polynomial with real coefficients, and set

$$F(x) = f(x) - f''(x) + f^{(4)}(x) - f^{(6)}(x) + \cdots.$$

Clearly, the sum on the right terminates after finitely many terms, and $F(x)$ is a polynomial.

We have the relation

$$\frac{d}{dx}(F'(x)\sin x - F(x)\cos x) = (F''(x) + F(x))\sin x = f(x)\sin x,$$

which, if we integrate, gives

$$\int_0^{\pi} f(x)\sin x\, dx = F(0) + F(\pi). \tag{31}$$

The identity (31) is the analogue of Hermite's identity for the function $\sin x$; it is satisfied for any polynomial $f(x)$ with real coefficients.

We suppose that the theorem is false, i.e., that π can be written in the form a/b, $a, b \in \mathbf{N}$. In (31) we set

$$f(x) = \frac{b^n}{n!}x^n(\pi - x)^n = \frac{1}{n!}x^n(a - bx)^n,$$

where n is a large natural number.

We shall show that with this choice of $f(x)$ the right side of (31) is a natural number, but the left side satisfies the inequality

$$0 < \int_0^{\pi} f(x) \sin x \, dx < 1. \tag{32}$$

This leads to a contradiction, proving the theorem.

Since $f(x)\sin x > 0$ on the interval $0 < x < \pi$ and the integrand in (32) is continuous, the first inequality in (32) obviously holds. Next, we have $x^n(\pi - x)^n \le \pi^{2n}$ for $0 \le x \le \pi$. Hence,

$$\int_0^{\pi} f(x) \sin x \, dx \le \frac{b^n \pi^{2n}}{n!} \int_0^{\pi} \sin x \, dx = 2\frac{(a^2/b)^n}{n!} = 2\frac{C^n}{n!}.$$

Since $\lim_{n\to\infty} C^n/n! = 0$, it follows that there exists $n_0 \in \mathbf{N}$ such that $2C^n/n! < 1$ for $n \ge n_0$, in which case the inequality on the right in (32) is fulfilled.

The polynomial $f(x)$ has 0 as a root of multiplicity n. Hence,

$$f(0) = f'(0) = \cdots = f^{(n-1)}(0) = 0.$$

By Lemma 1, the l-th derivative of $x^n(a - bx)^n$ has all coefficients divisible by $l!$. This implies that all of the derivatives of $f(x)$ of order $l \ge n$ have integer coefficients. Hence, $f^{(n)}(0), \dots, f^{(2n)}(0)$ are integers, and so

$$f^{(l)}(0) \in \mathbf{Z}, \qquad l = 0, 1, 2, \dots.$$

We have

$$f(x) = f(\pi - x), \qquad f^{(l)}(x) = (-1)^l f^{(l)}(\pi - x), \qquad l = 1, 2, \dots,$$

$$f^{(l)}(\pi) = (-1)^l f^{(l)}(0) \in \mathbf{Z}, \qquad l = 0, 1, 2, \dots.$$

Hence,

$$(F(0) + F(\pi)) \in \mathbf{Z}. \tag{33}$$

The relations (32) and (33) contradict (31), and the theorem is proved.

Theorem 5 (Lindemann). *π is transcendental.*

Proof. The proof that follows is based on the equation $e^{\pi i} + 1 = 0$ and Hermite's identity (17). We suppose that the theorem is false, i.e., that π is an algebraic number. Then $\gamma = \pi i$ is also algebraic. Let $\nu = \deg \gamma$, and let $\gamma = \gamma_1, \dots, \gamma_\nu$ be the conjugates of γ. Since $e^\gamma + 1 = 0$, we have

$$\prod_{i=1}^{\nu}(1 + e^{\gamma_i}) = 0.$$

Expanding this product, we obtain

$$\prod_{i=1}^{\nu}(1 + e^{\gamma_i}) = \sum_{\epsilon_1=0}^{1} \cdots \sum_{\epsilon_\nu=0}^{1} e^{\epsilon_1\gamma_1+\cdots+\epsilon_\nu\gamma_\nu} = 0. \tag{34}$$

The exponents inside the multiple sum in (34) include some which are nonzero, e.g., when $\epsilon_1 = 1$ and $\epsilon_2 = \cdots = \epsilon_\nu = 0$, and also some which are zero, e.g., when $\epsilon_1 = \cdots = \epsilon_\nu = 0$. Suppose that there are precisely m nonzero exponents and $a = 2^\nu - m$ which are zero, $a \geq 1$. Then, if we let $\alpha_1, \ldots, \alpha_m$ denote the nonzero exponents, we can rewrite (34) as follows:

$$a + e^{\alpha_1} + \cdots + e^{\alpha_m} = 0, \qquad a \geq 1. \tag{35}$$

We now show that the numbers $\alpha_1, \ldots, \alpha_m$ are the set of roots of a polynomial $\psi(x) \in \mathbf{Z}[x]$ of degree m. To see this, we observe that the polynomial

$$\varphi(x) = \prod_{\epsilon_1=0}^{1} \cdots \prod_{\epsilon_\nu=0}^{1} (x - (\epsilon_1\gamma_1 + \cdots + \epsilon_\nu\gamma_\nu)),$$

considered as a polynomial in $\gamma_1, \ldots, \gamma_\nu$ with coefficients in $\mathbf{Z}[x]$, is symmetric in $\gamma_1, \ldots, \gamma_\nu$. Hence, by Lemma 1 of Chapter 1, $\varphi(x)$ is in $\mathbf{Q}[x]$. The roots of the degree 2^ν polynomial $\varphi(x)$ are $\alpha_1, \ldots, \alpha_m$ and 0 with multiplicity a. Hence, the degree m polynomial $x^{-a}\varphi(x) \in \mathbf{Q}[x]$ has precisely the number $\alpha_1, \ldots, \alpha_m$ as its roots. If we let $r \in \mathbf{N}$ be the least common denominator of the coefficients of this polynomial, then the polynomial

$$\psi(x) = \frac{r}{x^a}\varphi(x) = b_m x^m + \cdots + b_1 x + b_0 \in \mathbf{Z}[x], \qquad b_m > 0, \quad b_0 \neq 0,$$

also has precisely $\alpha_1, \ldots, \alpha_m$ as its roots.

In Hermite's identity (17) we successively set $x = \alpha_1, \ldots, \alpha_m$. If we add the resulting equations and use (35), we obtain:

$$-aF(0) - \sum_{k=1}^{m} F(\alpha_k) = \sum_{k=1}^{m} e^{\alpha_k} \int_0^{\alpha_k} f(t)e^{-t}dt. \tag{36}$$

The rest of the proof of Theorem 5 is similar to the proof of Theorem 3. In (36) we set

$$\begin{aligned} f(x) &= \frac{1}{(n-1)!} b_m^{mn-1} x^{n-1} \psi^n(x) = \\ &= \frac{1}{(n-1)!} b_m^{(m+1)n-1} x^{n-1} (x-\alpha_1)^n \cdots (x-\alpha_m)^n, \end{aligned} \tag{37}$$

where n is a large natural number. We shall show that with this choice of $f(x)$ the equality (36) leads to a contradiction.

As in the case of Theorem 3, we obtain equalities similar to (23), (24) and (26):

$$f^{(l)}(0) = 0, \qquad l = 0, 1, \ldots, n-2, \qquad f^{(n-1)}(0) = b_m^{mn-1} b_0^n,$$

$$F(0) = \sum_{l=n-1}^{(m+1)n-1} f^{(l)}(0) = b_m^{mn-1} b_0^n + nA, \qquad A \in \mathbf{Z}. \tag{38}$$

Since α_k is a root of $f(x)$ of multiplicity n, we obtain equations similar to (25):

$$f^{(l)}(\alpha_k) = 0, \qquad l = 0, 1, \ldots, n-1; \;\; k = 1, \ldots, m. \tag{39}$$

By Lemma 1, the l-th derivative of $x^{n-1}\psi^n(x)$ has integer coefficients which are all divisible by $n!$. Hence, for $l > n$ the coefficients of $f^{(l)}(x)$ are integers divisible by $b_m^{mn-1} n$. Then, from (39), we have

$$\begin{gathered} F(\alpha_k) = \sum_{l=n}^{(m+1)n-1} f^{(l)}(\alpha_k) = n b_m^{mn-1} \Phi(\alpha_k), \\ k = 1, \ldots, m, \qquad \Phi(z) \in \mathbf{Z}[z]. \end{gathered} \tag{40}$$

The numbers $\beta_k = b_m \alpha_k$, $k = 1, \ldots, m$, are algebraic integers which make up the complete set of roots of a degree m polynomial in $\mathbf{Z}[x]$ with leading coefficient 1. Furthermore,

$$b_m^{mn-1} \Phi(\alpha_k) = H(\beta_k), \qquad H(x) \in \mathbf{Z}[x].$$

Hence, by Lemma 1 of Chapter 1,

$$\sum_{k=1}^{m} b_m^{mn-1} \Phi(\alpha_k) = \sum_{k=1}^{m} H(\beta_k) = B, \qquad B \in \mathbf{Z}. \tag{41}$$

From (38), (40) and (41) we find that

$$aF(0) + \sum_{k=1}^{m} F(\alpha_k) = ab_0^n b_m^{mn-1} + n(aA + B). \tag{42}$$

Now let n be any natural number satisfying the conditions

$$(n, b_0 b_m) = 1, \qquad n > a. \tag{43}$$

Then the right side of (42) is an integer which is not divisible by n, and so is nonzero. Hence,

$$\left| aF(0) + \sum_{k=1}^{m} F(\alpha_k) \right| \geq 1. \tag{44}$$

We now find an upper bound for the right side of (36). Suppose that all of the points $\alpha_1, \ldots, \alpha_m$ are contained in the circle $|x| \leq R$. We denote

$$\max_{|x| \leq R} |b_m^m \psi(x)| = C,$$

where C does not depend on n. Then

$$\max_{|x| \leq R} |f(x)| \leq \frac{R^{n-1} C^n}{(n-1)!}.$$

Hence, there exists an n_0 such that for any $n \geq n_0$ which satisfies (43) we have the inequalities

$$\left| \sum_{k=1}^{m} e^{\alpha_k} \int_0^{\alpha_k} f(x) e^{-x} dx \right| \leq \sum_{k=1}^{m} \left| \int_0^{\alpha_k} |f(x)| |e^{(\alpha_k - x)}| dx \right| \leq$$
$$\leq \frac{R^{n-1} e^R}{(n-1)!} C^n \sum_{k=1}^{m} \left| \int_0^{\alpha_k} dx \right| \leq m e^R \frac{(RC)^n}{(n-1)!} < 1. \tag{45}$$

The inequalities (44) and (45), along with (36), lead to the contradiction $1 < 1$. The theorem is proved.

A consequence of Theorem 5 is the impossibility of squaring the circle. Namely, in 1837, P. L. Wantzel [86:1] showed that the line segments which can be constructed by ruler and compass are precisely those whose lengths can be expressed in terms of roots of quadratic equations with rational coefficients, roots of quadratic equations whose coefficients are roots of quadratic equations with rational coefficients, and so on, i.e., they can be expressed in terms of numbers obtained by successively solving a series of quadratic equations. The set of such numbers is a subset of the field **A**. Since we proved that π is transcendental, it follows that it is impossible to construct a segment of length $\sqrt{\pi}$ using a ruler and compass.

§ 3. Transcendence of the values of the exponential function at algebraic points

Theorem 6 (Lindemann). *If ξ is a nonzero algebraic number, then e^ξ is transcendental.*

This theorem can be proved using Hermite's identity, much in the same way as Theorem 5, but with somewhat more complicated terms in the transformations. The proof given below is based on a different idea, which originated in the work of Gel'fond [24:1,2] on Hilbert's seventh problem.

In this proof one uses the expansion of the function $e^{\xi z}$ in a Newton interpolation series to construct a sequence of polynomials $P_n(x, y)$ with integer coefficients for which $P_n(\xi, e^\xi)$ decreases rapidly as n increases. But if we suppose that both ξ and e^ξ are algebraic, and if we use the lower bound in Theorem 11 of Chapter 1 for a polynomial in two algebraic variables with integer coefficients, then we obtain a lower bound for $|P_n(\xi, e^\xi)|$ which contradicts the earlier upper bound. This will imply that e^ξ cannot be algebraic.

Before proving Theorem 6, we study the interpolation series for the function $e^{\xi z}$.

Suppose that $f(z)$ is an analytic function in the region D, and $z_1, \ldots, z_n$ are a fixed n-tuple of points of D (some of which may be the same). We define:

$$F_0(t) = 1, \qquad F_k(t) = (t - z_1) \cdots (t - z_k), \qquad k = 1, \ldots, n.$$

Let $z \in D$. We have the following obvious identity for each value of k, $k = 1, \ldots, n$:

$$\frac{1}{t - z}\left(1 - \frac{z - z_k}{t - z_k}\right) = \frac{1}{t - z_k}.$$

Multiplying both sides by $F_{k-1}(z)/F_{k-1}(t)$, we obtain

$$\frac{1}{t - z}\left(\frac{F_{k-1}(z)}{F_{k-1}(t)} - \frac{F_k(z)}{F_k(t)}\right) = \frac{F_{k-1}(z)}{F_k(t)}, \qquad k = 1, \ldots, n.$$

Taking the sum of these identities for $k = 1, \ldots, n$ and using the fact that $F_0(t) = 1$, we find that

$$\frac{1}{t-z} - \frac{F_n(z)}{F_n(t)(t-z)} = \sum_{k=1}^{n} \frac{F_{k-1}(z)}{F_k(t)},$$

or

$$\frac{1}{t-z} = \sum_{k=1}^{n} \frac{F_{k-1}(z)}{F_k(t)} + \frac{F_n(z)}{F_n(t)(t-z)}. \tag{46}$$

Let C be a simple closed contour bounding a region lying entirely in D and containing the points $z_1, \ldots, z_n$ and z. If we multiply both sides of (46) by $\frac{1}{2\pi i} f(t)$ and integrate over C in the positive direction, using Cauchy's formula, we obtain:

$$\begin{aligned} f(z) &= \frac{1}{2\pi i} \int_C \frac{f(t)}{t-z} dt = \\ &= \sum_{k=1}^{n} F_{k-1}(z) \cdot \frac{1}{2\pi i} \int_C \frac{f(t)}{F_k(t)} dt + \frac{1}{2\pi i} \int_C \frac{F_n(z) f(t)}{F_n(t)(t-z)} dt. \end{aligned} \tag{47}$$

We denote:

$$A_{k-1} = \frac{1}{2\pi i} \int_C \frac{f(t)}{F_k(t)} dt, \qquad k = 1, \ldots, n, \tag{48}$$

$$R_n(z) = \frac{1}{2\pi i} \int_C \frac{F_n(z) f(t)}{F_n(t)(t-z)} dt. \tag{49}$$

From (47), (48) and (49) it follows that

$$f(z) = \sum_{k=0}^{n-1} A_k F_k(z) + R_n(z), \qquad z \in D. \tag{50}$$

The expansion (50) is called the Newton interpolation formula for $f(z)$ with interpolation points $z_1, \ldots, z_n$.

Now suppose that we are given an infinite sequence $z_1, \ldots, z_n, \ldots$ of points in D. We assume that $\lim_{n\to\infty} R_n(z) = 0$ for all z in a region G, where $G \subset D$. Then

$$f(z) = \sum_{n=0}^{\infty} A_n F_n(z) = \sum_{n=0}^{\infty} A_n (z-z_1)\cdots(z-z_n), \qquad z \in G. \tag{51}$$

The expansion (51) is called the Newton interpolation series for $f(z)$ in the region G with interpolation points $z_1, \dots, z_n, \dots$. Note that if all of the interpolation points are the same, then this is a Taylor series.

Let m be a natural number. For our interpolation points we choose the infinite sequence

$$z_1, z_2, \dots, z_n, \dots \tag{52}$$

with period m, where

$$\begin{aligned} z_n &= n \qquad \text{for} \qquad n = 1, \dots, m, \\ z_{n+m} &= z_n \qquad \text{for all} \qquad n \geq 1. \end{aligned} \tag{53}$$

We expand the function $f(z) = e^{\xi z}$, where ξ is a nonzero complex number, in a Newton series with interpolation points (52).

Let R be any number such that $R > m$. We now estimate the remainder term (49) in Newton's formula for the function $e^{\xi z}$ and for $n > 2R$:

$$R_n(z) = \frac{1}{2\pi i}\int_C \frac{(z - z_1)\cdots(z - z_n)e^{\xi t}}{(t - z_1)\cdots(t - z_n)(t - z)}dt, \qquad |z| \leq R, \tag{54}$$

where C is the circle $|t| = n$.

Since (53) implies that $1 \leq z_k \leq m$ for $k = 1, \dots, n$, we have

$$|z - z_k| \leq |z| + |z_k| \leq R + m$$

and

$$\left|\prod_{k=1}^{n}(z - z_k)\right| \leq (R + m)^n \tag{55}$$

for all z such that $|z| \leq R$.

Next, since $n > 2R > 2m$, it follows that on the circle $|t| = n$ we have

$$|t - z_k| \geq |t| - |z_k| \geq n - m > \frac{n}{2},$$

$$|t - z| \geq |t| - |z| \geq n - R > \frac{n}{2},$$

and

$$\left|(t - z)\prod_{k=1}^{n}(t - z_k)\right| > \left(\frac{n}{2}\right)^{n+1}. \tag{56}$$

Using the inequalities (55), (56) and $|e^{\xi t}| \leq e^{|\xi| n}$ (since $|t| = n$), we obtain the following bound for the integral (54):

$$\begin{aligned} |R_n(z)| \leq & \frac{1}{2\pi} 2\pi n \frac{e^{|\xi| n}(R+m)^n}{(n/2)^{n+1}} = \\ = & \frac{2^{n+1} e^{|\xi| n}(R+m)^n}{n^n}, \qquad |z| \leq R. \end{aligned} \tag{57}$$

The numbers R and m are fixed. Hence, from (57) we conclude that

$$\lim_{n\to\infty} R_n(z) = 0 \tag{58}$$

for all z such that $|z| \leq R$.

We can choose R to be arbitrarily large. Hence, (58) holds for any complex number z. Thus, for any z the function $e^{\xi z}$ can be represented by the Newton series (51) with interpolation points (52) given by the formulas (53). That is, we have

$$e^{\xi z} = \sum_{n=0}^{\infty} A_n (z - z_1) \cdots (z - z_n), \tag{59}$$

where

$$A_n = \frac{1}{2\pi i} \int_C \frac{e^{\xi t}}{(t - z_1) \cdots (t - z_{n+1})} dt, \qquad n = 0, 1, 2, \ldots. \tag{60}$$

If we choose C to be the circle $|t| = n$, where $n > 2m$, we obtain the following upper bound for A_n using the same procedure as for the integral (54) above:

$$|A_n| \leq \frac{1}{2\pi} 2\pi n \frac{e^{|\xi| n}}{(n/2)^{n+1}} = \frac{e^{|\xi| n + (n+1)\ln 2}}{n^n} < e^{\gamma n} n^{-n}, \tag{61}$$

where $\gamma > 0$ depends only on ξ.

In the proof of Theorem 6 we shall also want a lower bound for A_n. For this we need a lemma.

Using the formulas (53) for the sequence of interpolation points (52), we can write $F_{n+1}(t)$ as follows:

$$F_{n+1}(t) = (t - z_1) \cdots (t - z_{n+1}) = \prod_{k=1}^{m} (t - k)^{n_k + 1}, \tag{62}$$

where the integers n_k depend on n and satisfy the conditions

$$n_1 + \cdots + n_m + m = n + 1,$$

$$n_1 - 1 \leq n_m \leq n_{m-1} \leq \cdots \leq n_1 \leq \frac{n}{m}. \tag{63}$$

Hence, the equation (60) takes the form

$$A_n = \frac{1}{2\pi i} \int_C \frac{e^{\xi t}}{F_{n+1}(t)} dt = \frac{1}{2\pi i} \int_C \frac{e^{\xi t}}{(t-1)^{n_1+1} \cdots (t-m)^{n_m+1}} dt. \tag{64}$$

Let M be the least common multiple of $1, \ldots, m$, and set

$$r = n_1 = \max_{1 \leq k \leq m} n_k \leq \frac{n}{m}. \tag{65}$$

Lemma 3. *Let ξ be a nonzero complex number, let $n \geq m$ be a natural number, let A_n be defined by (64) and r by (65). Then there exists a polynomial*

$$P_n = P_n(x, y) \in \mathbf{Z}[x, y] \tag{66}$$

such that

$$\deg_x P_n \leq r, \qquad \deg_y P_n \leq m, \tag{67}$$

$$H = H_{P_n} \leq r!(2M)^n, \tag{68}$$

and

$$r! M^n A_n = P_n(\xi, \eta), \qquad \eta = e^{\xi}. \tag{69}$$

Proof. By Cauchy's theorem, from (64) we have

$$A_n = \sum_{k=1}^{m} \frac{1}{2\pi i} \int_{\Gamma_k} \frac{e^{\xi t}}{(t-1)^{n_1+1} \cdots (t-m)^{n_m+1}} dt, \tag{70}$$

where Γ_k is the circle $|t - k| = 1/2$ traversed in the positive direction.

We expand $e^{\xi t}$ in a Taylor series in powers of $(t - k)$:

$$e^{\xi t} = \eta^k e^{\xi(t-k)} = \eta^k \sum_{l=0}^{\infty} \frac{\xi^l}{l!}(t-k)^l, \qquad \eta = e^{\xi}.$$

Then

$$e^{\xi t} = \eta^k \sum_{l=0}^{n_k} \frac{\xi^l}{l!}(t-k)^l + H_k(t),$$

where $H_k(t)$ is an entire function having an n_{k+1}-order zero at the point $t = k$. Hence, by Cauchy's theorem,

$$\int_{\Gamma_k} \frac{H_k(t)}{(t-1)^{n_1+1}\dots(t-m)^{n_m+1}}dt = 0,$$

and

$$\frac{1}{2\pi i}\int_{\Gamma_k} \frac{e^{\xi t}}{(t-1)^{n_1+1}\dots(t-m)^{n_m+1}}dt = \\ = \sum_{l=0}^{n_k} \frac{\eta^k \xi^l}{l!}\frac{1}{2\pi i}\int_{\Gamma_k} \frac{(t-k)^l}{(t-1)^{n_1+1}\dots(t-m)^{n_m+1}}dt. \tag{71}$$

For $k = 1, \dots, m$ we denote

$$a_{k,l} = \frac{1}{2\pi i}\int_{\Gamma_k} \frac{(t-k)^l}{(t-1)^{n_1+1}\dots(t-m)^{n_m+1}}dt, \qquad l = 0, 1, \dots, n_k. \tag{72}$$

We now prove that all $a_{k,l} \in \mathbf{Q}$ and $M^n a_{k,l} \in \mathbf{Z}$.

The number $a_{k,l}$ is equal to the residue at $t = k$ of the integrand in (72), i.e., it is the coefficient of $(t-k)^{-1}$ in the Laurent expansion of this function in powers of $(t-k)$. We now find this expansion.

Let $s \in \mathbf{N}$, $1 \le s \le m$, $s \ne k$. We consider the function

$$\frac{1}{t-s} = \frac{1}{(t-k)-(s-k)} = -\frac{1}{s-k}\frac{1}{1-(t-k)/(s-k)}. \tag{73}$$

We set

$$t - k = Mu, \tag{74}$$

and we expand the function (73) in powers of u. We obtain:

$$\frac{1}{t-s} = -\frac{1}{s-k}\sum_{\nu=0}^{\infty}\left(\frac{M}{s-k}\right)^{\nu}u^{\nu} = -\frac{1}{M}\sum_{\nu=0}^{\infty}\left(\frac{M}{s-k}\right)^{\nu+1}u^{\nu} = \\ = \frac{1}{M}\sum_{\nu=0}^{\infty} b_\nu u^\nu, \qquad b_\nu = -\left(\frac{M}{s-k}\right)^{\nu+1}. \tag{75}$$

The series (75) is absolutely convergent in the disc $|u| < |s-k|/M$.

Since M is the least common multiple of $1, \dots, m$, and since $1 \le |s-k| \le m-1$, it follows that all of the numbers $M/(s-k)$ are integers; hence, $b_\nu \in \mathbf{Z}$, $\nu = 0, 1, 2, \dots$.

The function

$$\prod_{\substack{s=1\\ s\neq k}}^{m} \frac{1}{(t-s)^{n_s+1}}, \qquad 1 \le k \le m,$$

is a product of $n - n_k$ factors of the form $1/(t-s)$, $s \neq k$, some of which are the same. If we multiply together all of the series (75) corresponding to these functions $1/(t-s)$, by (74) we obtain:

$$\prod_{\substack{s=1\\ s\neq k}}^{m} \frac{1}{(t-s)^{n_s+1}} = \frac{1}{M^{n-n_k}} \sum_{\nu=0}^{\infty} c_\nu u^\nu = \frac{1}{M^{n-n_k}} \sum_{\nu=0}^{\infty} \frac{c_\nu}{M^\nu}(t-k)^\nu, \tag{76}$$

where all $c_\nu \in \mathbf{Z}$. By the theorem on products of series, the series in (76) converge, respectively, in the discs $|u| < 1/M$ and $|t-k| < 1$.

By the residue theorem, we find from (72) and (76) that

$$a_{k,l} = \frac{c_{n_k-l}}{M^{n-l}}, \qquad l = 0, 1, \dots, n_k; \quad k = 1, \dots, m. \tag{77}$$

From (77) we conclude that $a_{k,l} \in \mathbf{Q}$ and $M^n a_{k,l} \in \mathbf{Z}$.

From (70), (71) and (72) we obtain

$$A_n = \sum_{k=1}^{m} \sum_{l=0}^{n_k} \frac{a_{k,l} \xi^l \eta^k}{l!}, \tag{78}$$

which, by (77), implies that there exists a polynomial $P_n(x, y)$ (66) whose degrees in x and y satisfy (67) and such that (69) holds.

We estimate $a_{k,l}$, using (72). If $t \in \Gamma_k$, then $|t-k| = 1/2$, and for $s \neq k$ we have $|t-s| \geq 1/2$. Hence,

$$|a_{k,l}| \le \frac{1}{2\pi} \pi \frac{1}{(1/2)^{n-l+1}} \le 2^n, \qquad l = 0, 1, \dots, n_k; \quad k = 1, \dots, m. \tag{79}$$

From (69), (78) and (79) we conclude that the height of $P_n(x, y)$ satisfies the inequality (68).

Proof of Theorem 6. We suppose that the theorem is false, i.e., that ξ and $\eta = e^\xi$ are algebraic numbers, $\xi \neq 0$. Let $h \geq 1$ be the degree of an algebraic number field containing ξ and η.

We consider the Newton interpolation series (59) for $e^{\xi z}$ with interpolation points (52) given by (53), where we set $m = h + 1$. By what was proved above, the coefficients A_n in this expansion satisfy the upper bound (61); by Lemma 3, the equality (69) holds, where $P_n(x,y) \in \mathbf{Z}[x,y]$ and we have the equalities (67) and (68) for the degrees in x and y and the height of $P_n(x,y)$.

Let c, c_1 and c_2 denote positive constants which depend only on ξ and η.

By Theorem 11 of Chapter 1, the assumptions that $\xi \in \mathbf{A}$ and $\eta = e^{\xi} \in \mathbf{A}$ imply either that $P_n(\xi,\eta) = 0$ or else that

$$|P_n(\xi,\eta)| \geq \frac{c^k}{H^{h-1}} = e^{k \ln c - (h-1) \ln H}, \tag{80}$$

where $k \leq r + m$, and $\ln H < r \ln r + n \ln(2M)$, as follows from (67) and (68).

Since $m = h + 1$, and since (65) implies that $r \leq n/m$, it follows that from (80) we have the bound

$$|P_n(\xi,\eta)| > e^{-\frac{m-2}{m} n \ln n - c_1 n}. \tag{81}$$

From (69) and (61) we conclude that

$$|P_n(\xi,\eta)| < e^{\gamma n - n \ln n + n \ln M + r \ln r} < e^{-\frac{m-1}{m} n \ln n + c_2 n}. \tag{82}$$

For n sufficiently large the inequalities (81) and (82) contradict one another. Consequently, there exists $n_0 \in \mathbf{N}$ such that $A_n = 0$ for all $n \geq n_0$, and so the function $e^{\xi z}$ must be a polynomial, which is impossible. This contradiction completes the proof of Theorem 6.

Lindemann's theorem on the linear independence over $\mathbf{A}$ of the values at algebraic points of the exponential function will be proved by another method, using a construction of a set of linear forms in the exponential expressions with coefficients in $\mathbf{A}[z]$. We shall see that this proof is closely related to the method in § 3. It is worthwhile for the reader to become familiar with this new method, since it will later be generalized considerably and applied to a broader class of functions. We shall begin our description of the method with the simplest case — approximation of the exponential function by rational functions.

§ 4. Approximation of e^z by rational functions

Let $F(z)$ be a function which is analytic at the point $z = 0$. We shall let $\operatorname{ord} F(z)$ denote the order of zero of $F(z)$ at $z = 0$.

We are interested in the problem of approximating e^z by rational functions P/Q, where $P = P(z)$ and $Q = Q(z)$ are polynomials in $\mathbf{C}[z]$ whose degrees are at most n and which satisfy $\operatorname{ord}(e^z + P/Q) \geq N$ for some large integer N. Then we can write

$$R = Qe^z + P = \gamma z^N + \cdots. \tag{83}$$

We write the polynomials P and Q in (83) with undetermined coefficients and set the coefficients of z^k, $k = 0, 1, \ldots, 2n$, equal to zero. As a result we obtain a system of $2n+1$ homogeneous linear equations in $2n+2$ unknowns. This system has a nontrivial solution. Consequently, we can take $N = 2n+1$ in (83). Below we shall show that in this case $\gamma \neq 0$.

In order to find formulas for the polynomials P and Q, we consider the differential operator $D = d/dz$. Let

$$\varphi(u) = a_0 + a_1 u + a_2 u^2 + \cdots \tag{84}$$

be a power series with real coefficients, and let $f = f(z)$ be an arbitrary function of z. We define the operator $\varphi(D)$ by setting

$$\varphi(D)f(z) = \sum_{n=0}^{\infty} a_n D^n f(z) = \sum_{n=0}^{\infty} a_n \frac{d^n f(z)}{dz^n}, \tag{85}$$

where we assume that $f(z)$ is differentiable the requisite number of times and that the series on the right is either finite or else a convergent infinite series. In the applications below, these conditions of differentiability and convergence will always hold, since we shall be applying the operator D in only two cases: when $f(z)$ is a polynomial, and when $\varphi(z)$ is a polynomial and $f(z)$ has derivatives of every order. In these cases (85) is a finite sum.

We note some obvious properties of the operators $\varphi(D)$:

$$\begin{aligned} (\varphi_1(D) + \varphi_2(D))f &= \varphi_1(D)f + \varphi_2(D)f, \\ (\varphi_2(D)\varphi_1(D))f &= \varphi_2(D)(\varphi_1(D)f). \end{aligned} \tag{86}$$

Furthermore, if $a_0 \neq 0$ in (84), then we can form the series

$$\varphi^{-1}(u) = b_0 + b_1 u + b_2 u^2 + \cdots, \qquad b_0 = a_0^{-1},$$

and then, if f is a polynomial, we have:

$$\varphi^{-1}(D)(\varphi(D)f) = f.$$

If the power series $\varphi(u)$ and the polynomial f have integer coefficients, then $\varphi(D)f$ also has integer coefficients, and if we have $a_0 = \pm 1$ in (84), then $\varphi^{-1}(D)f$ has integer coefficients as well.

We denote

$$Jf = \int_0^z f(t)dt.$$

Then

$$DJf = f(z), \qquad JDf = f(z) - f(0).$$

We define

$$J^1 f = Jf, \qquad J^{n+1} f = J(J^n f), \qquad n = 1, 2, \ldots.$$

By induction, we then find that

$$J^{n+1} f = \int_0^z \frac{(z-t)^n}{n!} f(t)dt, \qquad n = 0, 1, 2, \ldots. \tag{87}$$

In what follows we shall apply differential operators to the function $f(z) = e^{\lambda z} P(z)$, where λ is a constant and $P = P(z) \in \mathbf{C}[z]$. Since

$$D(e^{\lambda z} P) = e^{\lambda z}(\lambda + D)P,$$

it follows that

$$D^n(e^{\lambda z} P) = e^{\lambda z}(\lambda + D)^n P, \qquad n = 1, 2, \ldots, \tag{88}$$

where

$$(\lambda + D)^n P = Q(z) \in \mathbf{C}[z]. \tag{89}$$

Conversely, if $\lambda \neq 0$ and $Q = Q(z)$ is any given polynomial, then (89) implies that

$$P = (\lambda + D)^{-n} Q$$

is the unique polynomial which solves the differential equation

$$D^n(e^{\lambda z} P) = e^{\lambda z} Q.$$

If $\lambda = \pm 1$, then $P \in \mathbf{Z}[z]$ whenever $Q \in \mathbf{Z}[z]$.

Lemma 4. *For any n, $n \geq 0$, there exist unique degree n polynomials $P = P(z)$ and $Q = Q(z)$ in $\mathbf{Z}[z]$ such that*

$$\begin{aligned} R = R(z) &= Q(z)e^z + P(z) = c_0 z^{2n+1} + \cdots, \\ c_0 &= \frac{1}{(n+1)\cdots(2n+1)}. \end{aligned} \tag{90}$$

The polynomials P and Q are given by the formulas

$$P(z) = (-1 + D)^{-n-1} z^n, \qquad Q(z) = (1 + D)^{-n-1} z^n, \tag{91}$$

and $R(z)$ can be represented by the integral

$$R(z) = \frac{z^{2n+1}}{n!} \int_0^1 t^n (1-t)^n e^{tz}\, dt, \qquad n = 0, 1, 2, \ldots, \tag{92}$$

or equivalently

$$R(z) = \frac{z^{2n+1}}{n!} e^{z/2} \int_0^1 t^n (1-t)^n \cosh\big((t - \frac{1}{2})z\big)\, dt, \qquad n = 0, 1, 2, \ldots. \tag{93}$$

Proof. Above we showed that there exist polynomials P and Q of degree at most n such that (83) holds with $N = 2n+1$. To find P and Q we differentiate this equation $(n+1)$ times. Then

$$D^{n+1}(Qe^z) = D^{n+1}R,$$

or, by (88),

$$e^z(1+D)^{n+1}Q = D^{n+1}R = \gamma_0 z^n + \cdots, \qquad \gamma_0 = (2n+1)\cdots(n+1)\gamma. \tag{94}$$

From this we obtain

$$(1+D)^{n+1}Q = e^{-z}(\gamma_0 z^n + \cdots) = \gamma_0 z^n + \cdots = \gamma_0 z^n, \tag{95}$$

because $(1+D)^{n+1}Q$ is a polynomial of degree at most n; hence,

$$Q = \gamma_0 (1+D)^{-n-1} z^n. \tag{96}$$

We similarly find P:

$$\begin{aligned} D^{n+1}(Pe^{-z}) &= D^{n+1}(Re^{-z}), \\ e^{-z}(-1+D)^{n+1}P &= \gamma_0 z^n + \cdots, \\ (-1+D)^{n+1}P &= e^z(\gamma_0 z^n + \cdots) = \gamma_0 z^n + \cdots = \gamma_0 z^n, \\ P &= \gamma_0(-1+D)^{-n-1} z^n. \end{aligned} \tag{97}$$

From (94), (96) and (97) it follows that $\gamma \neq 0$. The equations (96) and (97) show that P and Q are unique up to the constant factor γ_0. Setting $\gamma_0 = 1$, we obtain (90)-(91) with P and Q in $\mathbf{Z}[z]$.

From (94) and (95) we find that

$$D^{n+1}R = z^n e^z.$$

Since R and the first n derivatives of R vanish at $z = 0$, it follows from the definition of the symbol J and the formula (87) that we have

$$R = J^{n+1} D^{n+1} R = \frac{1}{n!} \int_0^z (z-t)^n t^n e^t \, dt,$$

from which we obtain (92) by a change of variables.

If we replace t by $1-t$ in (92), we find that

$$R(z) = \frac{z^{2n+1}}{n!} \int_0^1 t^n (1-t)^n e^{(1-t)z} \, dt, \tag{98}$$

and then (92) and (98) imply (93). The lemma is proved.

Lemma 4 will be applied to investigate the arithmetic properties of the exponential function. We first prove that e^k *is irrational for any* $k \in \mathbf{N}$.

Suppose the contrary, i.e., that $e^k = p/q$, $p, q \in \mathbf{N}$. In (90) we set $z = k$. Then $P(k)$ and $Q(k)$ are integers. Hence, $qR(k) = m \in \mathbf{Z}$.

From (92) we find that for all $z \in \mathbf{C}$

$$|R(z)| \le \frac{|z|^{2n+1}}{n!} e^{|z|},$$

and for all $x \in \mathbf{R}$, $x > 0$,

$$R(x) > 0.$$

These inequalities show that the following inequalities hold for all sufficiently large natural numbers n:

$$0 < m \le q \frac{k^{2n+1}}{n!} e^k < 1,$$

and this is a contradiction.

Thus, e^k is irrational for any $k \in \mathbf{N}$, and hence for any $k \in \mathbf{Z}$, $k \ne 0$. We note that in fact e^a *is irrational for any* $a \in \mathbf{Q}$, $a \ne 0$, since if $a = k/l$ with k and l nonzero integers, then $e^k = (e^a)^l$. If e^a were rational, then e^k would also be rational.

We next prove that π *and* π^2 *are irrational*.

By Lemma 4, the polynomials $P(z)$ and $Q(z)$ in (90) are uniquely determined. We replace z by $-z$ in (90) and multiply by e^z. Then

$$P(-z)e^z + Q(-z) = e^z R(-z) = -c_0 z^{2n+1} + \cdots,$$

and hence

$$P(-z) = -Q(z). \tag{99}$$

We set $z = \pi i$ in (90) and (93). Using (99), we obtain

$$e^{\frac{\pi i}{2}} = i, \qquad \cosh\big((t - \tfrac{1}{2})\pi i\big) = \cos\big((t - \tfrac{1}{2})\pi\big) = \sin \pi t,$$

$$P(\pi i) + P(-\pi i) = R(\pi i), \tag{100}$$

$$R(\pi i) = (-1)^{n+1} \frac{\pi^{2n+1}}{n!} \int_0^1 t^n (1-t)^n \sin \pi t \, dt. \tag{101}$$

Since the integrand in (101) is positive for $0 < t < 1$, we have

$$R(\pi i) \neq 0. \tag{102}$$

The polynomial $P(z) + P(-z)$ has only even powers of z, and so it is a polynomial in $\mathbf{Z}[z^2]$ of degree $s = [n/2]$.

We now suppose that our claim is false, i.e., that $\pi^2 = p/q$, $p, q \in \mathbf{N}$. Then from (100) we conclude that $q^s R(\pi i) = m \in \mathbf{Z}$. From (101) and (102) we obtain the inequality

$$0 < |m| \leq \frac{q^{n/2} \pi^{2n+1}}{n!} < 1,$$

which holds for all n sufficiently large. But this inequality is impossible, since $m \in \mathbf{Z}$. The contradiction proves that π^2 is irrational, and so π is also irrational.

§ 5. Linear approximating forms for $e^{\rho_1 z}, \dots, e^{\rho_m z}$

We now generalize Lemma 4 to the case of several exponential functions.

Lemma 5. *Let $\rho_1, \dots, \rho_m$ be distinct complex numbers, $m \geq 1$, let $n_1, \dots, n_m \in \mathbf{Z}^+$, and set*

$$\sum_{k=1}^{m} (n_k + 1) = N + 1. \tag{103}$$

Then there exists an m-tuple of polynomials $P_1 = P_1(z), \dots, P_m = P_m(z)$, unique up to a constant factor, which have degrees $n_1, \dots, n_m$, respectively, and *have the property that the function*

$$R = R(z) = \sum_{k=1}^{m} P_k(z)e^{\rho_k z} \tag{104}$$

has ord $R = N$. *The polynomials* P_k *have the form*

$$P_k(z) = \sum_{\substack{s=1 \\ s\neq k}}^{m} (\rho_k - \rho_s + D)^{-n_s-1} \frac{z^{n_k}}{n_k!}, \qquad k = 1, \dots, m, \tag{105}$$

and $R(z)$ *can be represented by the following integral for* $z = x$, $x > 0$:

$$R(x) = \int_{\substack{t_1+\cdots+t_m=x \\ t_1\geq 0,\dots,t_m\geq 0}} \cdots \int \prod_{k=1}^{m} \Big(\frac{t_k^{n_k}}{n_k!} e^{\rho_k t_k} \Big) dt_1 \dots dt_{m-1}. \tag{106}$$

Remark. In what follows we shall only need the expression (106) for $R(z)$. However, it is possible to obtain an expression for $R(z)$ which holds for any complex z. In fact, if we replace t_k by $t_k x$ in (106), then this change of variables leads to the relation

$$R(x) = x^N \int_{\substack{t_1+\cdots+t_m=1 \\ t_1\geq 0,\dots,t_m\geq 0}} \cdots \int \prod_{k=1}^{m} \Big(\frac{t_k^{n_k}}{n_k!} e^{\rho_k t_k x} \Big) dt_1 \dots dt_{m-1},$$

which, one easily sees, remains valid if x is replaced by a complex number z.

Proof of Lemma 5. We regard $P_1, \dots, P_m$ as polynomials of degree $n_1, \dots, n_m$, respectively, with undetermined coefficients. We substitute them on the right in (104) and set the coefficients of z^k, $k = 0, 1, \dots, N-1$, equal to zero. As a result we obtain a system of N homogeneous linear equations in the $(n_1+1)+\cdots+(n_m+1)$ unknown coefficients of $P_1, \dots, P_m$. By (103), the number of unknowns in this system is greater than the number of equations. Hence it has a nontrivial solution.

We have thereby shown that there exist polynomials $P_1, \dots, P_m$, not all zero, of degree no greater than $n_1, \dots, n_m$, respectively, such that the power series for the function R in (104) has the form

$$R = c\frac{z^N}{N!} + \cdots \tag{107}$$

for some constant c. We shall later show that the constant c in (107) is nonzero and that, for any fixed value of c, the solution $P_1, \dots, P_m$ is uniquely determined.

We now use induction on m to prove that (105) holds. For $m = 1$ we have

$$P_1 = c\frac{z^{n_1}}{n_1!}, \tag{108}$$

so that $c \neq 0$.

Now suppose that (105) has been proved in the case when m is replaced by $m-1$, $m > 1$. We prove that it then holds for m as well.

If $m > 1$, we write $N = N_m$ for clarity. Using (88), we find from (104) that

$$\begin{aligned} D^{n_m+1}\left(Re^{-\rho_m z}\right) &= \sum_{k=1}^{m} D^{n_m+1}\left(e^{(\rho_k-\rho_m)z}P_k\right) = \\ &= \sum_{k=1}^{m-1} e^{(\rho_k-\rho_m)z}(\rho_k - \rho_m + D)^{n_k+1}P_k, \end{aligned} \tag{109}$$

and from (107) we obtain the following relation, using (103):

$$\begin{aligned} D^{n_m+1}\left(Re^{-\rho_m z}\right) &= cD^{n_m+1}\left(\frac{z^{N_m}}{N_m!}e^{-\rho_m z} + \cdots\right) = \\ &= c\frac{z^{N_{m-1}}}{N_{m-1}!} + \cdots. \end{aligned} \tag{110}$$

The polynomials

$$Q_k = (\rho_k - \rho_m + D)^{n_m+1}P_k, \qquad k = 1, \ldots, m-1, \tag{111}$$

have the same degrees as the corresponding P_k, and so they are not all zero. Hence, (109) and (110) show that the function

$$\begin{aligned} H &= H(z) = D^{n_m+1}\left(Re^{-\rho_m z}\right) = \\ &= Q_1 e^{(\rho_1-\rho_m)z} + \cdots + Q_{m-1}e^{(\rho_{m-1}-\rho_m)z} = c\frac{z^{N_{m-1}}}{N_{m-1}!} + \cdots \end{aligned} \tag{112}$$

satisfies the conditions of the lemma with m replaced by $m-1$ and $\rho_1, \ldots, \rho_m$ replaced by $\rho_1 - \rho_m, \ldots, \rho_{m-1} - \rho_m$, with some constant c.

From (111) we find that

$$P_k = (\rho_k - \rho_m + D)^{-n_m-1}Q_k, \qquad k = 1, \ldots, m-1.$$

Hence, using (108) and the induction assumption and setting $c = 1$, we obtain:

$$P_1 = \prod_{s=2}^{m}(\rho_1 - \rho_s + D)^{-n_s-1}\frac{z^{n_1}}{n_1!}.$$

If we change the indexing of $\rho_1, \dots, \rho_m$ and use what was proved for P_1, we obtain (105) for all $k = 1, \dots, m$. By induction, then, (105) holds for any m.

We now proceed to the proof of (106). From (112) and (87) we obtain

$$R(z) = e^{\rho_m z} J^{n_m+1} H = e^{\rho_m z} \int_0^z \frac{(z-t)^{n_m}}{n_m!} H(t)dt. \tag{113}$$

We use (113) and induction to prove (106). For $m = 2$, from (113) and (108) we have:

$$R(x) = e^{\rho_2 x} \int_0^x \frac{(x-t)^{n_2}}{n_2!} e^{(\rho_1-\rho_2)t} \frac{t^{n_1}}{n!} dt =$$
$$= \int_{\substack{t_1+t_2=x\\ t_1\geq 0, t_2\geq 0}} \frac{t_1^{n_1} t_2^{n_2}}{n_1! n_2!} e^{\rho_1 t_1 + \rho_2 t_2} dt_1, \qquad x > 0.$$

We now suppose that (106) with m replaced by $m-1$ holds for any $\rho_1, \dots$ $\dots, \rho_{m-1}$. Then we have

$$H(t) = \int_{\substack{t_1+\dots+t_{m-1}=t\\ t_1>0,\dots,t_{m-1}>0}} \cdots \int \prod_{k=1}^{m-1} \frac{t_k^{n_k}}{n_k!} e^{(\rho_k - \rho_m)t_k} dt_1 \dots dt_{m-2}, \qquad t > 0.$$

Substituting this formula for $H(t)$ in (113), with z replaced by x in the latter equality, and setting $t_m = x - t$, we find that (106) holds for m; by induction, it holds for all m.

We shall call (106) a linear approximating form for the functions $e^{\rho_1 z}, \dots$ $\dots, e^{\rho_m z}$.

We note that Lemma 4 follows from Lemma 5 if we set $m = 2$, $n_1 = n_2 = n$, $\rho_1 = 1$ and $\rho_2 = 0$.

Corollary of Lemma 5. *Under the conditions of Lemma 5, for arbitrary complex $\rho_1, \dots, \rho_m$ the formula (106) implies the upper bound*

$$|R(1)| \leq \frac{e^{|\rho_1|+\dots+|\rho_m|}}{n_1! \cdots n_m!}, \tag{114}$$

and for arbitrary real $\rho_1, \dots, \rho_m$ it implies the lower bound

$$R(1) > 0. \tag{115}$$

Lemma 6. *Under the conditions of Lemma 5 with $m \geq 2$, the polynomials P_k in (105) satisfy the inequalities*

$$|P_k(1)| \leq (2M+2)^N, \qquad k = 1, \dots, m, \tag{116}$$

where

$$M = \max_{1 \le i < j \le m} \frac{1}{|\rho_j - \rho_i|}.$$

If, in addition, $\rho_1, \ldots, \rho_m \in \mathbf{A}$, *then*

$$n_k! q^N P_k(1) \in \mathbf{Z_A}, \qquad k = 1, \ldots, m, \tag{117}$$

where q *is a natural number such that*

$$\frac{q}{\rho_i - \rho_j} \in \mathbf{Z_A}, \qquad i = 1, \ldots, m; \ \ j = 1, \ldots, m, \ \ i \ne j.$$

Proof. From (105) we have for any $x > 0$:

$$|P_k(x)| \le (M^{-1} - D)^{n_k - N} \frac{x^{n_k}}{n_k!}, \qquad k = 1, \ldots, m, \tag{118}$$

since the right sides of these inequalities majorize $P_k(x)$ as polynomials in x.

Using the expansion

$$\begin{aligned}(\omega + D)^{-n-1} &= \omega^{-n-1} \sum_{l=0}^{\infty} \binom{-n-1}{l} \omega^{-l} D^l = \\ &= \omega^{-n-1} \sum_{l=0}^{\infty} (-1)^l \binom{n+1}{l} \omega^{-l} D^l, \qquad \omega \ne 0,\end{aligned} \tag{119}$$

we find that

$$\begin{aligned}(M^{-1} - D)^{n_k - N} \frac{x^{n_k}}{n_k!} &= M^{N-n_k} \sum_{l=0}^{n_k} \binom{N - n_k + l - 1}{l} M^l D^l \frac{x^{n_k}}{n_k!} \le \\ &\le \sum_{l=0}^{n_k} \binom{N}{l} M^{N-n_k+l} x^{n_k - l} \le (1+1)^N (M + x)^N = \\ &= (2M + 2x)^N, \qquad x > 0, \qquad k = 1, \ldots, m.\end{aligned} \tag{120}$$

From (118) and (120) we obtain (116).

If $\rho_1, \ldots, \rho_m \in \mathbf{A}$, then from (105) we have $P_k(1) \in \mathbf{A}$, $k = 1, \ldots, m$, and then we obtain (117) from (105) and (119).

We now use the linear approximating form constructed above to give *a new proof of Theorem 6 on the transcendence of* e^{ξ} *for* $\xi \in \mathbf{A} \cap \mathbf{R}$, $\xi \ne 0$. To do this, we suppose the contrary: $e^{\xi} \in \mathbf{A}$. Let $\mathbf{K}$ be the algebraic number field generated by ξ and e^{ξ}, and set $h = [\mathbf{K} : \mathbf{Q}]$. We choose $m = h + 1$, $\rho_k = (k-1)\xi$, $k = 1, \ldots, m$, and $n_1 = n_2 = \cdots = n_m = n > 1$, where n is sufficiently large.

In what follows let $c_1, c_2, \dots$ denote positive constants which do not depend on n.

From (105) it follows that the numbers $P_k(1)$ are in $\mathbf{K}$. If we use (105) and (116) for the conjugates of $P_k(1)$ in $\mathbf{K}$, we obtain:

$$\overline{|P_k(1)|} < c_1^n, \qquad k = 1, \dots, m, \tag{121}$$

since $N = m(n+1) - 1 < c_2 n$.

Since $e^{\xi} \in \mathbf{A}$, it follows by our choice of $\rho_1, \dots, \rho_m$ that the number

$$R(1) = P_1(1)e^{\rho_1} + \cdots + P_m(1)e^{\rho_m} \tag{122}$$

is also in $\mathbf{K}$.

From (117), using (103), we see that there exists $g \in \mathbf{N}$, namely

$$g = c_3^n n!, \tag{123}$$

such that

$$\delta = gR(1) \in \mathbf{Z_K}, \tag{124}$$

and from (115) we have $\delta \neq 0$. Hence,

$$|\mathcal{N}(\delta)| \geq 1. \tag{125}$$

On the other hand, by (114) along with (123) and (124), we have

$$|\delta| < c_4^n (n!)^{-h}, \tag{126}$$

since $m = h + 1$.

We note that the last bound may fail to hold for some of the numbers $\delta_2, \dots, \delta_k$ which are conjugate to $\delta = \delta_1$ in the field $\mathbf{K}$, since, if $\rho_{k,1}, \dots, \rho_{k,h}$ are the conjugates of $\rho_k = \rho_{k,1}$, then $e^{\rho_{k,2}}, \dots, e^{\rho_{k,h}}$ are not necessarily conjugates of e^{ρ_k}, and in fact they could even be transcendental numbers. However, (121) along with (122) and (123) imply the bound

$$\overline{|\delta|} < c_5^n n!. \tag{127}$$

From (126) and (127) we conclude that

$$|\mathcal{N}(\delta)| < c_4^n (n!)^{-h} (c_5^n n!)^{h-1} = \frac{c_6^n}{n!} < 1$$

for all n sufficiently large. The last inequality contradicts (125), and this proves our claim.

§ 6. A set of linear approximating forms

In the proof of transcendence of e^{ξ} at the end of § 5, an essential point was the nonvanishing of $\delta \in \mathbf{Z_K}$. This fact followed from (115). However, if $\rho_1, \ldots, \rho_m$ are arbitrary distinct complex numbers, then it no longer follows from the argument in § 5 that we have $R(1) \neq 0$. We must modify the proof so that the latter holds in that case.

We fix any k in the range $1, \ldots, m$, and we set $n_i = n$, $i = 1, \ldots, k$, $n \geq 1$, and $n_i = n - 1$, $i = k+1, \ldots, m$. We let R_k denote the corresponding approximating form from Lemma 5:

$$R_k = R_k(z) = \sum_{i=1}^{m} P_{k,i}(z) e^{\rho_i z}, \qquad k = 1, \ldots, m. \tag{128}$$

Thus, the polynomials $P_{k,i}(z)$ have degrees n_i equal to either n or $n-1$, depending on whether $k \geq i$ or $k < i$.

We consider the determinant of the system of linear forms (128):

$$\Delta = \Delta(z) = |P_{k,i}|_{k,i=1,\ldots,m}.$$

It is a sum of $m!$ terms, which are polynomials in z. All of these polynomials have degree less than mn, except for the one which is obtained by multiplying together the entries on the main diagonal; that polynomial has degree equal to mn. Hence, $\deg \Delta = mn$.

We let $\Delta_1, \ldots, \Delta_m$ denote the cofactors of the elements in the first column of Δ. Then

$$\Delta = (\Delta_1 R_1 + \cdots + \Delta_m R_m) e^{-\rho_1 z}. \tag{129}$$

Since $\operatorname{ord} R_k = (n_1 + \cdots + n_m) + m - 1 = mn + k - 1 \geq mn$, it follows from (129) that Δ has a zero of order at least mn at $z = 0$. But its degree is mn; thus,

$$\Delta(z) = \gamma z^{mn}, \qquad \Delta(1) = \gamma \neq 0.$$

From (129) we then find that at least one of the numbers $R_k(1)$, $1 \leq k \leq m$, is nonzero. This observation enables us to *extend the proof in § 5 of the transcendence of e^{ξ} to the case of a complex algebraic number ξ.* We have thereby proved Theorem 6 in full by a different method than before, based on the construction of linear approximating forms for the functions $e^{\rho_1 z}, \ldots, e^{\rho_m z}$.

§ 7. Lindemann's Theorem

We now use the method developed in § § 5 and 6 to prove Lindemann's theorem. We first prove a lemma.

Let $m \in \mathbf{N}$ and $N \in \mathbf{Z}^+$. We let $\mu^0_{N,m}$ denote the number of different m-tuples $(k_1, \ldots, k_m)$ of nonnegative integers which satisfy the condition $k_1 + \cdots + k_m = N$, and let $\mu_{N,m}$ denote the number of m-tuples which satisfy the inequality $k_1 + \cdots + k_m \leq N$.

Lemma 7. *For any* $m \in \mathbf{N}$ *and* $N \in \mathbf{Z}^+$,

$$\mu^0_{N,m} = \binom{N+m-1}{m-1} = \frac{(N+m-1)!}{N!(m-1)!}; \tag{130}$$

$$\mu_{N,m} = \binom{N+m}{m} = \frac{(N+m)!}{N!m!}. \tag{131}$$

Proof. The number of solutions of the equation $k_1 + \cdots + k_m = N$ in nonnegative integers $k_1, \ldots, k_m$ is equal to the coefficient of x^N in the series

$$(1 + x + x^2 + \cdots)^m = (1-x)^{-m}.$$

But

$$(1-x)^{-m} = \sum_{k=0}^{\infty} (-1)^k \frac{(-m)(-m-1)\cdots(-m-k+1)}{k!} x^k =$$
$$= \sum_{k=0}^{\infty} \frac{(k+m-1)!}{k!(m-1)!} x^k,$$

from which (130) follows. Next, we note that

$$\mu^0_{N,m} = \mu_{N,m-1}, \qquad N \geq 0, \qquad m \geq 2. \tag{132}$$

To see this, we set $k_m = 0, 1, \ldots, N$ successively in the equation $k_1 + \cdots + k_{m-1} = N - k_m$, and for each value of k_m we count the number of $(m-1)$-tuples $(k_1, \ldots, k_{m-1})$, $k_i \in \mathbf{Z}^+$, $i = 1, \ldots, m-1$; we find that

$$\mu^0_{N,m} = \sum_{k=0}^{N} \mu^0_{k,m-1} = \mu_{N,m-1}. \tag{133}$$

Now (131) follows from (130) and (132).

Lindemann's Theorem. *If $\xi_1, \dots, \xi_l$, $l \geq 1$, are algebraic numbers which are linearly independent over* $\mathbf{Q}$, *then the numbers $e^{\xi_1}, \dots, e^{\xi_l}$ are algebraically independent.*

Proof. Suppose the contrary, i.e., that $e^{\xi_1}, \dots, e^{\xi_l}$ are algebraically dependent. Then there exists a polynomial

$$P = P(z_1, \dots, z_l) \in \mathbf{Z_A}[z_1, \dots, z_l],$$

$$P \not\equiv 0, \qquad \deg_{\mathbf{Z}} P = \nu, \qquad \nu \geq 1,$$

such that

$$P\big(e^{\xi_1}, \dots, e^{\xi_l}\big) = 0. \tag{134}$$

Let $\mathbf{K}$ denote an algebraic number field containing $\xi_1, \dots, \xi_l$ and all of the coefficients of P, and let $h = [\mathbf{K} : \mathbf{Q}]$. We set $m = \mu_{s,l}$, where $s \in \mathbf{N}$, $s \geq \nu$, and we consider the set of m products of powers

$$z_1^{k_1} \cdots z_l^{k_l}, \qquad k_j \in \mathbf{Z}^+, \ \ j = 1, \dots, l, \ \ k_1 + \cdots + k_l \leq s. \tag{135}$$

We number these products in some arbitrary manner, denoting them $Z_1, \dots, Z_m$. We also consider the set of $r = \mu_{s-\nu,l}$ products of powers

$$z_1^{k_1} \cdots z_l^{k_l}, \qquad k_j \in \mathbf{Z}^+, \ \ j = 1, \dots, 1, \ \ k_1 + \cdots + k_l \leq s - \nu. \tag{136}$$

We number them in an arbitrary manner, denoting them $Y_{m-r+1}, \dots, Y_m$. Then

$$Y_k P = \sum_{i=1}^{m} a_{k,i} Z_i, \qquad k = m - r + 1, \dots, m, \tag{137}$$

where the $a_{k,i}$ are either zero or else the corresponding coefficients of P.

The right sides of (137) are linearly independent as linear forms in $Z_1, \dots, Z_m$ with constant coefficients, since the $Y_k P$ are polynomials in $z_1, \dots, z_l$ which are not related by any homogeneous linear equation with constant coefficients.

We denote

$$\rho_i = \sum_{j=1}^{l} k_j \xi_j, \qquad i = 1, \dots, m, \tag{138}$$

where $(k_1, \dots, k_l)$ is the set of exponents corresponding to the monomial $Z_i = z_1^{k_1} \cdots z_l^{k_l}$. Then $\rho_1, \dots, \rho_m$ are distinct numbers in $\mathbf{K}$, since $\xi_1, \dots, \xi_l$ are linearly independent over $\mathbf{Q}$ and the l-tuples $(k_1, \dots, k_l)$ corresponding to different i are distinct.

If we set $z_j = e^{\xi_j}$, $j = 1, \dots, l$, and use (137) and (138) along with (134), then we find that

$$\sum_{i=1}^{m} a_{k,i} e^{\rho_i} = 0, \qquad k = m - r + 1, \dots, m, \qquad a_{k,i} \in \mathbf{Z_K}. \tag{139}$$

We now consider the m approximating linear forms (128) (see § 6) for the functions $e^{\rho_1 z}, \dots, e^{\rho_m z}$:

$$R_k(z) = \sum_{i=1}^{m} P_{k,i}(z) e^{\rho_i z}, \qquad k = 1, \dots, m. \tag{140}$$

The determinant $\Delta(1) = |P_{k,i}(1)|$ of this system of linear forms at the point $z = 1$ is nonzero. Hence, the numerical linear forms in the numbers $e^{\rho_1}, \dots, e^{\rho_m}$ with coefficients in $\mathbf{Z_K}$ which are obtained by setting $z = 1$ in (140), are linearly independent; and we can choose $m - r$ of them with $k = k_1, \dots, k_{m-r}$ such that they, together with the forms (139), are linearly independent.

We denote

$$P_{k_j,i}(1) = a_{j,i}, \qquad R_{k_j}(1) = b_j,$$

$$j = 1, \dots, m - r, \qquad i = 1, \dots, m.$$

Then

$$\sum_{i=1}^{m} a_{k,i} e^{\rho_i} = b_k, \qquad k = 1, \dots, m, \tag{141}$$

where, by (139), we have $b_k = 0$, $k = m - r + 1, \dots, m$.

Let δ be the determinant of the system of linear forms (141), $\delta \neq 0$, and let $\delta_1, \dots, \delta_m$ be the cofactors of the elements in the first column of δ. From (141) we find that

$$\delta = (\delta_1 b_1 + \dots + \delta_{m-r} b_{m-r}) e^{-\rho_1} \neq 0. \tag{142}$$

We now make use of the properties of the linear forms (128). Since n_1 is equal to n or $n-1$, it follows by (117) that there exists an integer $g \in \mathbf{N}$, $g \leq c_7^n n!$, such that $g a_{k,i} \in \mathbf{Z_K}$ for $k = 1, \dots, m-r$, $i = 1, \dots, m$. Consequently, $g^{m-r}\delta \in \mathbf{Z_K}$; hence, $|\mathcal{N}(g^{m-r}\delta)| \geq 1$, and

$$|\mathcal{N}(\delta)| \geq g^{-h(m-r)} \geq c_8^n (n!)^{-h(m-r)}. \tag{143}$$

In addition, it follows from (116) that

$$\overline{|a_{k,i}|} < c_9^n, \qquad k = 1, \ldots, m; \quad i = 1, \ldots, m, \tag{144}$$

and then

$$\overline{|\delta|} < c_{10}^n. \tag{145}$$

On the other hand, by (114), we have

$$|b_k| < c_{11} n^{m-1} (n!)^{-m} < c_{12}^n (n!)^{-m}, \qquad k = 1, \ldots, m - r, \tag{146}$$

and, using (144), (146) and (142), we find that

$$|\delta| < c_{13}^n (n!)^{-m}. \tag{147}$$

From (145) and (147) we obtain:

$$|\mathcal{N}(\delta)| < c_{14}^n (n!)^{-m}. \tag{148}$$

Now let $n \to \infty$. Comparing the bounds (143) and (148), we find that $m \leq h(m - r)$, i.e.,

$$r \leq \left(1 - \frac{1}{h}\right) m. \tag{149}$$

But, by Lemma 7,

$$\begin{aligned} m &= \frac{(s+l)!}{s!\,l!} = \frac{1}{l!} s^l + O(s^{l-1}); \\ r &= \frac{(s-\nu+l)!}{(s-\nu)!\,l!} = \frac{1}{l!} s^l + O(s^{l-1}), \end{aligned} \tag{150}$$

and then (149) and (150) contradict one another for s sufficiently large. This contradiction proves Lindemann's theorem.

We note that, if $\xi_1, \ldots, \xi_l$ are linearly dependent over $\mathbf{Q}$, then $e^{\xi_1}, \ldots, e^{\xi_l}$ are algebraically dependent. To see this, suppose that

$$k_1 \xi_1 + \cdots + k_l \xi_l = 0, \qquad k_i \in \mathbf{Z}, \qquad i = 1, \ldots, l.$$

If we set

$$\beta_i = e^{\xi_i}, \qquad i = 1, \ldots, l,$$

then we have

$$\beta_1^{k_1} \cdots \beta_l^{k_l} = e^{k_1 \xi_1 + \cdots + k_l \xi_l} = 1,$$

which means that the β_i are algebraically dependent.

Lindemann's theorem can be restated in the following equivalent form.

Lindemann's Theorem. *If* $\alpha_1, \ldots, \alpha_k$, $k \geq 1$, *are distinct algebraic numbers, and if* $c_1, \ldots, c_k$ *are algebraic numbers which are not all zero, then*

$$c_1 e^{\alpha_1} + \cdots + c_k e^{\alpha_k} \neq 0.$$

We prove that the two statements of Lindemann's theorem are equivalent. This is obvious when $k = 1$. Suppose $k \geq 2$.

1. Suppose the first version of the theorem holds. In the hypothesis of the second version of the theorem, suppose that the numbers $\alpha_1, \ldots, \alpha_k$ include exactly l which are linearly independent over $\mathbf{Q}$, where $l \leq k$. We let $\beta_1, \ldots, \beta_l$ denote a maximal set of linearly independent α_i. Then

$$\alpha_i = a_{i,1}\beta_1 + \cdots + a_{i,l}\beta_l, \qquad i = 1, \ldots, k, \qquad a_{i,j} \in \mathbf{Q}. \tag{151}$$

Letting $q \in \mathbf{N}$ denote the least common denominator of all of the $a_{i,j}$, we set

$$q^{-1}\beta_i = \xi_i, \qquad qa_{i,j} = b_{i,j}, \qquad i = 1, \ldots, k; \ \ j = 1, \ldots, l. \tag{152}$$

We consider the rational function

$$Q(z_1, \ldots, z_l) = \sum_{i=1}^{k} c_i z_1^{b_{i,1}} \cdots z_l^{b_{i,l}} \tag{153}$$

with arbitrary algebraic coefficients, not all zero. We have $Q(z_1, \ldots, z_l) \not\equiv 0$, because the l-tuples $(b_{i,1}, \ldots, b_{i,l})$ are distinct. In (153) we set $z_j = e^{\xi_j}$, $j = 1, \ldots, l$. Using the first version of Lindemann's theorem along with (152), we obtain:

$$Q\left(e^{\xi_1}, \ldots, e^{\xi_l}\right) = \sum_{i=1}^{k} c_i e^{b_{i,1}\xi_1 + \cdots + b_{i,l}\xi_l} = c_1 e^{\alpha_1} + \cdots + c_k e^{\alpha_k} \neq 0$$

for any $c_i \in \mathbf{A}$ which are not all zero.

2. Now suppose that the second version of the theorem holds. We prove that the first version also holds. In fact, otherwise there would exist a polynomial

$$P = P(z_1, \ldots, z_l) \in \mathbf{A}[z_1, \ldots, z_l], \qquad P \not\equiv 0,$$

such that

$$P\left(e^{\xi_1},\dots,e^{\xi_l}\right)=\sum_{(k_1,\dots,k_l)} c_{k_1,\dots,k_l}e^{k_1\xi_1+\dots+k_l\xi_l}=0, \qquad (154)$$
$$c_{k_1,\dots,k_l}\in \mathbf{A}\setminus\{0\},$$

where the summation is taken over a finite number of distinct l-tuples $(k_1,\dots,k_l)$, $k_i\in\mathbf{Z}^+$, $i=1,\dots,l$. But (154) contradicts the second version of Lindemann's theorem, since the exponents $k_1\xi_1+\dots+k_l\xi_l$ are distinct because $\xi_1,\dots,\xi_l$ are $\mathbf{Q}$-linearly independent.

This completes the proof that the two statements of Lindemann's theorem are equivalent.

§ 8. Linear approximating forms and the Newton interpolation series for the exponential function

In this section we explain the connection between the methods in § § 5, 6 and 3.

Let ρ_k, n_k and N have the same meaning as in § 5. We consider the complex integral

$$J(z)=\frac{1}{2\pi i}\int_C \frac{e^{zt}}{F_N(t)}dt, \qquad F_N(t)=\prod_{k=1}^{m}(t-\rho_k)^{n_k+1}, \qquad (155)$$

where C is a simple closed contour traversed in the positive direction which contains $\rho_1,\dots,\rho_m$ in its interior. Using the equality

$$e^{zt}=e^{z\rho_k}e^{z(t-\rho_k)}=e^{z\rho_k}\sum_{l=0}^{\infty}\frac{z^l(t-\rho_k)^l}{l!},$$

we easily see that the residue of the integrand at $t=\rho_k$ is equal to $Q_ke^{\rho_kz}$, where $Q_k=Q_k(z)$ is a polynomial in z with $\deg Q_k\le n_k$. Then, by the residue theorem,

$$J(z)=\sum_{k=1}^{m}Q_k(z)e^{\rho_kz}. \qquad (156)$$

On the other hand,

$$J(z)=\sum_{l=0}^{\infty}a_l\frac{z^l}{l!}, \qquad a_l=\frac{1}{2\pi i}\int_C\frac{t^l}{F_N(t)}dt. \qquad (157)$$

We choose a contour C which contains all of the discs $|t| \le |\rho_k|$, $k = 1, \ldots, m$. Expanding $1/F_N(t)$ in a Laurent series, we obtain:

$$\frac{1}{F_N(t)} = t^{-N-1} \prod_{k=1}^{m} \left(1 - \frac{\rho_k}{t}\right)^{-n_k-1} = t^{-N-1} + \cdots.$$

Hence, from (157) we have

$$a_l = 0, \qquad l = 0, 1, \ldots, N-1, \qquad a_N = 1,$$

$$J(z) = \frac{z^N}{N!} + \cdots.$$

Then, by the uniqueness of the linear form $R(z)$ in (104), which was proved in § 6, it follows that

$$\begin{aligned} Q_k(z) &= P_k(z), \quad k = 1, \ldots, m, \\ J(z) &= R(z) = \frac{1}{2\pi i} \int_C \frac{e^{zt}}{F_N(t)} dt. \end{aligned} \tag{158}$$

We have thus obtained a representation for $R(z)$ as a complex integral.

The equalities (156) and (158) show that the linear approximating form (104) is the coefficient A_N (see (48)) in the Newton interpolation series for the function $f(\zeta) = e^{z\zeta}$ with the periodic sequence of interpolation points

$$\begin{aligned} z_N &= \rho_N \qquad \text{for } N = 1, \ldots, m, \\ z_{N+m} &= z_N \qquad \text{for all } N \ge 1. \end{aligned}$$

The case considered in § 3 occurs when $z = \xi$, $\rho_k = k$, $k = 1, \ldots, m$.

In § § 4–7 we proved theorems on irrationality, transcendence and algebraic independence of values of the exponential function by means of certain functional linear approximating forms and the numerical linear approximating forms in the numbers under consideration that are obtained from the functional forms. Many other problems of transcendental number theory can be proved similarly, by constructing linear forms with integer or algebraic integer coefficients in the numbers being studied.

To prove that a number α is irrational, one must show that the linear form in the numbers α and 1

$$L = a_1 \alpha + a_0$$

is nonzero for any $a_0, a_1 \in \mathbf{Z}$ which are not both zero. To prove linear independence of $\alpha_1, \ldots, \alpha_m$ over $\mathbf{Q}$ (or over an algebraic number field $\mathbf{K}$), one must show that the linear form

$$L = a_1\alpha_1 + \cdots + a_m\alpha_m$$

is nonzero for any $a_1, \ldots, a_m \in \mathbf{Z}$ which are not all zero. To prove that a number α is transcendental, one considers the linear form

$$L = a_n\alpha^n + \cdots + a_1\alpha + a_0$$

in the powers of α with coefficients in $\mathbf{Z}$ (or in $\mathbf{Z}_\mathbf{A}$); one must prove that $|L| > 0$ for all $n \in \mathbf{N}$ and for any $a_0, a_1, \ldots, a_n \in \mathbf{Z}$ (or $\in \mathbf{Z}_\mathbf{A}$) which are not all zero. Finally, to prove that the numbers $\alpha_1, \ldots, \alpha_m$ are algebraically independent, one must consider linear forms with $\mathbf{Z}$- (or $\mathbf{Z}_\mathbf{A}$-) coefficients in the products of powers

$$\alpha_1^{k_1} \cdots \alpha_m^{k_m}, \qquad 0 \le k_i \le N_i, \qquad i = 1, \ldots, m,$$

for all $N_1, \ldots, N_m \in \mathbf{N}$.

A central ingredient in the method that will be developed in the following chapters is the construction of the corresponding linear approximating forms in the functions and values of the functions under consideration.

Remarks

The proof of irrationality of e by means of the series (1), which is contained in many textbooks, appeared even in Fourier's 1815 course in analysis [20:1]. In [41:3,4], Liouville proved that the numbers e and e^2 cannot be quadratic irrationalities. By adopting a more complicated version of the elementary method, Hurwitz [29:1] proved that e cannot be a root of a cubic equation with coefficients in $\mathbf{Z}$.

We note that Lambert's proof in [36:1] implies a stronger assertion than what was stated in the Introduction: the number $(\tan a)/a$ is irrational for any nonzero a for which $a^2 \in \mathbf{Q}$.

In [38:1] Legendre proved that π^2 is irrational.

For many years after that, there were no successful attempts to obtain results on the arithmetic nature of e and π. Euler, Lambert, Legendre and others conjectured that e and π are not algebraic numbers.

After Hermite's proof [27:1] that e is transcendental and Lindemann's proof [40:1,2] that π is transcendental, and after the general theorems on arithmetic properties of the values of the exponential function, many well-known mathematicians published papers improving and simplifying the proofs of Hermite and Lindemann, without fundamentally changing their methods. We indicate some of these authors: Vahlen [80:1], Veblen [82:1], Hilbert [28:1], Hurwitz [29:2], Markov

[46:1,2], Mertens [49:1], Rouché [64:1], Sylvester [77:1], Schottky [69:1]. Especially worthy of note is Weierstrass's paper [87:1], containing substantial research on Lindemann's general theorem.

More complete information on work related to the method of Hermite and Lindemann can be found in Koksma's book [31:1], and a detailed survey of papers on the problem of squaring the circle is contained in Rudio's book [65:1]. In addition, a survey of work related to the Hermite-Lindemann method is contained in the article [18:1] by Fel'dman and Shidlovskii.

One can find different proofs of the transcendence of e and π, the transcendence of e^{ξ} for $\xi \in \mathbf{A}$, and of Lindemann's Theorem in the books by Baker [2:5], Waldschmidt [84:1], Gel'fond [24:8], Siegel [72:4], Le Veque [39:1], Lang [37:1,3], Fel'dman [17:8], Shidlovskii [70:35], Schneider [68:6]. In Mahler's monograph [42:7] an entire chapter is devoted to an exposition of various proofs of the transcendence of e and π.

The proof of Theorem 6 is a reworking of Gel'fond's proof in [24:9], Ch. 2 § 4. The exposition in § 4–7 largely follows Siegel's book [72:4].

Chapter 3
Transcendence and algebraic independence of the values of E-functions which are not connected by algebraic equations over the field of rational functions

§ 1. E-functions

An analytic function

$$f(z) = \sum_{n=0}^{\infty} c_n \frac{z^n}{n!}$$

is called an *E-function* if it satisfies the following conditions:

1°. $c_n \in \mathbf{K}$, $n = 0, 1, 2, \ldots$, where $\mathbf{K}$ is an algebraic number field.

2°. For any $\epsilon > 0$,

$$\overline{|c_n|} = O(n^{\epsilon n}), \qquad n \to \infty.$$

3°. There exists a sequence $\{q_n\}$, $q_n \in \mathbf{N}$, such that

$$q_n c_k \in \mathbf{Z_K}, \qquad k = 0, 1, \ldots, n; \quad n = 1, 2, \ldots, \tag{1}$$

and

$$q_n = O(n^{\epsilon n}), \qquad n \to \infty. \tag{2}$$

From condition 2° it follows that E-functions are entire functions.

There always exists a sequence $\{q_n\}$ satisfying (1). Namely, for each c_n we can find $r_n \in \mathbf{N}$ such that $r_n c_n \in \mathbf{Z_K}$. Then the sequence $q_n = r_0 \cdots r_n$ satisfies (1). Thus, the essence of condition 3° is the requirement (2).

The simplest examples of E-functions are: a constant function with algebraic value, a polynomial with algebraic coefficients, the functions e^z, $\sin z$, and $\cos z$. They obviously satisfy $1° - 3°$.

We consider some properties of the class of E-functions:

1) *the derivative of an E-function is an E-function;*

2) *if $f(z)$ is an E-function, then so is $\int_0^z f(t)dt$;*

3) *if $f(z)$ is an E-function and $\lambda \in \mathbf{A}$, then $f(\lambda z)$ is an E-function;*

4) *the sum or product of a finite number of E-functions is an E-function.*

The first three of these properties are obvious. We now prove the fourth.

It is enough to prove that the sum or product of two E-functions is an E-function. Let

$$f_1(z) = \sum_{n=0}^{\infty} \frac{a_n}{n!} z^n, \qquad f_2(z) = \sum_{n=0}^{\infty} \frac{b_n}{n!} z^n.$$

We may assume that all of the coefficients a_n and b_n belong to a single algebraic number field $\mathbf{K}$. By the definition of an E-function, for any $\epsilon > 0$ we have:

$$1^\circ. \qquad a_n \in \mathbf{K}, \qquad b_n \in \mathbf{K}, \qquad n = 0, 1, 2, \ldots. \tag{3}$$

$$2^\circ. \qquad \overline{|a_n|} = O\left(n^{\frac{\epsilon}{3}n}\right), \qquad \overline{|b_n|} = O\left(n^{\frac{\epsilon}{3}n}\right), \qquad n \to \infty. \tag{4}$$

3°. There exist sequences $\{q_n'\}$ and $\{q_n''\}$ such that

$$q_n' a_k \in \mathbf{Z_K}, \qquad q_n'' b_k \in \mathbf{Z_K}, \qquad k = 0, 1, \ldots, n; \quad n = 1, 2, \ldots, \tag{5}$$

$$q_n' = O\left(n^{\frac{\epsilon}{2}n}\right), \qquad q_n'' = O\left(n^{\frac{\epsilon}{2}n}\right), \qquad n \to \infty. \tag{6}$$

We consider the functions

$$F_1(z) = f_1(z) + f_2(z) = \sum_{n=0}^{\infty} \frac{c_n}{n!} z^n,$$

$$F_2(z) = f_1(z) f_2(z) = \sum_{n=0}^{\infty} \frac{d_n}{n!} z^n.$$

Then

$$c_n = a_n + b_n, \qquad d_n = \sum_{l=0}^{n} \frac{n!}{l!(n-l)!} a_l b_{n-l}.$$

We shall prove that $F_1(z)$ and $F_2(z)$ are E-functions.

1°. By condition (3), we have $c_n \in \mathbf{K}$, $d_n \in \mathbf{K}$, $n = 0, 1, 2, \ldots$, since the sums and products of numbers in $\mathbf{K}$ are in $\mathbf{K}$.

2°. Using (4), we find that

$$\overline{|c_n|} \le \overline{|a_n|} + \overline{|b_n|} = O\left(n^{\frac{\epsilon}{3}n}\right) + O\left(n^{\frac{\epsilon}{3}n}\right) = O(n^{\epsilon n}),$$

$$\overline{|d_n|} \le \sum_{l=0}^{n} \frac{n!}{l!(n-l)!} \overline{|a_l|}\, \overline{|b_{n-l}|} = 2^n O\left(n^{\frac{\epsilon}{3}n}\right) O\left(n^{\frac{\epsilon}{3}n}\right) = O(n^{\epsilon n}).$$

3°. We set $q_n = q'_n q''_n$. Using (5), we obtain

$$q_n c_k = q_n(a_k + b_k) \in \mathbf{Z_K},$$

$$q_n d_k = \sum_{l=0}^{k} \frac{k!}{l!(k-l)!}(q'_n a_l)(q''_n b_{k-l}) \in \mathbf{Z_K},$$

$$k = 0, 1, \ldots, n; \quad n = 1, 2, \ldots,$$

and, by (6), we have

$$q_n = q'_n q''_n = O(n^{\frac{\epsilon}{2}n})O(n^{\frac{\epsilon}{2}n}) = O(n^{\epsilon n}).$$

From property 4) we have:

5) *any polynomial in z and a finite number of E-functions with algebraic coefficients is an E-function.*

Combining the above properties of E-functions, we obtain the following result: *the E-functions form a ring of functions which is closed under the operations of differentiation, integration from 0 to z, and the change of variables from z to λz, where $\lambda \in \mathbf{A}$.*

An E-function will be called a KE-function if its power series coefficients belong to the algebraic number field **K**. In particular, when $\mathbf{K} = \mathbf{I}$ is a quadratic imaginary field, we call it an IE-function; and when $\mathbf{K} = \mathbf{Q}$ we call it a QE-function.

§ 2. The First Fundamental Theorem

Let **V** be a field, and let **W** be a commutative ring or field containing **V**. We say that the elements $w_1, \ldots, w_m \in \mathbf{W}$ are *algebraically independent over* **V** if they are not connected by an algebraic equation with coefficients in **V**, i.e., if $P(w_1, \ldots, w_m) \neq 0$ for any nonzero polynomial $P = P(z_1, \ldots, z_m) \in \mathbf{V}[z_1, \ldots, z_m]$. If this is not the case, then we say that $w_1, \ldots, w_m$ are *algebraically dependent over* **V**.

If P is taken to be a homogeneous polynomial in these definitions, then we say that $w_1, \ldots, w_m$ are *homogeneously algebraically independent over* **V** or *homogeneously algebraically dependent over* **V**, respectively.

In the case $\mathbf{V} = \mathbf{A}$ and $\mathbf{W} = \mathbf{C}$, we shall simply say: *the complex numbers $\alpha_1, \ldots, \alpha_m$ are algebraically independent* or *algebraically dependent* (*homogeneously algebraically independent* or *homogeneously algebraically dependent*). This terminology was used in the Introduction and in Chapter 2.

Lemma 1. *If* $\alpha_1, \ldots, \alpha_m \in \mathbf{C}$ *are algebraically dependent (homogeneously algebraically dependent), then they are algebraically dependent over* $\mathbf{Q}$ *(homogeneously algebraically dependent over* $\mathbf{Q}$*).*

Proof. We prove the first case. The homogeneous case is proved similarly.

There exists a nonzero polynomial $P = P(z_1, \ldots, z_m) \in \mathbf{A}[z_1, \ldots, z_m]$ such that $P(\alpha_1, \ldots, \alpha_m) = 0$. We consider an algebraic number field $\mathbf{K} = \mathbf{Q}(\theta)$, $[\mathbf{K} : \mathbf{Q}] = h$, which contains all of the coefficients of P. We write these coefficients as polynomials in $\mathbf{Q}[\theta]$. Then $P(z_1, \ldots, z_m) = H(z_1, \ldots, z_m, \theta)$, where $H(z_1, \ldots, z_m, y) \in \mathbf{Q}[z_1, \ldots, z_m, y]$.

Let $\theta = \theta_1, \ldots, \theta_h$ be the conjugates of θ. Since $H(z_1, \ldots, z_m, \theta)$ is nonzero as a polynomial in $z_1, \ldots, z_m$, it follows that the same is true of $H(z_1, \ldots, z_m, \theta_i)$ for $i = 1, \ldots, h$, since if $H(z_1, \ldots, z_m, y)$ vanished at $y = \theta_j$ for some j, then it would also vanish at $y = \theta_i$ for $i = 1, \ldots, h$, in particular, at $y = \theta$. Next, we observe that

$$F = F(z_1, \ldots, z_m; \theta_1, \ldots, \theta_h) = \prod_{i=1}^{h} H(z_1, \ldots, z_m, \theta_i), \qquad F \not\equiv 0,$$

is a symmetric polynomial in $\theta_1, \ldots, \theta_h$ with coefficients in $\mathbf{Q}[z_1, \ldots, z_m]$. By Lemma 1 of Chapter 1, we have: $F = F(z_1, \ldots, z_m) \in \mathbf{Q}[z_1, \ldots, z_m]$. Since $H(z_1, \ldots, z_m, \theta)$ is a factor of F, and since $H(\alpha_1, \ldots, \alpha_m, \theta) = 0$, it follows that $F(\alpha_1, \ldots, \alpha_m) = 0$, and this proves the lemma.

We shall also be considering the case when $\mathbf{V} = \mathbf{C}(z)$ and $\mathbf{W}$ is a ring containing $\mathbf{C}(z)$ and the functions $f_1(z), \ldots, f_m(z)$. In this case we say that the functions $f_1(z), \ldots, f_m(z)$ are *algebraically independent or dependent (homogeneously algebraically independent or dependent) over the field of rational functions.*

In this book we shall always be considering some set of analytic functions $f_1(z), \ldots, f_m(z)$ which form a solution of a system of homogeneous linear first order differential equations

$$y_k' = \sum_{i=1}^{m} Q_{k,i} y_i, \qquad k = 1, \ldots, m, \qquad Q_{k,i} \in \mathbf{C}(z), \tag{7}$$

or a system of non-homogeneous linear differential equations

$$y_k' = Q_{k,0} + \sum_{i=1}^{m} Q_{k,i} y_i, \qquad k = 1, \ldots, m, \qquad Q_{k,i} \in \mathbf{C}(z). \tag{8}$$

We let $T = T(z) \in \mathbf{C}[z]$ denote a polynomial which is a least common denominator of all of the rational functions $Q_{k,i}$ in (7) or (8), i.e., it has the property that all $TQ_{k,i}$ are in $\mathbf{C}[z]$.

First Fundamental Theorem. *Let*

$$f_1(z), \dots, f_m(z), \qquad m \geq 2, \tag{9}$$

be a set of E-functions which form a solution of the set of m homogeneous linear differential equations (7) and are homogeneously algebraically independent over $\mathbf{C}(z)$, and let $\xi \in \mathbf{A}$, $\xi T(\xi) \neq 0$. Then the numbers $f_1(\xi), \dots, f_m(\xi)$ are homogeneously algebraically independent.

In order to prove the First Fundamental Theorem we shall demonstrate some auxiliary lemmas. We shall show that, if the conditions in the theorem hold, then all $Q_{k,i}$ in (7) are in $\mathbf{K}(z)$, where $\mathbf{K}$ is an algebraic number field containing all of the power series coefficients of the functions in (9).

Lemma 2. *Suppose that $\mathbf{K}$ is an algebraic number field, $m \geq 1$, and the set of functions*

$$\varphi_k(z) = \sum_{n=0}^{\infty} a_{k,n} z^n, \qquad k = 1, \dots, m, \qquad a_{k,n} \in \mathbf{K}, \tag{10}$$

are analytic in a region containing the point $z = 0$ and are connected by the algebraic equation

$$P(z, \varphi_1(z), \dots, \varphi_m(z)) = 0, \tag{11}$$

$$P = P(z, z_1, \dots, z_m) \in \mathbf{C}[z, z_1, \dots, z_m], \qquad P \not\equiv 0,$$

$$\deg_z P = s, \qquad \deg_{\mathbf{z}} P = N, \qquad \mathbf{z} = (z_1, \dots, z_m).$$

Then there exists a polynomial

$$P^* = P^*(z, z_1, \dots, z_m) \in \mathbf{Z_K}[z, z_1, \dots, z_m], \qquad P^* \not\equiv 0,$$

$$\deg_z P^* \leq s, \qquad \deg_{\mathbf{z}} P^* \leq N,$$

such that

$$P^*(z, \varphi_1(z), \dots, \varphi_m(z)) = 0.$$

Proof. We take the coefficients of P to be indeterminates, and in the power series in z on the left in (11) we set the coefficient of each power of z equal to zero. This gives us a system of countably many homogeneous linear equations with coefficients in $\mathbf{K}$ with a finite number of unknowns — the undetermined coefficients. We choose a maximum linearly independent subsystem of equations.

By the hypothesis of the lemma, it has a nontrivial solution. Then, since the system is homogeneous, the desired coefficients can be chosen in $\mathbf{Z}_{\mathbf{K}}$. As a result, we obtain a polynomial P^* satisfying the conclusion of the lemma.

We note that in Lemma 2, if we regard P and P^* as polynomials in $z_1, \ldots, z_m$ with coefficients which are polynomials in z, then, in general, the sets of zeros of the coefficients of P and P^* may be different.

Lemma 3. *Suppose that the m-tuple of functions (9), $m \geq 2$, are analytic in a region containing the point $z = 0$, form a solution of the system of homogeneous linear differential equations (7), and are linearly independent over $\mathbf{C}(z)$. Then the system (7) is uniquely determined,*

$$Q_{k,i} \in \mathbf{K}(z), \qquad k, i = 1, \ldots, m, \tag{12}$$

and the polynomial $T = T(z)$ can be chosen so that

$$T \in \mathbf{Z}_{\mathbf{K}}[z], \qquad TQ_{k,i} \in \mathbf{Z}_{\mathbf{K}}[z], \qquad k, i = 1, \ldots, m. \tag{13}$$

Proof. We first prove that, under the conditions in Lemma 3, the system (7) is uniquely determined. Suppose that the functions (9) satisfy another system of differential equations

$$y_k' = \sum_{i=1}^{m} Q_{k,i}^* y_i, \qquad Q_{k,i}^* \in \mathbf{C}(z), \qquad k = 1, \ldots, m.$$

We substitute $y_i = f_i(z)$, $i = 1, \ldots, m$, in both of the systems, and we equate the right sides of the corresponding equations of the two systems. We obtain:

$$\sum_{i=1}^{m} (Q_{k,i}^* - Q_{k,i}) f_i(z) = 0, \qquad k = 1, \ldots, m.$$

Because the functions (9) are linearly independent over $\mathbf{C}(z)$, this gives us

$$Q_{k,i}^* = Q_{k,i}, \qquad k, i = 1, \ldots, m.$$

Next, if we apply Lemma 2 (with m replaced by $m+1$) to each of the equations

$$Tf_k'(z) = \sum_{i=1}^{m} TQ_{k,i} f_i(z), \qquad 1 \leq k \leq m,$$

we find that (13) holds, and hence (12) is satisfied as well.

Corollary. *Under the conditions of the First Fundamental Theorem, the coefficients in (7) satisfy the conditions (12) and (13).*

§ 3. Some properties of linear and fractional-linear forms

We consider m functions

$$f_k(z) = \sum_{\nu=0}^{\infty} c_{k,\nu} z^\nu, \qquad k = 1, \ldots, m, \quad m \geq 2, \quad c_{k,\nu} \in \mathbf{C}, \tag{14}$$

which are analytic in a region containing the point $z = 0$; m polynomials

$$P_k \in \mathbf{C}[z], \qquad \deg P_k \leq n, \qquad n \in \mathbf{Z}^+, \tag{15}$$

with undetermined coefficients; and a linear form

$$R(z) = \sum_{k=1}^{m} P_k f_k(z) = \sum_{\nu=0}^{\infty} \gamma_\nu z^\nu. \tag{16}$$

There are $m(n+1)$ undetermined coefficients of the polynomials (15). If we require that the form $R(z)$ satisfies: $\gamma_\nu = 0$, $\nu = 0, 1, \ldots, m(n+1) - 2$, then we obtain $m(n+1) - 1$ homogeneous linear equations in the $m(n+1)$ undetermined coefficients of the polynomials (15). Since the number of unknowns is greater than the number of equations, the system has a nontrivial solution. Consequently, we can find $m(n+1)$ coefficients for the polynomials (15), not all of them zero, such that the form $R(z)$ satisfies: $\operatorname{ord} R(z) \geq m(n+1) - 1$.

We show that, if the functions (14) are not all identically zero, and if the degrees of the polynomials (15) are bounded, then it is not possible to modify the coefficients of the polynomials in such a way as to obtain a linear form $R(z)$ which is not identically zero and which has arbitrarily high order at $z = 0$.

Lemma 4. *Let $f_1(z), \ldots, f_m(z)$ be an m-tuple of functions (14) which are analytic in a region containing the point $z = 0$, and let $n \in \mathbf{Z}^+$. Then there exists $N \in \mathbf{N}$, depending on the functions (14) and the integer n, such that for any choice of polynomials (15) the linear form $R(z)$ in (16) either is identically zero or else satisfies the relation* $\operatorname{ord} R(z) \leq N$.

Proof. We first prove the lemma in the case when all $P_1, \ldots, P_m \in \mathbf{C}$.

We may suppose that at least one of the functions (14) is not identically zero. We consider the set $\{R(z)\}$ of all possible linear forms (16) for which $R(z) \not\equiv 0$. Then $\{R(z)\} \neq \emptyset$. If we have a set of forms $R_1(z), \ldots, R_k(z)$ for which the numbers $\operatorname{ord} R_1(z), \ldots, \operatorname{ord} R_k(z)$ are distinct, then clearly these forms are linearly independent over $\mathbf{C}$. Since any $m+1$ forms of the form (16) are linearly dependent over $\mathbf{C}$, it follows that the set $\{R(z)\}$ must include at most m forms $R(z)$ for which the numbers $\operatorname{ord} R(z)$ are distinct. If we denote $N = \max \operatorname{ord} R(z)$, we obtain the lemma in this case.

Now suppose that $P_k \in \mathbf{C}[z]$, $\deg P_k \leq n$, $n \in \mathbf{N}$, $k = 1, \ldots, m$. We consider the $m(n+1)$ functions

$$f_{k,\nu}(z) = z^\nu f_k(z), \qquad k = 1, \ldots, m; \quad \nu = 0, 1, \ldots, n,$$

and we apply the assertion proved above to these functions, where m is replaced by $m(n+1)$. We then find that, for any $b_{k,\nu} \in \mathbf{C}$, $k = 1, \ldots, m$, $\nu = 0, 1, \ldots, n$, the linear form

$$R(z) = \sum_{k=1}^{m} \sum_{\nu=0}^{n} b_{k,\nu} z^\nu f_k(z) = \sum_{k=1}^{m} P_k f_k(z)$$

either is identically zero or else satisfies $\operatorname{ord} R(z) \leq N$, where N depends only on the functions (14) and the integer n.

We shall always assume that a rational function in $\mathbf{C}(z)$ is represented as a fraction with relatively prime polynomials in the numerator and denominator. By the *degree of a rational function* we mean the sum of the degrees of the numerator and denominator. In particular, if the rational function is a polynomial, then this coincides with the usual definition of the degree of a polynomial. The degree of a rational function has the following obvious properties:

1. *The degree of a product of rational functions does not exceed the sum of the degrees of the factors.*

2. *The degree of a quotient of two rational functions does not exceed the sum of the degrees of the two functions.*

3. *The degree of a sum of rational functions does not exceed twice the sum of the degrees of the summands.*

Lemma 5. *Let*

$$\varphi_1(z), \ldots, \varphi_s(z); \quad \psi_1(z), \ldots, \psi_m(z) \tag{17}$$

be analytic functions in some region such that for at least one value of i

$$\psi_i(z) \not\equiv 0, \qquad 1 \leq i \leq m. \tag{18}$$

Then there exists $N_0 \in \mathbf{N}$, depending only on the functions (17), such that there does not exist an $(s+m)$-tuple of complex numbers

$$\alpha_1, \ldots, \alpha_s; \quad \beta_1, \ldots, \beta_m, \tag{19}$$

with

$$\sum_{i=1}^{m} \beta_i \psi_i(z) \not\equiv 0, \tag{20}$$

such that the quotient

$$\omega = \sum_{i=1}^{s} \alpha_i \varphi_i(z) / \sum_{i=1}^{m} \beta_i \psi_i(z) \tag{21}$$

is a rational function of z of degree greater than N_0.

Proof. Suppose that we have an $(s+m)$-tuple (19) such that (20) holds and the function (21) belongs to $\mathbf{C}(z)$. Let r be the $\mathbf{C}(z)$-*rank* of the set of functions (17) (the maximal number of functions which are linearly independent over $\mathbf{C}(z)$). By (18), we have $1 \le r \le m+s$. We choose any r functions from (17) which are linearly independent over $\mathbf{C}(z)$, and we denote them $g_1(z), \dots, g_r(z)$. Then each of the functions in (17) can be expressed uniquely as a linear combination of $g_1(z), \dots, g_r(z)$ with coefficients in $\mathbf{C}(z)$:

$$\varphi_i(z) = \sum_{j=1}^{r} A_{i,j} g_j(z), \qquad i = 1, \dots, s, \qquad A_{i,j} \in \mathbf{C}(z), \tag{22}$$

$$\psi_i(z) = \sum_{j=1}^{r} B_{i,j} g_j(z), \qquad i = 1, \dots, m, \qquad B_{i,j} \in \mathbf{C}(z). \tag{23}$$

Let N denote the maximum of the degrees of the $(s+m)r$ rational functions $A_{i,j}$ and $B_{i,j}$. Clearly, N depends only on the functions (17).

We substitute the expansions (22) and (23) for $\varphi_i(z)$ and $\psi_i(z)$ in the right side of (21). We obtain:

$$\omega = \frac{\sum_{i=1}^{s} \sum_{j=1}^{r} \alpha_i A_{i,j} g_j(z)}{\sum_{i=1}^{m} \sum_{j=1}^{r} \beta_i B_{i,j} g_j(z)} = \frac{\sum_{j=1}^{r} C_j g_j(z)}{\sum_{j=1}^{r} D_j g_j(z)}, \tag{24}$$

where

$$C_j = \sum_{i=1}^{s} \alpha_i A_{i,j}, \qquad j = 1, \dots, r, \tag{25}$$

$$D_j = \sum_{i=1}^{m} \beta_i B_{i,j}, \qquad j = 1, \dots, r. \tag{26}$$

From (20) it follows that

$$\sum_{j=1}^{r} D_j g_j(z) \not\equiv 0.$$

Hence we have $D_j \not\equiv 0$ for at least one value of j.

Using the properties of the degree of a rational function, from (25) and (26) we conclude that the degrees of all of the rational functions C_j and D_j are bounded by $2sN$ and $2mN$, respectively.

From (24) we obtain:

$$\sum_{j=1}^{r} (C_j - \omega D_j) g_j(z) = 0. \tag{27}$$

Since the functions $g_1(z), \ldots, g_r(z)$ are linearly independent over $\mathbf{C}(z)$, it follows from (27) that

$$C_j - \omega D_j = 0, \qquad j = 1, \ldots, r. \tag{28}$$

We now choose j so that $D_j \not\equiv 0$. Then for this j the equality (28) gives us

$$\omega = C_j / D_j. \tag{29}$$

The degree of the right side of (29) is at most $N_0 = 2sN + 2mN = 2(s+m)N$. The lemma is proved.

§ 4. Properties of linear forms in functions which satisfy a system of homogeneous linear differential equations

Let $\mathbf{M}$ denote the module of linear forms in the variables $y_1, \ldots, y_m$, $m \geq 2$, over the ring $\mathbf{C}[z]$. An element $R \in \mathbf{M}$ has the form

$$R = \sum_{i=1}^{m} P_i y_i, \qquad P_i = P_i(z) \in \mathbf{C}[z], \qquad i = 1, \ldots, m. \tag{30}$$

We consider a system of first order homogeneous linear differential equations

$$y_k' = \sum_{i=1}^{m} Q_{k,i} y_i, \qquad k = 1, \ldots, m, \qquad m \geq 2, \tag{31}$$

$$Q_{k,i} = Q_{k,i}(z) \in \mathbf{C}(z).$$

The polynomial $T = T(z)$ for the system (31) will have the same meaning as in § 2, i.e., all $TQ_{k,i} \in \mathbf{C}[z]$.

On $\mathbf{M}$ we define the differential operator

$$D = \frac{\partial}{\partial z} + \sum_{k=1}^{m}\Big(\sum_{i=1}^{m} Q_{k,i}y_i\Big)\frac{\partial}{\partial y_k}, \tag{32}$$

corresponding to the system of differential equations (31). If $R \in \mathbf{M}$, then obviously also $TDR \in \mathbf{M}$. Thus, we have $TD : \mathbf{M} \to \mathbf{M}$.

Let $y_1, \ldots, y_m$ be any solution of the system of differential equations (31). We substitute this solution in the form (30) in place of the variables. This makes R into a function of z, and the derivative R' with respect to z is a linear form in $y_1, \ldots, y_m, y_1', \ldots, y_m'$ with coefficients in $\mathbf{C}[z]$. If in R' we replace the derivatives $y_1', \ldots, y_m'$ by the right sides of the corresponding differential equations (31), then we find that TR' is a linear form in $y_1, \ldots, y_m$ with coefficients in $\mathbf{C}[z]$. Consequently, in this situation applying the operator D to the form R amounts to computing the total derivative of R with respect to z and then replacing $y_1', \ldots, y_m'$ by the right sides of the differential equations (31). Thus, if $y_1, \ldots, y_m$ is a solution of (31), then

$$DR = \frac{d}{dz}R = R'. \tag{33}$$

In what follows, with the above assumptions concerning $y_1, \ldots, y_m$, we shall agree to let R' denote the result of differentiating R as a function of z and then replacing $y_1', \ldots, y_m'$ by the right sides of (31).

We consider an arbitrary linear form in $\mathbf{M}$

$$R_1 = \sum_{i=1}^{m} P_{1,i}y_i, \qquad P_{1,i} \in \mathbf{C}[z], \tag{34}$$

and we set

$$R_k = TDR_{k-1}, \qquad k = 2, 3, \ldots. \tag{35}$$

Using the above argument, we see that $R_k \in \mathbf{M}$, $k = 2, 3, \ldots$, and so we can write

$$R_k = \sum_{i=1}^{m} P_{k,i}y_i, \qquad k = 1, 2, \ldots, \qquad P_{k,i} \in \mathbf{C}[z]. \tag{36}$$

Using (32) and (35), it is easy to see that for $k > 1$ the polynomials $P_{k,i}$ can be determined recursively from the formulas

$$P_{k,i} = T\Big(P'_{k-1,i} + \sum_{j=1}^{m} P_{k-1,j}Q_{j,i}\Big), \qquad i = 1,\dots,m. \tag{37}$$

If we substitute an arbitrary solution of the system of differential equations (31) in place of the variables $y_1,\dots,y_m$ in the linear form (34), then, according to (33), the equations (35) take the form

$$R_k = TR'_{k-1}, \qquad k = 2,3,\dots. \tag{38}$$

By the *rank of a set of linear forms* with coefficients in a ring or field $\mathbf{V}$ we mean the maximum number of forms in the set which are linearly independent over $\mathbf{V}$.

Lemma 6. *If $R_1 \in \mathbf{M}$ is an arbitrary linear form (34), then the rank of the set of linear forms $R_1, R_2, \dots$ in (34)–(36) is equal to l, $0 \le l \le m$, if and only if the forms $R_1,\dots,R_l$ are linearly independent over $\mathbf{C}[z]$ while the forms $R_1,\dots,R_{l+1}$ are linearly dependent over $\mathbf{C}[z]$.*

Proof. Suppose that $R_1,\dots,R_l$ are linearly independent, but $R_1,\dots, R_{l+1}$ are linearly dependent over $\mathbf{C}[z]$. Then the latter $l+1$ forms are connected by a homogeneous linear equation with coefficients in $\mathbf{C}[z]$ in which R_{l+1} occurs. If we apply the operator TD to both sides of this equation $k-1$ times, where $k \ge 2$, we obtain a sequence of homogeneous linear equations for $R_1,\dots,R_{l+k}$, from which we can deduce a linear homogeneous equation for $R_1,\dots,R_l, R_{l+k}$ with coefficients in $\mathbf{C}[z]$ in which R_{l+k} occurs. Since this holds for any $k = 1,2,\dots$, it follows that the set of linear forms $R_1, R_2,\dots$ has rank l.

Conversely, if the set of forms $R_1, R_2,\dots$ has rank l, then the forms $R_1,\dots,R_l$ are linearly independent, since otherwise, by what was just proved, the rank of the forms $R_1, R_2,\dots$ would be less than l. The lemma is proved.

We consider the first m of the linear forms (36):

$$R_k = \sum_{i=1}^{m} P_{k,i}y_i, \qquad k = 1,\dots,m, \qquad P_{k,i} \in \mathbf{C}[z]. \tag{39}$$

We let $\Delta = \Delta(z)$ denote the determinant of the set of linear forms (39):

$$\Delta = \Delta(z) = |P_{k,i}|_{k,i=1,\dots,m}, \tag{40}$$

and we let $\Delta_{k,i}$ denote the cofactor of $P_{k,i}$ in this determinant.

For fixed j, $1 \leq j \leq m$, we multiply both sides of each of the m equations in (39) by $\Delta_{k,j}$ for $k = 1, \ldots, m$, respectively. Adding the resulting equalities, we obtain

$$\sum_{k=1}^{m} \Delta_{k,j} R_k = \sum_{i=1}^{m} \Big(\sum_{k=1}^{m} \Delta_{k,j} P_{k,i} \Big) y_i. \tag{41}$$

If we let $\delta_{i,j}$ denote the determinant obtained from Δ by replacing its j-th column by its i-th column, then

$$\sum_{k=1}^{m} \Delta_{k,i} P_{k,j} = \delta_{i,j} = \begin{cases} 0, & \text{if } i \neq j, \\ \Delta, & \text{if } i = j. \end{cases}$$

Hence, from (41) we obtain

$$\Delta y_j = \sum_{k=1}^{m} \Delta_{k,j} R_k, \qquad 1 \leq j \leq m, \tag{42}$$

identically in $y_1, \ldots, y_m$ and z.

The determinant Δ is nonzero if and only if the forms $R_1, \ldots, R_m$ are linearly independent, i.e., have rank m.

If the linear form R_1 vanishes when we substitute a nontrivial solution $y_1, \ldots$ $\ldots, y_m$ of the system (31), then it follows from (38) that all of the forms R_1, $R_2, \ldots$ also vanish for this solution. From (42) we then find that $\Delta \equiv 0$, since $y_1, \ldots, y_m$ is a nontrivial solution, and so $y_j \not\equiv 0$ for at least one value of j. This means that the forms $R_1, \ldots, R_m$ are linearly dependent over $\mathbf{C}[z]$. It is not hard to see that the converse also holds. This follows, in particular, from the following more general fact.

Lemma 7. *Let R_1 be any linear form of the form (34), and suppose that the set of forms $R_1, \ldots, R_m$ in (39) has rank l, where $1 \leq l < m$. Then it is possible to choose $\mu = m - l$ linearly independent solutions of the system of differential equations (31)*

$$y_{1,s}, \ldots, y_{m,s}, \qquad s = 1, \ldots, \mu, \tag{43}$$

in such a way that the forms $R_1, R_2, \ldots$ vanish when any of the μ solutions (43) is substituted in place of the variables $y_1, \ldots, y_m$.

Proof. The system of linear forms (39) has rank l. By Lemma 6, this means that the linear forms $R_1, \ldots, R_l$ are linearly independent, and the forms $R_1, \ldots$ $\ldots, R_l, R_{l+1}$ are linearly dependent. Thus, we can find $A_1, \ldots, A_{l+1} \in \mathbf{C}[z]$ such that the following equation holds identically in the variables $y_1, \ldots, y_m$ and z:

$$A_1 R_1 + \cdots + A_{l+1} R_{l+1} \equiv 0, \qquad A_{l+1} \not\equiv 0. \tag{44}$$

If we substitute an arbitrary solution $y_1, \ldots, y_m$ of (31) into the form R_1 and use (38), we can rewrite the identity (44) as follows:

$$\begin{aligned} &B_l R_1^{(l)} + \cdots + B_1 R_1' + B_0 R_1 \equiv 0, \\ &B_i \in \mathbf{C}[z], \qquad i = 1, \ldots, l, \qquad B_l = A_{l+1} T^l \not\equiv 0. \end{aligned} \tag{45}$$

Since (44) and (45) hold identically in $y_1, \ldots, y_m$ and z, it follows that the linear form R_1 satisfies the l-th order linear differential equation (45), where $l < m$, for any solution $y_1, \ldots, y_m$ of (31).

Let $\mathbf{V}$ be the $\mathbf{C}$-vector space of solutions of the system of linear differential equations (31), and let $\mathbf{U}$ be the $\mathbf{C}$-vector space of solutions of the differential equation (45). The dimensions $\dim \mathbf{V}$ and $\dim \mathbf{U}$ satisfy the equalities $\dim \mathbf{V} = m$ and $\dim \mathbf{U} = l$. Above we showed that the linear form R_1 in (34) can be used to obtain a map $\varphi : \mathbf{V} \to \mathbf{U}$ which is a $\mathbf{C}$-vector space homomorphism. Let $\operatorname{Ker} \varphi$ denote the *kernel of the map* φ (the preimage of zero), and let $\operatorname{Im} \varphi$ denote the *image of the map* φ. Then, by a well-known theorem of linear algebra (see [37:3], p. 107), we have

$$\dim \mathbf{V} = \dim \operatorname{Ker} \varphi + \dim \operatorname{Im} \varphi.$$

Since $\dim \operatorname{Im} \varphi \leq l$, we then find that $\dim \operatorname{Ker} \varphi \geq \mu = m - l$. We choose any μ linearly independent solutions of the system (31) in the subspace of solutions $\operatorname{Ker} \varphi$. Denoting them $y_{i,s}$, $i = 1, \ldots, m$, $s = 1, \ldots, \mu$, we obtain:

$$R_{k,s} = \sum_{i=1}^{m} P_{k,i} y_{i,s} = 0, \qquad k = 1, \ldots, m; \quad s = 1, \ldots, \mu. \tag{46}$$

The lemma is proved.

It is possible to prove that $\dim \operatorname{Ker} \varphi = \mu$. That is, μ is the maximum number of linearly independent solutions of (31) such that the linear form R_1 vanishes when the solution is substituted for the variables.

§ 5. Order of zero of a linear form at $z = 0$

Lemma 8. *Suppose that the m-tuple of functions*

$$f_1(z), \dots, f_m(z), \qquad m \geq 2, \tag{47}$$

are analytic in a region containing the point $z = 0$, form a solution of the system of homogeneous linear differential equations (31), and are linearly independent over $\mathbf{C}(z)$; set

$$p = \min_{1 \leq i \leq m} \operatorname{ord} f_i(z), \tag{48}$$

and

$$q = \max\Big(\deg T, \max_{\substack{1 \leq i \leq m \\ 1 \leq k \leq m}} \deg TQ_{k,i}\Big). \tag{49}$$

Furthermore, let $n \in \mathbf{N}$; let

$$\begin{gathered} R_1 = \sum_{i=1}^{m} P_{1,i} y_i, \qquad R_1 \not\equiv 0, \\ P_{1,i} \in \mathbf{C}[z], \qquad \deg P_{1,i} \leq n, \qquad i = 1, \dots, m, \end{gathered} \tag{50}$$

be any linear form in $\mathbf{M}$; and let l be the rank of the set of linear forms (39) obtained from R_1 using (35). Then there exists a constant r_0, depending only on the functions (47), such that the following relation holds for $y_i = f_i(z)$, $i = 1, \dots, m$:

$$\operatorname{ord} R_1(z) = \operatorname{ord} \sum_{i=1}^{m} P_{1,i} f_i(z) \leq ln + r_0. \tag{51}$$

If $l = m$, then r_0 in (51) can be replaced by r_1, where

$$r_1 = q\frac{(m-1)m}{2} + m + p - 1. \tag{52}$$

Proof. The set of forms (39) has rank l, where $l \leq m$. Since $R_1 \not\equiv 0$, it follows that $l > 0$.

We consider the linear forms

$$R_k = \sum_{i=1}^{m} P_{k,i} y_i, \qquad k = 1, \dots, l, \tag{53}$$

and the rectangular coefficient matrix for these forms

$$\|P_{k,i}\|_{k=1,\dots,l;\,i=1,\dots,m}. \tag{54}$$

By Lemma 6, the forms (53) are linearly independent, and so the matrix (54) has rank l and contains at least one nonzero minor of order l. Without loss of generality we may assume that this nonzero minor can be chosen to be the determinant

$$\Delta_0 = \Delta_0(z) = |P_{k,i}|_{k,i=1,\dots,l} \not\equiv 0, \tag{55}$$

since we can always renumber the functions (47).

If $l < m$, then each column of the matrix (54) with index $j > l$ is a linear combination of the first l columns; hence,

$$P_{k,j} = \sum_{i=1}^{l} P_{k,i} D_{i,j}, \qquad D_{i,j} \in \mathbf{C}(z), \qquad k = 1,\dots,l;\quad j = l+1,\dots,m. \tag{56}$$

Using (56), from (53) we find that

$$\begin{aligned} R_k &= \sum_{i=1}^{m} P_{k,i} y_i = \sum_{i=1}^{l} P_{k,i} y_i + \sum_{j=l+1}^{m} P_{k,j} y_j = \\ &= \sum_{i=1}^{l} P_{k,i} y_i + \sum_{j=l+1}^{m} \sum_{i=1}^{l} P_{k,i} D_{i,j} y_j = \sum_{i=1}^{l} P_{k,i} \Big(y_i + \sum_{j=l+1}^{m} D_{i,j} y_j \Big). \end{aligned}$$

Thus, the linear forms (53) can be represented in the form

$$R_k = \sum_{i=1}^{l} P_{k,i} u_i, \qquad k = 1,\dots,l, \tag{57}$$

where

$$u_i = y_i + \sum_{j=l+1}^{m} D_{i,j} y_j, \qquad i = 1,\dots,l, \qquad l \le m. \tag{58}$$

On the other hand, if $l = m$, then we take $u_i = y_i$, in accordance with (58).

If $l < m$, then, taking (55) into account, we see from (56) that the rational functions $D_{i,j}$, $i = 1,\dots,l$, $j = l+1,\dots,m$, are uniquely determined for a fixed linear form R_1. We now derive formulas which express these functions in terms of the solutions of the system (31).

Since l is in the interval $1 \leq l < m$, by Lemma 7 we can choose a fundamental system of solutions

$$\|y_{i,s}\|_{i,s=1,\ldots,m}, \tag{59}$$

of (31) in such a way that (46) holds. Hence, if we denote

$$u_{i,s} = y_{i,s} + \sum_{j=l+1}^{m} D_{i,j} y_{j,s}, \qquad i = 1, \ldots, l; \quad s = 1, \ldots, \mu, \tag{60}$$

then the following relations hold, by (46), (57), (58) and (60):

$$R_{k,s} = \sum_{i=1}^{l} P_{k,i} u_{i,s} = 0, \qquad k = 1, \ldots, l; \quad s = 1, \ldots, \mu. \tag{61}$$

For each fixed s, $1 \leq s \leq \mu$, we consider the system of l homogeneous linear equations (61) in the l unknowns $u_{1,s}, \ldots, u_{l,s}$. Because of the condition (55), we can be sure that this system has only the trivial solution. According to (60), this means that

$$u_{i,s} = y_{i,s} + \sum_{j=l+1}^{m} D_{i,j} y_{j,s} = 0, \qquad i = 1, \ldots, l; \quad s = 1, \ldots, \mu. \tag{62}$$

For each fixed i, $1 \leq i \leq l$, (62) is a system of μ homogeneous linear equations in the μ functions $D_{i,j}$, $j = l + 1, \ldots, m$. This system has a unique solution for any i, provided that its determinant

$$\lambda = |y_{j,s}|_{j=l+1,\ldots,m;s=1,\ldots,\mu} \tag{63}$$

is not identically zero as a function of z.

We suppose the contrary, i.e., that $\lambda \equiv 0$, and we consider the determinant σ of the matrix (59).

We fix any i, $1 \leq i \leq l$, and we add the i-th row of σ and the $(l+1)$-th,$\ldots$, m-th rows, after first multiplying the j-th row by $D_{i,j}$ for $j = l + 1, \ldots, m$. By (62), the first μ entries in the i-th row of σ then become zero. If we do this for all i, $i = 1, \ldots, l$, we find that σ contains an $l \times \mu$ bloc of zeros in its upper-left corner. Thus, in the first μ columns of σ we have this bloc entirely of zeros, under which we have a square bloc with μ rows and columns having determinant $\lambda \equiv 0$. Hence, by Laplace's theorem, the determinant σ of the matrix (59) is identically zero as a function of z. But the determinant of a matrix of fundamental solutions of a system of homogeneous linear differential equations cannot be identically zero. This contradiction shows that $\lambda \not\equiv 0$; consequently, all of the functions $D_{i,j}$, $i = 1, \ldots, l$, $j = l + 1, \ldots, m$, are uniquely determined by (62).

We consider the rectangular matrix

$$\|y_{i,s}\|_{i=1,\dots,l;s=1,\dots,\mu}, \tag{64}$$

and we let $\lambda_{i,j}$ denote the determinant obtained from λ (see (63)) if we replace the row $y_{j,1},\dots,y_{j,\mu}$, where $l+1 \le j \le m$, by the row $y_{i,1},\dots,y_{i,\mu}$, $1 \le i \le l$, of the matrix (64). Then from (62) we find that

$$D_{i,j} = -\frac{\lambda_{i,j}}{\lambda}, \qquad \lambda \not\equiv 0, \qquad i = 1,\dots,l; \;\; j = l+1,\dots,m. \tag{65}$$

The equations (65) express the functions $D_{i,j}$ in terms of the entries of the matrix (59), and this matrix depends on the choice of the form R_1. Thus, we shall transform the right sides of the equations in (65) in such a way that they contain only functions of a fixed fundamental set of solutions of (31). Let

$$\|y^*_{i,k}\|_{i,k=1,\dots,m} \tag{66}$$

be the matrix of an arbitrary fixed fundamental set of solutions.

In the matrix

$$\|y^*_{j,k}\|_{j=l+1,\dots,m;k=1,\dots,m} \tag{67}$$

there are

$$m_1 = \binom{m}{m-l} = \frac{m!}{l!(m-l)!}$$

$\mu \times \mu$ minors. We number them arbitrarily, and denote them

$$\psi_1(z),\dots,\psi_{m_1}(z). \tag{68}$$

If we replace the row $y^*_{j,1},\dots,y^*_{j,m}$ in (67), where $l+1 \le j \le m$, by the row $y^*_{i,1},\dots,y^*_{i,m}$ in (66), where $1 \le i \le l$, then we can form m_1 minors of dimension $\mu \times \mu$ in the resulting matrix. We number them arbitrarily, and denote them

$$\varphi_{i,j,1}(z),\dots,\varphi_{i,j,m_1}(z), \qquad 1 \le i \le l, \;\; l+1 \le j \le m. \tag{69}$$

The matrix (59) differs from the fixed matrix (66) by a matrix factor

$$\|c_{k,s}\|_{k,s=1,\dots,m}, \qquad c_{k,s} \in \mathbf{C}, \tag{70}$$

where the square matrix (70) has the property that its determinant is non-zero. Hence,

$$\begin{aligned}\|y_{i,s}\|_{i,s=1,\ldots,m} &= \|y^*_{i,k}\|_{i,k=1,\ldots,m}\ \|c_{k,s}\|_{k,s=1,\ldots,m} = \\ &= \Big\|\sum_{k=1}^{m} y^*_{i,k}c_{k,s}\Big\|_{i,s=1,\ldots,m}.\end{aligned} \tag{71}$$

We again consider (65). Using (71), we find the following expression for the denominator on the right in (65):

$$\lambda = |y_{j,s}| = \Big|\sum_{k=1}^{m} y^*_{j,k}c_{k,s}\Big|_{j=l+1,\ldots,m;s=1,\ldots,\mu}. \tag{72}$$

The determinant on the right in (72) is equal to the determinant of the product of the two rectangular matrices

$$\|y^*_{j,k}\|_{j=l+1,\ldots,m;k=1,\ldots,m} \tag{73}$$

and

$$\|c_{k,s}\|_{k=1,\ldots,m;s=1,\ldots,\mu}. \tag{74}$$

By the Binet–Cauchy formula (see [23:1], p. 9), this determinant is equal to the sum of all possible products of a minor of maximal size $\mu \times \mu$ in the matrix (73) with a corresponding minor of the same size in the matrix (74). Hence, using the notation (68), we have

$$\lambda = \sum_{k=1}^{m_1} \beta_k\psi_k(z), \qquad \lambda \not\equiv 0,$$

where the constants $\beta_1, \ldots, \beta_m$ are homogeneous forms of dimension μ in the entries of the matrix (74).

Similarly, using the notation (69), we obtain

$$-\lambda_{i,j} = \sum_{k=1}^{m_1} \alpha_{i,j,k}\psi_{i,j,k}(z), \qquad i = 1, \ldots, l; \;\; j = l+1, \ldots, m,$$

where the constants $\alpha_{i,j,k}$ are also homogeneous forms of dimension μ in the entries of the matrix (74).

Now the equations (65) can be rewritten as follows:

$$D_{i,j} = \frac{\sum_{k=1}^{m_1} \alpha_{i,j,k}\varphi_{i,j,k}(z)}{\sum_{k=1}^{m_1} \beta_k\psi_k(z)}, \qquad \sum_{k=1}^{m_1} \beta_k\psi_k(z) \not\equiv 0, \tag{75}$$

$$i = 1, \dots, l; \quad j = l+1, \dots, m.$$

Applying Lemma 5 to each of the equalities in (75), we find that the degree of each of the rational functions $D_{i,j}$ is no greater than a certain number v, depending only on the functions (68) and (69) (which are minors of the fixed matrix (66) of a fundamental set of solutions of (31)).

If we consider all linear forms R_1 which satisfy (50) for variable n, in general we obtain different sets of rational functions $D_{i,j}$, determined from (62) using (75) for different values of l, different ways of numbering the functions (47), and different sets of functions (68) and (69). Here there are finitely many different sets of functions (68) and (69) and corresponding equations (75). Consequently, by what was proved above, there exists a number v_0 such that, for any linear form R_1 and any numbering of the functions (47), the degree of each of the rational functions $D_{i,j}$ in (56) is no greater than v_0. Here v_0 depends only on the system of differential equations (31). And because the functions (47) are linearly independent over $\mathbf{C}(z)$, it follows by Lemma 3 that the system (31) is uniquely determined by these functions, and so v_0 depends only on the set of functions (47).

We substitute the functions (47) in place of $y_1, \dots, y_m$ in the linear forms (53). Then the equations (58) take the form

$$u_i = f_i(z) + \sum_{j=l+1}^{m} D_{i,j} f_j(z), \qquad i = 1, \dots, l, \qquad l \leq m. \tag{76}$$

We let $T_i = T_i(z)$ denote a polynomial which serves as least common denominator of all the rational functions $D_{i,j}$, $j = l+1, \dots, m$, in (76); if $l = m$, we set $T_i = 1$. The expressions

$$H_i = T_i u_i, \qquad i = 1, \dots, l, \qquad l \leq m, \tag{77}$$

are linear forms in the functions (47) with coefficients in $\mathbf{C}[z]$.

Since the functions (47) are linearly independent over $\mathbf{C}(z)$, it follows from (76) that none of the functions in (77) is identically zero as a function of z. We denote $\tau = \min_{1 \leq i \leq l} \operatorname{ord} H_i$. If $l = m$, we have $\tau = p$.

By what was proved above, the degrees of the polynomial coefficients of the linear forms in (77) are bounded by μv_0. Hence, if we apply Lemma 4 to each of these forms, we find that there exists a number τ_0, depending only on the functions (47), such that, for any n and any of our linear forms R_1, the number τ corresponding to the functions (77) satisfies the inequality $\tau \leq \tau_0$. If $l = m$, we have $\tau = p$.

We consider the determinant Δ_0 (see (55)) of the set of linear forms (57). We let $\Delta_{k,i}^0$ denote the cofactor of $P_{k,i}$ in Δ_0. In the same way that (42) was obtained, we find the following equations for the system of linear forms (57):

$$\Delta_0(z)u_j = \sum_{k=1}^{l} \Delta^0_{k,j} R_k(z), \qquad 1 \le j \le l. \tag{78}$$

We multiply both sides of these equations by the corresponding T_j. Using the notation (77), we obtain:

$$\Delta_0(z)H_j = T_j \sum_{k=1}^{l} \Delta^0_{k,j} R_k(z), \qquad 1 \le j \le l. \tag{79}$$

We choose j, $1 \le j \le l$, in such a way that $\operatorname{ord} H_j = \tau$, and we consider the j-th equation in (78).

From (37) it follows that $\deg P_{k,i} \le n + q(k-1)$, $i = 1, \dots, l$. Hence, $\deg \Delta_0(z) \le ln + q\frac{l(l-1)}{2}$. Since $\operatorname{ord} H_j = \tau$, it follows that

$$\operatorname{ord} \Delta_0(z)H_j \le ln + q\frac{(l-1)l}{2} + \tau. \tag{80}$$

Let $\operatorname{ord} R_1(z) = \nu$. Then from (38) we find that $\operatorname{ord} R_k(z) \ge \nu - k + 1$, $k = 1, \dots, l$. Consequently,

$$\operatorname{ord} \sum_{k=1}^{l} \Delta^0_{k,i} R_k(z) \ge \nu - l + 1. \tag{81}$$

The inequality (51) follows from (79) and (80)–(81), if we set

$$r_0 = q\frac{(m-1)m}{2} + m - 1 + \tau_0. \tag{82}$$

Since we have $\tau = p$ when $l = m$, in that case we can replace r_0 by r_1 in the inequality (51). The lemma is proved.

§ 6. The determinant of a set of linear forms

We introduce the following notation:

$$n_0 = 2(r_0 - m + 1) = q(m-1)m + 2\tau_0, \tag{83}$$

$$s \in \mathbf{N}, \qquad 0 \le s < \left[\frac{n}{2}\right], \tag{84}$$

$$t = q\frac{(m-1)m}{2} + s + p, \tag{85}$$

where r_0 and n were determined in Lemma 8.

Lemma 9. *Suppose that the conditions of Lemma 8 are fulfilled, and that the linear form R_1 (50) satisfies the following condition for $y = f_i(z)$, $i = 1, \dots, m$:*

$$\operatorname{ord} R_1(z) \geq m(n+1) - s - 1. \tag{86}$$

Then for $n \geq n_0$ the linear forms (39) obtained from R_1 by means of (35) are linearly independent, and the determinant $\Delta(z)$ (see (40)) of the forms (39) has the form

$$\begin{gathered} \Delta(z) = z^{mn-s-p}\Delta_1(z), \qquad \Delta_1(z) \not\equiv 0, \\ \Delta_1(z) \in \mathbf{C}[z], \qquad \deg \Delta_1(z) = \kappa, \qquad 0 \leq \kappa \leq t. \end{gathered} \tag{87}$$

Proof. As in Lemma 8, we suppose that the linear forms (39) have rank l, $1 \leq l \leq m$; then from (51) and (86) we have

$$m(n+1) - s - 1 \leq ln + r_0,$$

from which, by (84), we obtain:

$$(m-l)n \leq r_0 + s + 1 - m \leq \frac{n}{2} + r_0 - m. \tag{88}$$

If $n \geq n_0$ and $l < m$, then it follows from (83) that we have a contradiction in (88). Thus, $l = m$ and $\Delta_0(z) = \Delta(z)$.

We again consider the equation in (78) for $l = m$:

$$\Delta(z) f_j(z) = \sum_{k=1}^{m} \Delta_{k,j} R_k(z), \qquad \Delta(z) \not\equiv 0, \qquad \operatorname{ord} f_j(z) = p. \tag{89}$$

From (38) and (86) we conclude that $\operatorname{ord} R_k(z) \geq m(n+1) - s - k$. Then from (89) we obtain the equations in (87), where

$$\begin{aligned} \kappa = \deg \Delta_1(z) &= \deg \Delta(z) - (mn - s - p) \leq \\ &\leq \left(mn + q\frac{(m-1)m}{2}\right) - (mn - s - p) = q\frac{(m-1)m}{2} + s + p = t. \end{aligned}$$

The lemma is proved.

§ 7. Passing to linearly independent numerical linear forms

Lemma 10 (C. L. Siegel). *Suppose that the functions (47) are analytic in a region containing the point $z = 0$, form a solution of the system of homogeneous linear differential equations (31), and are linearly independent over* $\mathbf{C}(z)$. *Let p and q be given by (48) and (49), and let n_0, s and t satisfy (83)–(85). Further suppose that R_1 is a linear form (50) such that $n \geq n_0$ and (86) holds for $y_i = f_i(z)$, $i = 1, \ldots, m$; let $R_{k+1} = TDR_k$, $k = 1, 2, \ldots$, $\alpha \in \mathbf{C}$, and $\alpha T(\alpha) \neq 0$.*

Then the coefficient matrix

$$\|P_{k,i}(\alpha)\|_{i=1,\ldots,m;k=1,\ldots,m+t}, \tag{90}$$

of the linear forms

$$R_k(\alpha) = \sum_{i=1}^{m} P_{k,i}(\alpha) f_i(\alpha), \qquad k = 1, \ldots, m+t, \tag{91}$$

has rank m, and thus m linearly independent forms can be chosen from among the forms (91).

Proof. Because all of the conditions of Lemma 9 are fulfilled, it follows that (87) holds. Since $\alpha T(\alpha) \neq 0$, we can say that if $\Delta(z)$ has a τ-th order zero at $z = \alpha$, then, by (87),

$$0 \leq \tau \leq t. \tag{92}$$

We consider the linear forms $R_1, \ldots, R_m$ in $\mathbf{M}$ of the form (36) with variables $y_1, \ldots, y_m$ which are constructed from R by means of (35), and also the relations (42), which hold identically in $y_1, \ldots, y_m$ and z. We apply the operator TD to both sides of (42). Using (35), we obtain the following relations, which hold identically in $y_1, \ldots, y_m$ and z:

$$T(z)\Delta'(z)y_j + \Delta(z)\mathcal{L}_{1,j,0} = \sum_{k=1}^{m+1} M_{1,j,k}(z) R_k, \qquad 1 \leq j \leq m,$$

where $\mathcal{L}_{1,j,0}$ is some linear form in $y_1, \ldots, y_m$ with coefficients in $\mathbf{C}[z]$, and all $M_{1,j,k}(k) \in \mathbf{C}[z]$.

If we repeat this process τ times, we obtain the following relations, which hold identically in $y_1, \dots, y_m$ and z:

$$T^{\tau}(z)\Delta^{(\tau)}(z)y_j + \sum_{k=0}^{\tau-1} \Delta^{(k)}(z)\mathcal{L}_{\tau,j,k} = \sum_{k=1}^{m+\tau} M_{\tau,j,k}(z)R_k =$$
$$= \sum_{k=1}^{m+\tau} M_{\tau,j,k}(z)\big(P_{k,1}(z)y_1 + \dots + P_{k,m}(z)y_m\big), \quad 1 \le j \le m, \tag{93}$$

where $\mathcal{L}_{\tau,j,k}$ are linear forms in $y_1, \dots, y_m$ with coefficients in $\mathbf{C}[z]$, and $M_{\tau,j,k} \in \mathbf{C}[z]$.

By assumption, the following conditions are fulfilled:

$$\Delta^{(k)}(\alpha) = 0, \qquad k = 0, 1, \dots, \tau - 1,$$

$$T^{\tau}(\alpha)\Delta^{(\tau)}(\alpha) = \beta \neq 0. \tag{94}$$

We set $z = \alpha$ in (93) and denote $M_{\tau,j,k}(\alpha) = \beta_{j,k}$. Then, by (94), the identities in (93) take the form

$$\beta y_j = \sum_{k=1}^{m+\tau} \beta_{j,k}\big(P_{k,1}(\alpha)y_1 + \dots + P_{k,m}(\alpha)y_m\big),$$
$$\beta \neq 0, \qquad 1 \le j \le m. \tag{95}$$

From (95) it follows that the variables $y_1, \dots, y_m$ can be represented as linear combinations of the $m + \tau$ linear forms

$$P_{k,1}(\alpha)y_1 + \dots + P_{k,m}(\alpha)y_m, \qquad k = 1, \dots, m + \tau.$$

Because of (92), this implies that the matrix (91) has rank m. The lemma is proved.

§ 8. Auxiliary lemmas on solutions of systems of homogeneous linear equations

Lemma 11 (C. L. Siegel). *Suppose that the linear forms*

$$L_i = \sum_{j=1}^{q} a_{i,j}x_j, \qquad i = 1, \dots, p, \quad p < q, \quad q \ge 2, \tag{96}$$

satisfy the conditions

$$a_{i,j} \in \mathbf{Z}, \quad |a_{i,j}| \le A, \quad A \in \mathbf{N}, \quad i = 1, \ldots, p, \quad j = 1, \ldots, q. \tag{97}$$

Then there exists a nontrivial solution $x_1, \ldots, x_q$ *of the system of equations*

$$L_i = 0, \qquad i = 1, \ldots, p, \tag{98}$$

such that

$$x_j \in \mathbf{Z}, \qquad |x_j| < 1 + (qA)^{\frac{p}{q-p}}, \qquad j = 1, \ldots, q. \tag{99}$$

Proof. We use the Dirichlet pigeon-hole principle. Let $H \in \mathbf{N}$. If we let each of the unknowns $x_1, \ldots, x_q$ independently take on the $2H + 1$ values $0, \pm 1, \pm 2, \ldots \ldots, \pm H$ in all of the forms (96), we obtain $(2H + 1)^q$ points $L = (L_1, \ldots, L_p)$ in p-dimensional space $\mathbf{R}^p$. Because of (97), we know that all of these points lie in the hypercube

$$-qAH \le L_i \le qAH, \qquad i = 1, \ldots, p. \tag{100}$$

This hypercube has exactly $(2qAH + 1)^p$ points with coordinates in $\mathbf{Z}$. If H is chosen so that

$$(2qAH + 1)^p < (2H + 1)^q, \tag{101}$$

then the number of points we have with integer coordinates will be greater than the total number of integer points in the hypercube (100). Thus, one can find two of our q-tuples $x'_1, \ldots, x'_q$ and $x''_1, \ldots, x''_q$ which give the same point $L = (L_1, \ldots, L_p)$. If we denote

$$x_j = x'_j - x''_j, \qquad j = 1, \ldots, q,$$

then the q-tuple $x_1, \ldots, x_q$ is a solution of (98) which satisfies the conditions

$$|x_j| \le |x'_j| + |x''_j| \le 2H, \qquad j = 1, \ldots, q. \tag{102}$$

We now choose $2H$ equal to the even number which lies in the following interval of length 2:

$$(qA)^{\frac{p}{q-p}} - 1 \le 2H < (qH)^{\frac{p}{q-p}} + 1. \tag{103}$$

Then (101) holds, because the inequalities $q \ge 2$ and (103) imply that

$$(2qAH + 1)^p < (qA)^p(2H + 1)^p \le (2H + 1)^{q-p}(2H + 1)^p = (2H + 1)^q.$$

Using (102) and (103), we find that

$$|x_j| \le 2H < (qA)^{\frac{p}{q-p}} + 1, \qquad j = 1, \dots, q.$$

This shows that (99) is fulfilled. The lemma is proved.

By a *basis of the algebraic number field* $\mathbf{K} = \mathbf{Q}(\theta)$, $[\mathbf{K} : \mathbf{Q}] = h$, we mean a basis of the ring of integers $\mathbf{Z}_\mathbf{K}$of this field, i.e., an h-tuple $\omega_1, \dots, \omega_h \in \mathbf{Z}_\mathbf{K}$ of algebraic integers which are $\mathbf{Q}$-linearly independent and have the property that any $\alpha \in \mathbf{Z}$ can be expressed uniquely in the form

$$\alpha = b_1\omega_1 + \dots + b_h\omega_h, \qquad b_i \in \mathbf{Z}, \qquad i = 1, \dots, h. \tag{104}$$

The *discriminant of the field* $\mathbf{K}$ is defined to be:

$$d = \begin{vmatrix} \omega_1^{(1)} & \dots & \omega_h^{(1)} \\ \omega_1^{(2)} & \dots & \omega_h^{(2)} \\ \dots & \dots & \dots \\ \omega_1^{(h)} & \dots & \omega_h^{(h)} \end{vmatrix}^2 \neq 0,$$

where $\omega_1^{(j)}, \dots, \omega_h^{(j)}$ are the conjugates of $\omega_1, \dots, \omega_h$, respectively, in the field $\mathbf{K}$.

Lemma 12. *Suppose that*

$$\omega_1, \dots, \omega_h \tag{105}$$

is a basis of the algebraic number field $\mathbf{K}$, $[\mathbf{K} : \mathbf{Q}] = h$; α *is any element of* $\mathbf{Z}_\mathbf{K}$; *and (104) holds. Then*

$$|b_i| < \gamma\overline{|\alpha|}, \qquad \gamma > 0, \qquad i = 1, \dots, h, \tag{106}$$

where the constant γ *depends only on the choice of basis (105).*

Proof. By the definition of a basis, any $\alpha \in \mathbf{Z}_\mathbf{K}$ can be expressed uniquely in the form (104). Let $\alpha^{(j)}$, $j = 1, \dots, h$, be the conjugates of α, and let $\omega_1^{(j)}, \dots, \omega_h^{(j)}$, $j = 1, \dots, h$, be the conjugates of the basis elements (105) in the field $\mathbf{K}$. The following equations hold in the h conjugate fields $\mathbf{K}^{(j)}$, $j = 1, \dots, h$:

$$\alpha^{(j)} = b_1\omega_1^{(j)} + \dots + b_h\omega_h^{(j)}, \qquad j = 1, \dots, h. \tag{107}$$

We regard (107) as a system of h linear equations in the h unknowns $b_1, \dots, b_h$. The determinant of this system is $\pm\sqrt{d} \neq 0$, where d is the discriminant of $\mathbf{K}$. If ω_0 is the maximum modulus of all of the $(h-1) \times (h-1)$ minors in the determinant of (107), then the solution of this system satisfies the inequalities

$$|b_i| < \frac{h\omega_0 \overline{|\alpha|}}{|d|} = \gamma \overline{|\alpha|}, \qquad i = 1, \ldots, h,$$

where γ depends only on the choice of basis (105). The lemma is proved.

Lemma 13 (C. L. Siegel). *Suppose that the linear forms (96) satisfy the conditions*

$$\begin{aligned} a_{i,j} \in \mathbf{Z_K}, \qquad \overline{|a_{i,j}|} \le A, \qquad A \in \mathbf{R}, \qquad A > 0, \\ i = 1, \ldots, p, \quad j = 1, \ldots, q. \end{aligned} \tag{108}$$

Then there exists a nontrivial solution $x_1, \ldots, x_q$ of the system (98) such that

$$x_j \in \mathbf{Z_K}, \qquad \overline{|x_j|} < c(cqA)^{\frac{p}{q-p}}, \qquad c > 0, \qquad j = 1, \ldots, q, \tag{109}$$

where the constant c depends only on the field **K**.

Proof. Let (105) be a basis of **K**. Then every $\alpha \in \mathbf{Z_K}$ can be expressed uniquely in the form (104), where, by Lemma 12, the numbers $b_1, \ldots, b_h$ satisfy the inequalities (106) with γ a constant depending only on the choice of basis (105). We set

$$\begin{aligned} x_j = x_{j,1}\omega_1 + \cdots + x_{j,h}\omega_h, \qquad j = 1, \ldots, q, \\ x_{j,k} \in \mathbf{Z}, \qquad j = 1, \ldots, q, \quad k = 1, \ldots, h, \end{aligned} \tag{110}$$

and we substitute the right sides of the equations (110) in place of $x_1, \ldots, x_q$ in (98). We obtain

$$L_i = \sum_{j=1}^{q} \sum_{k=1}^{h} a_{i,j}\omega_k x_{j,k}, \qquad i = 1, \ldots, p. \tag{111}$$

We next represent the pqh products $a_{i,j}\omega_k$ as linear combinations of the basis elements (105) with coefficients in **Z**, and we substitute these linear combinations in (111). Because of the **Q**-linear independence of the basis elements, this system can then be split up into ph homogeneous linear equations with coefficients in **Z** in the qh unknowns $x_{1,1}, \ldots, x_{q,h}$, since the coefficient of each $\omega_1, \ldots, \omega_h$ in each equation (111) must be zero.

By (106) and (108), the coefficients in these ph linear equations have absolute value less than $\gamma \max \overline{|a_{i,j}\omega_k|}$, and hence less than $\gamma_1 A$, where γ_1 is a positive constant which depends only on the choice of basis (105). If we apply Lemma 11 to our system of ph homogeneous linear equations with coefficients in **Z** and

qh unknowns, we find a solution $x_{1,1}, \ldots, x_{q,h} \in \mathbf{Z}$ such that

$$|x_{j,k}| < 1 + (\gamma_1 hqA)^{\frac{p}{q-p}}, \qquad j = 1, \ldots, q, \quad k = 1, \ldots, h. \tag{112}$$

Using (110), from (112) we find that

$$\overline{|x_j|} < h\big(1 + (\gamma_1 hqA)^{\frac{p}{q-p}}\big) \max_{1 \le k \le h} \overline{|\omega_k|} < c(cqA)^{\frac{p}{q-p}},$$

$$c > 0, \qquad j = 1, \ldots, m,$$

where c clearly depends only on $\mathbf{K}$.

§ 9. Functional linear approximating forms

The functions in the lemmas proved above belong to a broader class of functions than the E-functions. Hence, those lemmas can be used when generalizing the method of this book to other classes of functions.

In all of the lemmas proved below which lead up to the proof of the First Fundamental Theorem (except for Lemma 18), we shall restrict our attention to the class of E-functions, since the proofs will make use of the arithmetic properties of the power series coefficients. These results can be proved again when the method is applied to other classes of functions, such as G-functions (see, for example, [55:1], [22:6]).

We consider KE-functions $f_1(z), \ldots, f_m(z)$, $m \ge 2$, and a linear form

$$R = \sum_{i=1}^{m} P_i f_i(z), \qquad P_i \in \mathbf{C}[z], \qquad \deg P_i \le n, \qquad i = 1, \ldots, m.$$

In § 3 we showed that it is possible to choose coefficients of the polynomials P_i, not all zero, in such a way that $\operatorname{ord} R \ge m(n+1) - 1$. These coefficients are found from a system of $m(n+1) - 1$ homogeneous linear equations obtained by setting the coefficients of z^ν, $\nu = 0, 1, \ldots, m(n+1) - 2$, equal to zero in the power series for the function R. The system always has a nontrivial solution, since it has more unknowns than equations. The coefficients of the system belong to the field $\mathbf{K}$, since the power series coefficients of our KE-functions belong to $\mathbf{K}$; and because the equations are homogeneous, the coefficients of P_i can be chosen to belong to $\mathbf{Z}_{\mathbf{K}}$.

In this situation, Lemma 13 gives us an upper bound for the moduli of the coefficients of the polynomials P_i and their conjugates in $\mathbf{K}$. However, the magnitude of this estimate is not good enough to obtain the arithmetic results which interest us, since the fraction p/q is close to 1. But if we weaken somewhat the requirement on the order of zero of R at $z = 0$, so that this order of zero is not the greatest possible but is sufficiently great, then we can use Lemma 13 to obtain a bound for the moduli of the coefficients of the P_i and their conjugates which is suitable for our applications.

Lemma 14. *Suppose that $f_1(z), \ldots, f_m(z)$, $m \geq 2$, is a set of KE-functions, $n \in \mathbf{N}$, $\epsilon \in \mathbf{R}$, $0 < \epsilon < 1/2$. Then there exist m polynomials*

$$P_k = P_k(z) = \sum_{l=0}^{n} b_{k,l} z^l, \qquad k = 1, \ldots, m, \tag{113}$$

with the following properties:

1) all $b_{k,l} \in \mathbf{Z_K}$, they are not all zero, and they satisfy

$$\overline{|b_{k,l}|} = O\left(n^{(1+\epsilon)n}\right), \qquad k = 1, \ldots, m; \;\; l = 0, 1, \ldots, n, \tag{114}$$

as $n \to \infty$, uniformly in k and l;

2) the linear form

$$R = \sum_{k=1}^{m} P_k(z) f_k(z) = \sum_{\nu=\tau}^{\infty} a_\nu \frac{z^\nu}{\nu!} \tag{115}$$

satisfies the property

$$\operatorname{ord} R \geq \tau, \qquad \tau = m(n+1) - [\epsilon n] - 1; \tag{116}$$

3) the coefficients in the series (115) satisfy the following condition for all n sufficiently large, uniformly in ν:

$$\overline{|a_\nu|} = \nu^{\epsilon\nu} O(n^n), \qquad \nu \geq \tau. \tag{117}$$

Proof. We denote

$$f_k(z) = \sum_{\nu=0}^{\infty} c_{k,\nu} \frac{z^\nu}{\nu!}, \qquad k = 1, \ldots, m, \tag{118}$$

$$P_k(z) = n! \sum_{\nu=0}^{n} B_{k,\nu} \frac{z^\nu}{\nu!}, \qquad k = 1, \ldots, m, \tag{119}$$

where the $B_{k,\nu}$ are undetermined coefficients in $\mathbf{Z_K}$. Then $P_k(z) \in \mathbf{Z_K}[z]$. We further set

$$P_k(z) f_k(z) = n! \sum_{\nu=0}^{\infty} d_{k,\nu} \frac{z^\nu}{\nu!}, \qquad k = 1, \ldots, m, \tag{120}$$

where

$$d_{k,\nu} = \sum_{\rho=0}^{\min(n,\nu)} \binom{\nu}{\rho} B_{k,\rho} c_{k,\nu-\rho}. \tag{121}$$

From (115) and (120) we obtain

$$a_\nu = n! \sum_{k=1}^{m} d_{k,\nu}, \qquad \nu = 0, 1, 2, \ldots. \tag{122}$$

From the definition of an E-function it follows that there exists a sequence $\{q_\nu\}$, $q_\nu \in \mathbf{N}$, such that

$$q_\nu c_{k,l} \in \mathbf{Z_K}, \qquad l = 0, 1, \ldots, \nu; \quad k = 1, \ldots, m; \quad \nu = 1, 2, \ldots,$$

where

$$q_\nu = O\left(\nu^{\epsilon_1 \nu}\right), \qquad \epsilon_1 = \frac{\epsilon^2}{18m^2}, \qquad \nu = 0, 1, 2, \ldots, \tag{123}$$

and also

$$\overline{|c_{k,\nu}|} = O\left(\nu^{\epsilon_1 \nu}\right), \qquad \nu = 0, 1, 2, \ldots; \quad k = 1, \ldots, m. \tag{124}$$

In order for (116) to be fulfilled, we require that the following equalities hold:

$$\frac{q_\nu a_\nu}{n!} = \sum_{k=1}^{m} \sum_{\rho=0}^{\min(n,\nu)} \binom{\nu}{\rho} q_\nu c_{k,\nu-\rho} B_{k,\rho} = 0, \qquad \nu = 0, 1, \ldots, \tau - 1. \tag{125}$$

These equalities form a system of τ homogeneous linear equations in the $m(n+1)$ unknowns $B_{k,\rho}$. By (123) and (124), the coefficients of the system (125) satisfy the condition

$$\overline{\left|q_\nu \binom{\nu}{\rho} c_{k,\nu-\rho}\right|} = O\left(\nu^{\epsilon_1 \nu}\right) 2^\nu O\left(\nu^{\epsilon_1 \nu}\right) = O\left(\nu^{\frac{\epsilon^2}{8m^2}\nu}\right) =$$

$$= O\left((mn+m)^{\frac{\epsilon^2}{8m^2}m(n+1)}\right) = \tag{126}$$

$$= O\left(n^{\frac{\epsilon^2}{3m}n}\right), \qquad \nu \le \tau.$$

If we use Lemma 13 with the values

$$p = m(n+1) - [\epsilon n] - 1, \qquad q = m(n+1), \qquad A = O\left(n^{\frac{\epsilon^2}{3m}n}\right),$$

we find that the $B_{k,\nu}$ are in $\mathbf{Z_K}$, are not all zero, and satisfy:

$$\overline{|B_{k,\nu}|} < c(cqA)^{p/(q-p)} = c\left(cm(n+1)\, O\left(n^{\frac{\epsilon^2}{3m}n}\right)\right)^{2m/\epsilon} = \tag{127}$$

$$= O\left(n^{\epsilon n}\right), \qquad \nu = 0, 1, \ldots, n; \quad k = 1, \ldots, m,$$

since

$$\frac{p}{q-p} = \frac{m(n+1) - [\epsilon n] - 1}{[\epsilon n] + 1} < \frac{m(n+1)}{\epsilon n} \le \frac{2m}{\epsilon}.$$

Since the coefficients $b_{k,\nu}$ of the polynomials (113) are equal to $(n!/\nu!)B_{k,\nu}$, by (119), and since $n! \le n^n$, it follows that (127) implies the bound (114).

The relation (117) follows from (121) and (122), in view of (124) and (127). In fact, we have

$$|a_\nu| = n! \left|\sum_{k=1}^{m} d_{k,\nu}\right| = n! \left|\sum_{k=1}^{m} \sum_{\rho=0}^{\min(n,\nu)} \binom{\nu}{\rho} B_{k,\rho} c_{k,\nu-\rho}\right| =$$

$$= m2^\nu O\left(n^{\epsilon n}\right) O\left(\nu^{\frac{\epsilon^2}{18m^2}\nu}\right) O\left(n^n\right) = \nu^{\epsilon \nu} O\left(n^n\right), \qquad \nu \ge \tau.$$

The linear form (115) constructed in Lemma 14, which has a high order zero at $z = 0$, approximates zero well in the algebraic sense. Linear forms in various functions which have this property are called *functional linear approximating forms* or simply *approximating forms.*

Suppose that our set of KE-functions forms a solution of the system (7) of homogeneous linear differential equations. We set $R_1 = R$, where R is the linear approximating form (115) which satisfies the conditions of Lemma 14. Then, if we use (35) (or (38)), we can successively obtain linear forms R_1, $R_2, \ldots$ (see (36)), in which $y_i = f_i(z)$, $i = 1, \ldots, m$. If our KE-functions are linearly independent over $\mathbf{C}(z)$, then, by Lemma 3, the coefficients of the system of differential equations (7) satisfy the conditions (12) and (13); it then follows from (37) that all of the linear forms R_1, $R_2, \ldots$, have coefficients $P_{k,i}(z) \in \mathbf{Z_K}[z]$.

In Lemma 9 we set $s = [\epsilon n]$, $0 < \epsilon < 1/2$. The lemma then tells us that for $n \geq n_0$ the linear forms $R_1, \ldots, R_m$ are linearly independent, and the determinant formed from their coefficients $P_{k,i}$ has the form (87), where $\Delta_1(z) \in \mathbf{Z_K}[z]$. The coefficients of the polynomials $P_{k,i}$ and their conjugates in $\mathbf{K}$, and also the power series coefficients for the linear forms $R_1, R_2, \ldots$, can be bounded using (37) along with (114) and (117).

Each of the linear forms $R_1, \ldots, R_m$ has a high order zero at $z = 0$. Thus, from a single functional linear approximating form (115) one can use (35) to obtain a sequence of m linearly independent functional linear approximating forms.

§ 10. Numerical linear approximating forms

In § 9 we constructed m linearly independent linear approximating forms from a set of m KE-functions which satisfy a system (7) of linear differential equations. We now proceed to the arithmetic part of the method, where the basic problem is to construct a set of linearly independent numerical linear approximating forms in the values of our functions. Using Lemma 10, we do this by going from the functional linear approximating forms to a set of m linearly independent numerical forms which are obtained from a large (but not too large) number of the functional forms.

If R_1 is the linear form constructed in Lemma 14, and if $R_{k+1}(z) = TR'_k(z)$, $k = 1, 2, \ldots$, $\alpha \in \mathbf{C}$, $\alpha T(\alpha) \neq 0$, then, by Lemma 10, the linear forms

$$R_k(\alpha), \qquad k = 1, \ldots, m+t, \qquad t = q\frac{(m-1)m}{2} + [\epsilon n] + p, \tag{128}$$

include m linearly independent forms.

We now prove some bounds for $|R_k(\xi)|$ and $\overline{|P_{k,i}(\xi)|}$ at $\xi \in \mathbf{A}$ which we shall later need.

Lemma 15. *Suppose that the KE-functions $f_1(z), \ldots, f_m(z)$, $m \geq 2$, form a solution of the system (7) of linear differential equations and are linearly independent over $\mathbf{C}(z)$, and let $\xi \in \mathbf{K}$. Suppose that R_1 is the linear form constructed in Lemma 14 for any n, $n \geq n_0$, with ϵ replaced by $\epsilon_1 = \epsilon/6(m+1)$ for some ϵ, $0 < \epsilon < 1$, and let t_1 be the number obtained from t by replacing ϵ by ϵ_1 in (128). Then the following bounds for the linear forms $R_k(z)$ and their coefficients are satisfied as* *n increases:*

$$|R_k(\xi)| = O\left(n^{-(m-1-\epsilon/2)n}\right), \qquad k \leq m + t_1, \tag{129}$$

$$\overline{|P_{k,i}(\xi)|} = O\left(n^{(1+\epsilon/2)n}\right), \qquad k \leq m + t_1; \quad i = 1, \ldots, m. \tag{130}$$

Proof. If the coefficients of the power series

$$\varphi(z) = \sum_{\nu=0}^{\infty} a_\nu z^\nu, \qquad \psi(z) = \sum_{\nu=0}^{\infty} b_\nu z^\nu$$

satisfy the conditions $|a_\nu| \leq b_\nu$, $b_\nu \geq 0$, $\nu = 0, 1, 2, \ldots$, then we agree to write $\varphi(z) \ll \psi(z)$. We further introduce the notation

$$\varphi^*(z) = \sum_{\nu=0}^{\infty} |a_\nu| z^\nu.$$

Then it is obvious that $\varphi(z) \ll \varphi^*(z)$.

We have

$$T(z) \ll c(1+z)^q, \qquad T(z)Q_{k,i}(z) \ll c(1+z)^q, \tag{131}$$

where $c > 0$ is a constant which does not depend on n, and q, as above, is the maximum degree of all of the polynomials T and $TQ_{k,i}$.

Since

$$R_1 = \sum_{i=1}^{m} P_{1,i}(z) f_i(z) = \sum_{\nu=0}^{\infty} a_\nu \frac{z^\nu}{\nu!}$$

is the linear form constructed in Lemma 14, it follows by (116) and (117) that

$$a_\nu = 0, \qquad \nu = 0, 1, 2, \ldots, \tau - 1,$$

$$|a_\nu| = \nu^{\epsilon_1 \nu} O\left(n^n\right), \qquad \nu \geq \tau,$$

where

$$\tau = m(n+1) - [\epsilon_1 n] - 1.$$

Hence

$$R_1(z) \ll R_1^*(z) = \sum_{\nu=T}^{\infty} |a_\nu| \frac{z^\nu}{\nu!} = \sum_{\nu=T}^{\infty} \nu^{\epsilon_1 \nu} \frac{z^\nu}{\nu!} O\left(n^n\right). \tag{132}$$

We prove the following two relations by induction:

$$R_{k+1}(z) \ll c^k (1+z)^{kq} \prod_{\nu=0}^{k-1} \left(\nu q + \frac{d}{dz}\right) R_1^*(z), \qquad k = 0, 1, 2, \dots, \tag{133}$$

$$P_{k+1,i}(z) \ll c^k (1+z)^{kq+n} \prod_{\nu=0}^{k-1} (\nu q + m + n)\, O\left(n^{(1+\epsilon_1)n}\right), \qquad i = 1, \dots, m; \quad k = 0, 1, 2, \dots. \tag{134}$$

The first of these relations remains true if the coefficients of the polynomials $P_{k+1,i}(z)$ are replaced by their conjugates in **K** with any fixed index. (We set $\prod_{\nu=0}^{k-1} \varphi_\nu = 1$.)

By Lemma 14, (133) and (134) hold for $k = 0$. We suppose that they hold for some $k - 1$, $k \geq 1$, and we prove that they are then also fulfilled for k. Using (131) and the induction assumption, we obtain:

$$R_{k+1}(z) = T(z) R_k'(z) \ll c(1+z)^q c^{k-1} \Big((k-1) q (1+z)^{(k-1)q-1} + \\ + (1+z)^{(k-1)q} \frac{d}{dz} \Big) \prod_{\nu=0}^{k-2} \left(\nu q + \frac{d}{dz}\right) R_1^*(z) \ll \\ \ll c^k (1+z)^{kq} \prod_{\nu=0}^{k-1} \left(\nu q + \frac{d}{dz}\right) R_1^*(z).$$

$$
\begin{aligned}
P_{k+1,i}(z) = T(z)\left(P'_{k,i}(z) + \sum_{j=1}^{m} P_{k,j}(z)Q_{j,i}(z)\right) \ll \\
\ll c(1+z)^q c^{k-1}\prod_{\nu=0}^{k-2}(\nu q+m+n)\, O\left(n^{(1+\epsilon_1)n}\right) \times \\
\times \left(m+\frac{d}{dz}\right)(1+z)^{(k-1)q+n} \ll \\
\ll c^k(1+z)^{kq+n}\prod_{\nu=0}^{k-1}(\nu q+m+n)\, O\left(n^{(1+\epsilon_1)n}\right).
\end{aligned}
$$

We hence conclude that (133) and (134) hold for any k, $k \in \mathbf{Z}^+$.

Now suppose that k satisfies the inequality $k \le m + t_1 = \epsilon n + O(1)$. Then (134) is easily seen to imply the bound (130). Furthermore, from (132) and (133) we find that

$$
\begin{aligned}
\prod_{\nu=0}^{k-1}\left(\nu q+\frac{d}{dz}\right) R_1^*(z) \ll O\left(n^{2\epsilon_1 n}\right)\left(1+\frac{d}{dz}\right)^k R_1^*(z) = \\
= O\left(n^{(1+2\epsilon_1)n}\right)\left(1+\frac{d}{dz}\right)^k \sum_{\nu=\tau}^{\infty} \nu^{\epsilon_1\nu}\frac{z^\nu}{\nu!} = \\
= O\left(n^{(1+2\epsilon_1)n}\right)\sum_{\rho=0}^{k}\sum_{\nu=\tau}^{\infty}\binom{k}{\rho}\nu^{\epsilon_1\nu}\frac{z^{\nu-\rho}}{(\nu-\rho)!} \ll \\
\ll O\left(n^{(1+2\epsilon_1)n}\right)2^k\sum_{\nu=\tau-k}^{\infty}(\nu+k)^{\epsilon_1(\nu+k)}\frac{z^\nu}{\nu!} \ll \\
\ll O\left(n^{(1+2\epsilon_1)n}\right)2^k\sum_{\nu=\tau-k}^{\infty}(2\nu)^{2\epsilon_1\nu}\frac{z^\nu}{\nu!},
\end{aligned}
$$

since $\tau \ge 2k$, and hence $\nu \ge \tau - k \ge k$ for n sufficiently large. If we set $z = \xi \in \mathbf{K}$ and estimate the value of the last series on the right using the order of its first term, we obtain:

$$
2^k\sum_{\nu=\tau-k}^{\infty}(2\nu)^{2\epsilon_1\nu}\frac{\xi^\nu}{\nu!} = O\left(n^{3\epsilon_1 mn-mn}\right).
$$

Hence,

$$\left(\prod_{\nu=0}^{k-1}\left(\nu q+\frac{d}{dz}\right)R_1^*(z)\right)_{z=\xi} = O\left(n^{(1+2\epsilon_1)n}\right)O\left(n^{3\epsilon_1 mn-mn}\right) = \\ = O\left(n^{-(m-1-(2+3m)\epsilon_1)n}\right). \tag{135}$$

The bounds (133) and (135) give us (129).

Lemma 16. *Suppose that the KE-functions $f_1(z),\ldots,f_m(z)$, $m\geq 2$, form a solution of the system (7) of linear differential equations and are linearly independent over $\mathbf{C}(z)$, and let $\xi\in\mathbf{K}$, $\xi T(\xi)\neq 0$. Then for any ϵ, $0<\epsilon<1$, and for any $n\in\mathbf{N}$, $n\geq n_0$, where n_0 is given by (83), there exists a set of m linearly independent linear forms in the numbers $f_1(\xi),\ldots,f_m(\xi)$*

$$L_k=\sum_{i=1}^m a_{k,i}f_i(\xi),\qquad k=1,\ldots,m,\qquad a_{k,i}\in\mathbf{Z}_\mathbf{K}, \tag{136}$$

such that

$$L_k=O\left(n^{-(m-1-\epsilon)n}\right),\qquad k=1,\ldots,m, \tag{137}$$

$$\overline{|a_{k,i}|}=O\left(n^{(1+\epsilon)n}\right),\qquad k,i=1,\ldots,m, \tag{138}$$

where the bound (137) remains valid if the number ξ and all of the coefficients of $R_k(z)$ are replaced by their conjugates in the field $\mathbf{K}_i$ conjugate to $\mathbf{K}$, for any $i=1,\ldots,h$.

Proof. Let n and ϵ satisfy the conditions of the lemma, and let the form $R_1(z)$ be constructed in accordance with Lemma 14 with our chosen n and with ϵ replaced by $\epsilon_1=\epsilon/(6(m+1))$. We consider the forms $R_k(z)$, $k=1,2,\ldots$, where $R_k=TR'_{k-1}$. By Lemma 10 with t replaced by $t_1=p+q(m-1)m/2+[\epsilon_1 n]$, we find that the linear forms

$$R_k(\xi)=\sum_{i=1}^m P_{k,i}(\xi)f_i(\xi),\qquad k=1,\ldots,m+t_1, \tag{139}$$

include m linearly independent forms. According to Lemma 15, those forms satisfy (129) and (130).

Suppose that the forms

$$R_s(\xi) = \sum_{i=1}^{m} P_{s,i}(\xi) f_i(\xi), \qquad s = s_1, \dots, s_m, \tag{140}$$
$$s_k \le m + t_1, \qquad k = 1, \dots, m,$$

are linearly independent. We choose $a \in \mathbf{N}$ so that $a\xi \in \mathbf{Z_K}$. From the conditions in Lemma 14 and the equalities (37) it follows that

$$\deg P_{k,i}(z) \le n + (k-1)q, \qquad i = 1, \dots, m; \quad k = 1, 2, \dots.$$

Hence, if we denote

$$g = n + q(m + t_1 - 1) = O(n), \tag{141}$$

we have

$$a_{k,i} = a^g P_{s_k,i}(\xi) \in \mathbf{Z_K}, \qquad i, k = 1, \dots, m,$$

since the coefficients of $P_{k,i}(z)$ are numbers in $\mathbf{Z_K}$.

We set

$$L_k = a^g R_{s_k}(\xi) = \sum_{i=1}^{m} a_{k,i} f_i(\xi), \qquad k = 1, \dots, m. \tag{142}$$

Then the linear forms (142) will be linearly independent. In view of the bounds (129), (130) and (141), we have (137) and (138). The lemma is proved.

§ 11. Rank of the m-tuple $f_1(\xi), \dots, f_m(\xi)$

We say that a set of numbers $\alpha_1, \dots, \alpha_m$ has *rank ρ over the field* $\mathbf{K}$ if ρ is the maximum number of $\mathbf{K}$-linearly independent α_i that can be found among $\alpha_1, \dots, \alpha_m$. If an m-tuple has rank ρ over $\mathbf{K}$, then $\rho \le m$, and if $\rho < m$, then the numbers are connected by exactly $m - \rho$ linearly independent homogeneous linear equations with coefficients in $\mathbf{K}$.

Lemma 17. *Suppose that the KE-functions $f_1(z), \dots, f_m(z)$, $m \ge 2$, form a solution of the system (7) of homogeneous linear differential equations and are linearly independent over $\mathbf{C}(z)$, and let $\xi \in \mathbf{K} = \mathbf{Q}(\theta)$, $\xi T(\xi) \ne 0$, and $h = [\mathbf{K} : \mathbf{Q}]$. Then the rank ρ over $\mathbf{K}$ of the m numbers*

$$f_1(\xi), \dots, f_m(\xi) \tag{143}$$

satisfies the inequality $\rho \geq m/h$, *and if* $\theta \notin \mathbf{R}$, *then it satisfies* $\rho \geq 2m/h$. *In particular, if* $\mathbf{K} = \mathbf{I}$, *where* $\mathbf{I}$ *is an imaginary quadratic field, then* $\rho = m$.

Proof. Since ξ is not a singular point for the system (7), it follows that some of the numbers (143) are nonzero. Thus, $\rho \geq 1$. If $\rho = m$, then the lemma holds. So we suppose that $1 \leq \rho < m$. Then the numbers (143) are connected by exactly $m - \rho$ homogeneous linear equations

$$L_k^0 = \sum_{i=1}^{m} c_{k,i} f_i(\xi) = 0, \qquad k = \rho + 1, \ldots, m, \qquad c_{k,i} \in \mathbf{Z_K}, \tag{144}$$

where $L_{\rho+1}^0, \ldots, L_m^0$ are linear forms in the numbers (143) which are linearly independent over $\mathbf{K}$. We denote $c = \max \overline{|c_{k,i}|}$. Then

$$\overline{|c_{k,i}|} \leq c, \qquad k = \rho + 1, \ldots, m; \quad i = 1, \ldots, m. \tag{145}$$

According to Lemma 16, for any ϵ, $0 < \epsilon < 1$, and for any $n \geq n_0$ there exist m linearly independent linear forms (136) in the numbers (143) such that

$$L_k = O\left(n^{-(m-1-\epsilon/2)n}\right), \qquad k = 1, \ldots, m, \tag{146}$$

$$\overline{|a_{k,i}|} = O\left(n^{(1+\epsilon/2)n}\right), \qquad k, i = 1, \ldots, m. \tag{147}$$

Hence, it is possible to choose ρ of these forms which, along with the forms (144), are all linearly independent. To be definite, suppose that the ρ forms are

$$L_k = \sum_{i=1}^{m} a_{k,i} f_i(\xi), \qquad k = 1, \ldots, \rho. \tag{148}$$

We let δ denote the determinant of the $m \times m$ coefficient matrix of the linear forms (144) and (148). Since all of the entries in this matrix are in $\mathbf{Z_K}$, it follows that $\delta \in \mathbf{Z_K}$. We let $\delta_{i,k}$ denote the cofactor of the entry in the k-th row and i-th column of δ.

If we multiply both sides of each of the equations in (144) and (146) by $\delta_{k,j}$ with j fixed, $1 \leq j \leq m$, and add the equations obtained in this way for $k = 1, \ldots, m$, we obtain the following equations from (144) (compare with the equations (42) for the determinant Δ):

$$\delta f_j(\xi) = \sum_{k=1}^{\rho} \delta_{k,j} L_k(\xi), \qquad 1 \leq j \leq m. \tag{149}$$

Using (145) and (147), we find that

$$|\delta_{k,i}| = m^m c^{m-\rho} O\left(n^{(1+\epsilon/2)n(\rho-1)}\right) = O\left(n^{(1+\epsilon)(\rho-1)n}\right), \quad k = 1, \ldots, \rho; \quad i = 1, \ldots, m, \tag{150}$$

and similarly,

$$\overline{|\delta|} = O\left(n^{(1+\epsilon)\rho n}\right). \tag{151}$$

We choose j so that $f_j(\xi) \neq 0$. Then from the j-th equation in (149), along with the bounds (146) and (150), we find that

$$|\delta| \leq \frac{1}{|f_j(\xi)|} \sum_{k=1}^{\rho} |\delta_{k,j}||L_k(\xi)| = O\left(n^{(1+\epsilon)(\rho-1)n}\right) O\left(n^{-(m-1-\epsilon/2)n}\right),$$

from which we obtain:

$$\delta = O\left(n^{((1+\epsilon)\rho-m)n}\right). \tag{152}$$

From (151) and (152) we obtain the bound

$$\mathcal{N}(\delta) = O\left(n^{((1+\epsilon)\rho h-m)n}\right), \tag{153}$$

where $\mathcal{N}(\delta)$ is the norm of an element $\delta \in \mathbf{Z_K}$. But the norm of a number in $\mathbf{Z_K}$ belongs to $\mathbf{Z}$. Since the forms (144) and (148) are linearly independent, we have $\delta \neq 0$, and hence $\mathcal{N}(\delta) \neq 0$. Thus,

$$|\mathcal{N}(\delta)| \geq 1. \tag{154}$$

From the conditions (153) and (154) we obtain the inequality

$$(1+\epsilon)\rho h - m \geq 0,$$

which, since ϵ is arbitrary, implies that $\rho h - m \geq 0$, i.e., $\rho \geq m/h$.

Now suppose that $\theta \notin \mathbf{R}$. Then the conjugates of δ in $\mathbf{K}$ include a conjugate $\overline{\delta}$ for which $|\overline{\delta}| = |\delta|$. Hence, instead of (153), the bounds (151) and (152) lead to the result

$$\mathcal{N}(\delta) = O\left(n^{((1+\epsilon)\rho h-2m)n}\right),$$

from which it follows that $\rho \geq 2m/h$. Finally, when $\mathbf{K} = \mathbf{I}$, we have $h = 2$, and so $\rho \geq m$; but since $\rho \leq m$, we conclude that $\rho = m$. The lemma is proved.

§ 12. Proof of the First Fundamental Theorem

In order to prove the First Fundamental Theorem, we must show that there is no homogeneous polynomial

$$P = P(x_1, \dots, x_m) \in \mathbf{A}[x_1, \dots, x_m], \qquad P \not\equiv 0, \qquad \deg_{\mathbf{x}} P \le k, \qquad k \ge 1, \tag{155}$$

which vanishes when each x_i is replaced by $f_i(\xi)$, $i = 1, \dots, m$. This means that we have to solve a problem which is nonlinear in the functions $f_1(z), \dots, f_m(z)$. But all of the basic lemmas in our method are linear in the functions under consideration. It turns out, however, that a nonlinear problem is easily reduced to a linear problem by means of the lemma below, in which we show that, if instead of $f_1(z), \dots, f_m(z)$ we consider the set of all monomials in these functions of total degree N, then this new collection of functions forms a solution of a system of homogeneous linear differential equations with coefficients in $\mathbf{C}(z)$, and hence the method described above can be applied.

Lemma 18. *Suppose that the m-tuple of analytic functions $f_1(z), \dots, f_m(z)$, $m \ge 2$, forms a solution of the system (7) of homogeneous linear differential equations. Then for any $N \in \mathbf{N}$ the set of $\mu^0_{N,m} = (N + m - 1)!/N!(m - 1)!$ monomials in these functions*

$$v_{k_1,\dots,k_m} = f_1^{k_1}(z) \cdots f_m^{k_m}(z),$$

$$k_i \in \mathbf{Z}^+, \qquad i = 1, \dots, m, \qquad k_1 + \dots + k_m = N, \tag{156}$$

forms a solution of a system of homogeneous linear differential equations of the form (7), in which m is replaced by $\mu^0_{N,m}$ and the coefficients are $\mathbf{Z}$-linear combinations of the coefficients in the original system (7), and hence these coefficients do not have any poles other than the poles of the system (7).

Proof. Differentiating the functions in (156), we obtain

$$v'_{k_1,\dots,k_m} = \sum_{i=1}^{m} k_i f_1^{k_1}(z) \cdots f_m^{k_m}(z) \frac{f_i'(z)}{f_i(z)}, \qquad k_1 + \dots + k_m = N.$$

If we replace $f_i'(z)$ here by the right-hand side of the corresponding equation in (7) with y_i set equal to $f_i(z)$, then we conclude that the lemma is true.

Lemma 18 still holds if we consider an m-tuple of functions $f_1(z), \dots, f_m(z)$, $m \ge 1$, which form a solution of a non-homogeneous system (8) of linear differential equations and the set of $\mu_{N,m} = (N+m)!/N!m!$ monomials

$$v_{k_1,\dots,k_m} = f_1^{k_1}(z)\cdots f_m^{k_m}(z),$$

$$k_i \in \mathbf{Z}^+, \qquad i = 1,\dots,m, \qquad k_1+\dots+k_m \le N.$$

Corollary. *If in Lemma 18 we further assume that $f_1(z),\dots,f_m(z)$ are KE-functions which satisfy the conditions of the First Fundamental Theorem, then, by Lemma 3, the set of $\mu^0_{N,m}$ monomials (156) forms a solution of a system of linear differential equations of the form (7), in which all $Q_{k,i} \in \mathbf{K}(z)$. The polynomial $T = T(z)$ for this system coincides with the corresponding polynomial $T(z)$ for the original system (7), and we have $T \in \mathbf{Z_K}[z]$, $TQ_{k,i} \in \mathbf{Z_K}[z]$, $k, i = 1,\dots,m$.*

We now proceed to the proof of the First Fundamental Theorem. Let ξ be any number in $\mathbf{A}$, $\xi T(\xi) \ne 0$. We suppose the contrary of the theorem: under the conditions of the theorem, the numbers $f_1(\xi),\dots,f_m(\xi)$ are homogeneously algebraically independent. Then there exists a homogeneous polynomial (155) such that

$$P(f_1(\xi),\dots,f_m(\xi)) = 0. \tag{157}$$

We shall suppose that $\mathbf{K}$ is an algebraic number field which contains ξ, the coefficients of the power series of all of the KE-functions (9), and all of the coefficients of P. According to Lemma 18 and its corollary, $\mathbf{K}$ contains the coefficients of all of the polynomials T and $TQ_{k,i}$ in the system of differential equations satisfied by the monomials (156) with $N \in \mathbf{N}$ arbitrary.

We consider the $\mu^0_{N-k,m}$ expressions

$$f_1^{k_1}(z)\cdots f_m^{k_m}(z)\,P(f_1(z),\dots,f_m(z)), \qquad k_1+\dots+k_m = N-k. \tag{158}$$

Each of these expressions is a linear form in the variables (156) with coefficients in $\mathbf{K}$. These linear forms are linearly independent, since, if we order the monomials (156) lexicographically with respect to powers of $f_m(z),\dots, f_1(z)$, then the leading terms of the forms (158) are distinct.

In (158) we set $z = \xi$. Using (157), we obtain $\mu^0_{N-k,m}$ linearly independent homogeneous linear equations for the $\mu^0_{N,m}$ numbers

$$f_1^{k_1}(\xi)\cdots f_m^{k_m}(\xi), \qquad k_i \ge 0,\ \ i = 1,\dots,m,\ \ k_1+\dots+k_m = N, \tag{159}$$

with coefficients in $\mathbf{K}$.

On the other hand, by assumption the functions (9) are homogeneously algebraically independent over $\mathbf{C}(z)$; hence, for any $N \in \mathbf{N}$ the functions (156) are linearly independent over $\mathbf{C}(z)$. By Lemma 18 and its corollary, the functions

(156) satisfy the hypothesis of Lemma 17 with m replaced by $\mu^0_{N,m}$. Since ξ is not a singular point of the system of differential equations (7), it follows from Lemma 18 that ξ is not a singular point of the system of differential equations satisfied by the functions (156). Hence, by Lemma 17, the rank over $\mathbf{K}$ of the set of numbers (159) is not less than $\mu^0_{N,m}/h$, where $h = [\mathbf{K} : \mathbf{Q}]$. This means that

$$\mu^0_{N,m} - \mu^0_{N-k,m} \geq \frac{\mu^0_{N,m}}{h}. \tag{160}$$

But the estimates

$$\mu^0_{N,m} = \frac{(N+m-1)!}{N!(m-1)!} = \frac{1}{(m-1)!}N^{m-1} + O(N^{m-2}),$$

$$\mu^0_{N,m} - \mu^0_{N-k,m} = \frac{(N+m-1)!}{N!(m-1)!} - \frac{(N-k+m-1)!}{(N-k)!(m-1)!} = O(N^{m-2})$$

imply that (160) is impossible for N sufficiently large. This contradiction completes the proof of the First Fundamental Theorem.

The next lemma shows that in the First Fundamental Theorem the hypothesis that the KE-functions (9) are homogeneously algebraically independent over $\mathbf{C}(z)$ is not only sufficient, but necessary as well.

Lemma 19. *Suppose that the functions $f_1(z), \dots, f_m(z)$, $m \geq 1$, are analytic in a region containing $z = 0$, have power series coefficients in z belonging to the algebraic number field $\mathbf{K}$, and are homogeneously algebraically dependent (algebraically dependent) over $\mathbf{C}(z)$, and let $\xi \in \mathbf{K}$. Then the numbers $f_1(\xi), \dots, f_m(\xi)$ are homogeneously algebraically dependent (algebraically dependent) over $\mathbf{K}$.*

Proof. We consider the homogeneous case; the non-homogeneous case is similar. By assumption, there exists a polynomial homogeneous in $z_1, \dots, z_m$

$$P = P(z, z_1, \dots, z_m) \in \mathbf{C}[z, z_1, \dots, z_m], \qquad P \not\equiv 0,$$

$$\deg_{\mathbf{z}} P \leq k, \qquad k \geq 1, \qquad \mathbf{z} = (z_1, \dots, z_m),$$

such that

$$P(z, f_1(z), \dots, f_m(z)) = 0. \tag{161}$$

By Lemma 2, we may assume that

$$P = P(z, z_1, \dots, z_m) \in \mathbf{K}[z, z_1, \dots, z_m], \qquad P \not\equiv 0.$$

Suppose that P is an irreducible polynomial. Then the equation (161) can be written as follows:

$$P(z, f_1(z), \dots, f_m(z)) = \sum_{k_1+\dots+k_m=k} P_{k_1,\dots,k_m}(z) f_1^{k_1}(z) \cdots f_m^{k_m}(z) = 0, \quad (162)$$

where the summation is taken over a set of m-tuples $(k_1, \dots, k_m)$ in which $k_1, \dots$ $\dots, k_m \in \mathbf{Z}^+$, and the polynomials $P_{k_1,\dots,k_m}(z) \not\equiv 0$ do not all have a factor in common.

In (162) we set $z = \xi$. Then this equation becomes an algebraic relation with coefficients in $\mathbf{K}$ between the numbers $f_1(\xi), \dots, f_m(\xi)$. It is nontrivial, since at least one of the $P_{k_1,\dots,k_m}(\xi)$ is nonzero (because the polynomials $P_{k_1,\dots,k_m}(z)$ do not all have a factor in common). This proves that $f_1(\xi), \dots, f_m(\xi)$ are homogeneously algebraically dependent over $\mathbf{K}$. The lemma is proved.

§ 13. Consequences of the First Fundamental Theorem

In this section we describe some corollaries of the First Fundamental Theorem. *Suppose that the E-functions $f_1(z), \dots, f_m(z)$ and the number $\xi \in \mathbf{A}$ satisfy the hypothesis of the First Fundamental Theorem. In that case:*

1°. *All of the numbers $f_1(\xi), \dots, f_m(\xi)$ are nonzero; consequently, all of the zeros of each of the functions $f_1(z), \dots, f_m(z)$ which are not zeros of the polynomial $zT(z)$, are transcendental.*

2°. *For each s, $1 \leq s \leq m$, the $m-1$ numbers $f_k(\xi)/f_s(\xi)$, $k = 1, \dots, m$, $k \neq s$, are algebraically independent.*

3°. *At least $m-1$ of the numbers $f_1(\xi), \dots, f_m(\xi)$ are transcendental.*

4°. *If one of the numbers $f_1(\xi), \dots, f_m(\xi)$ is algebraic, then the remaining $m-1$ of them are algebraically independent.*

5°. *If $P(z, z_1, \dots, z_m) \in \mathbf{A}[z, z_1, \dots, z_m]$ is an irreducible polynomial which is homogeneous of degree k, $k \geq 1$, in $z_1, \dots, z_m$, then all of the zeros of the function $P(z, f_1(z), \dots, f_m(z))$ which are not zeros of the polynomial $zT(z)$, are transcendental.*

Suppose that $f(z)$ is an E-function which satisfies the m-th order homogeneous linear differential equation

$$\begin{aligned} &P_m(z) y^{(m)} + \dots + P_1(z) y' + P_0(z) y = 0, \qquad m \geq 2, \\ &\qquad\qquad P_k(z) \in \mathbf{C}[z], \qquad k = 0, 1, \dots, m. \end{aligned} \quad (163)$$

We denote

$$f_1(z) = f(z), \qquad f_k(z) = f^{(k-1)}(z), \qquad k = 2, \dots, m. \quad (164)$$

Then the m E-functions (164) form a solution of the system of first order homogeneous linear differential equations

$$\begin{aligned} y_k' &= y_{k+1}, \qquad k = 1, \dots, m-1, \\ y_m' &= -\frac{P_0(z)}{P_m(z)} y_1 - \dots - \frac{P_{m-1}(z)}{P_m(z)} y_m. \end{aligned} \tag{165}$$

If the functions (164) are homogeneously algebraically independent over $\mathbf{C}(z)$, then the First Fundamental Theorem can be applied, giving the following result.

Theorem 1. *Suppose that $f(z)$ is an E-function which satisfies the m-th order homogeneous linear differential equation (163) and does not satisfy any homogeneous algebraic differential equation with coefficients in $\mathbf{C}(z)$ of order less than m, and let $\xi \in \mathbf{A}$, $\xi P_m(\xi) \neq 0$. Then the numbers $f(\xi), f'(\xi), \dots, f^{(m-1)}(\xi)$ are homogeneously algebraically independent.*

From Theorem 1 we obtain corollaries similar to corollaries 1°-5° of the First Fundamental Theorem. We take special note of the case $m = 2$.

Theorem 2. *Suppose that $f(z)$ is an E-function which satisfies the second order homogeneous linear differential equation*

$$P_2(z)y'' + P_1(z)y' + P_0(z)y = 0, \qquad P_k \in \mathbf{C}[z], \qquad k = 0, 1, 2 \tag{166}$$

and does not satisfy any first order homogeneous algebraic differential equation with coefficients in $\mathbf{C}[z]$, and let $\xi \in \mathbf{A}$, $\xi P_2(\xi) \neq 0$. Then the logarithmic derivative $f'(\xi)/f(\xi)$ is transcendental, at least one of the numbers $f(\xi)$ and $f'(\xi)$ is transcendental, and all of the zeros of $f(z)$ and $f'(z)$ which are not zeros of $zP_2(z)$ are transcendental.

A simple application of Theorem 2 is the case of the E-function $f(z) = \sin z$, which satisfies the differential equation $y'' + y = 0$. If $f(z)$ satisfied a first order homogeneous algebraic differential equation with coefficients in $\mathbf{C}(z)$, that would mean that $\tan z$ is an algebraic function. But that is impossible, since it has infinitely many zeros, and an algebraic function has only finitely many zeros. Thus, for any $\xi \in \mathrm{A}$, $\xi \neq 0$, the number $\tan \xi$ is transcendental, and at least one of the numbers $\sin \xi$, $\cos \xi$ is transcendental. But because $\sin^2 \xi + \cos^2 \xi = 1$, this means that both $\sin \xi$ and $\cos \xi$ are transcendental. In addition, since π is a zero of $\sin z$, it is also transcendental.

We consider an m-tuple of E-functions $f_1(z), \dots, f_m(z)$ which form a solution of the system of linear differential equations

$$y_k' = Q_{k,0} + \sum_{i=1}^{m} Q_{k,i} y_i, \qquad k = 1, \dots, m, \qquad Q_{k,i} \in \mathbf{C}(z), \tag{167}$$

which may possibly be homogeneous, i.e., if $Q_{k,0} = 0$, $k = 1, \dots, m$. We add the function $f_0(z) = 1$ to our set of functions. Then the $(m+1)$-tuple of functions $f_0(z), f_1(z), \dots, f_m(z)$ satisfies a system of homogeneous linear differential equations of the form (7) with m replaced by $m+1$, regardless of whether (167) is a homogeneous or non-homogeneous system.

Notice that the functions $f_0(z), f_1(z), \dots, f_m(z)$ are homogeneously algebraically independent over $\mathbf{C}(z)$ if and only if the functions $f_1(z), \dots, f_m(z)$ are algebraically independent over $\mathbf{C}(z)$; and the numbers $f_0(\xi), f_1(\xi), \dots, f_m(\xi)$ are homogeneously algebraically independent if and only if the numbers $f_1(\xi), \dots, f_m(\xi)$ are algebraically independent. Thus, if we apply the First Fundamental Theorem to the functions $f_0(z), f_1(z), \dots, f_m(z)$, we obtain the following result from corollary 2°.

Second Fundamental Theorem. *Suppose that the E-functions $f_1(z), \dots, f_m(z)$ form a solution of the system of linear differential equations (167) and are algebraically independent over $\mathbf{C}(z)$, and let $\xi \in \mathbf{A}$, $\xi T(\xi) \neq 0$. Then the numbers $f_1(\xi), \dots, f_m(\xi)$ are algebraically independent.*

We now give some corollaries of the Second Fundamental Theorem.

Suppose that the E-functions $f_1(z), \dots, f_m(z)$ and the number $\xi \in \mathbf{A}$ satisfy the hypothesis of the Second Fundamental Theorem. Then:

1°. *Each of the numbers $f_1(\xi), \dots, f_m(\xi)$ is transcendental.*

2°. *All of the zeros and all of the A-points (the points z where the function takes the algebraic value A) of the functions $f_1(z), \dots, f_m(z)$ which are not zeros of the polynomial $zT(z)$, are transcendental.*

3°. *If $P(z, z_1, \dots, z_m) \in \mathbf{A}[z, z_1, \dots, z_m]$ is an irreducible polynomial of degree k in $z_1, \dots, z_m$, where $k \geq 1$, then all of the zeros and algebraic A-points of the function $P(z, f_1(z), \dots, f_m(z))$ which are not zeros of $zT(z)$, are transcendental.*

If $f(z)$ is an E-function which satisfies the m-th order linear differential equation

$$P_m(z)y^{(m)} + \dots + P_1(z)y' + P_0(z)y = Q(z), \qquad m \geq 1,$$

$$P_k(z) \in \mathbf{C}[z], \qquad k = 0, 1, \dots, m, \qquad Q(z) \in \mathbf{C}[z], \tag{168}$$

then, using the notation (164), we find that the E-functions in (164) form a solution of a system of linear differential equations similar to (165) but with an additional term $Q(z)/P_m(z)$ on the right in the last equation. Thus, if we apply the Second Fundamental Theorem to this set of E-functions, we obtain the following result.

Theorem 3. *Suppose that $f(z)$ is an E-function satisfying the m-th order linear differential equation (168) and not satisfying any algebraic differential equation with coefficients in $\mathbf{C}[z]$ of order less than m (in the case $m = 1$ this means that*

$f(z)$ is not an algebraic function), and let $\xi \in \mathbf{A}$, $\xi P_m(\xi) \neq 0$. Then the numbers $f(\xi), f'(\xi), \ldots, f^{(m-1)}(\xi)$ are algebraically independent.

Theorem 3 has corollaries analogous to corollaries 1°-3° of the Second Fundamental Theorem.

The case $m = 1$ is of special interest in Theorem 3. In that case the result is:

Theorem 4. *If $f(z)$ is a transcendental E-function which satisfies the first order linear differential equation*

$$P_1(z)y' + P_0(z)y = Q(z), \qquad P_0(z), P_1(z), Q(z) \in \mathbf{C}[z], \tag{169}$$

and if $\xi \in \mathbf{A}$, $\xi P_1(\xi) \neq 0$, then $f(\xi)$ is a transcendental number.

The simplest example of Theorem 4 is given by the exponential function $y = e^z$, which satisfies the differential equation $y' = y$. Hence, e^ξ is a transcendental number for any $\xi \in \mathbf{A}$, $\xi \neq 0$. In particular, e itself is a transcendental number, and so is $\ln \xi$ for any $\xi \in \mathbf{A}$, $\xi \neq 1$.

We now show that Lindemann's theorem is a simple consequence of the Second Fundamental Theorem.

Lemma 20. *If $\beta_1, \ldots, \beta_n$, $n \geq 1$, are arbitrary distinct complex numbers, then the functions $e^{\beta_1 z}, \ldots, e^{\beta_n z}$ are linearly independent over $\mathbf{C}(z)$.*

Proof. The lemma is trivial for $n = 1$. Suppose that it holds for $n = k$, $k \geq 1$; we prove that it also holds for $n = k+1$. This will prove the lemma for all $n \geq 1$.

Suppose that the claim is false, i.e., that the following equation holds:

$$P_1 e^{\beta_1 z} + \cdots + P_k e^{\beta_k z} + P_{k+1} e^{\beta_{k+1} z} = 0.$$

$$P_i \not\equiv 0, \qquad P_i \in \mathbf{C}[z], \qquad i = 1, \ldots, k+1.$$

Then

$$P_1 e^{(\beta_1 - \beta_{k+1})z} + \cdots + P_k e^{(\beta_k - \beta_{k+1})z} + P_{k+1} = 0.$$

If $\deg P_{k+1} = s$, then, differentiating the last equation $s+1$ times, we obtain an equation which contradicts the induction assumption, since all of the differences $\beta_i - \beta_{k+1}$, $i = 1, \ldots, k$, are distinct.

Lemma 21. *If the complex numbers $\alpha_1, \ldots, \alpha_m$, $m \geq 1$, are linearly independent over $\mathbf{Q}$, then the functions $e^{\alpha_1 z}, \ldots, e^{\alpha_m z}$ are algebraically independent over $\mathbf{C}(z)$.*

Proof. Suppose the lemma is false. Then there exists a polynomial

$$P = P(z, z_1, \ldots, z_m) \in \mathbf{C}[z, z_1, \ldots, z_m],$$

$$P \not\equiv 0, \qquad \deg_{\mathbf{z}} P = k, \qquad \mathbf{z} = (z_1, \ldots, z_m),$$

such that

$$P\left(z, e^{\alpha_1 z}, \ldots, e^{\alpha_m z}\right) = 0.$$

This equation can be written in the form

$$P\left(z, e^{\alpha_1 z}, \ldots, e^{\alpha_m z}\right) = \sum_{k_1+\cdots+k_m \le k} P_{k_1,\ldots,k_m}(z) e^{(k_1\alpha_1+\cdots+k_m\alpha_m)z} = 0,$$

$$P_{k_1,\ldots,k_m}(z) \in \mathbf{C}[z], \qquad P_{k_1,\ldots,k_m(z)} \not\equiv 0,$$

where the summation is taken over some set of m-tuples of positive integers $(k_1, \ldots, k_m)$. By the hypothesis of the lemma, the exponents $k_1\alpha_1 + \cdots + k_m\alpha_m$ in the last equation are distinct. Hence, by Lemma 20, this equation is impossible, and the lemma is proved.

Lemma 21 along with the Second Fundamental Theorem imply Lindemann's theorem in one of its equivalent formulations.

One case which is of special interest occurs when we have a set of IE-functions and consider their values at numbers in **I**. Since $\mathbf{Q} \subset \mathbf{I}$, this includes the case of a set of QE-functions and their values at rational points. In the imaginary quadratic case, the First and Second Fundamental Theorems, along with all of the above corollaries, follow immediately from Lemmas 1, 17 and 18. Moreover, all of these results are a consequence of the more general theorem below, which also follows from those three lemmas.

Theorem 5. *Suppose that the IE-functions $f_1(z), \ldots, f_m(z)$, $m \ge 2$, form a solution of the system of homogeneous linear differential equations (7) and are not connected by a homogeneous algebraic equation of degree k with coefficients in* $\mathbf{C}(z)$, *where $k \in \mathbf{N}$; and let $\xi \in \mathbf{I}$, $\xi T(\xi) \ne 0$. Then the numbers $f_1(\xi), \ldots, f_m(\xi)$ are not connected by a homogeneous algebraic equation of degree k with coefficients in* **I**. *In particular, for $k = 1$, the numbers $f_1(\xi), \ldots, f_m(\xi)$ are linearly independent over* **I**, *and none of them is zero.*

If the functions $f_1(z), \ldots, f_m(z)$ are homogeneously algebraically independent over $\mathbf{C}(z)$, then Theorem 5 holds for any k. Hence, in this case the numbers $f_1(\xi), \ldots, f_m(\xi)$ are homogeneously algebraically independent over **I**, and so, by Lemma 1, they are homogeneously algebraically independent. This shows that in the case of IE-functions Theorem 5 contains the result in the First Fundamental Theorem. The following result follows from Theorem 5.

Theorem 6. *Suppose that the IE-functions $f_1(z), \ldots, f_m(z)$, $m \ge 1$, form a solution of the system of linear differential equations (167) and are not connected by* *an algebraic equation of degree k ($k \in \mathbf{N}$) with coefficients in $\mathbf{C}(z)$, and let $\xi \in \mathbf{I}$, $\xi T(\xi) \neq 0$. Then the numbers $f_1(\xi), \ldots, f_m(\xi)$ are not connected by an algebraic equation of degree k with coefficients in $\mathbf{I}$. In particular, for $k = 1$, the numbers $f_1(\xi), \ldots, f_m(\xi)$, along with 1, are linearly independent over $\mathbf{I}$, and none of them belongs to $\mathbf{I}$ (hence they are irrational).*

Theorem 6 implies the Second Fundamental Theorem for IE-functions. Theorems 5 and 6 can easily be adopted to the situation when we have an IE-function $f(z)$ which satisfies the linear differential equation (163) or (168), and we consider $f(z)$ along with its successive derivatives. We state the result which corresponds to the linear case of Theorem 6.

Theorem 7. *Suppose that $f(z)$ is an IE-function which satisfies an m-th order linear differential equation (168), $m \geq 1$, and does not satisfy any linear differential equation of the same type of order less than m (if $m = 1$, we suppose that it does not belong to $\mathbf{I}(z)$); and let ξ be any number in $\mathbf{I}$ with $\xi P_m(\xi) \neq 0$. Then the numbers $f(\xi), f'(\xi), \ldots, f^{(m-1)}(\xi)$, along with 1, are linearly independent over $\mathbf{I}$; in particular, none of them belongs to $\mathbf{I}$, and hence none of them is rational.*

It is an interesting problem to extend Theorem 7 to an arbitrary algebraic number field $\mathbf{K}$. If this could be done, then transcendence of each of the numbers $f(\xi), f'(\xi), \ldots, f^{(m-1)}(\xi)$, $\xi \in \mathbf{A}$, $\xi P_m(\xi) \neq 0$, would follow from linear independence over $\mathbf{C}(z)$ of the corresponding functions, along with 1.

We note that it is much simpler to apply the general theorems in Chapter 3 on algebraic independence of the values of E-functions to concrete E-functions than it is to apply Siegel's general theorem [72:4] or the fundamental theorem of [70:1,9]. Even for the simplest functions it is difficult to verify the conditions in the latter theorems. It is a simpler problem to prove that a set of functions satisfying some linear differential equations is algebraically independent over $\mathbf{C}(z)$. There are several methods which can be used to do this for certain classes of functions which are solutions of linear differential equations of arbitrary order.

Remarks

In Chapter 3 we have given the proof of algebraic independence of the values of E-functions in a modern form. Some of the lemmas used in this proof are due to Siegel, some are generalizations of lemmas of Siegel, and some of the lemmas arose in the course of generalizing the method.

Siegel first published his basic method in 1929 in [72:3], and he published his general theorem in 1949 in [72:4]. In this theorem the functions under consideration form a solution of a system of homogeneous linear differential equations. The possibilities of applying the theorem to concrete functions were limited: it could be applied to a set of E-functions each of which satisfies a first or second order homogeneous linear differential equation.

In 1954 in [70:1,9] a theorem was proved which is more general than Siegel's theorem. It can be applied to solutions of non-homogeneous linear differential equations, in particular, to certain sets of E-functions which satisfy 3-rd and 4-th order linear differential equations. The hypothesis of this theorem, as in the case of Siegel's theorem, is only a sufficient condition for the conclusion to hold.

The fundamental theorems in Chapter 3 were first announced in 1955 in [70:2,3], and their proofs were published in 1959 in [70:8]. In 1966 in [70:19] the basic lemmas in the method were refined. This made it possible to obtain a best possible result in the case of the values of IE-functions at points in **I** in the fundamental Lemma 17 on the rank of the numbers $f_1(\xi), \ldots, f_m(\xi)$, thereby proving Theorems 5–7.

The most important step in generalizing Siegel's method to obtain the fundamental theorems in Chapter 3 is the proof of Lemmas 8 and 9, which enable one to construct a set of linearly independent functional approximating forms under the conditions in the fundamental theorems. The proof of these lemmas is based on the auxiliary lemmas 4–7.

The basic results of Chapter 3 are contained in reports at the Third All-Union Mathematical Congress (Moscow, 1956) [70:7], the Fourth All-Union Mathematical Congress (Leningrad, 1961) [70:15], and the International Congress of Mathematicians (Moscow, 1966) [70:17,22].

After Siegel's monograph [72:4] appeared, his method was included in books by Gel'fond [24:8] and Schneider [68:6].

The First and Second Fundamental Theorems are contained in Mahler's book [42:7]. These theorems were also the subject of his lectures [42:6]. The Second Fundamental Theorem can be found in books by Lang [37:1] and Baker [2:5]. The basic results of Chapter 3 are contained in Shidlovskii's book [70:35].

For surveys of results connected with the method considered in Chapter 3, see the articles by Fel'dman and Shidlovskii [18:1] and Shidlovskii [70:25,26].

In Pershikova's paper [58:1], a general theorem is proved which gives algebraic independence of the values of a subclass of E-functions.

Chapter 4
Transcendence and algebraic independence of the values of E-functions which are connected by algebraic equations over the field of rational functions

§ 1. Rank of the m-tuple $f_1(\xi), \dots, f_m(\xi)$

In this chapter we generalize the method in Chapter 3 in such a way that it can be applied to obtain arithmetic information on the values of E-functions which are algebraically dependent over $\mathbf{C}(z)$. As in Chapter 3, the linear case plays a basic role. What we have to do is generalize Lemma 17 of Chapter 3 on the rank of the m-tuple $f_1(\xi), \dots, f_m(\xi)$ to include the possibility that the E-functions $f_1(z), \dots, f_m(z)$ are linearly dependent over $\mathbf{C}(z)$.

We consider a set of functions

$$f_1(z), \dots, f_m(z), \qquad m \geq 2, \tag{1}$$

which are analytic in a region. We suppose that these functions are linearly dependent over $\mathbf{C}(z)$, and that their *rank over* $\mathbf{C}(z)$ (the maximum number of linearly independent functions) is equal to r, $1 \leq r \leq m-1$. To be definite, we suppose that the functions

$$f_1(z), \dots, f_r(z) \tag{2}$$

are linearly independent over $\mathbf{C}(z)$. Then

$$f_j(z) = \sum_{i=1}^{r} B_{j,i} f_i(z), \qquad j = r+1, \dots, m, \qquad B_{j,i} \in \mathbf{C}(z). \tag{3}$$

Suppose that the functions (1) form a solution of the system of homogeneous linear differential equations

$$y_k' = \sum_{i=1}^{m} Q_{k,i} y_i, \qquad k = 1, \dots, m, \qquad Q_{k,i} \in \mathbf{C}(z), \tag{4}$$

and we define the polynomial $T = T(z)$ in the same way as in Chapter 3. We write the first r equations in (4) as follows:

$$y'_k = \sum_{i=1}^{r} Q_{k,i} y_i + \sum_{j=r+1}^{m} Q_{k,i} y_j, \qquad k = 1, \dots, r.$$

From these equations along with (3) we find that the functions (2) satisfy the system of homogeneous linear differential equations

$$y'_k = \sum_{i=1}^{r} \left(Q_{k,i} + \sum_{j=r+1}^{m} Q_{k,j} B_{j,i} \right) y_i, \qquad k = 1, \dots, r. \tag{5}$$

If the functions in (1) are E-functions, then we can apply Lemma 17 of Chapter 3 to the linearly independent functions (2), which satisfy the system of differential equations (5), thereby obtaining arithmetic results about their values at any $\xi \in \mathbf{A}$ not equal to zero or a singular point of the system (5). However, the singular points of this system may include not only the singular points of (4), but also the poles of the functions $B_{j,i}$, which may be unknown to us. This makes it more complicated to apply Lemma 17 to our functions. But it is not hard to overcome this difficulty using the lemmas proved below.

Lemma 1. *Suppose that the m-tuple of functions (1) has rank r over $\mathbf{C}(z)$, $1 \le r \le m-1$, and α is any fixed complex number. Then it is possible to choose a subset of r $\mathbf{C}(z)$-linearly independent functions (2) (perhaps changing the numbering in (1)) in such a way that the equations (3) expressing $f_{r+1}(z), \dots, f_m(z)$ in terms of the functions (2) have the property that the point $z = \alpha$ is not a pole of any of the functions $B_{j,i}$.*

Proof. We use induction on the number m of functions in (1). If $m = 2$, r must equal 1, and the functions (1) are connected by a unique (up to constant multiple) linear equation

$$A_1 f_1(z) + A_2 f_2(z) = 0,$$

in which $A_1 = A_1(z)$ and $A_2 = A_2(z)$ are relatively prime polynomials in $\mathbf{C}[z]$. Then at least one of these polynomials is nonzero for $z = \alpha$. Changing the numbering of the functions (1), if necessary, we may assume that $A_2(\alpha) \neq 0$, in which case

$$f_2(z) = B f_1(z), \qquad f_1(z) \not\equiv 0, \qquad B \in \mathbf{C}(z),$$

where B does not have a pole at $z = \alpha$. Thus, the lemma holds for $m = 2$.

We now suppose that the lemma holds for $m-1$ functions, $m \geq 3$, for any r, $1 \leq r \leq m-2$. We shall prove that it then holds for m functions for any r, $1 \leq r \leq m-1$.

In this case the functions (1) are connected by precisely $m - r \geq 1$ linearly independent linear equations over $\mathbf{C}(z)$. We choose any one of them:

$$\sum_{i=1}^{m} A_i f_i(z) = 0, \tag{6}$$

where $A_1, \ldots, A_m$ are polynomials in $\mathbf{C}[z]$ not all having a common factor. Next, if we argue as in the case $m = 2$, perhaps renumbering the functions (1), we obtain the following equation from (6):

$$f_m(z) = \sum_{i=1}^{m-1} B^*_{m,i} f_i(z), \qquad B^*_{m,i} \in \mathbf{C}(z), \qquad i = 1, \ldots, m, \tag{7}$$

in which none of the functions $B^*_{m,i}$ has a pole at the point $z = \alpha$.

If $r = m - 1$, then the functions $f_1(z), \ldots, f_{m-1}(z)$ are linearly independent over $\mathbf{C}(z)$, and the lemma holds. On the other hand, if $r < m - 1$, then these functions are connected by exactly $m - r - 1$ linearly independent linear equations over $\mathbf{C}(z)$; then, by the induction assumption, we have the following equations (after renumbering the functions, if necessary):

$$f_j(z) = \sum_{i=1}^{r} B_{j,i} f_i(z), \qquad j = r+1, \ldots, m-1, \qquad B_{j,i} \in \mathbf{C}(z), \tag{8}$$

where none of the $B_{j,i}$ has a pole at $z = \alpha$, and the functions in (2) are linearly independent over $\mathbf{C}(z)$.

If we substitute the right sides in (8) in place of $f_{r+1}(z), \ldots, f_{m-1}(z)$ in (7), we obtain the relation

$$f_m(z) = \sum_{i=1}^{r} B_{m,i} f_i(z), \qquad B_{m,i} \in \mathbf{C}(z), \tag{9}$$

in which none of the $B_{m,i}$ has a pole at $z = \alpha$.

Equations (8) and (9) show that the lemma holds for m functions and any value of r, $1 \leq r \leq m-1$. By induction, this completes the proof of the lemma.

Lemma 2. *Suppose that the functions (1) form a solution of the system (4) of homogeneous linear differential equations and have rank r over the field $\mathbf{C}(z)$, $1 \leq r \leq m-1$; let α be any fixed complex number such that $\alpha T(\alpha) \neq 0$. Then one can choose r functions (2) from among the functions (1) (changing the numbering,* *if necessary) in such a way that these r functions are linearly independent over* $\mathbf{C}(z)$ *and form a solution of a system of homogeneous linear differential equations*

$$y_k' = \sum_{i=1}^{r} Q_{k,i}^* y_i, \qquad k = 1, \ldots, r, \qquad Q_{k,i}^* \in \mathbf{C}(z), \tag{10}$$

in which the point $z = \alpha$ is not a pole of any of the functions $Q_{k,i}^*$.

Proof. Lemma 2 follows from Lemma 1 if we choose functions (2) satisfying Lemma 1, use (3) to form the system of differential equations (5), and denote

$$Q_{k,i}^* = Q_{k,i} + \sum_{j=r+1}^{m} Q_{k,j} B_{j,i}, \qquad k, i = 1, \ldots, r.$$

Lemma 3. *Suppose that the set of KE-functions (1) forms a solution of the system (4) of homogeneous linear differential equations and has rank r over* $\mathbf{C}(z)$*, $0 \leq r \leq m$; let $\xi \in \mathbf{K} = \mathbf{Q}(\theta)$, $\xi T(\xi) \neq 0$, and $h = [\mathbf{K} : \mathbf{Q}]$. Then the rank ρ of the set of numbers $f_1(\xi), \ldots, f_m(\xi)$ over the field* $\mathbf{K}$ *satisfies the bound $\rho \geq r/h$, and if $\theta \notin \mathbf{R}$, then it satisfies $\rho \geq 2r/h$. In particular, if $\mathbf{K} = \mathbf{I}$, then $\rho = r$.*

Proof. The lemma holds for $r = m$, since in that case it is the same as Lemma 17 of Chapter 3. The lemma is trivial for $r = 0$. So we suppose that $1 \leq r \leq m - 1$. According to Lemma 2, given $\xi \in \mathbf{K}$, we can choose r KE-functions (2) from among the functions (1) which are $\mathbf{C}(z)$-linearly independent and form a solution of the system (10) of homogeneous linear differential equations, in which the point $z = \xi$ is not a pole of any of the functions $Q_{k,i}^*$. Hence, the KE-functions (2) satisfy all of the conditions in Lemma 17 of Chapter 3 with m replaced by r. If we let ρ^* denote the rank over $\mathbf{K}$ of the set of numbers $f_1(\xi), \ldots, f_r(\xi)$, it then follows from Lemma 17 that $\rho^* \geq r/h$, and if $\theta \notin \mathbf{R}$, then $\rho^* \geq 2r/h$. In particular, for $\mathbf{K} = \mathbf{I}$ we have $\rho^* = r$. Since $\rho \geq \rho^*$, the lemma is proved.

In Chapter 4 we use Lemma 3 to prove several algebraic independence theorems for the values of a subset of E-functions in the case when the basic set of E-functions under consideration is algebraically dependent over $\mathbf{C}(z)$.

§ 2. Some lemmas

In Chapter 3 we proved the First Fundamental Theorem (the homogeneous case), which established homogeneous algebraic independence of the values of a set of E-functions. The First Fundmental Theorem was then seen to imply the Second

Fundamental Theorem (non-homogeneous case), which asserted the algebraic independence of the values of a set of functions. Most of the theorems and lemmas proved in this book will follow a similar pattern. Hence, for brevity of exposition, every pair of theorems or lemmas (one homogeneous and one non-homogeneous) will be stated as a single theorem which includes both cases. In most such situations, the non-homogeneous case will be a consequence of the homogeneous case if one replaces m by $m+1$ and adds the function $f_0(z) \equiv 1$ to the m-tuple of functions under consideration. When this is the case, we shall give only the proof of the homogeneous part of the theorem or lemma.

In the homogeneous case our set of E-functions (1) will satisfy a system (4) of homogeneous linear differential equations; in the non-homogeneous case it will satisfy a system of linear differential equations

$$y_k' = Q_{k,0} + \sum_{i=1}^{m} Q_{k,0} y_i, \qquad k = 1, \ldots, m, \qquad Q_{k,i} \in \mathbf{C}(z), \tag{11}$$

which may in fact be a homogeneous system, i.e., if $Q_{k,0} \equiv 0$, $k = 1, \ldots, m$. The polynomial $T(z)$ will be defined in the same way for the system (11) as it was for the system (4).

In the statement of the First Fundamental Theorem in Chapter 3, we assumed that all of the coefficients $Q_{k,i}$ belong to $\mathbf{C}(z)$ in the system of differential equations (3.7) satisfied by the E-functions (3.9). We then showed, as a consequence of Lemma 3 of Chapter 3, that these coefficients are all in $\mathbf{K}(z)$. This resulted from the $\mathbf{C}(z)$-linear independence of the functions (3.9). The same thing happened with the other theorems in Chapter 3, both in the homogeneous and non-homogeneous cases.

In the theorems in Chapter 4 and some of the theorems in later chapters, the KE-functions (1) under consideration will be algebraically dependent over $\mathbf{C}(z)$. In particular, they may even be linearly dependent. In these theorems as well, we shall assume that the coefficients $Q_{k,i}$ belong to $\mathbf{C}(z)$ in the corresponding system of differential equations (4) or (11). Then Lemma 2 of Chapter 3 can be used to modify the $Q_{k,i}$ in such a way that they belong to $\mathbf{K}(z)$. However, in general this may change the poles of these functions. Hence, it is conceivable that we would have to require that $Q_{k,i} \in \mathbf{K}(z)$ in the statements of the theorems. But in actual fact it turns out that this is not necessary.

Namely, in the proofs of the theorems the lemmas in our method will be applied not to the functions (1), but rather to bases over $\mathbf{C}(z)$ of certain sets of monomials in these functions, which also satisfy systems of linear differential equations. These bases will be linearly independent over $\mathbf{C}(z)$; hence, all of the coefficients of the systems of linear differential equations which they satisfy will belong to $\mathbf{K}(z)$, by Lemma 3 of Chapter 3.

The statements of many of the theorems proved in the book will involve algebraic equations over $\mathbf{C}(z)$ for the KE-functions (1) in which the left sides are polynomials in the functions (1) with coefficients in $\mathbf{C}(z)$. In most of these theorems, we can use Lemmas 4 and 5 below in order to reduce to the case when these polynomials are assumed to belong to $\mathbf{Z_K}[z]$.

Lemma 4. *Suppose that*

$$\varphi_k(z) = \sum_{n=0}^{\infty} a_{k,n} z^n, \qquad k = 1, \dots, m, \qquad m \geq 2, \qquad a_{k,n} \in \mathbf{K}, \tag{12}$$

is a set of functions which is analytic in a region containing the point $z = 0$ and has $\mathbf{C}(z)$-rank $m-1$, and suppose that these functions are connected by a homogeneous linear equation

$$\sum_{i=1}^{m} A_i(z)\varphi_i(z) = 0, \qquad A_i(z) \in \mathbf{C}[z], \tag{13}$$

where

$$A_1(z), \dots, A_m(z) \tag{14}$$

do not have any factor in common. Then there exists $a \in \mathbf{C}$ such that

$$aA_i(z) \in \mathbf{Z_K}[z], \qquad i = 1, \dots, m. \tag{15}$$

Proof. By assumption, the polynomials (14) in equation (13) are uniquely determined up to a constant factor. On the other hand, by Lemma 2 of Chapter 3, the coefficients of the polynomials (14) can be chosen in $\mathbf{Z_K}$. Hence, there exists $a \in \mathbf{C}$ such that (15) holds.

Lemma 5. *Suppose that the functions (12) are connected by an algebraic equation*

$$B(z, \varphi_1(z), \dots, \varphi_m(z)) = \sum A_{k_1,\dots,k_m}(z)\varphi_1^{k_1}(z) \cdots \varphi_m^{k_m}(z) = 0,$$

$$B = B(z, z_1, \dots, z_m) \in \mathbf{C}[z, z_1, \dots, z_m], \tag{16}$$

$$A_{k_1,\dots,k_m}(z) \in \mathbf{C}[z], \qquad A_{k_1,\dots,k_m}(z) \not\equiv 0,$$

where B is a primitive polynomial in the m variables $z_1, \dots, z_m$ over $\mathbf{C}[z]$, and the summation is taken over k ($k \geq 2$ finite) different m-tuples $(k_1, \dots, k_m)$ with $k_1, \dots, k_m \in \mathbf{Z}^+$. Furthermore, suppose that exactly $k-1$ of the monomials in the functions (12) which occur in (16) are linearly independent over $\mathbf{C}(z)$. Then there exists $a \in \mathbf{C}$ such that all $aA_{k_1,\dots,k_m}(z) \in \mathbf{Z_K}[z]$.

Lemma 5 follows immediately from Lemma 4 and the primitivity of the polynomial B.

§ 3. Estimate for the dimension of a vector space spanned by monomials in elements of a field extension

Let $\mathbf{V}$ be a field, and let $\mathbf{W}$ be a field extension of $\mathbf{V}$. If $\mathbf{U}$ is a subset of $\mathbf{W}$ and $l \in \mathbf{Z}^+$ is the maximum number of elements of $\mathbf{U}$ which are algebraically independent over $\mathbf{V}$, then l is called the *transcendence degree of the set* $\mathbf{U}$ *over* $\mathbf{V}$ and is denoted $\deg \mathrm{tr}_{\mathbf{V}} \mathbf{U}$. Similarly, if $\mathbf{U} \subset \mathbf{W}$ and l is the maximum number of elements of $\mathbf{U}$ which are homogeneously algebraically independent over $\mathbf{V}$, then l is called the *homogeneous transcendence degree of the set* $\mathbf{U}$ *over* $\mathbf{V}$ and is denoted $\deg \mathrm{tr}^0_{\mathbf{V}} \mathbf{U}$.

If $\mathbf{V} = \mathbf{A}$, $\mathbf{W} = \mathbf{C}$ and $\mathbf{U} \subset \mathbf{C}$, then we say more briefly "the transcendence degree (homogeneous transcendence degree) of the numbers $\mathbf{U}$," and in that case we write $\deg \mathrm{tr}\, \mathbf{U}$ (resp. $\deg \mathrm{tr}^0 \mathbf{U}$).

An element $w \in \mathbf{W}$ is said to be algebraically dependent on $u_1, \ldots, u_n \in \mathbf{W}$ over $\mathbf{V}$ if w is algebraic over the field $\mathbf{V}(u_1, \ldots, u_n)$, i.e., if w satisfies an algebraic equation

$$a_k w^k + \cdots + a_1 w + a_0 = 0, \qquad k \geq 1, \tag{17}$$

whose coefficients $a_0, a_1, \ldots, a_k$ belong to $\mathbf{V}[u_1, \ldots, u_n]$ and are not all zero.

The notion of algebraic dependence has properties similar to those of linear dependence (see [83:1]).

We say that *the set* $\{w_1, \ldots, w_m\}$ *depends algebraically over* $\mathbf{V}$ *on the elements* $\{u_1, \ldots, u_n\}$ if each element in the former set depends algebraically over $\mathbf{V}$ on the elements of the second set. If the two sets $\{w_1, \ldots, w_m\}$ and $\{u_1, \ldots, u_n\}$ each depend algebraically over $\mathbf{V}$ on one another, then we say that they are *algebraically equivalent sets*. If two sets are algebraically equivalent over $\mathbf{V}$, then they have the same transcendence degree over $\mathbf{V}$.

We introduce analogous definitions in the homogeneous case. An element $w \in \mathbf{W}$ is said to be *homogeneously algebraically dependent on* $u_1, \ldots, u_n \in \mathbf{W}$ *over* $\mathbf{V}$ if w satisfies an algebraic equation (17) whose left side is a homogeneous polynomial in $u_1, \ldots, u_n, w$ of total degree s in these variables and whose coefficients a_l, $0 \leq l \leq k$, are homogeneous polynomials in $\mathbf{V}[u_1, \ldots, u_n]$, not all zero, with $\deg_{\mathbf{u}} a_l = s - l$. If the two sets $\{w_1, \ldots, w_m\}$ and $\{u_1, \ldots, u_n\}$ each depend homogeneously algebraically over $\mathbf{V}$ on one another, then we say that they are *homogeneously algebraically equivalent sets*. If two sets are homogeneously algebraically equivalent over $\mathbf{V}$, then they have the same homogeneous transcendence degree over $\mathbf{V}$.

Suppose that

$$u_m, \dots, u_1 \in \mathbf{W}, \qquad m \geq 1. \tag{18}$$

Then any set D which consists of monomials in the elements (18) of the form

$$u_m^{k_m} \cdots u_1^{k_1}, \qquad k_i \in \mathbf{Z}^+, \qquad i = 1, \dots, m,$$

will always be assumed to be ordered lexicographically according to powers of $u_m, \dots, u_1$. Any polynomial in the variables (18) or any algebraic equation involving these variables (or other variables) with coefficients in some field will be assumed to be written with its terms in lexicographical order according to powers of $u_m, \dots, u_1$ (or other variables).

In this section we shall agree to write any algebraic relation over $\mathbf{V}$ between the elements (18) in such a way that its leading term has coefficient 1.

In what follows, the elements (18) will be either numbers or else certain functions (1).

Let L_N^0 denote the set of monomials

$$\begin{aligned} &u_m^{k_m} \cdots u_1^{k_1}, \qquad k_1 + \cdots + k_m = N, \qquad N \in \mathbf{N}, \\ &\qquad\qquad\quad k_i \in \mathbf{Z}^+, \qquad i = 1, \dots, m, \qquad m \geq 1, \end{aligned} \tag{19}$$

where $(k_m, \dots, k_1)$ run through all m-tuples satisfying the conditions indicated. We let $\mathbf{L}_N^0$ denote the vector space over $\mathbf{V}$ spanned by the elements (19) of L_N^0.

Similarly, we let L_N denote the set of monomials

$$\begin{aligned} &u_m^{k_m} \cdots u_1^{k_1}, \qquad k_1 + \cdots + k_m \leq N, \qquad N \in \mathbf{N}, \\ &\qquad\qquad\quad k_i \in \mathbf{Z}^+, \qquad i = 1, \dots, m, \qquad m \geq 1, \end{aligned} \tag{20}$$

and we let $\mathbf{L}_N$ denote the $\mathbf{V}$-vector space spanned by the elements (20) of L_N.

We further denote

$$r_{N,m,l}^0 = \dim \mathbf{L}_N^0 \qquad (r_{N,m,l} = \dim \mathbf{L}_N).$$

Lemma 6. *If* $m \geq 1$ *and*

$$\deg \mathbf{tr}_{\mathbf{V}}^0 \{u_m, \dots, u_1\} = l, \qquad 1 \leq l \leq m,$$

$$(\deg \mathbf{tr}_{\mathbf{V}} \{u_m, \dots, u_1\} = l, \qquad 0 \leq l \leq m),$$

then there exist positive constants c_1 and c_2, depending only on m, the elements (18) and the field $\mathbf{V}$, such that

$$c_1 N^{l-1} \le r^0_{N,m,l} \le c_2 N^{l-1}, \qquad N = 1,2,\ldots \tag{21}$$

$$(c_1 N^{l} \le r_{N,m,l} \le c_2 N^{l}, \qquad N = 1,2,\ldots). \tag{22}$$

The constants c_1 and c_2 can be chosen as follows:

$$c_1 = \frac{1}{(l-1)!}, \qquad c_2 = l(m-l+1)(lk)^{m-l},$$
$$\left(c_1 = \frac{1}{l!}, \qquad c_2 = (l+1)(m-l+1)((l+1)k)^{m-l}\right),$$

where k is determined in the proof of the lemma.

Proof. We shall consider the homogeneous case; the non-homogeneous case follows from it if we replace m by $m+1$, l by $l+1$, and add the element $u_0 \equiv 1$. We first prove the lower bound in (21). We may assume that the elements $u_1, \ldots, u_l$ are homogeneously algebraically independent over $\mathbf{V}$. Hence, the monomials

$$u_l^{k_l} \cdots u_1^{k_1}, \qquad k_i \in \mathbf{Z}, \qquad i = 1,\ldots,l, \qquad k_1 + \cdots + k_l = N,$$

are linearly independent over $\mathbf{V}$. By Lemma 7 of Chapter 2, there are $\mu^0_{N,l} = (N+l-1)!/N!(l-1)!$ of these monomials; hence,

$$r^0_{N,m,l} \ge \frac{1}{(l-1)!}(N+1)\cdots(N+l-1) \ge \frac{N^{l-1}}{(l-1)!}, \qquad N = 1,2,\ldots,$$

which establishes the lower bound.

We next prove the upper bound in (21). If $l = m$, then

$$\begin{aligned} r^0_{N,m,m} &= \frac{1}{(m-1)!}(N+1)\cdots(N+m-1) = \\ &= m\frac{N+1}{2}\cdots\frac{N+m-1}{m} \le mN^{m-1}, \qquad N = 1,2,\ldots, \end{aligned} \tag{23}$$

and the upper bound is established.

Now suppose that $1 \le l < m$ (hence $m \ge 2$). Here l is the maximum number of elements in (18) which are homogeneously algebraically independent over $\mathbf{V}$. We take $u_{i_1}, \ldots, u_{i_l}$ to be any fixed set of elements in (18) which are homogeneously algebraically independent over $\mathbf{V}$, and we let $u_{i_{l+1}}, \ldots, u_{i_m}$ be the rest of the elements in (18). Then $u_{i_{l+1}}, \dots, u_{i_m}$ are connected with $u_{i_1}, \dots, u_{i_l}$ by algebraic equations

$$P_\nu(u_{i_1}, \dots, u_{i_l}, u_{i_\nu}) = 0, \tag{24}$$

$$P_\nu(x_1, \dots, x_l, x_\nu) \in \mathbf{V}[x_1, \dots, x_l, x_\nu], \qquad \nu = l+1, \dots, m,$$

where P_ν is an irreducible homogeneous polynomial written in the standard way, i.e., with its terms in lexicographic order and with leading coefficient 1.

We consider the set of all possible equations of the form (24) for different choices of l homogeneously algebraically independent elements among $u_1, \dots$ $\dots, u_m$. We discard duplicate equations, and denote the remaining distinct equations

$$Q_1 = 0, \quad \dots, \quad Q_s = 0; \tag{25}$$

we let A_1 denote the set of leading terms

$$u_{i_{1.1}}^{\kappa_{i_{1.1}}} \cdots u_{i_{l+1.1}}^{\kappa_{i_{l+1.1}}}, \quad \dots, \quad u_{i_{1.s}}^{\kappa_{i_{1.s}}} \cdots u_{i_{l+1.s}}^{\kappa_{i_{l+1.s}}},$$

$$i_{1,j} > i_{2,j} > \cdots > i_{l+1,j}, \qquad j = 1, \dots, s;$$

and we set

$$k = \max(\kappa_{i_{1.1}}, \dots, \kappa_{i_{l+1.1}}, \dots, \kappa_{i_{1.s}}, \dots, \kappa_{i_{l+1.s}}).$$

Let A_2 be the subset of the monomials $u_m^{k_m} \cdots u_1^{k_1}$ in L_N^0 (see (19)) for which as least $m - l$ different exponents satisfy the inequality $k_i < k$. We fix each of the possible $m - l$ values of $i_{l+1}, \dots, i_m$ and corresponding exponents $k_{i_{l+1}} < k, \dots, k_{i_m} < k$, and we let all of the other exponents $k_{i_1}, \dots, k_{i_l}$ run through all values in $\mathbf{Z}^+$ such that

$$k_{i_1} + \cdots + k_{i_l} = N - k_{i_{l+1}} - \cdots - k_{i_m}.$$

Then, using Lemma 7 of Chapter 2, we find that the number of elements in the set A_2 is at most

$$\binom{m}{m-l} \sum_{\substack{k_{i_{l+1}}+\cdots+k_{i_m}\le N\\ 0\le k_{i_j}<k,\ j=l+1,\dots,m}} \frac{(N-k_{i_{l+1}}-\cdots-k_{i_m}+l-1)!}{(N-k_{i_{l+1}}-\cdots-k_{i_m})!(l-1)!} <$$

$$< \frac{m!}{(m-l)!l!}k^{m-l}\frac{(N+l-1)!}{N!(l-1)!} = \tag{26}$$

$$= l(m-l+1)\frac{l+1}{2}\cdots\frac{l+(m-l)}{m-l+1}k^{m-l}\frac{N+1}{2}\cdots\frac{N+l-1}{l} \le$$

$$\le l(m-l+1)(lk)^{m-l}N^{l-1}.$$

We now prove that each of the elements (19) in L_N^0 can be expressed as a $\mathbf{V}$-linear combination of the elements of A_2. In fact, any element $u_m^{k_m}\cdots u_1^{k_1}$ in L_N^0 not in A_2 (if such monomials exist) has at least $l+1$ exponents with different indices $i_1,\dots,i_{l+1}$ such that $k_{i_1}\ge k,\dots,k_{i_{l+1}}\ge k$. But any $l+1$ elements from among $u_1,\dots,u_m$ are algebraically dependent over $\mathbf{V}$. Hence, by the definition of the set A_1, we have the equality

$$u_m^{k_m}\cdots u_1^{k_1} = u_{i_{1,j}}^{\kappa_{i_{1,j}}}\cdots u_{i_{l+1,j}}^{\kappa_{i_{l+1,j}}} u_m^{k_m^*}\cdots u_1^{k_1^*}, \tag{27}$$

where $u_{i_{1,j}}^{\kappa_{i_{1,j}}}\cdots u_{i_{l+1,j}}^{\kappa_{i_{l+1,j}}}$ is the leading term in the corresponding algebraic equation $Q_j=0$, $1\le j\le s$, in (25) which connects the elements $u_{i_1},\dots,u_{i_{l+1}}$ over $\mathbf{V}$, and $k_m^*,\dots,k_1^*\in\mathbf{Z}^+$.

Using the equation $Q_j=0$, we can represent the product $u_{i_{1,j}}^{\kappa_{i_{1,j}}}\cdots u_{i_{l+1,j}}^{\kappa_{i_{l+1,j}}}$ as a $\mathbf{V}$-linear combination of the monomials (18) of lower order. We substitute this linear combination of the monomials (18) in place of that product in the right side of (26). We then have a linear equation for $u_m^{k_m}\cdots u_1^{k_1}$ in terms of lower order elements of L_N^0:

$$u_m^{k_m}\cdots u_1^{k_1} = a_1 u_m^{k_{m,1}}\cdots u_1^{k_{1,1}}+\cdots+a_t u_m^{k_{m,t}}\cdots u_1^{k_{1,t}}, \qquad a_i\in\mathbf{V},\quad i=1,\dots,t. \tag{28}$$

On the right side of (28) each of the monomials $u_m^{k_{m,i}}\cdots u_1^{k_{1,i}}$, $1\le i\le m$, which does not belong to A_2 can be expressed linearly in terms of lower order monomials, as indicated above. We continue this process until all of the monomials in terms of which $u_m^{k_m}\cdots u_1^{k_1}$ is expressed as a $\mathbf{V}$-linear combination belong to A_2. We have thus shown that any element of L_N^0 is a $\mathbf{V}$-linear combination of elements of A_2.

From this it follows that A_2 contains a basis of the vector space $\mathbf{L}_N^0$. Then the upper bound in (21) follows from (23) and (26). The lemma is proved.

Remark. Roughly speaking, we can say that the number k in the expression for c_2 is the maximum degree of all of the polynomials on the left in the equations in (25).

§ 4. The Third Fundamental Theorem

Third Fundamental Theorem. *Suppose that the E-functions*

$$f_1(z), \dots, f_m(z), \tag{29}$$

$m \geq 2$ $(m \geq 1)$, *form a solution of the system (4) of m homogeneous linear differential equations (the system (11) of m linear differential equations) and have homogeneous transcendence degree (transcendence degree) over $\mathbf{C}(z)$ equal to l, $0 \leq l \leq m$. Let $\xi \in \mathbf{A}$, $\xi T(\xi) \neq 0$. Then the homogeneous transcendence degree (transcendence degree) of the set of numbers*

$$f_1(\xi), \dots, f_m(\xi) \tag{30}$$

is also equal to l.

Proof. Let $\mathbf{K}$ be an algebraic number field containing ξ and all of the coefficients of the power series in z of the E-functions (29), and set $h = [\mathbf{K} : \mathbf{Q}]$. We denote $l' = \deg\operatorname{tr}^0\{f_1(\xi), \dots, f_m(\xi)\}$, $0 \leq l' \leq m$. If $l' \geq 1$, then we can choose l' homogeneously algebraically independent numbers in (30). Then, by Lemma 19 of Chapter 3, the corresponding functions from (29) will be homogeneously algebraically independent over $\mathbf{C}(z)$. This implies that

$$l' \leq l, \tag{31}$$

which is trivially true for $l' = 0$ as well.

We now suppose that $l \geq 1$. Choosing $N \in \mathbf{N}$ large, we consider the set of monomials in the functions (29)

$$f_m^{k_m}(z) \cdots f_1^{k_1}(z), \qquad k_1 + \cdots + k_m = N,$$

$$k_i \in \mathbf{Z}^+, \qquad i = 1, \dots, m, \qquad m \geq 2, \tag{32}$$

and we let r_N^0 denote the rank over $\mathbf{C}(z)$ of this set of functions.

In Lemma 6, setting $\mathbf{V} = \mathbf{C}(z)$, $u_i = f_i(z)$, $i = 1, \dots, m$, we find that r_N^0 satisfies the inequalities

$$c_1 N^{l-1} < r_N^0 < c_2 N^{l-1}, \qquad N = 1, 2, \dots, \tag{33}$$

where c_1 and c_2 are positive constants not depending on N.

By Lemma 18 of Chapter 3, the E-functions (32) and the number ξ satisfy all of the conditions of Lemma 3 with m replaced by $\mu_{N,m}^0$. Hence, letting ρ_N^0 denote the **K**-rank of the set of numbers

$$f_m^{k_m}(\xi) \cdots f_1^{k_1}(\xi), \qquad k_1 + \cdots + k_m = N,$$

$$k_i \in \mathbf{Z}^+, \qquad i = 1, \dots, m, \qquad m \geq 2, \tag{34}$$

we find by Lemma 3 that

$$\rho_N^0 \geq r_N^0 / h.$$

From this inequality and (33) it follows that

$$\rho_N^0 \geq \frac{c_1}{h} N^{l-1}, \qquad c_1 > 0. \tag{35}$$

On the other hand, if we set $\mathbf{V} = \mathbf{K}$, $u_i = f_i(\xi)$, $i = 1, \dots, m$, in Lemma 6, we obtain the following inequality for the rank of the numbers (34):

$$c_3 N^{l'-1} < \rho_N^0 < c_4 N^{l'-1}, \qquad N = 1, 2, \dots,$$

where c_3 and c_4 are positive constants not depending on N. For N sufficiently large, these inequalities along with (35) imply that

$$l' \geq l \tag{36}$$

(this also holds for $l = 0$). From (31) and (36) we obtain $l' = l$, and the theorem is proved.

In the case $l = m$ the Third Fundamental Theorem becomes the First (and Second) Fundamental Theorems.

The Third Fundamental Theorem in the homogeneous case gives us a series of corollaries similar to those of the First Fundamental Theorem.

Suppose that the E-functions (29) and the number $\xi \in \mathbf{A}$ satisfy the conditions of the homogeneous case of the Third Fundamental Theorem. Then:

1°. *At least l of the numbers in (30) are nonzero.*

2°. *For every s, $1 \leq s \leq m$, if $f_s(\xi) \neq 0$, then among the numbers*

$$f_k(\xi)/f_s(\xi), \qquad k = 1, \dots, m, \qquad k \neq s,$$

there are $l-1$ algebraically independent numbers.

3°. *There are at least $l-1$ transcendental numbers among the numbers in (30).*

4°. *If one of the numbers in (30) is a nonzero algebraic number, then the remaining numbers include $l-1$ algebraically independent numbers.*

As in the case of the First and Second Fundamental Theorems, the Third Fundamental Theorem can be carried over to the case of an E-function which satisfies a homogeneous linear differential equation

$$P_m(z)y^{(m)} + \cdots + P_1(z)y' + P_0(z)y = 0, \qquad m \geq 2,$$

$$P_k(z) \in \mathbf{C}[z], \qquad k = 0, 1, \ldots, m, \tag{37}$$

or a linear differential equation

$$P_m(z)y^{(m)} + \cdots + P_1(z)y' + P_0(z)y = Q(z), \qquad m \geq 1,$$

$$P_k(z) \in \mathbf{C}[z], \qquad k = 0, 1, \ldots, m, \qquad Q(z) \in \mathbf{C}[z]. \tag{38}$$

Here the algebraic independence condition for $f(z)$ and its successive derivatives takes on a somewhat different form.

Theorem 1. *Suppose that the E-function $f(z)$ is a solution of the m-th order homogeneous linear differential equation (37) (linear differential equation (38)), $m \geq 2$ ($m \geq 1$). Further suppose that $f(z)$ satisfies a homogeneous algebraic differential equation (an algebraic differential equation) with coefficients in $\mathbf{C}[z]$ of order l, $1 \leq l \leq m$, but none of lower order (if $l = 1$, one supposes, respectively, that $f(z) \not\equiv 0$ or that $f(z)$ is not a rational function). Let $\xi \in \mathbf{A}$, $\xi T(\xi) \neq 0$. Then the homogeneous transcendence degree (transcendence degree) of the set of numbers $f(\xi), f'(\xi), \ldots, f^{(m-1)}(\xi)$ is equal to l.*

Theorem 1 is a consequence of the Third Fundamental Theorem and the following lemma.

Lemma 7. *Suppose that the analytic function $f(z)$ is a solution of the m-th order homogeneous linear differential equation (37) (the linear differential equation (38)), $m \geq 2$ ($m \geq 1$). Then the homogeneous transcendence degree (transcendence degree) over $\mathbf{C}(z)$ of the set of functions $f(z), f'(z), \ldots, f^{(m-1)}(z)$ is equal to l, $1 \leq l \leq m$, if and only if $f(z)$ satisfies a homogeneous algebraic differential equation (an algebraic differential equation) with coefficients in $\mathbf{C}[z]$ of order l and none of lower order (if $l = 1$, one assumes, respectively, that $f(z) \not\equiv 0$ or that $f(z)$ is not a rational function).*

Proof. We consider the non-homogeneous case; the homogeneous case is similar. If $l = m$, then the lemma holds. Suppose that $1 \leq l < m$.

1. Suppose that the function $y = f(z)$ is a solution of the differential equation

$$P = Q_k(y^{(l)})^k + \cdots + Q_1 y^{(l)} + Q_0 = 0, \qquad Q_0 Q_k \not\equiv 0, \qquad k \geq 1,$$

$$Q_i \in \mathbf{C}\left[z, y, y', \ldots, y^{(l-1)}\right], \qquad i = 0, 1, \ldots, k, \tag{39}$$

where P is an irreducible polynomial in $l+2$ variables, and $f(z)$ does not satisfy an algebraic differential equation with coefficients in $\mathbf{C}[z]$ of order less than l. Then $\deg \operatorname{tr}_{\mathbf{C}(z)} \left\{y, y', \ldots, y^{(l-1)}\right\} = l$. If $l = m - 1$, then from (39) we conclude that the function $y^{(l)}$ depends algebraically over $\mathbf{C}(z)$ on the functions $y, y', \ldots, y^{(l-1)}$.

On the other hand, if $l < m - 1$, then, differentiating both sides of (39), we obtain

$$(kQ_k(y^{(l)})^{k-1} + \cdots + Q_1)y^{(l+1)} + (Q_k'(y^{(l)})^k + \cdots + Q_1' y^{(l)} + Q_0') = 0. \tag{40}$$

We now prove that

$$R = R(z, y, y', \ldots, y^{(l)}) = kQ_k(y^{(l)})^{k-1} + \cdots + Q_1 \neq 0, \qquad kQ_k \not\equiv 0.$$

In fact, if this were not the case, then R would be divisible as a polynomial in $l + 2$ variables by the irreducible polynomial P, since otherwise we could eliminate $y^{(l)}$ from the two equations $P = 0$ and $R = 0$, obtaining an algebraic equation with coefficients in $\mathbf{C}[z]$ relating the functions $y, y', \ldots, y^{(l-1)}$, and this is impossible. On the other hand, R cannot be divisible by P, since $R \not\equiv 0$ as a polynomial in $z, y, y', \ldots, y^{(l)}$, and it has degree in $y^{(l)}$ which is less than the degree of P in $y^{(l)}$. Consequently, $R \not\equiv 0$ as a function of z.

It now follows from (39) and (40) that $y^{(l)}$ and $y^{(l+1)}$ depend algebraically over $\mathbf{C}(z)$ on the functions $y, y', \ldots, y^{(l-1)}$.

If $l < m - 2$, then, differentiating both sides of (40), we similarly find that $y^{(l)}, y^{(l+1)}, y^{(l+2)}$ depend algebraically over $\mathbf{C}(z)$ on the functions $y, y', \ldots, y^{(l-1)}$. If we repeat this argument $m - l$ times, we can conclude that the functions $y^{(l)}, \ldots, y^{(m-1)}$ all depend algebraically over $\mathbf{C}(z)$ on $y, y', \ldots, y^{(l-1)}$. This implies that the two sets of functions $\{y, y', \ldots, y^{(m-1)}\}$ and $\{y, y', \ldots, y^{(l-1)}\}$ are algebraically equivalent over $\mathbf{C}(z)$. Hence,

$$\deg \operatorname{tr}_{\mathbf{C}(z)} \left\{y, y', \ldots, y^{(m-1)}\right\} = \deg \operatorname{tr}_{\mathbf{C}(z)} \left\{y, y', \ldots, y^{(l-1)}\right\} = l.$$

2. Now let $\deg \operatorname{tr}_{\mathbf{C}(z)} \{y, y', \ldots, y^{(m-1)}\} = l$, $1 \leq l < m$. Then y is not an algebraic function. Suppose that y satisfies an s-th order algebraic differential equation with coefficients in $\mathbf{C}[z]$ and does not satisfy a lower order differential equation of that type. We have $1 \leq s \leq l$, and, by what we proved in case 1, $s = l$. The lemma is proved.

We shall use the Third Fundamental Theorem to prove several theorems giving arithmetic properties of the values of E-functions at algebraic points.

§ 5. Transcendence of the values of E-functions connected by arbitrary algebraic equations over $\mathbf{C}(z)$

Let Ω be a subset of the complex numbers. We shall say that a statement holds *for almost all* $\alpha \in \Omega$ if it holds for all $\alpha \in \Omega$ except for a finite (or empty) subset.

Theorem 2. *Suppose that the E-functions*

$$f_1(z), \ldots, f_m(z), \qquad m \geq 1, \tag{41}$$

form a solution of the system (11) of linear differential equations. Then for almost all $\xi \in \mathbf{A}$ *each of the numbers* $f_1(\xi), \ldots, f_m(\xi)$ *is transcendental if the corresponding function in (41) is transcendental.*

The proof of this theorem uses the following lemma.

Lemma 8. *If the polynomials*

$$H_i = H_i(y, z) \in \mathbf{C}[y, z], \qquad i = 1, \ldots, k, \qquad k \geq 2,$$

do not all have a common factor, and if $f(z)$ *is an analytic function in some region, then the set of common zeroes of all of the functions*

$$H_i(f(z), z), \qquad i = 1, \ldots, k$$

is finite (possibly empty).

Proof. If any of the polynomials H_i does not depend on y and is nonzero, then the lemma is clearly true. Now suppose that all of the nonzero polynomials depend on y. We consider two cases.

1) $k = 2$. Then both of the polynomials H_1 and H_2 are nonzero. We regard them as polynomials in y with coefficients in $\mathbf{C}[z]$, and we let $R(z)$ be the resultant of the two polynomials. Then there exist $P_1(y, z), P_2(y, z) \in \mathbf{C}[y, z]$ such that

$$R(z) = P_1(y, z)H_1(y, z) + P_2(y, z)H_2(y, z), \qquad R(z) \in \mathbf{C}[z].$$

Since H_1 and H_2 are relatively prime, it follows that $R(z) \not\equiv 0$, and the lemma follows from the last equation with y set equal to $f(z)$.

2) We consider the set of all nonzero polynomials in the set $H_1, \ldots, H_k$. There are at least two, and they do not have any factor in common. We write each of

these polynomials as a product of powers of irreducible polynomials in $\mathbf{C}[y, z]$. Let $P_1 = P_1(y, z), \ldots, P_n = P_n(y, z)$ be the set of all pairwise relatively prime irreducible factors which occur in the factorization of our polynomials. Then $n \geq 2$.

If we have the relations

$$H_1(f(\alpha), \alpha) = 0, \ \ldots, \ H_k(f(\alpha), \alpha) = 0 \tag{42}$$

for some $z = \alpha$, then this means that at least two of the numbers $P_1(f(\alpha), \alpha), \ldots$ $\ldots, P_n(f(\alpha), \alpha)$ vanish, since $H_1, \ldots, H_k$ have no common factor. But by case 1), any two fixed polynomials P_j, $1 \leq j \leq n$, can vanish simultaneously for only finitely many values of z. Consequently, if there is any value of z for which any pair of the polynomials $P_1, \ldots, P_n$ vanish simultaneously, then there is only a finite set of such values. This implies that the equalities (42) can occur simultaneously for only finitely many values of α. The lemma is proved.

Proof of Theorem 2. Any algebraic E-function must be a polynomial. Hence, we may assume that all of the functions in (41) are transcendental, since otherwise the subset of those which are transcendental forms a solution of a system of linear differential equations similar to (11), and we could restrict ourselves to those transcendental functions.

Suppose that $\deg \operatorname{tr}_{\mathbf{C}(z)}\{f_1(z), \ldots, f_m(z)\} = l$, $1 \leq l \leq m$. If $l = m$ (in particular, if $m = 1$), then Theorem 2 follows from the Second Fundamental Theorem. Hence, we shall consider the case when $1 \leq l \leq m - 1$ and $m > 1$. It is sufficient to prove the conclusion of the theorem for any of the functions in (41) which is transcendental. To be definite, let us suppose that this is $f_m(z)$.

We now define a finite set $A_m \in \mathbf{A}$, which we call the set of exceptional points for $f_m(z)$. We shall prove that $f_m(\xi)$ is transcendental for any $\xi \in \mathbf{A} \setminus A_m$. This will prove the theorem.

Two cases are possible:

1. $\deg \operatorname{tr}_{\mathbf{C}(z)}\{f_1(z), \ldots, f_{m-1}(z)\} = l - 1$ with $l > 1$. We let A_m be the set consisting of zero and the zeros of the polynomial $T(z)$ for the system (11) of differential equations. If we apply Lemma 19 of Chapter 3 to all l-tuples of functions taken from among the functions

$$f_1(z), \ldots, f_{m-1}(z), \tag{43}$$

we find that for any $\xi \in \mathbf{A}$ the transcendence degree of the set of numbers $f_1(\xi), \ldots, f_{m-1}(\xi)$ is less than l. Then, the Third Fundamental Theorem tells us that $f_m(\xi)$ is transcendental for any $\xi \in \mathbf{A} \setminus A_m$.

2. $\deg \operatorname{tr}_{\mathbf{C}(z)}\{f_1(z), \ldots, f_{m-1}(z)\} = l$. Then it is possible to find l of the functions in (43) which are algebraically independent over $\mathbf{C}(z)$. To be definite, suppose that $f_1(z), \ldots, f_l(z)$ is some such l-tuple of algebraically independent functions (where the functions in (43) may be renumbered).

In this case the functions $f_1(z), \dots, f_l(z), f_m(z)$ are algebraically dependent over $\mathbf{C}(z)$, and so are connected by an algebraic equation

$$P = P(z, f_1(z), \dots, f_l(z), f_m(z)) = 0, \tag{44}$$

where P is an irreducible polynomial in $l+2$ variables whose coefficients may be assumed to lie in $\mathbf{A}$, by Lemma 5, and which contains $f_m(z)$ and at least one of the functions $f_1(z), \dots, f_l(z)$ (since $f_m(z)$ is a transcendental function).

We regard P as a polynomial in $f_1(z), \dots, f_l(z)$ with coefficients in $\mathbf{A}[f_m(z), z]$. At least two of these coefficients are nonzero, and the coefficients have no common factor. By Lemma 8, the set A_P of points $\xi \in \mathbf{A}$ which are common zeros of these coefficients must be a finite set.

We now define the set A_m as follows. It contains 0, all of the zeros of $T(z)$, and also all of the points in the different sets A_P corresponding to the different l-tuples of functions in (43) which are algebraically independent over $\mathbf{C}(z)$.

Let $\xi \in \mathbf{A}$, $\xi T(\xi) \neq 0$. Suppose that $f_m(\xi) \in \mathbf{A}$. Since the set of functions (41) has transcendence degree l, it follows by the Third Fundamental Theorem that one can find l of the numbers

$$f_1(\xi), \dots, f_{m-1}(\xi)$$

which are algebraically independent. Without loss of generality (after possibly renumbering the functions in (43)), we may suppose that $f_1(\xi), \dots, f_l(\xi)$ is such an algebraically independent set. By Lemma 19 of Chapter 3, the corresponding functions $f_1(z), \dots, f_l(z)$ are algebraically independent over $\mathbf{C}(z)$, and are connected with $f_m(z)$ by an algebraic equation (44), as above. We again regard P as a polynomial in $f_1(z), \dots, f_l(z)$ with coefficients in $\mathbf{A}[f_m(z), z]$, and we set $z = \xi$ in (44). Then the coefficients of P are numbers in $\mathbf{A}$, and since the set $f_1(\xi), \dots, f_l(\xi)$ is algebraically independent, it follows that all of these coefficients vanish. Hence, $\xi \in A_P$, and so $\xi \in A_m$. This implies that $f_m(\xi)$ is a transcendental number for all $\xi \in \mathbf{A} \setminus A_m$. The theorem is proved.

Corollary. *Under the conditions of Theorem 2, almost all algebraic A-points of each of the transcendental E-functions in (41) are transcendental numbers. In particular, almost all of the zeros of these functions are transcendental.*

Just as the Second Fundamental Theorem implied Theorem 3 of Chapter 3, Theorem 2 above implies

Theorem 3. *If $f(z)$ is a transcendental E-function which is a solution of the linear differential equation (38), then for almost all $z \in \mathbf{A}$ the value at z of the function f or any of its derivatives is a transcendental number.*

From Theorem 3 we also obtain a corollary similar to the corollary of Theorem 2, but for a function $f(z)$ and its successive derivatives.

Theorem 4. *Suppose that each of the E-functions in (41) is a solution of a linear differential equation with coefficients in* $\mathbf{C}[z]$, *let*

$$P = P(z, f_1(z), \ldots, f_m(z)) \in \mathbf{A}[z, f_1(z), \ldots, f_m(z)],$$

and suppose that P *is transcendental as a function of* z. *Then for almost all* $\xi \in \mathbf{A}$

$$P(\xi, f_1(\xi), \ldots, f_m(\xi))$$

is a transcendental number. In particular, almost all of the zeros and algebraic A-points of the function $P(z, f_1(z), \ldots, f_m(z))$ *are transcendental.*

Theorem 4 is a consequence of Theorem 2 and the following lemma.

Lemma 9. *If each of the* m *analytic functions in (41) is a solution of a linear differential equation with coefficients in* $\mathbf{C}[z]$, *then any polynomial*

$$P = P(z, f_1(z), \ldots, f_m(z)) \in \mathbf{C}[z, f_1(z), \ldots, f_m(z)] \tag{45}$$

is also a solution of a linear differential equation with coefficients in $\mathbf{C}[z]$.

Proof. It suffices to consider the case when P contains at least one of the functions in (41). Suppose that $f_k(z)$ is a solution of an n_k-th order linear differential equation, $k = 1, \ldots, m$. We set $M = n_1 + \cdots + n_m$, and we consider the set of M E-functions

$$f_k^{(l)}(z), \qquad l = 0, 1, \ldots, n_k - 1; \quad k = 1, \ldots, m. \tag{46}$$

We write each of the differential equations satisfied by the functions in (41) as a system of first order linear differential equations, in the same way as this was done in § 13 of Chapter 3 for the equations (3.163) and (3.168). The m systems altogether form a system of M linear differential equations with coefficients in $\mathbf{C}[z]$ having the set of functions (46) as its solution.

Let N, $N \geq 1$, be the total degree of the polynomial P in (45) with respect to the variables (41). We consider the $\mu_{N,m}$ monomials in the functions (46) whose total degree does not exceed N. By Lemma 18 of Chapter 3, this set of functions satisfies a system of first order linear differential equations with coefficients in $\mathbf{C}(z)$, and the polynomial P is a linear form in certain of these functions.

We set $R_1 = P$. Starting with the linear form R_1 and using (3.38), we construct linear forms $R_2, R_3, \ldots$ in our set of monomials in the functions (46). Arguing as in Lemma 6 of Chapter 3, we find that the first l of these linear forms, where $l \leq \mu_{N,m} + 1$, are linearly dependent, and hence they are connected by a homogeneous linear equation with coeficients in $\mathbf{C}[z]$. If we replace the forms $R_2, R_3, \ldots$ in this equation by the successive derivatives of R_1 (using the relation $R_k = TR'_{k-1}$), we obtain a homogeneous linear differential equation with coefficients in $\mathbf{C}[z]$ having R_1 as a solution. This completes the proof of the lemma, and thus the proof of Theorem 4 as well.

Remark. Theorem 4 obviously holds if the E-functions in (41), either by themselves or together with some other E-functions, form a solution of a system of linear differential equations of the form (11).

§ 6. Algebraic independence of the values of E-functions which are connected by arbitrary algebraic equations over $\mathbf{C}(z)$

Theorem 5. *Suppose that the E-functions*

$$f_1(z), \dots, f_m(z), \tag{47}$$

where $m \geq 2$ ($m \geq 1$), form a solution of the system of homogeneous linear differential equations (4) (of linear differential equations (11)), and the functions $f_1(z), \dots, f_l(z)$, $1 \leq l \leq m$, are homogeneously algebraically independent (algebraically independent) over $\mathbf{C}(z)$. Then for almost all $\xi \in \mathbf{A}$ the numbers $f_1(\xi), \dots, f_l(\xi)$ are homogeneously algebraically independent (algebraically independent).

The proof of the theorem requires two lemmas.

Lemma 10. *Let $\alpha_1, \dots, \alpha_m$, $m \geq 2$, be complex numbers which are connected by homogeneous algebraic equations (by algebraic equations)*

$$B_\nu(\alpha_1, \dots, \alpha_\nu) = 0, \qquad \nu = l+1, \dots, m,$$

$$B_\nu = B_\nu(y_1, \dots, y_\nu) \in \mathbf{A}[y_1, \dots, y_\nu], \tag{48}$$

where $1 \leq l < m$, and B_ν contains y_ν and has the property that at least one of the polynomials in $y_1, \dots, y_{\nu-1}$ which appear as coefficients of powers of y_ν does not vanish when the numbers $\alpha_1, \dots, \alpha_{\nu-1}$ are substituted in place of $y_1, \dots, y_{\nu-1}$. Then

$$\begin{gathered} \deg \operatorname{tr}^0\{\alpha_1, \dots, \alpha_m\} = \deg \operatorname{tr}^0\{\alpha_1, \dots, \alpha_l\} \\ \left(\deg \operatorname{tr}\{\alpha_1, \dots, \alpha_m\} = \deg \operatorname{tr}\{\alpha_1, \dots, \alpha_l\}\right). \end{gathered} \tag{49}$$

Proof. From the equations (48) it follows that each of the numbers α_ν, $\nu = l+1, \dots, m$, is homogeneously algebraically dependent (algebraically dependent) on the numbers $\alpha_1, \ldots, \alpha_{\nu-1}$, and so is homogeneously algebraically dependent (algebraically dependent) on the numbers $\alpha_1, \ldots, \alpha_l$. Thus, the two sets $\alpha_1, \ldots, \alpha_m$ and $\alpha_1, \ldots, \alpha_l$ are homogeneously algebraically equivalent (algebraically equivalent). This implies the equality (49).

Lemma 11. *Suppose that the E-functions (47), $m \geq 2$ ($m \geq 1$), form a solution of the system of homogeneously linear differential equations (4) (of linear differential equations (11)), and let l, $1 \leq l \leq m-1$, be the homogeneous transcendence degree (transcendence degree) of this set of functions over $\mathbf{C}(z)$. Further suppose that the functions $f_1(z), \ldots, f_l(z)$ are homogeneously algebraically independent (algebraically independent) over $\mathbf{C}(z)$, and the functions $f_1(z), \ldots, f_m(z)$ are connected by homogeneous algebraic equations (arbitrary algebraic equations)*

$$P_\nu = P_\nu(f_1(z), \ldots, f_\nu(z)) = 0, \qquad \nu = l+1, \ldots, m, \tag{50}$$

where P_ν is a polynomial in ν variables with coefficients in $\mathbf{A}[z]$ in which $f_\nu(z)$ appears. Then, if the numbers $f_1(\xi), \ldots, f_l(\xi)$ are homogeneously algebraically dependent (algebraically dependent) at some point $\xi \in \mathbf{A}$ with $\xi T(\xi) \neq 0$, it follows that for at least one of the polynomials P_ν, regarded as a polynomial in $f_\nu(z)$ with coefficients in $\mathbf{A}[z, f_1(z), \ldots, f_{\nu-1}(z)]$, all of these coefficients vanish when we substitute $z = \xi$.

Proof. We suppose the contrary, i.e., that there is a point $\xi \in \mathbf{A}$, $\xi T(\xi) \neq 0$, such that $f_1(\xi), \ldots, f_l(\xi)$ are homogeneously algebraically dependent (algebraically dependent) but each of the polynomials P_ν in (50), regarded as a polynomial in $f_\nu(z)$ with coefficients in $\mathbf{A}[z, f_1(z), \ldots, f_{\nu-1}(z)]$, has at least one coefficient which does not vanish at $z = \xi$. Because

$$P_\nu(f_1(\xi), \ldots, f_\nu(\xi)) = 0, \qquad \nu = l+1, \ldots, m,$$

our supposition means that the two sets $f_1(\xi), \ldots, f_m(\xi)$ and $f_1(\xi), \ldots, f_l(\xi)$ satisfy the conditions of Lemma 10, according to which these sets have the same homogeneous transcendence degree (transcendence degree). But, by assumption, the numbers $f_1(\xi), \ldots, f_l(\xi)$ are homogeneously algebraically dependent (algebraically dependent). Consequently, the sets $f_1(\xi), \ldots, f_m(\xi)$ and $f_1(\xi), \ldots, f_l(\xi)$ have homogeneous transcendence degree (transcendence degree) less than l.

On the other hand, the E-functions in (47) satisfy the conditions of the Third Fundamental Theorem, from which it follows that the set $f_1(\xi), \ldots, f_m(\xi)$ has homogeneous transcendence degree (transcendence degree) equal to l. This contradiction proves the lemma.

Proof of Theorem 5. We may assume that the homogeneous transcendence degree (transcendence degree) of the set of functions (47) over $\mathbf{C}(z)$ is equal to l, and the functions $f_1(z), \ldots, f_l(z)$ are homogeneously algebraically independent (algebraically independent) over $\mathbf{C}(z)$. If $l = m$, then the theorem follows from the First and Second Fundamental Theorems. We now suppose that $1 \leq l \leq m-1$, and the functions $f_{l+1}(z), \ldots, f_m(z)$ are connected with the functions $f_1(z), \ldots, f_l(z)$ by the homogeneous algebraic equations (arbitrary algebraic equations)

$$P_\nu = P_\nu(f_1(z), \ldots, f_l(z), f_\nu(z)) = 0, \qquad \nu = l+1, \ldots, m, \tag{51}$$

where P_ν is an irreducible and primitive polynomial in $l+1$ variables with coefficients in $\mathbf{C}[z]$ which contains $f_\nu(z)$. By Lemma 5, we may assume that $P_\nu \in \mathbf{A}[z, f_1(z), \ldots, f_l(z), f_\nu(z)]$. The equations (51) are a special case of the equations (50) considered in Lemma 11.

Let $\xi \in \mathbf{A}$, $\xi T(\xi) \neq 0$. If the numbers $f_1(\xi), \ldots, f_l(\xi)$ are homogeneously algebraically dependent (algebraically dependent), then, by Lemma 11, at least one of the polynomials P_ν in (51), regarded as a function of $f_\nu(z)$ with coefficients in $\mathbf{A}[z, f_1(z), \ldots, f_l(z)]$, has all coefficients vanishing at $z = \xi$. But, by Theorem 4, for each of the polynomials P_ν it is impossible for any of the coefficients, let alone all of them, to vanish at more than a finite set of numbers in $\mathbf{A}$. Hence, at almost all algebraic points ξ the numbers $f_1(\xi), \ldots, f_l(\xi)$ are homogeneously algebraically independent (algebraically independent). The theorem is proved.

The homogeneous case of Theorem 5 implies a series of corollaries similar to those which followed from the First and Second Fundamental Theorem.

Suppose that the E-functions (47) satisfy the conditions of the homogeneous case of Theorem 5. Then:

1°. *For almost all $\xi \in \mathbf{A}$ the numbers $f_1(\xi), \ldots, f_l(\xi)$ are nonzero.*

2°. *For every s, $1 \leq s \leq l$, if $f_s(\xi) \neq 0$, then for almost all $\xi \in \mathbf{A}$ the ratios $f_k(\xi)/f_s(\xi)$, $k = 1, \ldots, l$, $k \neq s$, are algebraically independent.*

3°. *For almost all $\xi \in \mathbf{A}$ at least $l-1$ of the numbers $f_1(\xi), \ldots, f_l(\xi)$ are transcendental.*

§ 7. E-functions connected by special types of equations

The Third Fundamental Theorem and Theorems 1–5 establish "generic" transcendence and algebraic independence of the values of our E-functions at algebraic points when these functions are connected by algebraic equations over $\mathbf{C}(z)$. However, these theorems do not tell us exactly which are the exceptional algebraic points, where the transcendence or algebraic independence does not hold.

In these theorems the exceptional points include the number 0, the zeros of the polynomial $T(z)$, and possibly some other algebraic points. The proofs of Theorem 2 and Lemma 11 show that these points are determined by the algebraic equations which relate our functions over $\mathbf{C}(z)$, and these proofs indicate how one might attempt to find them.

We shall show that in certain cases, when the algebraic equations connecting the E-functions over $\mathbf{C}(z)$ have a special structure, one can determine the set of all exceptional points. In some cases we shall give a finite set of algebraic points which contains all of the exceptional points. Theorems will then be proved about the transcendence and algebraic independence of the values at the algebraic points of these E-functions which are connected by algebraic equations over $\mathbf{C}(z)$.

In the general case, assuming that we know the basic algebraic equations connecting the E-functions over $\mathbf{C}(z)$, we are not yet able to prove theorems of this type, even in the situation when the E-functions are only connected by a single algebraic equation.

We consider the simplest case, when our functions are connected by a special type of algebraic equation.

Theorem 6. *Suppose that the set of E-functions*

$$f_1(z), \dots, f_m(z), \qquad m \geq 2, \tag{52}$$

form a solution of the system of homogeneous linear differential equations (4) (of linear differential equations (11)) and has homogeneous transcendence degree (transcendence degree) over $\mathbf{C}(z)$ *equal to* l, $1 \leq l \leq m-1$. *Suppose that* $f_1(z), \dots, f_l(z)$ *are homogeneously algebraically independent (algebraically independent) over* $\mathbf{C}(z)$, *and the functions in (52) are connected by algebraic equations (50) in which* P_ν *is a homogeneous polynomial (an arbitrary polynomial) in* ν *variables with coefficients in* $\mathbf{A}[z]$ *and with leading coefficient of the form*

$$A_\nu(z) f_\nu^{s_\nu}(z), \qquad A_\nu(z) \in \mathbf{A}[z], \qquad A_\nu(z) \not\equiv 0, \qquad s_\nu \geq 1. \tag{53}$$

Further let $\xi \in \mathbf{A}$, $\xi T(\xi) A_{l+1}(\xi) \cdots A_m(\xi) \neq 0$. *Then the numbers*

$$f_1(\xi), \dots, f_l(\xi) \tag{54}$$

are homogeneously algebraically independent (algebraically independent), and in the homogeneous case with $l > 1$ *the numbers* $f_1(\xi)/f_l(\xi), \dots, f_{l-1}(\xi)/f_l(\xi)$ *are algebraically independent.*

Proof. We suppose the contrary: that the numbers (54) are homogeneously algebraically dependent (algebraically dependent). We regard each polynomial P_ν in (50) as a polynomial in $f_\nu(z)$ with coefficients in $\mathbf{A}[z, f_1(z), \dots, f_{\nu-1}(z)]$. We set $z = \xi$ in (50). Then, by Lemma 11, for at least one value of ν, $l+1 \leq \nu \leq m$, the coefficients of all powers of $f_\nu(z)$ in P_ν vanish; in particular, $A_\nu(\xi) = 0$. But this contradicts the hypothesis of the theorem. This contradiction proves the theorem.

The proofs of the homogeneous and non-homogeneous cases of Theorem 6 are similar. The non-homogeneous case is a consequence of the homogeneous case only when all of the leading terms in the non-homogeneous case are simply powers of the corresponding f_ν.

Under the conditions of Theorem 6, if the E-functions (52) are connected by algebraic equations (51) in which P_ν is a homogeneous (arbitrary) irreducible and primitive polynomial in $l+1$ variables with coefficients in $\mathbf{C}[z]$ containing $f_\nu(z)$, and if the leading coefficients of all of the polynomials P_ν in (51) have the form (53), then, by Lemma 5, we may assume that $P_\nu \in \mathbf{A}[z, f_1(z), \dots, f_l(z), f_\nu(z)]$; hence, the equations (51) are a special case of (50).

We note that, under the conditions of Theorem 6, there might be points ξ satisfying the condition $A_{l+1}(\xi), \dots, A_m(\xi) = 0$ for which the numbers (54) are algebraically independent.

§ 8. E-functions connected by algebraic equations with constant coefficients

Theorem 7. *Suppose that the set of E-functions*

$$f_1(z), \dots, f_m(z), \tag{55}$$

$m \geq 2$ $(m \geq 1)$, *forms a solution of the system of homogeneous linear differential equations (4) (of linear differential equations (11)), and has the same positive homogeneous transcendence degree (transcendence degree) both over* $\mathbf{C}(z)$ *and over* $\mathbf{C}$. *Further suppose that*

$$P = P(z_1, \dots, z_m) \in \mathbf{A}[z_1, \dots, z_m], \qquad \deg_{\mathbf{z}} P = s, \qquad s \geq 1, \tag{56}$$

is any homogeneous (arbitrary) polynomial such that

$$P(f_1(z), \dots, f_m(z)) \not\equiv 0 \tag{57}$$

as a function of z. *Let* $\xi \in \mathbf{A}$, $\xi T(\xi) \neq 0$. *Then*

$$P(f_1(\xi), \dots, f_m(\xi)) \neq 0. \tag{58}$$

To prove this theorem we shall need the following lemma.

Lemma 12. *Let $\varphi(z)$ be a nonnegative non-decreasing function of natural numbers N which increases no faster than a polynomial in N, i.e., it satisfies the conditions*

$$\begin{aligned} \varphi(N+1) &\geq \varphi(N), \qquad n = 1, 2, \ldots, \\ 0 \leq \varphi(N) &\leq cN^l, \qquad n = 1, 2, \ldots, \end{aligned} \tag{59}$$

for some positive l and c. Then for any $\epsilon > 0$ and any $s \in \mathbf{N}$ and for infinitely many values of N one has

$$\varphi(N) - \varphi(N-s) < \epsilon\varphi(N). \tag{60}$$

Proof. We suppose the contrary: that for some s and ϵ, $0 < \epsilon < 1$, the inequality (60) holds for only finitely many N (the lemma is trivial if $\epsilon \geq 1$). Then there exists a natural number N_0 such that for all $N \geq N_0$

$$\varphi(N) - \varphi(N-s) \geq \epsilon\varphi(N),$$

i.e.,

$$\varphi(N) \geq \frac{1}{1-\epsilon}\varphi(N-s).$$

If we use this inequality $[(N-N_0)/s]$ times and take into account the first condition in (59), we obtain the inequality

$$\varphi(N) \geq \left(\frac{1}{1-\epsilon}\right)^{[(N-N_0)/s]} \varphi(N_0),$$

which shows that $\varphi(N)$ grows at least as fast as an exponential function with base greater than 1. But this contradicts the second condition in (59). This proves the lemma.

Proof of Theorem 7. Let l denote the homogeneous transcendence degree over $\mathbf{C}$ of the set of functions (55). If $l = m$, then the theorem follows from the First Fundamental Theorem. So suppose that $1 \leq l < m$. We suppose that the claim in the theorem is false, i.e., that

$$P(f_1(\xi), \ldots, f_m(\xi)) = 0. \tag{61}$$

Let $\mathbf{K}$ denote an algebraic number field containing the number ξ, the coefficients of the power series in z of all of the E-functions (55), and all of the coefficients of the polynomial P. For every $N \in \mathbf{N}$ we consider the set of $\mu^0_{N,m} = (N+m-1)!/N!(m-1)!$ monomials in the functions in (55)

$$f_m^{k_m}(z) \cdots f_1^{k_1}(z), \qquad k_i \geq 0, \qquad i = 1, \ldots, m; \;\; k_1 + \cdots + k_m = N. \tag{62}$$

We let L_N^0 denote this set. Let r_n^0 and $\bar{r}_N^0$, respectively, denote the dimension of the vector space spanned by the functions in L_N^0 over $\mathbf{C}$ and the dimension of the space spanned by L_N^0 over $\mathbf{C}(z)$. By Lemma 6, there exist positive constants c_1, c_2, c_3 and c_4 such that

$$c_1 N^{l-1} < r_N^0 < c_2 N^{l-1} \tag{63}$$

and

$$c_3 N^{l-1} < \bar{r}_N^0 < c_4 N^{l-1}. \tag{64}$$

(We note that it is possible to prove that $r_N^0 = \bar{r}_N^0$.)

Since we have $1 \leq l < m$, it follows that, starting from some value of N, the functions in L_N^0 are linearly dependent over $\mathbf{C}$. We shall consider only such values of N. For each such N we choose from among the functions in L_N^0 any $\mathbf{C}$-basis for the $\mathbf{C}$-vector space spanned by those functions. The functions in the basis will be denoted

$$F_{N,i}(z), \qquad i = 1, \ldots, r_N^0, \tag{65}$$

numbered in increasing order. We then list the elements of L_N^0 not in the basis (65), denoting them

$$F_{N,j}(z), \qquad j = r_N^0 + 1, \ldots, \mu_{N,m}^0. \tag{66}$$

Each of the functions in (66) can be uniquely represented as a linear combination of the functions (65) with complex coefficients. By Lemma 5, these coefficients are in $\mathbf{K}$. Hence, we have equations of the form

$$F_{N,j}(z) - \sum_{i=1}^{r_N^0} \gamma_{N,j,i} F_{N,i}(z) = 0, \tag{67}$$

$$\gamma_{N,j,i} \in \mathbf{K}, \qquad j = r_N^0 + 1, \ldots, \mu_{N,m}^0.$$

For $N > s$ we consider the products

$$F_{N-s,j}(z) P(f_1(z), \ldots, f_m(z)), \qquad j = 1, \ldots, r_{N-s}^0.$$

If we multiply each term in P by $F_{N-s,j}(z)$, we see that all of these products are linear forms in various of the functions (65) and (66) with coefficients in **K**:

$$F_{N-s,j}(z)P(f_1(z),\dots,f_m(z)) = \sum_{i=1}^{\mu^0_{N,m}} \beta_{N,j,i}F_{N,i}(z), \tag{68}$$

$$\beta_{N,j,i} \in \mathbf{K}, \qquad j = 1,\dots,r^0_{N-s}.$$

We prove that the left sides of (67) and the products (68), regarded as linear forms in the functions (65) and (66) with coefficients in **K**, are linearly independent.

Suppose that this is not the case. Then we have the following equations, which hold identically in the variables (65) and (66):

$$\begin{aligned}\sum_{j=1}^{r^0_{N-s}} \alpha_j F_{N-s,j}(z)P(f_1(z),\dots,f_m(z)) &= \sum_{j=1}^{r^0_{N-s}} \alpha_j \sum_{i=1}^{\mu^0_{N,m}} \beta_{N,j,i}F_{N,i}(z) = \\ &= \sum_{j=r^0_N+1}^{\mu^0_{N,m}} \alpha_j \left(F_{N,j}(z) - \sum_{i=1}^{r^0_N} \gamma_{N,j,i}F_{N,i}(z) \right), \qquad \alpha_j \in \mathbf{K},\end{aligned} \tag{69}$$

where at least one of the α_j is nonzero.

Since the equations (67) are linearly independent over **K**, it follows from (69) that at least one of the numbers α_j, $j = 1,\dots,r^0_{N-s}$, is nonzero. From (67) and (69) we find that

$$\sum_{j=1}^{r^0_{N-s}} \alpha_j F_{N-s,j}(z)P(f_1(z),\dots,f_m(z)) = 0. \tag{70}$$

Since the condition (57) holds, it follows from (70) that we have

$$\sum_{j=1}^{r^0_{N-s}} \alpha_j F_{N-s,j}(z) = 0.$$

But this is a contradiction, since the functions $F_{N-s,j}(z)$, $j = 1,\dots,r^0_{N-s}$, are linearly independent over **C**, and at least one of the numbers α_j, $j = 1,\dots,r^0_{N-s}$, is nonzero. This contradiction shows that the linear forms (67) and (68) are linearly independent.

If we set $z = \xi$ in the equations (67) and the linear forms (68) and use (61), we obtain $\mu^0_{N,m} - r^0_N + r^0_{N-s}$ linear, homogeneous — and, by what was proved above, linearly independent — equations for the numbers $F_{N,i}(\xi)$, $i = 1, \dots, \mu^0_{N,m}$, with coefficients in **K**. Hence, letting ρ^0_N denote the **K**-rank of this set of numbers, we have

$$\rho^0_N \le r^0_N - r^0_{N-s}. \tag{71}$$

The integer r^0_N satisfies (63) as a function of N. Hence, Lemma 12 can be applied, according to which, for any $\epsilon > 0$ and for infinitely many values of N, we have the inequality

$$r^0_N - r^0_{N-s} < \epsilon r^0_N. \tag{72}$$

By Lemma 18 of Chapter 3, the set of functions (62) which make up the set L^0_N forms a solution of a system of homogeneous linear differential equations of the form (4) with m replaced by $\mu^0_{N,m}$. Thus, this set of functions satisfies all of the conditions of Lemma 3 with m replaced by $\mu^0_{N,m}$. According to that lemma,

$$\rho^0_N \ge \bar{r}^0_N / h, \qquad h = [\mathbf{K} : \mathbf{Q}]. \tag{73}$$

From (63), (64), (71), (72) and (73) we find that the following inequalities hold for infinitely many values of N:

$$0 < \frac{c_3}{h} N^{l-1} < \frac{\bar{r}^0_N}{h} \le \rho^0_N < \epsilon c_2 N^{l-1},$$

from which it follows that

$$0 < \frac{c_3}{h} < \epsilon c_2.$$

But these inequalities are impossible, since ϵ is arbitrary. This contradiction completes the proof of Theorem 7.

From Theorem 7 we obtain a series of corollaries about the arithmetic properties of the values of a set of E-functions which are connected by algebraic equations with constant coefficients.

Theorem 8. *Suppose that the E-functions (55), $m \ge 2$ ($m \ge 1$), form a solution of the system of homogeneous linear differential equations (4) (of linear differential equations (11)) and have homogeneous transcendence degree (transcendence degree) l, $1 \le l \le m$, both over $\mathbf{C}(z)$ and over $\mathbf{C}$. Suppose that the functions $f_1(z), \dots, f_l(z)$ are homogeneously algebraically independent (algebraically independent) over $\mathbf{C}(z)$. Further let $\xi \in \mathbf{A}$, $\xi T(\xi) \ne 0$. Then the numbers*

$f_1(\xi), \dots, f_l(\xi)$ are homogeneously algebraically independent (algebraically independent), and in the homogeneous case with $l > 1$ the numbers $f_1(\xi)/f_l(\xi), \dots$ $\dots, f_{l-1}(\xi)/f_l(\xi)$ are algebraically independent.

The method used to prove Theorem 7, and hence also Theorem 8, is not suitable for proving a similar theorem about algebraic independence of the values of E-functions (76) which are connected by arbitrary algebraic equations with coefficients in $\mathbf{C}(z)$. In that case the proof begins in the same way, but with $\mathbf{C}$ replaced by $\mathbf{C}(z)$. Here one can use Lemma 3 to choose the functions $F_{N,i}(z)$, $i = 1, \dots, r_N^0$, so that all of the $\gamma_{N,j,i}$ in (67) are functions in $\mathbf{K}(z)$ which have no poles other than the poles of the system of differential equations (4). Similarly, from (67) and (68) we obtain $\mu_{N,m}^0 - r_N^0 - r_{N-s}^0$ linearly independent linear forms in the $F_{N,i}(z)$, $i = 1, \dots, \mu_{N,m}^0$, with coefficients in $\mathbf{K}(z)$. However, when we set $z = \xi$ in them we can no longer claim that the resulting linear equations for $F_{N,i}(\xi)$, $i = 1, \dots, \mu_{N,m}^0$, with coefficients in $\mathbf{K}$, are linearly independent.

Theorem 7 immediately implies

Theorem 9. *Suppose that the E-functions (55), $m \geq 2$ ($m \geq 1$), form a solution of the system of homogeneous linear differential equations (4) (of linear differential equations (11)) and have homogeneous transcendence degree (transcendence degree) l, $1 \leq l \leq m$, both over $\mathbf{C}(z)$ and over $\mathbf{C}$. Let $\xi \in \mathbf{A}$, $\xi T(\xi) \neq 0$. Then:*

1) if the functions (55) are linearly independent (linearly independent along with the function $f(z) \equiv 1$) over $\mathbf{C}(z)$, then the numbers

$$f_1(\xi), \dots, f_m(\xi) \tag{74}$$

are linearly independent (linearly independent along with the number 1) over $\mathbf{A}$, each of the numbers in (74) is nonzero (is transcendental), and all of the zeros (algebraic A-points) of all of the functions in (55) which are not 0 or zeros of $T(z)$, are transcendental;

2) if the functions (55) are not connected by a homogeneous algebraic equation (an arbitrary algebraic equation) over $\mathbf{C}(z)$ of degree k, $k > 1$, then the numbers in (74) are not connected by a homogeneous algebraic equation (an arbitrary algebraic equation) of degree k with algebraic coefficients.

Theorem 9 contains the following theorem (as well as its homogeneous analog, which we omit).

Theorem 10. *Suppose that the E-function $f(z)$ satisfies the linear differential equation*

$$P_m(z)y^{(m)} + \dots + P_1(z)y' + P_0(z)y = Q(z), \qquad m \geq 1,$$

with coefficients in $\mathbf{C}[z]$ and does not satisfy a differential equation of the same type of order lower than m (if $m = 1$, it is not a rational function). Suppose *that the transcendence degrees of the set of functions $f(z), f'(z), \ldots, f^{(m-1)}(z)$ over $\mathbf{C}(z)$ and over $\mathbf{C}$ are the same. Let $\xi \in \mathbf{A}$, $\xi T(\xi) \neq 0$. Then the numbers $1, f(\xi), f'(\xi), \ldots, f^{(m-1)}(\xi)$ are linearly independent over $\mathbf{A}$, and all of the numbers $f^{(i)}(\xi)$, $i = 0, 1, \ldots$, are transcendental.*

It is natural to conjecture that Theorem 10 remains true without the assumption that the transcendence degrees of $f(z), f'(z), \ldots, f^{(m-1)}(z)$ over $\mathbf{C}(z)$ and over $\mathbf{C}$ are the same.

The simplest example of Theorem 9 is the set of functions $y_1 = \sin z$ and $y_2 = \cos z$. They form a solution of the differential equations $y_1' = y_2, y_2' = -y_1$, and they are connected by the equation $y_1^2 + y_2^2 = 1$. Theorem 9 then tells us that for any nonzero $\xi \in \mathbf{A}$ each of the numbers $\sin \xi$ and $\cos \xi$ is transcendental.

Suppose that the set of KE-functions (55) satisfies all of the conditions in Theorem 7. We consider the vector space $\mathbf{L}_N^0$ spanned over $\mathbf{C}(z)$ by the elements (62) of the set L_N^0. Then $\dim \mathbf{L}_N^0 = r_N^0$. Let $\xi \in \mathbf{A}$, $\xi T(\xi) \neq 0$. We let $L_{N,\xi}^0$ denote the set obtained from the elements of L_N^0 after substituting $z = \xi$, and we consider the vector space $\mathbf{L}_{N,\xi}^0$ spanned over $\mathbf{A}$ by the elements of $L_{N,\xi}^0$. We set $\rho_N^0 = \dim \mathbf{L}_{N,\xi}^0$. From Theorem 7 it obviously follows that $r_N^0 = \rho_N^0$ for any N. Moreover, any $\mathbf{C}(z)$-basis of $\mathbf{L}_N^0$ consisting of monomials in (62) becomes a basis of $\mathbf{L}_{N,\xi}^0$ when one substitutes $z = \xi$.

Similar statements hold under the conditions of the First Fundamental Theorem, when the set of KE-functions in (55) are algebraically independent over $\mathbf{C}(z)$, and $\xi \in \mathbf{A}$ with $\xi T(\xi) \neq 0$, since the theorem implies that the set of monomials in (62) with ξ substituted for z is homogeneously linearly independent over $\mathbf{A}$ for any N.

If the functions (55) are IE-functions which are connected by algebraic equations over $\mathbf{C}(z)$, and if $\xi \in \mathbf{I}$ with $\xi T(\xi) \neq 0$, then we have $r_N^0 = \rho_N^0$, where ρ_N^0 is the dimension over $\mathbf{I}$ of the vector space $\mathbf{L}_{N,\xi}^0$. This follows from Lemma 3. But in this case one cannot say that there is a correspondence between the bases of the vector spaces $\mathbf{L}_N^0$ and $\mathbf{L}_{N,\xi}^0$.

In the three cases discussed above, one has analogous statements concerning equality of the dimensions r_N and ρ_N and correspondence between the bases of the vector space spanned by the functions (62) of L_N for which the exponents satisfy the inequality $k_1 + \cdots + k_m \leq N$, and the vector space spanned by their values at the point $z = \xi$.

It is a very difficult problem to either prove or find a counterexample to the equality $r_N^0 = \rho_N^0$ ($r_N = \rho_N$), where ρ_N^0 and ρ_N are the dimensions of the vector spaces $\mathbf{L}_{N,\xi}^0$ and $\mathbf{L}_{N,\xi}$ over $\mathbf{A}$, or to the corresponding claim about correspondence of vector space bases, in the general case of KE-functions (55) having arbitrary transcendence degree over $\mathbf{C}(z)$, where we take $\xi \in \mathbf{A}$ with $\xi T(\xi) \neq 0$ and perhaps exclude a finite number of such ξ.

Here is an example which shows when one might expect to have $\rho_N^0 < r_N^0$.

Suppose that the KE-functions (55) have homogeneous transcendence degree $m-1$ over $\mathbf{C}(z)$, and are connected by the algebraic equation

$$P(z, f_1(z), \dots, f_m(z)) = 0,$$

$$P = P(z, z_1, \dots, z_m) \in \mathbf{K}[z, z_1, \dots, z_m],$$

where P is an irreducible polynomial in $m+1$ variables which is homogeneous in $z_1, \dots, z_m$. If $\xi \in \mathbf{A}$, $\xi T(\xi) \neq 0$, and the polynomial $P(\xi, z_1, \dots, z_m)$ is reducible over $\mathbf{A}$, then we have $r_N^0 > \rho_N^0$. But if $P(\xi, z_1, \dots, z_m)$ is irreducible over $\mathbf{A}$, then $r_N^0 = \rho_N^0$.

§ 9. E-functions which are connected by a single algebraic equation over C(z)

Suppose that the KE-functions

$$f_1(z), \dots, f_m(z), \qquad m \geq 2, \tag{75}$$

have homogeneous transcendence degree (transcendence degree) over $\mathbf{C}(z)$ equal to l, $1 \leq l < m$, and the functions

$$f_1(z), \dots, f_l(z) \tag{76}$$

are homogeneously algebraically independent (algebraically independent) over $\mathbf{C}(z)$.

Because of Lemma 5, we may assume that the functions in (75) are connected by algebraic equations

$$P_\nu = P_\nu(f_1(z), \dots, f_l(z), f_\nu(z)) = 0, \qquad \nu = l+1, \dots, m, \tag{77}$$

where P_ν is a homogeneous (an arbitrary) irreducible and primitive polynomial in $l+1$ variables with coefficients in $\mathbf{K}[z]$, and it contains $f_\nu(z)$.

We let $\Omega_\nu \subset \mathbf{A}$ denote the set of common zeros of all of the polynomials in $\mathbf{K}[z]$ which are coefficients in the terms of P_ν containing $f_\nu(z)$. If $\Omega_\nu \neq \emptyset$, then for $\xi \in \Omega_\nu$ we set $z = \xi$ in (77). Then the coefficients vanish in all of the terms of P_ν containing $f_\nu(z)$. But, since P_ν is a primitive polynomial, it follows that at least one of the coefficients of its terms which do not contain $f_\nu(z)$ must be nonzero at $z = \xi$. Hence, the equation $P_\nu = 0$ with $z = \xi$ implies that the numbers $f_1(\xi), \dots, f_l(\xi)$ are homogeneously algebraically independent (algebraically independent).

An example of this situation is when P_ν is a linear form in $f_1(z), \dots, f_l(z)$, $f_\nu(z)$, and the set Ω_ν consists of the zeros of the coefficient of $f_\nu(z)$.

We set

$$\Omega = \bigcup_{\nu=l+1}^{m} \Omega_\nu.$$

The above argument shows that, if we want to prove that the values of the E-functions (76) at points in **A** are homogeneously algebraically independent (algebraically independent), the exceptional points must include the set Ω. Here we note that the set of exceptional points may also include other $\xi \in \mathbf{A}$ for which $\xi T(\xi) \neq 0$.

Under the conditions of Theorem 6, if $\Omega \neq \emptyset$, then it consists of either all of the zeros or only some of the zeros of the polynomials $A_\nu(z)$, $\nu = l+1, \dots, m$. Under the conditions of Theorem 8, we have $\Omega = \emptyset$, and the set of exceptional points does not contain any $\xi \in \mathbf{A}$ for which $\xi T(\xi) \neq 0$.

A difficult and interesting problem is the following: under the above conditions, is it possible for the set of exceptional points for the functions (76) to contain $\xi \in \mathbf{A}$ such that $\xi T(\xi) \neq 0$ and $\xi \notin \Omega$? In general, this question has not been answered even in the case $l = m - 1$. We can prove only the following theorem.

Theorem 11. *Suppose that the IE-functions (75) form a solution of the system of homogeneous linear differential equations (4) (of linear differential equations (11)). Suppose that the functions*

$$f_1(z), \dots, f_{m-1}(z) \tag{78}$$

are homogeneously algebraically independent (algebraically independent) over $\mathbf{C}(z)$, *and the functions in (75) are connected by an algebraic equation*

$$P = P(f_1(z), \dots, f_m(z)) = 0, \tag{79}$$

in which P is a homogeneous (an arbitrary) irreducible and primitive polynomial in the variables (75) with coefficients in $\mathbf{C}[z]$. *Let $\xi \in \mathbf{I}$, $\xi T(\xi) \neq 0$, and $\xi \notin \Omega$. Then the numbers $f_1(\xi), \dots, f_{m-1}(\xi)$ are homogeneously algebraically independent (algebraically independent), and in the homogeneous case for $m > 2$ the numbers $f_1(\xi)/f_{m-1}(\xi), \dots, f_{m-2}(\xi)/f_{m-1}(\xi)$ are algebraically independent.*

Proof. As usual, we shall consider the homogeneous case. By Lemma 5, we may assume that $P \in \mathbf{I}[z, f_1(z), \dots, f_m(z)]$. By assumption, P contains $f_m(z)$. If P does not contain the functions (78), then $f_m(z) \equiv 0$, and the theorem follows from the First Fundamental Theorem.

Now suppose that P contains at least one of the functions (78), and that s is the total degree of P in the variables (75). Suppose that the theorem is false, i.e., that $\xi \in \mathbf{I}$, $\xi T(\xi) \neq 0$, $\xi \notin \Omega$, and the numbers $f_1(\xi), \ldots, f_{m-1}(\xi)$ are homogeneously algebraically dependent. Then, by Lemma 11, if we regard P as a polynomial in $f_m(z)$ with coefficients in $\mathbf{I}[z, f_1(z), \ldots, f_{m-1}(z)]$, all of the coefficients of the different powers of $f_m(z)$ vanish at $z = \xi$. By assumption, among these coefficients there is a polynomial

$$A = A[z, f_1(z), \ldots, f_{m-1}(z)] \in \mathbf{I}[z, f_1(z), \ldots, f_{m-1}(z)], \qquad A \not\equiv 0,$$

of degree k in the variables (78), where $1 \leq k < s$, which is not divisible by $z - \xi$. We have

$$A(\xi, f_1(\xi), \ldots, f_{m-1}(\xi)) = 0. \tag{80}$$

We prove that the equation (80) is impossible. In fact, by the hypothesis of the theorem, the set of monomials (62) with $N = k$ is linearly independent over $\mathbf{C}(z)$, since $k < s$. According to Lemma 18 of Chapter 3, this set of functions satisfies a system of homogeneous linear differential equations of the form (4) with m replaced by $\mu^0_{k,m}$, in which there are no singular points other than the singular points of the system (4). If we apply Lemma 3 to the set of IE-functions (62) with $N = k$, we find that the set of numbers

$$f_m^{k_m}(\xi) \cdots f_1^{k_1}(\xi), \qquad k_i \geq 0, \qquad i = 1, \ldots, m, \qquad k_1 + \cdots + k_m = k,$$

is linearly independent over $\mathbf{I}$. This means that (80) is impossible, and the theorem is proved.

Note that in Theorems 8 and 11 the set of exceptional points was specified exactly. It would be interesting to determine whether or not Theorem 11 generalizes to the case of a set of KE-functions (75), or to the case of a set of IE-functions or KE-functions (75) which are connected by an arbitrary number of algebraic equations (77) over $\mathbf{C}(z)$.

In the case of an arbitrary set of E-functions which are connected by a single algebraic equation over $\mathbf{C}(z)$, at present we can prove only the following result.

Theorem 12. *Suppose that the E-functions (72) form a solution of a system of homogeneous linear differential equations (4) (of linear differential equations (11)) and have homogeneous transcendence degree (transcendence degree) $m - 1$ over $\mathbf{C}(z)$. Suppose that the functions (75) are connected by an algebraic equation (79), in which P is a homogeneous (an arbitrary) irreducible and primitive polynomial in the variables (75) with coefficients in $\mathbf{C}[z]$, and its leading term has the form*

$$A(z) f_m^{\kappa_m}(z) \cdots f_{t+1}^{\kappa_{t+1}}(z), \qquad A(z) \in \mathbf{C}[z], \qquad A(z) \not\equiv 0, \tag{81}$$

where $1 \le t \le m-1$. *Let* $\xi \in \mathbf{A}$, $\xi T(\xi) \ne 0$. *Then the numbers*

$$f_1(\xi), \dots, f_t(\xi) \tag{82}$$

are homogeneously algebraically independent (algebraically independent), and in the homogeneous case for $t > 1$ *the numbers* $f_1(\xi)/f_t(\xi), \dots, f_{t-1}(\xi)/f_t(\xi)$ *are algebraically independent.*

Proof. We first consider the homogeneous case. Suppose that the theorem is false, i.e., that the numbers (82) are homogeneously algebraically dependent. Then they are connected by an algebraic equation

$$P_0(f_1(\xi), \dots, f_t(\xi)) = 0, \tag{83}$$

$$P_0 = P_0(z_1, \dots, z_t) \in \mathbf{A}[z_1, \dots, z_t], \qquad \deg_{\mathbf{z}} P_0 = k, \qquad k \ge 1,$$

where P_0 is an irreducible homogeneous polynomial.

Let $\mathbf{K}$ be an algebraic number field containing the number ξ, the coefficients of P_0, and the coefficients of the power series in z of all of the E-functions (75). By Lemma 5, we may suppose that all of the numerical coefficients of P in (79) are in $\mathbf{K}$.

For every $N \in \mathbf{N}$ we consider the set L_N^0 consisting of the $\mu_{N,m}^0 = (N + m - 1)!/N!(m-1)!$ monomials (62) in the functions (75). As above, we take the set L_N^0, its various subsets, and our polynomials in the functions (75) all to be written in lexicographical order with respect to powers of $f_m(z), \dots, f_1(z)$.

We compute the $\mathbf{C}(z)$-rank r_N^0 of the set L_N^0 in (62). Let $s = \deg P$ be the total degree in the variables (75). By (79), we can write $\mu_{N-s,m}^0$ homogeneous linear equations

$$f_m^{k_m}(z) \cdots f_1^{k_1}(z) P(f_m(z), \dots, f_1(z)) = 0, \tag{84}$$

$$k_1 + \dots + k_m = N - s, \qquad k_i \ge 0, \qquad i = 1, \dots, m,$$

connecting the functions in L_N^0 over $\mathbf{K}[z]$. In fact, if we multiply the terms of P by

$$f_m^{k_m}(z) \cdots f_1^{k_1}(z) \tag{85}$$

term-by-term, the left side of each of the equations in (84) will contain a linear form in the elements of L_N^0 with coefficients in $\mathbf{K}[z]$. On the other hand, if we regard the left sides of (84) as polynomials in the functions (75), we note that the leading term in each of these polynomials is the result of multiplying the product in (85) by the leading term of P. Consequently, the leading terms of all of the polynomials on the left in (84) are distinct, and so the linear equations (84) in the functions of L_N^0 (see (62)) are linearly independent over $\mathbf{C}(z)$.

Under the conditions of the theorem, the left side of any algebraic equation which connects the functions (75) over $\mathbf{C}(z)$ must be divisible by P as polynomials in m independent variables. Hence, any linear equation connecting the functions (62) over $\mathbf{C}(z)$ is a linear combination of the equations (84) with coefficients in $\mathbf{C}(z)$. This implies that for $N \geq s$ the $\mathbf{C}(z)$-rank of the set L_N^0 in (62) is equal to

$$r_N^0 = \mu_{N,m}^0 - \mu_{N-s,m}^0 = \frac{s}{(m-2)!} N^{m-2} + O(N^{m-3}). \tag{86}$$

We now estimate ρ_N^0, which denotes the $\mathbf{K}$-rank of the set of values of the functions (62) at the point $z = \xi$.

We divide the set L_N^0 into two subsets A_N and B_N. The set A_N consists of the elements of L_N^0 which are divisible by

$$f_m^{\kappa_m}(z) \cdots f_{t+1}^{\kappa_{t+1}}(z),$$

i.e., all monomials of the form

$$f_m^{k_m+\kappa_m}(z) \cdots f_{t+1}^{k_{t+1}+\kappa_{t+1}}(z) f_t^{k_t}(z) \cdots f_1^{k_1}(z), \qquad k_1 + \cdots + k_m = N - k.$$

The set B_N consists of the remaining elements of L_N^0. The set A_N has $\mu_{N-s,m}^0$ elements; hence B_N has $\mu_{N,m}^0 - \mu_{N-s,m}^0 = r_N^0$ elements.

We consider the set L_{N-k}^0, and we let $F_i(z)$, $i = 1, \ldots, r_{N-k}^0$, be all of the elements of B_{N-k} written in increasing order. Then the expressions

$$F_i(z) P_0(f_t(z), \ldots, f_1(z)), \qquad i = 1, \ldots, r_{N-k}^0, \tag{87}$$

are linear forms in the elements of B_N with coefficients in $\mathbf{K}$, since, by the definition of B_N, the product of any of the $F_i(z)$ by any monomial involving only $f_t(z), \ldots, f_1(z)$ with sum of the powers equal to k, is an element of B_N.

In (84) and (87) we set $z = \xi$. By (83), we obtain $\mu_{N-s,m}^0 + r_{N-k}^0$ homogeneous linear equations connecting the values at $z = \xi$ of the functions (62) over $\mathbf{K}$. These equations are linearly independent, since the leading terms on the left have different orders, because the condition $A(\xi) \neq 0$ ensures that the coefficients of the leading terms in (84) do not vanish at $z = \xi$. Consequently,

$$\rho_N^0 \leq \mu_{N,m}^0 - \mu_{N-s,m}^0 - r_{N-k}^0 = r_N^0 - r_{N-k}^0.$$

Using (86), we see that the last inequality gives

$$\rho_N^0 = O(N^{m-3}). \tag{88}$$

By Lemma 18 of Chapter 3, the E-functions in (62) form a solution of a system of homogeneous linear differential equations of the form (4) with m replaced by $\mu^0_{N,m}$, in which there are no singular points other than those of the system (4). Hence, Lemma 3 can be applied to these E-functions. According to Lemma 3, if we again use (86), we obtain the inequality

$$\rho^0_N \geq \frac{r^0_N}{h} = \frac{s}{h(m-2)!} N^{m-2} + O(N^{m-3}), \tag{89}$$

where h is the degree of the field **K**.

The estimates (88) and (89) are contradictory for N sufficiently large. This contradiction proves the theorem.

The above proof also goes through in the non-homogeneous case, if instead of the set L^0_N in (62) we consider the analogous set L_N of monomials in which the sum of the exponents satisfy the inequality $k_1 + \cdots + k_m \leq N$. The non-homogeneous case is a consequence of the homogeneous case only when the degree of the leading term in P is equal to the degree of P in the variables (75).

§ 10. Minimal equations

Let **V** be a field, let **W** be an extension field, and let

$$u_m, \ldots, u_1 \in \mathbf{W}, \qquad m \geq 1. \tag{90}$$

We consider an arbitrary nonempty set D of monomials in the elements (90)

$$u_m^{k_m} \cdots u_1^{k_1}, \qquad k_i \in \mathbf{Z}^+, \qquad i = 1, \ldots, m. \tag{91}$$

As in § 3, we write the elements of any such set in lexicographical order with respect to powers of $u_m, \ldots, u_1$.

An element $u_m^{\kappa_m} \cdots u_1^{\kappa_1}$ in D is said to be *minimal* if for any other element $u_m^{k_m} \cdots u_1^{k_1}$ in D there exists an i, $1 \leq i \leq m$, such that $\kappa_i < k_i$.

Lemma 13. *In any set D of monomials (91), the subset D_1 of minimal elements is finite and nonempty.*

Proof. Clearly, an element of D is a minimal element if its total degree $k_1 + \cdots + k_m$ is minimal. Now suppose that the lemma is false, and that D is an infinite set.

Let $u_m^{\bar{\kappa}_m} \cdots u_1^{\bar{\kappa}_1}$ be any minimal element such that the sum $\kappa = \bar{\kappa}_1 + \cdots + \bar{\kappa}_m$ is minimal. Then, by definition, any other minimal element $u_m^{\kappa_m} \cdots u_1^{\kappa_1}$ must have an exponent κ_i, $1 \le i \le m$, satisfying the inequalities

$$\kappa_i < \bar{\kappa}_i \le \kappa.$$

Hence, the set D_1 contains an infinite subset of minimal elements

$$u_m^{\kappa_{m,1}} \cdots u_1^{\kappa_{1,1}}, \; u_m^{\kappa_{m,2}} \cdots u_1^{\kappa_{1,2}}, \; \ldots, \; u_m^{\kappa_{m,n}} \cdots u_1^{\kappa_{1,n}}, \; \ldots,$$

whose exponents for some fixed value of i, $1 \le i \le m$, satisfy the inequalities

$$\kappa_{i,1} < \kappa, \quad \kappa_{i,2} < \kappa, \quad \ldots, \quad \kappa_{i,n} < \kappa, \quad \ldots.$$

This means that each of the exponents $\kappa_{i,1}, \kappa_{i,2}, \ldots, \kappa_{i,n}, \ldots$ is equal to one of the numbers $0, 1, \ldots, \kappa - 1$.

Consequently, there exists an infinite subset D_2 of the set of minimal elements

$$u_m^{\kappa^*_{m,1}} \cdots u_1^{\kappa^*_{1,1}}, \; u_m^{\kappa^*_{m,2}} \cdots u_1^{\kappa^*_{1,2}}, \; \ldots, \; u_m^{\kappa^*_{m,n}} \cdots u_1^{\kappa^*_{1,n}}, \; \ldots,$$

for which

$$\kappa^*_{i,1} = \kappa^*_{i,2} = \cdots = \kappa^*_{i,n} = \cdots.$$

If we repeat the above argument starting with the set D_2, we obtain an infinite subset D_3 of minimal elements

$$u_m^{\kappa^{**}_{m,1}} \cdots u_1^{\kappa^{**}_{1,1}}, \; u_m^{\kappa^{**}_{m,2}} \cdots u_1^{\kappa^{**}_{1,2}}, \; \ldots, \; u_m^{\kappa^{**}_{m,n}} \cdots u_1^{\kappa^{**}_{1,n}}, \; \ldots,$$

for which

$$\kappa^{**}_{i,1} = \kappa^{**}_{i,2} = \cdots = \kappa^{**}_{i,n} = \cdots, \qquad \kappa^{**}_{j,1} = \kappa^{**}_{j,2} = \cdots = \kappa^{**}_{j,n} = \cdots, \qquad i \ne j.$$

If we repeat the argument $m - 1$ times, we obtain an infinite subset D_{m-1} of minimal elements such that for some value of i, $1 \le i \le m$, any two elements $u_m^{\kappa_m} \cdots u_1^{\kappa_1}$ and $u_m^{\kappa^*_m} \cdots u_1^{\kappa^*_1}$ satisfy the condition:

$$\kappa_j = \kappa^*_j, \qquad j = 1, \ldots, m, \qquad j \ne i.$$

But this is impossible, since there can be only one minimal element in D_{m-1} with smallest exponent κ_i. This contradiction proves the lemma.

Lemma 14. *For any element $u_m^{k_m} \cdots u_1^{k_1}$ in D (see (91)), the subset D_1 of minimal elements contains an element $u_m^{\kappa_m} \cdots u_1^{\kappa_1}$ satisfying the condition*

$$k_i \ge \kappa_i, \qquad i = 1, \ldots, m.$$

Proof. Let $u_m^{k_m} \cdots u_1^{k_1}$ be a fixed element of D. We consider the subset D_0 of elements $u_m^{\bar{k}_m} \cdots u_1^{\bar{k}_1}$ in D for which $\bar{k}_i \leq k_i$, $i = 1, \ldots, m$.

We claim that every element $u_m^{\bar{k}_m} \cdots u_1^{\bar{k}_1} \in D_0$ for which $\bar{k}_1 + \cdots + \bar{k}_m$ is minimal, is a minimal element of D. In fact, if such an element were not a minimal element, then, by definition, the set D would contain an element $u_m^{k_m^*} \cdots u_1^{k_1^*}$ such that

$$k_i^* \leq \bar{k}_i, \qquad i = 1, \ldots, m.$$

Here for at least one value of i, $1 \leq i \leq m$, we have $k_i^* < \bar{k}_i$, since otherwise the element $u_m^{\bar{k}_m} \cdots u_1^{\bar{k}_1}$ would be minimal. Consequently,

$$k_1^* + \cdots + k_m^* < \bar{k}_1 + \cdots + \bar{k}_m. \tag{92}$$

However, $\bar{k}_i \leq k_i$, $i = 1, \ldots, m$. Hence, $u_m^{k_m^*} \cdots u_1^{k_1^*} \in D_0$, and then (92) contradicts our choice of the element $u_m^{\bar{k}_m} \cdots u_1^{\bar{k}_1}$.

Now the minimal element $u_m^{\bar{k}_m} \cdots u_1^{\bar{k}_1} \in D_0$ satisfies the lemma for the element $u_m^{k_m} \cdots u_1^{k_1}$.

Suppose that the elements in (90) are homogeneously algebraically dependent (algebraically dependent) over $\mathbf{V}$. We consider the set of all homogeneous (all arbitrary) algebraic equations connecting the elements (90) over $\mathbf{V}$. As in § 3, we agree to write the left sides of these equations in lexicographical order with respect to powers of the elements in (90), and to take the coefficient of the leading term to be 1.

Now let D be the set consisting of the leading terms of all of these equations. By Lemma 13, D contains a nonempty and finite subset D_1 consisting of all of the minimal elements

$$u_m^{\kappa_{m,1}} \cdots u_1^{\kappa_{1,1}}, \ \ldots, \ u_m^{\kappa_{m,\nu}} \cdots u_1^{\kappa_{1,\nu}}, \tag{93}$$

and, by Lemma 14, the exponents of any element $u_m^{k_m} \cdots u_1^{k_1} \in D$ satisfy the following inequalities for some value of j, $1 \leq j \leq \nu$:

$$k_i \geq \kappa_{i,j}, \qquad i = 1, \ldots, m. \tag{94}$$

We now consider any set of homogeneous (of arbitrary) algebraic equations between the elements (90) with coefficients in $\mathbf{V}$, such that each of the minimal elements in (93) occurs as a leading term in exactly one equation; this set of equations will be denoted

$$G_1 = 0, \ \cdots, \ G_\nu = 0, \qquad \deg_{\mathbf{u}} G_j = s_j, \tag{95}$$

where the equations are listed in decreasing order of their leading terms.

If $\deg \operatorname{tr}^0_{\mathbf{V}}\{u_m, \ldots, u_1\} = m - 1$ ($\deg \operatorname{tr}_{\mathbf{V}}\{u_m, \ldots, u_1\} = m - 1$), then the set (95) consists of a single irreducible equation. But if $\deg \operatorname{tr}^0_{\mathbf{V}}\{u_m, \ldots, u_1\} < m-1$ ($\deg \operatorname{tr}_{\mathbf{V}}\{u_m, \ldots, u_1\} < m - 1$), then this set is not the only set of equations satisfying the above conditions. In fact, let $G_j = 0$ be any of the equations in (95). We multiply the equations $G_{j+1} = 0, \ldots, G_\nu = 0$, respectively, by elements $a_{j+1}, \ldots, a_\nu \in \mathbf{V}$, add to the equation $G_j = 0$, and replace the equation $G_j = 0$ by this new equation. If we do this for some or all of the values of j, we obtain a new set of equations of the same type as (95).

We now show that, given the elements in (90), if we specify an additional condition, then there is a single uniquely determined set of equations among all sets of equations of the type (95).

We consider the set L^0_N (see (19)) (the set L_N in (20)) of monomials in the elements (90). We divide this set into two subsets A_N and B_N.

We shall carry out the argument only in the homogeneous case. The non-homogeneous case reduces to the homogeneous case if we replace m by $m+1$ and set $u_0 = 1$. Then the elements $u_m^{k_m} \cdots u_1^{k_1} \in L_N$ will take the form

$$u_m^{k_m} \cdots u_1^{k_1} = u_m^{k_m} \cdots u_1^{k_1} u_0^{k_0}, \qquad k_0 = N - (k_1 + \cdots + k_m).$$

For the minimal elements we also denote $\kappa_{0,j} = s_j - (\kappa_{1,j} + \cdots + \kappa_{m,j})$, $j = 1, \ldots, m$.

A monomial $u_m^{k_m} \cdots u_1^{k_1}$ belongs to B_N if for each j, $j = 1, \ldots, \nu$, its exponents include a value k_i such that

$$k_i < \kappa_{i,j}, \qquad 1 \le i \le m. \tag{96}$$

Otherwise, the monomial belongs to the set A_N.

If $A_N \ne \emptyset$, then for some j, $1 \le j \le \nu$, the exponents of a monomial $u_m^{k_m} \cdots u_1^{k_1}$ in A_N satisfy the inequalities (94).

We introduce the following notation for the minimal elements (93):

$$N^* = \min_{1 \le j \le \nu} \sum_{i=1}^{m} \kappa_{i,j}. \tag{97}$$

Then $A_N = \emptyset$ if $N < N^*$, and $A_N \ne \emptyset$ if $N \ge N^*$.

We note that for any $N \in \mathbf{N}$ the elements of B_N are linearly independent over $\mathbf{V}$. In fact, otherwise, by the definition of D, the leading term $u_m^{k_m} \cdots u_1^{k_1}$ of any linear equation which connects the elements of B_N over $\mathbf{V}$ must belong to D. Hence, by Lemma 14, for some value of j, $1 \le j \le \nu$, the exponents in this leading term must satisfy (94), and hence the monomial belongs to A_N, which is impossible.

Lemma 15. *If $N \geq N^*$, then every element of A_N is a linear combination of elements of B_N with coefficients in* **V**.

Proof. By (94), any element $u_m^{k_m} \cdots u_1^{k_1}$ in A_N can be written in the following form for some value of j, $1 \leq j \leq \nu$:

$$u_m^{k_m} \cdots u_1^{k_1} = u_m^{\kappa_{m,j}} \cdots u_1^{\kappa_{1,j}} u_m^{k_m^*} \cdots u_1^{k_1^*}, \qquad k_i^* \in \mathbf{Z}^+, \qquad i = 1, \ldots, m.$$

Hence, using the equation $G_j = 0$ in (95), we can express this element as a linear combination of lower order elements of L_N^0 (of L_N) with coefficients in **V**. If the right side of this representation contains elements of A_N, then we apply the above process to their highest order term. Repeating this procedure a finite number of times, we obtain the desired linear combination.

Lemma 16. *Given elements (90) which are homogeneously algebraically dependent (algebraically dependent) over* **V**, *there exists a unique set of homogeneous (of arbitrary) algebraic equations (95) with coefficients in* **V** *such that its set of leading terms coincides with the set of minimal elements (93), and every term $u_m^{k_m} \cdots u_1^{k_1}$ which is not a leading term has an exponent k_i satisfying (96) for any value of j, $j = 1, \ldots, \nu$.*

Proof. We consider any set of homogeneous (arbitrary) algebraic equations (95) such that each of the minimal elements in (93) occurs as the leading term inr exactly one equation. Using Lemma 15, in each equation $G_j = 0$ we replace all of the terms in the set A_{s_j} except for the leading term by linear combinations of elements of B_{s_j} with coefficients in **V**. As a result we obtain a set of equations which satisfy the lemma, since uniqueness follows from the linear independence of the elements of B_N for any $N \in \mathbf{N}$.

The set of homogeneous (arbitrary) algebraic equations over **V** connecting the elements (90) whose existence was established in Lemma 16, will be called *the set of homogeneous minimal equations (the set of minimal equations) over* **V** *for the elements (90).*

It is easy to see that the left side of each of the minimal equations over **V** for the elements (90) is an irreducible polynomial in the variables (90). In fact, if this were not the case, i.e., if the left side of one of the equations $G_j = 0$ were a reducible polynomial, then the leading term of this polynomial could not be a minimal element of the set D_1 of leading terms of all of the algebraic equations over **V** for the elements (90).

Lemma 17. *For any $N \in \mathbf{N}$ the subset B_N of L_N^0 (of L_N) forms a basis of the vector space $\mathbf{L}_N^0$ (the vector space $\mathbf{L}_N$).*

Proof. The elements of B_N are linearly independent over $\mathbf{V}$, and, by Lemma 15, every element of $\mathbf{L}_N^0$ (of $\mathbf{L}_N$) is a linear combination over $\mathbf{V}$ of elements of B_N. This implies the lemma.

We note that the definition of the set of minimal equations over $\mathbf{V}$ for the elements (90) depends on the given numbering of these elements. The set may change if those elements are re-numbered.

§ 11. Dimension of the vector spaces spanned by monomials in the elements of a field extension

We return to the question which was examined in § 3. Again let $\mathbf{V}$ be a field and $\mathbf{W}$ be an extension field, let L_N^0 and L_N be the sets (19) and (20) of monomials in elements $u_m, \dots, u_1 \in \mathbf{W}$, let $\mathbf{L}_N^0$ and $\mathbf{L}_N$ be the vector spaces over $\mathbf{V}$ spanned by the sets (19) and (20), and let $r_{N,m,l}^0 = \dim \mathbf{L}_N^0$, $r_{N,m,l} = \dim \mathbf{L}_N$. Earlier we established the inequalities (21) and (22) for $r_{N,m,l}^0$ and $r_{N,m,l}$; now we shall determine the actual form of these functions.

Theorem 13. *Suppose that $m \geq 1$, and the set of elements $u_m, \dots, u_1 \in \mathbf{W}$ has homogeneous transcendence degree (transcendence degree) over $\mathbf{V}$ equal to l, $1 \leq l \leq m$ $(0 \leq l \leq m)$. Then there exist numbers $\alpha_0, \alpha_1, \dots, \alpha_{l-1} \in \mathbf{Q}$, $\alpha_{l-1} > 0$, $(\beta_0, \beta_1, \dots, \beta_l \in \mathbf{Q}$, $\beta_l > 0)$ and $N_0 \in \mathbf{N}$, depending only on $u_m, \dots, u_1$ and the field $\mathbf{V}$, such that for all $N \geq N_0$*

$$r_{N,m,l}^0 = \alpha_{l-1} N^{l-1} + \dots + \alpha_1 N + \alpha_0 \tag{98}$$

(respectively,

$$r_{N,m,l} = \beta_l N^l + \dots + \beta_1 N + \beta_0.) \tag{99}$$

Proof. We shall treat the homogeneous case. According to Lemma 17, the subset B_N of L_N^0 considered in § 10 forms a basis of the vector space $\mathbf{L}_N^0$. Hence, $r_{N,m,l}^0$ is equal to the number of elements in B_N. In order to find this number, we represent B_N as a union of certain subsets.

We consider the set of all ν-tuples of nonzero exponents $\kappa_{i_1,1}, \dots, \kappa_{i_\nu,\nu}$, one exponent taken from each of the minimal elements in (93). Let us take any such ν-tuple. It might have exponents having the same first index $i_1, \dots, i_\nu$; in that case we remove from the ν-tuple all but one of the exponents with the same first index. The resulting t-tuple of exponents $(t \leq \nu)$ will be denoted

$$l_{i_1}, \dots, l_{i_t}. \tag{100}$$

Here t may be different for different ν-tuples; but always $t \leq \nu$. The number n of sets of exponents (100) clearly does not depend on N.

We re-number the sets of exponents (100) in an arbitrary way from 1 to n, and to the i-th set $(1 \leq i \leq n)$ we associate the set $B_{N,i}$ consisting of elements $u_m^{k_m} \cdots u_1^{k_1}$ in L_N^0 which satisfy the inequalities

$$k_{i_1} < l_{i_1}, \ \ldots, \ k_{i_t} < l_{i_t}. \tag{101}$$

Then any element $u_m^{k_m} \cdots u_1^{k_1}$ in B_N satisfies (101) for some choice of the set (100), and so it belongs to at least one of the subsets $B_{N,i}$, $1 \leq i \leq n$. Here the set (100) is obtained from the ν-tuple of exponents $\kappa_{i_1,1}, \ldots, \kappa_{i_\nu,\nu}$ in (96) in the manner described above.

On the other hand, any element $u_m^{k_m} \cdots u_1^{k_1}$ of the subset $B_{N,i}$, $1 \leq i \leq n$, obviously satisfies (96) for some choice of the exponents $\kappa_{i_1,1}, \ldots, \kappa_{i_\nu,\nu}$, and so it belongs to B_N. Thus,

$$B_N = \bigcup_{i=1}^{n} B_{N,i}.$$

Let B be any subset of B_N. We let $\Psi(B)$ denote the number of elements in B. Then $r_{N,m,l}^0 = \Psi(B_N)$. We have

$$\begin{aligned} r_{N,m,l}^0 = \Psi(B_N) = & \sum_{i=1}^{n} \Psi(B_{N,i}) - \sum_{1 \leq i < j \leq n} \Psi(B_{N,i} \cap B_{N,j}) + \\ & + \sum_{1 \leq i < j < k \leq n} \Psi(B_{N,i} \cup B_{N,j} \cup B_{N,k}) - \cdots + \\ & + (-1)^{n-1} \Psi\left(\bigcap_{i=1}^{n} B_{N,i}\right). \end{aligned} \tag{102}$$

We now find $\Psi(B_{N,i})$, which is the number of elements $u_m^{k_m} \cdots u_1^{k_1}$ in the subset $B_{N,i}$, $1 \leq i \leq n$. We set $N_i = l_{i_1} + \cdots + l_{i_t}$, where $l_{i_1}, \ldots, l_{i_t}$ is the set of exponents (100) corresponding to the set $B_{N,i}$. We set $N > N_i$, and we fix any nonnegative integers $k_{i_1}, \ldots, k_{i_t}$ satisfying (101); we let the remaining exponents $k_{i_{t+1}}, \ldots, k_{i_m}$ run through the nonnegative integers for which

$$k_{i_{t+1}} + \cdots + k_{i_m} = N - k_{i_1} - \cdots - k_{i_t}.$$

Then, by Lemma 7 of Chapter 2, the number of elements $u_m^{k_m} \cdots u_1^{k_1}$ satisfying these conditions is equal to

$$\frac{(N - k_{i_1} - \cdots - k_{i_t} + m - t - 1)!}{(N - k_{i_1} - \cdots - k_{i_t})!(m - t - 1)!}.$$

Hence, using the inequalities (101), we obtain

$$\Psi(B_{N,i}) = \sum_{k_{i_1}=0}^{l_{i_1}-1} \cdots \sum_{k_{i_t}=0}^{l_{i_t}-1} \frac{(N - k_{i_1} - \cdots - k_{i_t} + m - t - 1)!}{(N - k_{i_1} - \cdots - k_{i_t})!(m - t - 1)!} = P(N),$$

where $P(x) \in \mathbf{Q}[x]$, and the coefficients of the polynomial $P(x)$ does not depend on N.

The subset $B_{N,i} \cap B_{N,j}$, $1 \leq i < j \leq n$, is determined by the two sets (100) corresponding to $B_{N,i}$ and $B_{N,j}$. In these two sets (100) for each index we choose the smaller exponent with that index. This gives us a new set of exponents and new inequalities similar to (101); this determines the subset $B_{N,i} \cap B_{N,j}$. Thus, $\Psi(B_{N,i} \cap B_{N,j})$ is also given by a polynomial in $\mathbf{Q}[N]$.

We similarly find that all of the expressions

$$\Psi(B_{N,i} \cap B_{N,j} \cap B_{N,k}), \quad 1 \leq i < j < k \leq n, \qquad \ldots, \ \Psi\left(\bigcap_{i=1}^{n} B_{N,i}\right)$$

are polynomials in $\mathbf{Q}[N]$.

We set

$$N_0 = \sum_{j=1}^{\nu} \max_{1 \leq i \leq m} \kappa_{i,j}.$$

Then from (102) we find that for $N \geq N_0$ the expression for the number $r^0_{N,m,l}$ of elements in the basis B_N has the form of a polynomial in $\mathbf{Q}[N]$. By Lemma 6, this polynomial has degree $l - 1$, and since $r^0_{N,m,l} > 0$, it follows that $\alpha_{l-1} > 0$. This proves (98). The relation (99) is a consequence of (98).

If we know minimal equations for the elements in (90), then we can use them to obtain effective estimates for $r^0_{N,m,l}$ and $r_{N,m,l}$.

Lemma 18. *Lemma 6 remains true if k is replaced by*

$$\kappa = \max_{\substack{1 \leq i \leq m \\ 1 \leq j \leq \nu}} \kappa_{i,j},$$

where the $\kappa_{i,j}$ are the exponents in the leading terms of the minimal equations (93).

The proof of Lemma 18 is similar to the proof of Lemma 6, except that in place of the equations (25) one considers the minimal equations (95).

§ 12. Algebraic independence of the values of IE-functions

Let

$$f_1(z), \ldots, f_m(z), \qquad m \geq 2, \tag{103}$$

be a set of functions which are analytic in some region. We suppose that they are homogeneously algebraically independent (algebraically independent) over $\mathbf{C}(z)$. We set $\mathbf{V} = \mathbf{C}(z)$, and $u_i = f_i(z)$, $i = 1, \ldots, m$. Then all of the arguments in § § 10-11 apply to the functions (103), and all of the results in those sections hold in this case.

In this context, a set of minimal homogeneous equations (minimal equations) for $u_1, \ldots, u_m$ with leading coefficients (93) will be called a set of minimal homogeneous equations (minimal equations) for the functions (103) over $\mathbf{C}(z)$; and we agree to write the left side of each of these equations in the form of a primitive polynomial in $\mathbf{C}[z][f_1(z), \ldots, f_m(z)]$, which (by what was proved before) is irreducible.

We let Λ denote the set consisting of all of the zeros of the leading coefficients of the minimal equations for the functions (103) over $\mathbf{C}(z)$.

If all of the functions in (103) are analytic at $z = 0$, and if all of the coefficients of the power series in z of these functions belong to an algebraic number field $\mathbf{K}$, then, using Lemmas 5, 16 and 17, we may assume that all of the minimal equations for these functions are in $\mathbf{Z_K}[z][f_1(z), \ldots, f_m(z)]$; and then $\Lambda \subset \mathbf{A}$.

We consider the set L_N^0 of monomials in the functions (103) of the form

$$f_m^{k_m}(z) \cdots f_1^{k_1}(z), \qquad k_1 + \cdots + k_m = N, \tag{104}$$

$$k_i \in \mathbf{Z}^+, \qquad i = 1, \ldots, m, \qquad m \geq 2,$$

and also the set L_N of monomials of the form

$$f_m^{k_m}(z) \cdots f_1^{k_1}(z), \qquad k_1 + \cdots + k_m \leq N, \tag{105}$$

$$k_i \in \mathbf{Z}^+, \qquad i = 1, \ldots, m, \qquad m \geq 2.$$

Let r_N^0 and r_N denote the dimensions of the corresponding vector spaces $\mathbf{L}_N^0$ and $\mathbf{L}_N$, and let A_N and B_N be the subsets of L_N^0 and L_N considered in § 10.

For any $N \in \mathbf{N}$, we number the functions of B_N in increasing order, and denote them

$$F_{N,i}(z), \qquad i = 1, \dots, r_N^0 \qquad (i = 1, \dots, r_N), \tag{106}$$

and for $N \geq N^*$, where N^* was defined in (97), we number the functions of A_N similarly, and denote them

$$F_{N,j}(z), \qquad j = r_N^0 + 1, \dots, \mu_{N,m}^0 \qquad (j = r_N + 1, \dots, \mu_{N,m}). \tag{107}$$

Lemma 19. *Suppose that the functions (103) are analytic in a region, form a solution of the system of homogeneous linear differential equations (4) (of linear differential equations (11)), and are homogeneously algebraically dependent (algebraically dependent) over* $\mathbf{C}(z)$. *Then for any* $N \in \mathbf{N}$ *the functions (106) form a solution of a system of homogeneous linear differential equations*

$$y'_{N,j} = \sum_{i=1}^{M} Q_{N,j,i} y_{N,i}, \qquad Q_{N,j,i} \in \mathbf{C}(z), \qquad M = r_N^0 \ (M = r_N), \tag{108}$$

which does not have any singular points other than the singular points of the system (4) (the system (11)) and the points in the set Λ.

Proof. By Lemma 17, for any $N \in \mathbf{N}$ the functions in (106) form a basis of the vector space $\mathbf{L}_N^0$ (the vector space $\mathbf{L}_N$) which is spanned over $\mathbf{C}(z)$ by the elements of L_N^0 (of L_N); and by Lemma 15, for $N \geq N^*$ the functions in (107) can be represented uniquely in the form

$$F_{N,j} = \sum_{i=1}^{M} \gamma_{N,j,i} F_{N,i}(z), \qquad \gamma_{N,j,i} \in \mathbf{C}(z), \tag{109}$$

$$M = r_N^0 \ (M = r_N), \qquad j = r_N^0 + 1, \dots, \mu_{N,m}^0 \ (j = r_N + 1, \dots, \mu_{N,m}).$$

From the proof of Lemma 15 it follows that in (109) the right side is obtained from the left side by successively replacing certain monomials in the functions (103) by linear combinations of elements of L_N^0 (of L_N) whose coefficients are coefficients of the minimal equations (95). Hence, the functions $\gamma_{N,j,i}$ cannot have any poles except for points of Λ.

According to Lemma 18 of Chapter 3, the functions (104) in L_N^0 (the functions (105) in L_N) form a solution of a system of homogeneous linear differential equations of the form (4) with m replaced by $M = r_N^0$ ($M = r_N$). This implies the lemma for $N < N^*$, since $A_N = \emptyset$ and $r_N^0 = \mu_{N,m}^0$ ($r_N = \mu_{N,m}$). If $N \geq N^*$, then in our system of differential equations we keep only the equations whose left sides are the derivatives of the functions (106), and we transform the right side of those equations by substituting the right side of (109) in place of the corresponding functions (107) which occur there. As a result we obtain a system of homogeneous linear differential equations (108) whose coefficients satisfy the conditions in the lemma.

Theorem 14. *Suppose that the IE-functions (103) form a solution of the system of homogeneous linear differential equations (4) (of linear differential equations (11)), these functions have homogeneous transcendence degree (transcendence degree) l, $1 \leq l \leq m-1$, over $\mathbf{C}(z)$, and the functions*

$$f_1(z), \ldots, f_l(z) \tag{110}$$

are homogeneously algebraically independent (algebraically independent) over $\mathbf{C}(z)$. Further let

$$A_1(z) f_m^{\kappa_{m,1}}(z) \cdots f_1^{\kappa_{1,1}}(z), \ \ldots, \ A_\nu(z) f_m^{\kappa_{m,\nu}}(z) \cdots f_1^{\kappa_{1,\nu}}(z), \tag{111}$$

$$A_i(z) \in \mathbf{C}[z], \qquad i = 1, \ldots, \nu,$$

be the leading terms in the set of minimal homogeneous equations (minimal equations) for the functions (103) over $\mathbf{C}(z)$. Suppose that $\xi \in \mathbf{I}$, $\xi T(\xi) A_1(\xi) \cdots A_\nu(\xi) \neq$ $\neq 0$. Then the numbers

$$f_1(\xi), \ldots, f_l(\xi) \tag{112}$$

are homogeneously algebraically independent (algebraically independent), and in the homogeneous case with $l > 1$ the numbers

$$f_1(\xi)/f_l(\xi), \ \ldots, \ f_{l-1}(\xi)/f_l(\xi)$$

are algebraically independent.

Proof. It is sufficient to prove the theorem in the homogeneous case. Suppose that $\xi \in \mathbf{I}$ satisfies the conditions of the theorem. For any $N \in \mathbf{N}$ we consider the set L_N^0 in (104) and its subsets A_N and B_N determined by the minimal homogeneous equations with leading terms (111). By Lemma 19, the functions (106) in B_N form a solution of the system of homogeneous linear differential equations (108), for which ξ is not a singular point. By Lemma 17, these functions are linearly independent over $\mathbf{C}(z)$. Hence, by Lemma 3, for any $N \in \mathbf{N}$ the numbers $F_{N,i}(\xi)$, $i = 1, \ldots, r_N^0$, are linearly independent over $\mathbf{I}$.

The functions (110) are homogeneously algebraically independent over $\mathbf{C}(z)$. This implies that each of the leading terms of the minimal equations (111) — and hence any element of A_N for any N — contains at least one of the functions $f_{l+1}(z),\dots,f_m(z)$. Thus, for any N the set of all expressions

$$f_l^{k_l}(z)\cdots f_1^{k_1}(z),\qquad k_1+\dots+k_l=N,$$

occurs in B_N; by what was proved above, for any N the numbers

$$f_l^{k_l}(\xi)\cdots f_1^{k_1}(\xi),\qquad k_1+\dots+k_l=N,$$

are linearly independent over **I**. Then, by Lemma 1 of Chapter 3, the numbers (112) are homogeneously algebraically independent.

Theorem 14 gives a set of algebraic points ξ where the values of the functions (110) are algebraically independent. But it remains an open question whether some or all of the points in Λ are really exceptional points.

It would be very interesting to generalize Theorem 14 to the case of a set of KE-functions and a point $\xi\in\mathbf{A}$, or else to give a counterexample to the possibility of such a generalization. However, this seems to be a very difficult problem. At present, using the minimal equations, all we can prove is Theorem 15 below.

§ 13. Algebraic independence of the values of KE-functions

Theorem 15. *Suppose that the E-functions*

$$f_1(z),\dots,f_m(z),\qquad m\ge 3\ \ (m\ge 2),\tag{113}$$

form a solution of the system of homogeneous linear differential equations (4) (of linear differential equations (11)) and have homogeneous transcendence degree (transcendence degree) l, $2\le l\le m-1$ ($1\le l\le m-1$), over $\mathbf{C}(z)$. Suppose that the functions (110) are homogeneously algebraically independent (algebraically independent) over $\mathbf{C}(z)$. Further let

$$A_1(z)f_m^{\kappa_{m,1}}(z)\cdots f_{t+1}^{\kappa_{t+1,1}}(z),\ \dots,\ A_\nu(z)f_m^{\kappa_{m,\nu}}(z)\cdots f_{t+1}^{\kappa_{t+1,\nu}}(z),\tag{114}$$

$$1\le t\le l,\qquad A_i(z)\in\mathbf{C}[z],\qquad i=1,\dots,\nu,$$

be the leading terms of the set of minimal homogeneous equations (minimal equations) over $\mathbf{C}(z)$ for the functions (113). Suppose that $\xi\in\mathbf{A}$, $\xi T(\xi)A_1(\xi)\cdots$ $\cdots A_\nu(\xi)\ne 0$. Then the numbers

$$f_1(\xi),\dots,f_t(\xi)\tag{115}$$

are homogeneously algebraically independent (algebraically independent), and in the homogeneous case with $t > 1$ the numbers

$$f_1(\xi)/f_t(\xi), \dots, f_{t-1}(\xi)/f_t(\xi)$$

are algebraically independent.

Proof. We first consider the homogeneous case. Let $\xi \in \mathbf{A}$ satisfy the conditions of the theorem. Suppose that the theorem is false, i.e., that

$$P(f_1(\xi), \dots, f_t(\xi)) = 0, \tag{116}$$

$$P = P(z_1, \dots, z_t) \in \mathbf{A}[z_1, \dots, z_t], \qquad \deg_{\mathbf{z}} P = k.$$

We let $\mathbf{K}$ denote an algebraic number field containing ξ, the coefficients of P, and the power series coefficients in z of all of the E-functions (113).

For any $N \in \mathbf{N}$ we consider the set L_N^0 (see (104)) and the subset B_N determined by the minimal homogeneous equations with leading terms (114). Let r_N^0 be the rank of L_N^0 over $\mathbf{C}(z)$. By Lemma 17, for any $N \in \mathbf{N}$ the subset B_N is a basis of the vector space $\mathbf{L}_N^0$ spanned over $\mathbf{C}(z)$ by the elements of L_N^0. Hence, the set B_N has r_N^0 elements. By Theorem 13, for $N \geq N_0$ we have

$$r_N^0 = \alpha_{l-1} N^{l-1} + \dots + \alpha_1 N + \alpha_0, \qquad \alpha_{l-1} > 0. \tag{117}$$

For any $N \in \mathbf{N}$ let the functions (106) be the elements of B_N listed in increasing order. We estimate the rank ρ_N^0 over $\mathbf{K}$ of the set of numbers $F_{N,i}(\xi)$, $i = 1, \dots, r_N^0$. To do this we consider the values at $z = \xi$ of the elements of B_{N-k} and the equations

$$F_{N-k,i}(\xi) P(f_1(\xi), \dots, f_t(\xi)) = 0, \qquad i = 1, \dots, r_{N-k}^0. \tag{118}$$

By the definition of B_N and the hypothesis of the theorem, the element

$$f_m^{k_m}(z) \cdots f_1^{k_1}(z), \qquad k_1 + \dots + k_m = N,$$

belongs to B_N if for each j, $j = t+1, \dots, m$, it has an exponent k_i satisfying the inequality

$$k_i < \kappa_{i,j}, \qquad i \leq i \leq m. \tag{119}$$

All of the elements $F_{N-k,i}(z)$ satisfy (119). Hence, all products of the following form also satisfy (119):

$$F_{N-k,i}(z) f_t^{\bar{k}_t} \cdots f_1^{\bar{k}_1}(z), \qquad \bar{k}_1 + \dots + \bar{k}_t = k.$$

Hence, these products are elements of B_N. Thus, the left sides in (118) are linear forms in the elements of B_N with coefficients in **K**:

$$F_{N-k,i}(\xi)P(f_1(\xi),\ldots,f_t(\xi)) = \sum_{j=1}^{r_N^0} \alpha_{N,i,j}F_{N,j}(\xi) = 0, \tag{120}$$

$$i = 1,\ldots,r_{N-k}^0,$$

where the numbers $\alpha_{N,i,j}$ are either zero or else coefficients of P.

The linear equations (120) between the numbers $F_{N,j}(\xi)$ with coefficients in **K** are linearly independent. This follows because their terms of highest order are different, since they are obtained by multiplying $F_{N-k,j}(\xi)$ by the leading term of $P(f_1(\xi),\ldots,f_m(\xi))$, and so have different orders. Hence, the elements of B_N are connected by at least r_{N-k}^0 linearly independent linear equations with coefficients in **K**. This along with (117) implies that

$$\rho_N^0 \le r_N^0 - r_{N-k}^0 = O(N^{l-2}). \tag{121}$$

On the other hand, by Lemma 18, the functions (106) form a solution of the system of homogeneous linear differential equations (108), for which ξ is not a singular point. Since these functions are linearly independent over $\mathbf{C}(z)$, it follows from Lemma 3 and the formula (117) that the following relation holds for $N \ge N_0$:

$$\rho_N^0 \ge \frac{r_N^0}{h} = \frac{\alpha_{l-1}}{h}N^{l-1} + O(N^{l-2}), \qquad h = [\mathbf{K}:\mathbf{Q}]. \tag{122}$$

The inequalities (121) and (122) are contradictory for N sufficiently large. This contradiction proves the theorem in the homogeneous case.

The non-homogeneous case follows from the homogeneous case only when the degrees in $f_m(z),\ldots,f_{t+1}(z)$ of all of the leading terms of the minimal equations (114) are equal to the degrees of the corresponding minimal equations. In the general situation, the proof in the non-homogeneous case goes through in the same way as the proof in the homogeneous case. One must consider L_N in place of L_N^0, and examine the minimal non-homogeneous equations.

The most interesting case of Theorem 15 is when $t = l$. Then the numbers $f_1(\xi),\ldots,f_l(\xi)$ are algebraically independent.

We note that, under the conditions of Theorems 14 and 15 and the corresponding theorems in Chapters 11 and 12, if the functions $f_1(z),\ldots,f_l(z)$ are fixed, then we can obtain $(m-l)!$ different sets of minimal equations by changing the numbering of $f_{l+1}(z),\ldots,f_m(z)$. Then, in general, the set of points where the theorems give algebraic independence of the values of the functions may change.

Theorem 12 is a special case of Theorem 15.

The elements of an arbitrary set D (see (91)) can be ordered in different ways. A natural way to order them is as follows. We consider terms to have higher order if they have larger total degree in $u_m, \ldots, u_1$; and in a homogeneous set of products we order the terms lexicographically with respect to powers of $u_m, \ldots, u_1$. If D consists of products all of the same total degree, then this ordering is the same as what we had before. But if D contains products of different total degrees, then a leading term in the new ordering may, in general, be different from a leading term in the old ordering.

One can introduce minimal equations in the new ordering and prove properties for them which are similar to the results in § 10. Statements and proofs can be given analogous to Theorems 6, 12, 14 and 15. We note that now the set of points $\xi \in \mathbf{A}$ where one proves the values of the functions to be algebraically independent, may be different from before.

We also note that Theorems 6–9, 11, 12, 14 and 15 can be adapted to the situation when one has an E-function which satisfies a homogeneous linear differential equation (38) (a linear differential equation (39)), and one is interested in the sequence consisting of the function along with its successive derivatives.

Remarks

Several theorems on transcendence and algebraic independence of the values at algebraic points of E-functions which satisfy algebraic equations over $\mathbf{C}(z)$ were announced in 1955–56 in [70:3–5]. These theorems were proved in 1960 in [70:11] under the assumption that the functions are connected by a single algebraic equation over $\mathbf{C}(z)$. We note that the proof of Theorem 4 in that paper contains a mistake: it is correct only when the number m of functions is 2.

Proofs of the basic results of Chapter 4 appeared in 1962 in [70:14].

The auxiliary propositions in § 1, which make it possible to extend the method of Chapter 3 to sets of E-functions which are algebraically dependent over $\mathbf{C}(z)$, were first proved in [70:14]. The proofs were complicated, and the properties of a fundamental set of solutions of the system of differential equations satisfied by the E-functions were used to construct the necessary linear equations connecting the functions.

In 1967 in [42:4], Mahler gave a different proof of the most basic of these propositions. His proof is also complicated and makes use of the properties of differential equations. It can be found in his book [42:7].

The proofs of the lemmas in § 1 are simple, and they do not use the properties of a fundamental set of solutions of differential equations. These proofs were published in 1983 in [70:37].

The lemmas in § 3, 10 and 11 on the dimensions of vector spaces spanned over a field **V** by monomials in certain elements follow from results of Hilbert [28:3] published in 1890. The proofs given in this chapter were taken from Oleinikov's 1962 paper [56:2], which also includes a treatment of minimal equations.

We note that in the non-homogeneous case the minimal equations form a basis of the ideal in the polynomial ring $\mathbf{V}[x_1, \ldots, x_m]$ consisting of all polynomials $P(x_1, \ldots, x_m)$ such that $P(u_1, \ldots, u_m) = 0$. In the homogeneous case the minimal equations form a basis of a homogeneous ideal.

The non-homogeneous case of Theorem 14 was proved by Chirskii in 1973 in [12:1].

In connection with bounds for the dimension of the vector spaces $\mathbf{L}_N^0$ and $\mathbf{L}_N$ in § 3 and 11, we cite Nesterenko's 1984 article [52:7].

Theorem 11 was published in [70:38].

Chapter 5
Transcendence and algebraic independence of the values of E-functions which satisfy first order linear differential equations

§ 1. Hypergeometric E-functions

In Chapters 3 and 4 we proved general theorems about transcendence and algebraic independence of the values of E-functions at algebraic points. Some of these theorems were applied to the simplest concrete E-functions. In those cases it was not necessary to prove that the functions were E-functions, since that fact was obvious from the definition of an E-function. But when we apply the general theorems of Chapters 3 and 4 to more complicated E-functions we must concern ourselves with proving that they are E-functions which satisfy linear differential equations with coefficients in $\mathbf{C}(z)$.

The structure of the class $\mathbf{E}$ consisting of *all E-functions which satisfy linear differential equations with coefficients in* $\mathbf{C}(z)$ has not yet been definitively studied. This seems to be a very difficult problem. The best one can do is describe a rather large subclass of the class $\mathbf{E}$, which may perhaps be all of $\mathbf{E}$. We now define this subclass of E-functions, which includes many functions which are used in various branches of pure and applied mathematics.

Let $\alpha \in \mathbf{C}$. We set

$$[\alpha, 0] = 1, \qquad [\alpha, n] = \alpha(\alpha+1)\cdots(\alpha+n-1), \qquad n = 1, 2, \ldots.$$

If $a_i, b_j \in \mathbf{C}$, $b_j \neq 0, -1, -2, \ldots$, $i = 1, \ldots, l$, $j = 1, \ldots, m$, $t = m - l > 0$, then we define the function

$$y = f(z) = \sum_{n=0}^{\infty} \gamma_n \left(\frac{z}{t}\right)^{tn}, \tag{1}$$

$$\gamma_n = \frac{[a_1, n]\cdots[a_l, n]}{[b_1, n]\cdots[b_m, n]}, \qquad n = 0, 1, 2, \ldots.$$

A function of the form (1) will be called a *generalized hypergeometric function*, or simply a *hypergeometric function*.

Lemma 1. *The function (1) satisfies an m-th order linear differential equation*

$$y^{(m)} + Q_{m-1}y^{(m-1)} + \dots + Q_1 y' + Q_0 y = t^m(b_1 - 1)\cdots(b_m - 1)z^{-m}, \quad (2)$$

$$Q_i \in \mathbf{C}\left[\frac{1}{z}\right], \qquad i = 0, 1, \dots, m-1.$$

Proof. We denote $\delta = z\frac{d}{dz}$. We have

$$(\delta + ta)\left(\frac{z}{t}\right)^{tn} = t(n+a)\left(\frac{z}{t}\right)^{tn}, \qquad a \in \mathbf{C}, \qquad n = 0, 1, 2, \dots.$$

Hence, if we apply the differential operator $\delta + ta$ to the function (1), we obtain

$$\begin{aligned}(\delta + t(b_m - 1))\cdots(\delta + t(b_1 - 1))y &= \\ = t^m(b_1 - 1)\cdots(b_m - 1) + t^m \sum_{n=1}^{\infty} \frac{[a_1, n]\cdots[a_l, n]}{[b_1, n-1]\cdots[b_m, n-1]}\left(\frac{z}{t}\right)^{tn} &= \\ = t^m(b_1 - 1)\cdots(b_m - 1) + t^l z^t \sum_{n=0}^{\infty} \frac{[a_1, n+1]\cdots[a_l, n+1]}{[b_1, n]\cdots[b_m, n]}\left(\frac{z}{t}\right)^{tn}&, \end{aligned} \quad (3)$$

$$(\delta + ta_l)\cdots(\delta + ta_1)y = t^l \sum_{n=0}^{\infty} \frac{[a_1, n+1]\cdots[a_l, n+1]}{[b_1, n]\cdots[b_m, n]}\left(\frac{z}{t}\right)^{tn}. \quad (4)$$

From (3) and (4) we obtain the differential equation

$$\left(\prod_{i=1}^{m}(\delta + t(b_i - 1)) - z^t \prod_{i=1}^{l}(\delta + ta_i)\right) y = t^m(b_1 - 1)\cdots(b_m - 1), \quad (5)$$

from which the differential equation (2) follows.

The only singular point of the differential equation (2) in the finite plane is the point $z = 0$.

If one or more of the numbers $b_1, \dots, b_m$ is 1, then the right side of (2) vanishes, and then the function (1) is a solution of a homogeneous linear differential equation. On the other hand, if $b_i \neq 1$ for $i = 1, \dots, m$, then, if we replace l and m by $l + 1$ and $m + 1$ and set $a_{l+1} = b_{m+1} = 1$, then we see that the function (1) satisfies a homogeneous linear differential equation of the form (2) except of order $m + 1$.

We further note that, if all of the a_i and b_j are rational, then in (2) we have

$$Q_i \in \mathbf{Q}\left[\frac{1}{z}\right], \qquad i = 0, 1, \dots, m-1.$$

We denote

$$c_\nu = \begin{cases} \gamma_n \frac{(tn)!}{t^{tn}}, & \nu = tn, \quad n = 0, 1, 2, \dots, \\ 0, & \nu \neq tn, \quad n = 0, 1, 2, \dots. \end{cases} \tag{6}$$

Then the hypergeometric function (1) can be written in the form

$$y = f(z) = \sum_{\nu=0}^{\infty} c_\nu \frac{z^\nu}{\nu!}. \tag{7}$$

Lemma 2. *If*

$$a_i, b_j \in \mathbf{Q}, \qquad b_j \neq 0, -1, -2, \dots, \qquad i = 1, \dots, l, \ \ j = 1, \dots, m,$$

then there exists a constant $c > 0$ such that the coefficients of the function (1) satisfy the following conditions:

1°. $|c_\nu| = O(c^\nu)$. (8)

2°. *There exists a sequence $\{q_\nu\}$, $q_\nu \in \mathbf{N}$, such that*

$$q_\nu c_k \in \mathbf{Z}, \qquad k = 0, 1, \dots, \nu, \qquad q_\nu = O(c^\nu), \qquad \nu \to \infty. \tag{9}$$

Proof. From (6) we have

$$c_{tn} = \gamma_n \frac{(tn)!}{t^{tn}} = \frac{[a_1, n] \cdots [a_l, n][1, n] \cdots [1, n]}{[b_1, n] \cdots [b_l, n][b_{l+1}, n] \cdots [b_m, n]} \frac{(tn)!}{(n!)^t t^{tn}}. \tag{10}$$

We consider the ratio

$$\frac{[a, n]}{[b, n]} = \prod_{\nu=1}^{n} \left(\left(1 + \frac{a-1}{\nu}\right) \Big/ \left(1 + \frac{b-1}{\nu}\right) \right),$$

$$a, b \in \mathbf{Q}, \qquad b \neq 0, -1, -2, \dots.$$

Let $n_0 \geq \max(|a-1|, |b-1|)$, $n_0 \in \mathbf{N}$. Then for $n > n_0$

$$\frac{[a, n]}{[b, n]} = \left(\prod_{\nu=1}^{n_0} \left(\left(1 + \frac{a-1}{\nu}\right) \Big/ \left(1 + \frac{b-1}{\nu}\right) \right) \right) \times$$

$$\times \exp\left(\sum_{\nu=n_0+1}^{n} \left(\ln\left(1 + \frac{a-1}{\nu}\right) - \ln\left(1 + \frac{b-1}{\nu}\right) \right) \right).$$

If we use Taylor's formula for $\ln(1+x)$ along with the well-known relation

$$\sum_{\nu=1}^{n} \frac{1}{\nu} = \ln n + \gamma + o\left(\frac{1}{n}\right), \qquad \gamma = 0.577\cdots,$$

we obtain

$$\frac{[a,n]}{[b,n]} = O\left(n^{c_1}\right), \qquad c_1 > 0. \tag{11}$$

Furthermore,

$$\frac{(tn)!}{(n!)^t} \le (\underbrace{1+\cdots+1}_{t \text{ times}})^{tn} = t^{tn}. \tag{12}$$

From (11) and (12) it follows that there exists a constant $c > 0$ such that (8) holds.

We now prove that condition 2° is fulfilled. Since $(tn)!/(n!)^t \in \mathbf{N}$, it follows from (10) that it suffices to show that the least common denominator of the $n+1$ numbers

$$[a,k]/[b,k], \qquad k = 0, 1, \ldots, n, \tag{13}$$

is of order $O(c^n)$ as $n \to \infty$, for any $a, b \in \mathbf{Q}$ with $b \neq 0, -1, -2, \ldots$.

We set $a = \alpha/\beta$, $b = \gamma/\delta$, $\alpha, \gamma \in \mathbf{Z}$, $\beta, \delta \in \mathbf{N}$, $(\alpha, \beta) = 1$, $(\gamma, \delta) = 1$, and we denote

$$\frac{M_k}{N_k} = \frac{\beta^{2k}[a,k]}{\delta^k[b,k]} = \frac{\beta^k \alpha(\alpha+\beta)(\alpha+2\beta)\cdots(\alpha+(k-1)\beta)}{\gamma(\gamma+\delta)(\gamma+2\delta)\cdots(\gamma+(k-1)\delta)}.$$

Let p be any prime divisor of N_k. Then $(p, \delta) = 1$.

If $(l, \delta) = 1$, then $\gamma + \delta x$ runs through a complete set of residues modulo l when x runs through any l successive integers. Thus, for every $\nu = 1, 2, \ldots$, if x runs through p^ν successive integers, then exactly one of the p^ν values $\gamma + \delta x$ will be divisible by p^ν. This means that at least $[kp^{-\nu}]$ and at most $1 + [kp^{-\nu}]$ of the k factors

$$\gamma,\ \gamma+\delta,\ \gamma+2\delta,\ \ldots,\ \gamma+(k-1)\delta$$

in N_k are divisible by p^ν. But if $p^\nu > |\gamma| + (k-1)\delta$, then none of these factors are divisible by p^ν.

Suppose that N_k is divisible by p^s but not by p^{s+1}. Then

$$\sum_{\nu=1}^{\nu_k} [kp^{-\nu}] \le s \le \sum_{\nu=1}^{\nu_k} \left(1 + [kp^{-\nu}]\right),$$

$$\nu_k = [\ln(|\gamma| + (k-1)\delta)/\ln p].$$

We thus have

$$\left[\frac{k}{p}\right] \le s \le \left[\frac{k}{p}\right] + O\left(\frac{k}{p^2}\right) + O\left(\frac{\ln(k+1)}{\ln p}\right), \tag{14}$$

where

$$p \le |\gamma| + (k-1)\delta.$$

We now estimate the power $p^{s'}$ dividing the numerator M_k.

If $(p, \beta) = 1$, then, arguing as above, we obtain bounds similar to (14), from which we see that

$$s' \ge \left[\frac{k}{p}\right]. \tag{15}$$

This inequality also holds for primes which divide β, since M_k has the factor β^k.

We now cancel factors in the numerator and denominator of M_k/N_k, letting r_p denote the exponent with which the prime p occurs in the exact denominator of M_k/N_k. Then from (14) and (15) we find that

$$r_p = O\left(\frac{n}{p^2}\right) + O\left(\frac{\ln n}{\ln p}\right), \qquad k = 0, 1, \dots, n.$$

Hence, the least common denominator q_n of the fractions (13), which differ from M_k/N_k by the factor β^{2k}/δ^k, satisfies the estimate

$$\begin{aligned} \ln q_n &= O(n) + \sum_{p \le O(n)} \left(O\left(\frac{n}{p^2}\right) + O\left(\frac{\ln n}{\ln p}\right)\right) \ln p = \\ &= O(n) + \sum_{p \le O(n)} O\left(\frac{n \ln p}{p^2}\right) + O(\ln n) \sum_{p \le O(n)} 1 = \\ &= O(n) + \pi(O(n))\, O(\ln n), \end{aligned}$$

where $\pi(x)$ denotes the number of primes not exceeding x. Using the estimate

$$\pi(x) = O\left(\frac{x}{\ln x}\right), \qquad x \to \infty,$$

we obtain: $\ln q_n = O(n)$. Hence, in view of (10), we have (9) with some constant c. Choosing c to be the larger of the constants in (8) and (9), we obtain Lemma 2.

Since $O(c^n) = O(n^{\epsilon n})$ for any $\epsilon > 0$, the following lemma follows from Lemmas 1 and 2.

Lemma 3. *Any hypergeometric function (1) with rational a_i and b_j belongs to the class* **E**, *i.e., it is an E-function which satisfies a linear differential equation with coefficients in* $\mathbf{C}[z]$ *(and even in* $\mathbf{Q}[z]$*).*

Functions of the form (1) with rational a_i and b_j will be called *hypergeometric E-functions.*

From Lemmas 1–3, Lemma 9 of Chapter 4 and the properties of E-functions in § 1 of Chapter 3, we obtain the following result: *any polynomial with algebraic coefficients in z and a finite set of hypergeometric E-functions is an E-function in the class* **E**, *as is any function obtained from it by replacing z by λz for $\lambda \in \mathbf{A}$.*

In 1949 Siegel [72:4] asked the question: can any E-function in the class **E** be represented as such a polynomial? This problem is very difficult: no results have yet been obtained in the direction of a solution.

The example that follows shows that there is a much simpler problem of this sort which also has not yet been solved.

It is easy to show that an E-function which satisfies a first order homogeneous linear differential equation with coefficients in $\mathbf{C}[z]$ is the product of a polynomial with algebraic coefficients and a function of the form $e^{\alpha z}$, where $\alpha \in \mathbf{A}$. However, we have not yet been able to determine the form that an E-function must have if it satisfies a first order non-homogeneous linear differential equation of the same type.

One question which has been investigated is whether a hypergeometric function (1) with algebraic parameters a_i and b_j can be an E-function. It turns out that, except for a case having little interest, a function of the form (1) is an E-function only when all of the a_i and b_j are rational. We shall state the final result without proof.

We say that two sets of algebraic numbers $(a_1, \dots, a_l)$ and $(b_1, \dots, b_m)$ *satisfy condition E* if either all $a_i, b_j \in \mathbf{Q}$, or else there exists a partition of the parameters not in $\mathbf{Q}$ into pairs $(a_{i_1}, b_{j_1}), \dots, (a_{i_s}, b_{j_s})$ ($i_\lambda \neq i_\mu$ if $\lambda \neq \mu$, and $j_\nu \neq j_\eta$ if $\nu \neq \eta$) such that $(a_{i_k} - b_{j_k}) \in \mathbf{Z}^+$ for all $k = 1, \dots, s$.

In 1978, A. I. Galochkin [22:11] proved: *A hypergeometric function (1) with complex parameters $a_1, \dots, a_l$ and $b_1, \dots, b_m$ which are not equal to $0, -1, -2, \dots$ and have the property that $a_i \neq b_j$ for $i = 1, \dots, l$ and $j = 1, \dots, m$, is an E-function if and only if $a_i, b_j \in \mathbf{A}$ for $i = 1, \dots, l$, $j = 1, \dots, m$, and the sets of algebraic numbers $(a_1, \dots, a_l)$ and $(b_1, \dots, b_m)$ satisfy condition E.*

A corollary of this result is that for $l = 0$ the hypergeometric function (1) is an E-function if and only if $b_j \in \mathbf{Q}$, $b_j \neq 0, -1, -2, \dots$, $j = 1, \dots, m$.

Hypergeometric functions with irrational algebraic parameters are not E-functions (except in the case indicated above) because the least common multiple of the denominators of the first n power series coefficients increases much faster than is allowed for an E-function.

§ 2. The simplest hypergeometric E-functions

Theorem 4 of Chapter 3 concerns transcendence of the values of a transcendental E-function f(z) which satisfies a first order linear differential equation

$$P_1(z)y' + P_0(z)y = Q(z), \qquad P_0(z), P_1(z), Q(z) \in \mathbf{C}[z], \tag{16}$$

where the value can be taken at any point $\xi \in \mathbf{A}$ with $\xi P_1(\xi) \neq 0$. This theorem was an immediate consequence of the Second Fundamental Theorem in Chapter 3. It implies, in particular, Lindemann's result on transcendence of e^ξ for $\xi \in \mathbf{A}\setminus\{0\}$, since the function $y = e^z$ satisfies the differential equation $y' = y$.

We consider the hypergeometric function

$$\varphi_\lambda(z) = \sum_{n=0}^{\infty} \frac{z^n}{(\lambda+1)\cdots(\lambda+n)}, \qquad \lambda \neq -1, -2, \ldots, \tag{17}$$

which satisfies the first order linear differential equation

$$y' + \left(\frac{\lambda}{z} - 1\right) y = \frac{\lambda}{z}, \tag{18}$$

with $z = 0$ as its only singular point.

If $\lambda \in \mathbf{Q}$, then, by Lemma 3, the function (17) is a hypergeometric E-function. Since it is an entire function which is not a polynomial, it is a transcendental function. Hence, we can use Theorem 4 of Chapter 3 to obtain the following result.

Theorem 1. *If* $\lambda \in \mathbf{Q}$, $\lambda \neq -1, -2, \ldots$, *and if* $\xi \in \mathbf{A}\setminus\{0\}$, *then the number* $\varphi_\lambda(\xi)$ *is transcendental.*

There is a connection between the function (17) for any λ and the exponential function e^z. Namely, if $\lambda = 0$, then

$$\varphi_0(z) = e^z.$$

If $\lambda = k$, where $k \in \mathbf{N}$, then we have

$$\sum_{n=0}^{k-1} \frac{z^n}{n!} + \frac{z^k}{k!}\varphi_k(z) = e^z.$$

And for $\lambda \neq 0, -1, -2, \ldots$ we can integrate (18) to obtain

$$\varphi_\lambda(z) = \lambda z^{-\lambda} e^z \int z^{\lambda-1} e^{-z} dz + c z^{-\lambda} e^z, \tag{19}$$

where c is a constant.

We introduce the following notation. If the function $f(z)$ is analytic at $z = 0$,

$$f(z) = \sum_{n=k}^{\infty} a_n z^n, \qquad k \geq 0, \tag{20}$$

and if $\mu \in \mathbf{C}$, $\mu \neq -k, -(k+1), -(k+2), \ldots$, then the symbol

$$\int^z t^{\mu-1} f(t)dt$$

will denote the anti-derivative of $z^{\mu-1} f(z)$ which is obtained by multiplying the series (20) by $z^{\mu-1}$ and integrating term-by-term with zero constant of integration, i.e.,

$$\int^z t^{\mu-1} f(t)dt = \int^z t^{\mu-1} \left(\sum_{n=k}^{\infty} a_n t^n \right) dt = \sum_{n=k}^{\infty} \frac{a_n}{n+\mu} z^{n+\mu}. \tag{21}$$

In particular, for $\mu + k > 0$

$$\int^z t^{\mu-1} f(t)dt = \int_0^z t^{\mu-1} f(t)dt.$$

Using this notation, we can rewrite (19) in the following form:

$$\varphi_\lambda(z) = \lambda z^{-\lambda} e^z \int^z t^{\lambda-1} e^{-t} dt. \tag{22}$$

Theorem 2. *If* $\lambda \in \mathbf{Q} \setminus \mathbf{Z}$, *and if* $\xi \in \mathbf{A}$, $\eta \in \mathbf{A}$, $\xi\eta \neq 0$, *then the numbers* $\varphi_\lambda(\xi)$ *and* e^η *are algebraically independent.*

We prove two lemmas which are needed for this theorem.

Let $\mathcal{L}$ be a *field of analytic functions which is closed under the differentiation operation*, i.e., if $f(z) \in \mathcal{L}$, then $f'(z) \in \mathcal{L}$. In particular, $\mathcal{L}$ could be the field $\mathbf{C}(z)$.

We consider the system of linear differential equations

$$y_k' = Q_{k,0} + \sum_{i=1}^{m} Q_{k,i} y_i, \qquad k = 1, \ldots, m, \qquad Q_{k,i} \in \mathcal{L}, \tag{23}$$

and the associated differential operator

$$D = \frac{\partial}{\partial z} + \sum_{k=1}^{m} \left(Q_{k,i} + \sum_{i=1}^{m} Q_{k,i} y_i \right) \frac{\partial}{\partial y_k}. \tag{24}$$

Then if

$$P = P(y_1, \dots, y_m) \in \mathcal{L}[y_1, \dots, y_m], \tag{25}$$

it follows that

$$DP \in \mathcal{L}[y_1, \dots, y_m].$$

Consequently,

$$D: \ \mathcal{L}[y_1, \dots, y_m] \longrightarrow \mathcal{L}[y_1, \dots, y_m].$$

Lemma 4. *Suppose that the functions*

$$\varphi_1(z), \dots, \varphi_m(z), \qquad m \geq 1, \tag{26}$$

are analytic in some region, form a solution of the system of linear differential equations (23), and have transcendence degree $\deg \operatorname{tr}_{\mathcal{L}}\{\varphi_1(z), \dots, \varphi_m(z)\} = m - 1$. *Further suppose that the functions (26) are connected by an algebraic equation*

$$P(\varphi_1(z), \dots, \varphi_m(z)) = 0, \tag{27}$$

where $P = P(y_1, \dots, y_m)$ *is an irreducible polynomial satisfying (25). Then the polynomial* DP *is divisible by* P *as polynomials in* m *variables, and the quotient is a function* $\omega \in \mathcal{L}$, *i.e.,*

$$DP \equiv \omega P. \tag{28}$$

Proof. The polynomial P contains at least one of the variables $y_1, \dots, y_m$. To be definite, suppose that it contains y_m.

Since we have (27), and since the functions (26) form a solution of the system (23), it follows that for $y_k = \varphi_k(z)$, $k = 1, \dots, m$,

$$DP(\varphi_1(z), \dots, \varphi_m(z)) = \frac{d}{dz} P(\varphi_1(z), \dots, \varphi_m(z)) = 0. \tag{29}$$

We eliminate $\varphi_m(z)$ from the two equations (27) and (29), obtaining the resultant of the two polynomials:

$$R = R(\varphi_1(z), \dots, \varphi_{m-1}(z)) = 0.$$

The polynomial R cannot contain the functions $\varphi_1(z), \dots, \varphi_{m-1}(z)$, since if it did, then we would have an algebraic relation for them over the field $\mathcal{L}$, and this would contradict the assumption that the polynomials (26) have transcendence degree $m - 1$. Thus, $R \equiv 0$, and so DP is divisible by P as polynomials in m variables.

It follows from the form of the differential operator (24) that the degree of the polynomial DP in the variables $y_1, \ldots, y_m$ does not exceed the degree of P. Hence, the quotient when DP is divided by P is an element of $\mathcal{L}$. The lemma is proved.

Corollary. *Under the conditions of Lemma 4, if $\mathcal{L} = \mathbf{C}(z)$, the polynomial $T = T(z)$ for the system (23) has the same meaning as in Chapters 3 and 4, and*

$$P = P(z, y_1, \ldots, y_m) \in \mathbf{C}[z, y_1, \ldots, y_m],$$

then TDP is divisible by the irreducible polynomial P as polynomials in $m+1$ independent variables, and the quotient is a polynomial in $\mathbf{C}[z]$.

Let

$$D = \frac{\partial}{\partial z} + \left(\left(1 - \frac{\lambda}{z}\right) y + \frac{\lambda}{z}\right) \frac{\partial}{\partial y} \tag{30}$$

be the differential operator associated with the differential equation (18).

Lemma 5. *If $\lambda \in \mathbf{C} \setminus \mathbf{Z}$ and*

$$P = P(y, z) \in \mathbf{C}[y, z], \qquad P \not\equiv 0,$$

then zDP is divisible by P as polynomials in two independent variables if and only if

$$P = \sigma z^b, \qquad b \in \mathbf{Z}^+, \qquad \sigma \in \mathbf{C}, \qquad \sigma \neq 0. \tag{31}$$

Proof. Suppose that zDP is divisible by P. Since the degree of zDP in y is no greater than the degree of P, and the degree in z may be 1 greater than the degree of P, it follows that the quotient when zDP is divided by P is a polynomial in z of degree at most 1. Hence,

$$zDP \equiv (az + b)P, \qquad a, b \in \mathbf{C}.$$

Now let $y = \varphi_\lambda(z)$. Then

$$zDP(\varphi_\lambda(z), z) = z\frac{d}{dz}P(\varphi_\lambda(z), z) = (az + b)P(\varphi_\lambda(z), z). \tag{32}$$

If we integrate the differential equation (32), we obtain

$$P(\varphi_\lambda(z), z) = \sigma z^b e^{az}, \qquad \sigma \neq 0. \tag{33}$$

Since $P(\varphi_\lambda(z), z)$ is an entire function, it follows from (33) that $b \in \mathbf{Z}^+$. If P does not depend on y, then $a = 0$ and P has the form (31).

On the other hand, if P does depend on y, then we set

$$P = Q_l y^l + Q_{l-1} y^{l-1} + \cdots + Q_0, \tag{34}$$

$$Q_l \not\equiv 0, \qquad Q_0 \not\equiv 0, \qquad Q_0, Q_1, \ldots, Q_l \in \mathbf{C}[z].$$

Applying the differential operator (30) to P, we obtain

$$\begin{aligned} zDP = \big(zQ_l' + l(z-\lambda)Q_l\big)y^l + \\ + \big(l\lambda Q_l + zQ_{l-1}' + (l-1)(z-\lambda)Q_{l-1}\big)y^{l-1} + \cdots \\ \cdots + \big(\lambda Q_1 + zQ_0'\big). \end{aligned} \tag{35}$$

If we equate the coefficients of y^l and y^{l-1} in (32), and use (34) and (35), we find that

$$\begin{aligned} zQ_l' + l(z-\lambda)Q_l \equiv (az+b)Q_l, \\ l\lambda Q_l + zQ_{l-1}' + (l-1)(z-\lambda)Q_{l-1} \equiv (az+b)Q_{l-1}. \end{aligned} \tag{36}$$

Now dividing both sides of the first equation in (36) by Q_l, we obtain

$$\frac{zQ_l'}{Q_l} + l(z-\lambda) \equiv (az+b), \tag{37}$$

from which it follows that the polynomial zQ_l' is divisible by the polynomial Q_l, and

$$\frac{zQ_l'}{Q_l} = s, \qquad s = \deg Q_l.$$

Integrating, we obtain

$$Q_l = cz^s, \qquad c \in \mathbf{C}, \qquad c \neq 0.$$

The identity (37) now takes the form

$$s + l(z-\lambda) = az + b,$$

from which $a = l$, $b = s - l\lambda$. If we substitute these values of a, b, and Q_l in the second identity in (36), we arrive at the identity

$$l\lambda cz^s + zQ_{l-1}' - (z + s - \lambda)Q_{l-1} \equiv 0, \qquad cl\lambda \neq 0, \qquad s \in \mathbf{Z}^+. \tag{38}$$

From (38) we conclude that $Q_{l-1} \not\equiv 0$. Comparing the coefficients of the highest power of z in this identity, we find that $s = \deg Q_{l-1} + 1$. Let

$$Q_{l-1} = z^k Q, \qquad zQ'_{l-1} = kz^k Q + z^{k+1} Q',$$

where Q is a polynomial which is not divisible by z. Substituting these expressions for Q_{l-1} and zQ'_{l-1} in (38), we obtain:

$$l\lambda c z^s + z^{k+1} Q' - (z + s - \lambda - k) z^k Q \equiv 0, \qquad cl\lambda \neq 0,$$

from which it follows (since $k < s$) that $s - k - \lambda = 0$. But this contradicts the hypothesis of the lemma. Thus, P does not depend on y, and it has the form (31). Since the converse is obvious, the lemma is proved.

Lemma 6. *If $\lambda \in \mathbf{C} \setminus \mathbf{Z}$ and $\alpha \in \mathbf{C} \setminus \{0\}$, then the functions $\varphi_\lambda(z)$ and $e^{\alpha z}$ are algebraically independent over $\mathbf{C}(z)$.*

Proof. We set $y_1 = \varphi_\lambda(z)$ and $y_2 = e^{\alpha z}$. These functions satisfy the system of differential equations

$$y'_1 = \left(1 - \frac{\lambda}{z}\right) y_1 + \frac{\lambda}{z}, \qquad y'_2 = \alpha y_2. \tag{39}$$

We consider the differential operators associated with the system (39): the operator D in (30) with $y = y_1$, and

$$D_1 = D + \alpha y_2 \frac{\partial}{\partial y_2}. \tag{40}$$

The function y_1 is a transcendental entire function. We suppose that the lemma is false, i.e., that y_1 and y_2 are algebraically dependent over $\mathbf{C}(z)$. Then there exists an algebraic equation

$$P = P(z, y_1, y_2) = P_l y_2^l + \cdots + P_1 y_2 + P_0 = 0, \tag{41}$$

$$P_l \not\equiv 0, \quad P_0 \not\equiv 0, \quad l \geq 1, \quad P \in \mathbf{C}[z, y_1, y_2], \quad P_k \in \mathbf{C}[z, y_1], \quad k = 0, 1, \ldots, l,$$

where P is an irreducible polynomial.

It follows from the form of the differential operators (30) and (40) that $zDP_k \in \mathbf{C}[z, y_1]$, $k = 0, 1, \ldots, l$, and $zD_1 P \in \mathbf{C}[z, y_1, y_2]$. Thus, we have $\deg \operatorname{tr}_{\mathbf{C}(z)} \{y_1, y_2\} = 1$, and the functions y_1 and y_2 are connected by the equation (41). Hence, by the corollary of Lemma 4, $zD_1 P$ is divisible by P as polynomials in three independent variables, and there exists $\omega \in \mathbf{C}[z]$ such that

$$zD_1 P \equiv \omega P. \tag{42}$$

Since the degree of zD_1P in y_1 and y_2 does not exceed the degree of P, and since its degree in z may be at most 1 more than the degree of P, it follows that $\omega = az + b$, where $a, b \in \mathbf{C}$.

If we note that

$$D_1P(z, y_1, y_2) = \frac{d}{dz}P(z, y_1, y_2),$$

and apply the operator zD_1 to both sides of (41), we find that

$$zD_1P = z(DP_l + l\alpha P_l)y_2^l + \cdots + z(DP_1 + \alpha P_1)y_2 + zDP_0 \equiv 0. \qquad (43)$$

We equate the coefficients of powers of y_2 in both sides of the identity (42), using (41) and (43). This gives

$$z(DP_k + k\alpha P_k) \equiv (az + b)P_k, \qquad k = 0, 1, \ldots, l. \qquad (44)$$

From (44) it follows that for each k the polynomial zDP_k is divisible by the polynomial P_k; then, by Lemma 5, the quotients are equal to $b_k \in \mathbf{C}$, $k = 0, 1, \ldots, l$. If we divide both sides of (44) by P_0 for $k = 0$ and by P_l for $k = l$, and if we then equate coefficients of z, we find that $a = 0$, and then also $\alpha = 0$. But this contradicts the hypothesis of the lemma. The lemma is proved.

Proof of Theorem 2. In Lemma 6 we set $\alpha = \eta/\xi$. Applying the Second Fundamental Theorem of Chapter 3 and Lemma 6 to the functions $\varphi_\lambda(z)$ and $e^{(\eta/\xi)z}$, which satisfy the system of linear differential equations (39), we obtain Theorem 2.

Let $\lambda \in \mathbf{C} \setminus \mathbf{Z}$. We denote

$$\omega(z) = \sum_{n=0}^{\infty} \frac{z^n}{n!(\lambda + n)} = z^{-\lambda} \int^z t^{\lambda-1} e^t dt. \qquad (45)$$

From (22) and (45) we obtain

$$\omega(z) = \frac{1}{\lambda} e^z \varphi_\lambda(-z). \qquad (46)$$

If $\lambda \in \mathbf{Q}$, then $\omega(z)$ is a product of E-functions, and so itself is an E-function. By Lemma 6, the functions e^z and $\varphi_\lambda(-z)$ are algebraically independent over $\mathbf{C}(z)$. It then follows from (46) that $\deg \operatorname{tr}_{\mathbf{C}(z)}\{e^z, \omega(z), \varphi_\lambda(-z)\} = 2$, and that any two of the functions e^z, $\omega(z)$ and $\varphi_\lambda(-z)$ are algebraically independent over $\mathbf{C}(z)$. In particular, this proves the following result.

Lemma 7. *If $\lambda \in \mathbf{C} \setminus \mathbf{Z}$, then the functions $\omega(z)$ and e^z are algebraically independent over $\mathbf{C}(z)$.*

Theorem 3. *For any $\lambda \in \mathbf{C} \setminus \mathbf{Z}$ and $\xi \in \mathbf{A} \setminus \{0\}$, the numbers $\omega(\xi)$ and e^{ξ} are algebraically independent.*

Proof. By differentiating (45), we see that the functions $y_0 = e^z$ and $y_1 = \omega(z)$ satisfy the system of differential equations

$$y_0' = y_0, \qquad y_1' = \frac{-\lambda}{z} y_1 + \frac{1}{z} y_0.$$

According to Lemma 7, these functions are algebraically independent over $\mathbf{C}(z)$. If we now apply the Second Fundamental Theorem of Chapter 3, we obtain the theorem.

We give some corollaries of Theorem 3.

Let λ be a rational fraction, $\lambda > 0$. Making a change of variables in (45), we obtain

$$\psi(z) = \int_0^1 t^{\lambda-1} e^{-zt} dt = \omega(-z). \tag{47}$$

Theorem 3 implies that for any $\xi \in \mathbf{A} \setminus \{0\}$, the numbers $\psi(\xi)$ and e^{ξ} are algebraically independent.

Lemma 8. *Let $k \in \mathbf{N}$. If the functions $z, f_1(z), \dots, f_n(z)$, $n \geq 1$, are algebraically independent over $\mathbf{C}$, then so are the functions $z^{1/k}, f_1(z), \dots, f_n(z)$.*

Proof. Suppose that the lemma is false, i.e., there exists

$$P = P(z, z_1, \dots, z_n) \in \mathbf{C}[z, z_1, \dots, z_n], \qquad P \not\equiv 0,$$

such that

$$P\big(z^{1/k}, f_1(z), \dots, f_n(z)\big) = 0. \tag{48}$$

We replace $z^{1/k}$ by $\zeta_i z^{1/k}$ on the left in (48), where ζ_i, $i = 1, \dots, k$, are the k-th roots of unity. Then $P\big(\zeta_i z^{1/k}, z_1, \dots, z_n\big) \not\equiv 0$ for all i, and, by the theorem on symmetric polynomials, the product

$$\prod_{i=1}^{k} P\big(\zeta_i z^{1/k}, f_1(z), \dots, f_n(z)\big) = 0$$

is an algebraic equation over $\mathbf{C}$ between $z, f_1(z), \dots, f_n(z)$, which is impossible.

We consider the "incomplete gamma-function"

$$F(z) = \int_0^z t^{\lambda-1} e^{-t} dt = z^{\lambda} \omega(-z), \qquad \lambda \in \mathbf{Q} \setminus \mathbf{Z}, \qquad \lambda > 0.$$

If we use Lemma 8, we obtain from Theorem 3: *for any* $\xi \in \mathbf{A}$, $\xi \neq 0$, *the numbers* $F(\xi)$ *and* e^{ξ} *are algebraically independent, and all of the zeros of* $F(z)$ *are transcendental.*

We define the "incomplete probability integral"

$$\Phi(z) = \int_0^z e^{-t^2} dt = \frac{1}{2} \int_0^{z^2} \frac{e^{-u}}{\sqrt{u}} du.$$

Again using Lemma 8 and Theorem 3, we find that *for any* $\xi \in \mathbf{A} \setminus \{0\}$, *the numbers* $\Phi(\xi)$ *and* e^{ξ^2} *are algebraically independent.*

Similarly, if we define

$$\Phi_\lambda(z) = \int_0^z e^{-t^\lambda} dt = \frac{1}{\lambda} \int_0^{z^\lambda} u^{\frac{1}{\lambda}-1} e^{-u} du$$

for $1/\lambda$ a positive rational number, we find that *for any* $\xi \in \mathbf{A} \setminus \{0\}$, *the numbers* $\Phi_\lambda(\xi)$ *and* e^{ξ^λ} *are algebraically independent.*

§ 3. Sets of solutions of first order linear differential equations

In Chapter 3 we showed that the Second Fundamental Theorem easily implies Lindemann's theorem on algebraic independence of the numbers $e^{\xi_1}, \dots, e^{\xi_m}$ for algebraic $\xi_1, \dots, \xi_m$ which are linearly independent over $\mathbf{Q}$, and also the equivalent fact concerning linear independence over $\mathbf{A}$ of the numbers $e^{\alpha_1}, \dots, e^{\alpha_n}$ when $\alpha_1, \dots, \alpha_n$ are distinct elements of $\mathbf{A}$.

The following result is known for the functions (17) for different values of λ and different values of the argument z.

Theorem 4 [70:10]. *Suppose that* $\lambda_0 \in \mathbf{Z}^+$; $\lambda_1, \dots, \lambda_n$, $n \geq 0$, *are rational fractions such that* $(\lambda_{j_1} - \lambda_{j_2}) \notin \mathbf{Z}$, $1 \leq j_1 < j_2 \leq n$; $\alpha_1, \dots, \alpha_m$, $m \geq 1$, *are elements of* $\mathbf{A}$ *which are linearly independent over* $\mathbf{Q}$; *and* $\xi_1, \dots, \xi_m \in \mathbf{A}$ *are distinct and nonzero. Then the* $(n+1)m$ *numbers* $\varphi_{\lambda_0}(\alpha_i)$ *and* $\varphi_{\lambda_j}(\xi_i)$, $i = 1, \dots, m$, $j = 1, \dots, n$, *are algebraically independent.*

Lindemann's theorem follows from this theorem with $\lambda_0 = 0$ and $n = 0$.

Lindemann's theorem has been generalized to the case of any transcendental E-function which satisfies a first order linear differential equation with coefficients in $\mathbf{C}[z]$.

Theorem 5 [70:12]. *Suppose that* $f(z)$ *is a transcendental E-function which satisfies a first order linear differential equation (16). Let* $\xi_1, \dots, \xi_m$, $m \geq 1$, *be* *algebraic numbers which are linearly independent over* $\mathbf{Q}$ *and have the property that* $P_1(\xi_1)\cdots P_1(\xi_m) \neq 0$. *Then the numbers*

$$f(\xi_1), \ldots, f(\xi_m) \tag{49}$$

are algebraically independent. If none of the solutions of (16) are rational functions, then the numbers (49) are algebraically independent for any distinct algebraic $\xi_1, \ldots, \xi_m$ *for which* $\xi_1 \cdots \xi_m P_1(\xi_1) \cdots P_1(\xi_m) \neq 0$.

We shall not prove Theorems 4 or 5 directly. Below we shall prove a more general theorem (Theorem 6) about algebraic independence of the values at algebraic points of a set of E-functions each of which satisfies a first order linear differential equation with coefficients in $\mathbf{C}(z)$. This theorem subsumes all known theorems of this type; in particular, Theorems 1–5 are corollaries.

By the *characteristic polynomial of a first order differential equation (16)* we mean the polynomial

$$\chi(z) = -\int_0^z R(t)dt,$$

where the polynomial $R(t)$ is the quotient (without remainder) when P_0 is divided by P_1.

We consider the system of first order linear differential equations:

$$P_{1,i}(z)y_i' + P_{0,i}(z)y_i = Q_i(z), \qquad i = 1, \ldots, m, \tag{50}$$

$$P_{1,i}(z),\ P_{0,i}(z),\ Q_i(z) \in \mathbf{C}[z].$$

We let $\chi_i(z)$ denote the characteristic polynomial of the i-th equation in (50), and we let X denote the set of indices i, $1 \leq i \leq m$, such that the i-th equation in (50) has a rational function solution (the identically zero function is allowed).

Theorem 6 (V. Kh. Salikhov). *Suppose that the set of transcendental E-functions*

$$f_1(z), \ldots, f_m(z), \qquad m \geq 1, \tag{51}$$

forms a solution of the system of linear differential equations (50), and let $\xi \in \mathbf{A}$, $\xi P_{1,1}(\xi) \cdots P_{1,m}(\xi) \neq 0$. *Further suppose that the following conditions are fulfilled:*

1°. *For any* $i, j = 1, \ldots, m$, $i \neq j$, $m \geq 2$,

$$\exp\left(\int \left(\frac{P_{0,i}(z)}{P_{1,i}(z)} - \frac{P_{0,j}(z)}{P_{1,j}(z)}\right) dz\right) \notin \mathbf{C}(z).$$

2°. *The set of characteristic polynomials* $\chi_i(z)$, $i \in X$, *is linearly independent over* $\mathbf{Q}$.

Then the numbers $f_1(\xi), \ldots, f_m(\xi)$ are algebraically independent.

Theorem 6 follows from the Second Fundamental Theorem of Chapter 3 and the following theorem on algebraic independence over $\mathbf{C}(z)$ of the functions in (51).

Theorem 7 (V. Kh. Salikhov). *Suppose that the set of transcendental entire functions (51) forms a solution of the system of linear differential equations (50), and conditions 1° and 2° in the statement of Theorem 6 are satisfied. Then the functions (51) are algebraically independent over $\mathbf{C}(z)$.*

Thus, we must prove Theorem 7 in order to establish Theorems 4–6 above. In order to do this we first prove several lemmas.

§ 4. Some lemmas

We start by generalizing Lemmas 19 and 20 of Chapter 3, § 13.

Lemma 9. *If $P_i(z)$, $i = 1, \ldots, n$, $n \geq 1$, are any distinct polynomials in $\mathbf{C}[z]$ such that $P_i(0) = 0$, $i = 1, \ldots, n$, then the functions $e^{P_i(z)}$, $i = 1, \ldots, n$, are linearly independent over $\mathbf{C}[z]$.*

Proof. For $n = 1$ the lemma is trivial. Suppose that the lemma holds for some $n = k$, $k \geq 1$; we shall prove that it holds for $n = k + 1$. Then the lemma will hold for all n by induction.

Suppose the contrary, i.e., that there exists a relation

$$\sum_{i=1}^{k+1} A_i e^{P_i(z)} = 0, \qquad A_i \in \mathbf{C}[z], \qquad A_i \not\equiv 0, \qquad i = 1, \ldots, k+1.$$

From this equality we obtain

$$\sum_{i=1}^{k} A_i e^{R_i(z)} + A_{k+1} = 0, \qquad R_i(z) = P_i(z) - P_{k+1}(z), \tag{52}$$

where $R_i(0) = 0$, $\deg R_i(z) \geq 1$, $i = 1, \ldots, k$, and $R_i(z) \not\equiv R_j(z)$ for $i \neq j$.

If $\deg A_{k+1} = s$, then, differentiating (52) $s + 1$ times, we obtain

$$\sum_{i=1}^{k} B_i e^{R_i(z)} = 0, \qquad B_i \in \mathbf{C}[z]. \tag{53}$$

It is easy to verify that $\deg B_i = \deg(A_i(R_i')^{s+1}) \geq 1$, $i = 1, \ldots, k$. Hence, $B_i \not\equiv 0$, $i = 1, \ldots, k$; but then (53) is impossible, by the induction assumption. This contradiction proves the lemma.

Lemma 10. *If*

$$P_i(z) \in \mathbf{C}[z], \qquad P_i(0) = 0, \qquad i = 1, \ldots, m, \tag{54}$$

then the functions

$$y_i = e^{P_i(z)}, \qquad i = 1, \ldots, m, \tag{55}$$

are algebraically independent over $\mathbf{C}(z)$ *if and only if the polynomials (54) are linearly independent over* $\mathbf{Q}$.

Proof. If the polynomials (54) are linearly dependent over $\mathbf{Q}$, then

$$\sum_{i=1}^{m} k_i P_i(z) = 0, \qquad k_i \in \mathbf{Z}, \qquad \sum_{i=1}^{m} k_i^2 > 0.$$

Then

$$y_1^{k_1} \cdots y_m^{k_m} = 1,$$

from which it follows that the functions (55) are algebraically dependent over $\mathbf{C}(z)$.

Now suppose that the polynomials (54) are linearly independent over $\mathbf{Q}$. If

$$P = P(z, y_1, \ldots, y_m) \in \mathbf{C}[z, y_1, \ldots, y_m], \qquad P \not\equiv 0,$$

then this polynomial can be written in the form

$$P = \sum_{(k_1, \ldots, k_m)} A_{k_1, \ldots, k_m} y_1^{k_1} \cdots y_m^{k_m},$$

$$A_{k_1, \ldots, k_m} \in \mathbf{C}[z], \qquad A_{k_1, \ldots, k_m} \not\equiv 0,$$

where the summation is taken over a finite set of m-tuples $(k_1, \ldots, k_m)$ of non-negative integers. By Lemma 9, we have

$$P\left(z, e^{P_1(z)}, \ldots, e^{P_m(z)}\right) = \sum_{(k_1, \ldots, k_m)} A_{k_1, \ldots, k_m} e^{\sum_{i=1}^{m} k_i P_i(z)} \not\equiv 0, \tag{56}$$

since the linear independence of the polynomials (54) means that all of the sums $k_1 P_1(z) + \cdots + k_m P_m(z)$ corresponding to different m-tuples $(k_1, \ldots, k_m)$ are distinct. It follows from (56) that the functions (55) are algebraically independent. The lemma is proved.

Lemma 11. *The differential equation*

$$P_1(z)y' + P_0(z)y = 0, \qquad P_0(z),\ P_1(z) \in \mathbf{C}[z], \tag{57}$$

has a solution $y(z)$ which can be written in the form

$$y(z) = \frac{\varphi(z)}{H(z)}, \qquad H(z) \in \mathbf{C}[z], \qquad \varphi(z) \not\equiv 0, \tag{58}$$

where $\varphi(z)$ is an entire function, if and only if the partial fraction decomposition of $P_0(z)/P_1(z)$ has the form

$$\frac{P_0(z)}{P_1(z)} = R(z) + \sum_{i=1}^{k} \frac{s_i}{z - \alpha_i}, \qquad R(z) \in \mathbf{C}[z], \tag{59}$$

where $k \in \mathbf{Z}^+$, $\alpha_i \in \mathbf{C}$, $s_i \in \mathbf{Z}$, $i = 1, \ldots, k$. If the condition (59) is fulfilled, then the function $\varphi(z)$ has the form

$$\begin{aligned} \varphi(z) = A(z)e^{\chi(z)}, \qquad & A(z) \in \mathbf{C}[z], \qquad A(z) \not\equiv 0, \\ & \chi(z) = -\int_0^z R(t)dt. \end{aligned} \tag{60}$$

Proof. If we have the decomposition (59), then, integrating it and using the differential equation (57), we obtain (58), in which $\varphi(z)$ is an entire function of the form (60).

Conversely, if the function (58) satisfies the differential equation (57), then the singular points of $P_0(z)/P_1(z)$ are the zeros and poles of $\varphi(z)/H(z)$. All of the singular points must be first order poles with residues in $\mathbf{Z}$. Consequently, we have the decomposition (59). The lemma is proved.

Lemma 12. *Suppose that the functions (51), $m \geq 2$, are analytic in some region and form a solution of the system of linear differential equations (50). Further suppose that*

$$f_i(z) \notin \mathbf{C}(z), \qquad i = 1, \ldots, m, \tag{61}$$

and the system (50) satisfies condition $1°$ of Theorem 6. Then the functions

$$1, f_1(z), \ldots, f_m(z) \tag{62}$$

are linearly independent over $\mathbf{C}(z)$.

Proof. Suppose the contrary, i.e., that the functions (62) are linearly dependent over $\mathbf{C}(z)$, and let k, $k < m$, be the dimension of the $\mathbf{C}(z)$-vector space $\mathbf{L}$ spanned by the functions (62). Then $k \geq 2$, by (61). Re-numbering the functions (51), if necessary, we may assume that the functions

$$1, f_1(z), \dots, f_{k-1}(z) \tag{63}$$

form a basis of $\mathbf{L}$. Then the functions $1, f_1(z), \dots, f_k(z)$ are linearly dependent over $\mathbf{C}[z]$, and we have an equation

$$P_0 + \sum_{i=1}^{k} P_i f_i(z) = 0, \qquad P_i \in \mathbf{C}[z], \qquad i = 1, \dots, m, \tag{64}$$

with $P_k \not\equiv 0$. By (61), we have $P_k f_k(z) \notin \mathbf{C}(z)$. It then follows from (64) that there exists l, $1 \leq l \leq k-1$, such that $P_l \not\equiv 0$.

If we differentiate the equation (64) and use the differential equation (50), we obtain

$$P_0' + \sum_{i=1}^{k} \frac{P_i Q_i}{P_{1,i}} + \sum_{i=1}^{k} \left(P_i' - \frac{P_{0,i}}{P_{1,i}} P_i \right) f_i(z) = 0. \tag{65}$$

Since the functions (62) form a basis of the vector space $\mathbf{L}$, the corresponding coefficients of the linear equations (64) and (65) must be proportional. Hence, in particular, we have

$$\frac{P_l' - \frac{P_{0.l}}{P_{1.l}} P_l}{P_l} = \frac{P_k' - \frac{P_{0.k}}{P_{1.k}} P_k}{P_k},$$

i.e.,

$$\frac{P_{0,l}}{P_{1,l}} - \frac{P_{0,k}}{P_{1,k}} = \frac{P_l'}{P_l} - \frac{P_k'}{P_k}. \tag{66}$$

But the equality (66) contradicts condition 1°. This contradiction proves the lemma.

§ 5. Proof of the theorems

Proof of Theorem 7. If $m = 1$, the theorem is obvious. Suppose that $m \geq 2$. We suppose that the theorem is false, i.e., that the functions (51) are algebraically dependent over $\mathbf{C}(z)$. Let l be the number such that any $l-1$ of these functions are algebraically independent over $\mathbf{C}(z)$, but there is a set of l of them which are algebraically dependent over $\mathbf{C}(z)$. Since all of the functions in (51) are transcendental, it follows that l satisfies the inequalities $2 \leq l \leq m$. By possibly re-numbering the functions (51), we may suppose that the functions $f_1(z), \ldots, f_l(z)$ are algebraically dependent over $\mathbf{C}(z)$, the first ν differential equations in (50) have solutions in $\mathbf{C}(z)$, $0 \leq \nu \leq l$, and the next $l - \nu$ of those differential equations do not have solutions in $\mathbf{C}(z)$.

We consider the i-th equation in (50):

$$P_{1,i}y_i' + P_{0,i}y_i = Q_i, \qquad 0 \leq i \leq \nu. \tag{67}$$

Let G_i/H_i, with $G_i,\ H_i \in \mathbf{C}[z]$, satisfy (67). Then the function

$$y_i^* = f_i(z) - \frac{G_i}{H_i}, \qquad 1 \leq i \leq \nu, \tag{68}$$

satisfies the differential equation

$$P_{1,i}y_i' + P_{0,i}y_i = 0, \qquad 1 \leq i \leq \nu.$$

Applying Lemma 11 to the function (68), we find that

$$y_i^* = f_i(z) - \frac{G_i}{H_i} = \frac{A_i}{H_i}e^{\chi_i(z)}, \tag{69}$$

$$A_i(z) \in \mathbf{C}[z], \qquad A_i \not\equiv 0, \qquad 1 \leq i \leq \nu,$$

where $\chi_i(z)$ is the characteristic polynomial of the differential equation (67).

By assumption, there exists an irreducible polynomial

$$P = P(z, y_1, \ldots, y_l) \in \mathbf{C}[z, y_1, \ldots, y_l], \qquad P \not\equiv 0,$$

such that

$$P(z, f_1(z), \ldots, f_l(z)) = 0. \tag{70}$$

We make the following substitutions for the variables $y_1, \ldots, y_l$ in the polynomial P:

$$\begin{aligned} y_i &= \frac{A_i w_i + G_i}{H_i}, \qquad & i &= 1, \ldots, \nu, \\ y_i &= w_i, & i &= \nu + 1, \ldots, l. \end{aligned} \tag{71}$$

As a result we obtain a polynomial in $w_1, \ldots, w_l$ with coefficients in $\mathbf{C}(z)$. We can multiply it by some polynomial $R \in \mathbf{C}[z]$ so as to obtain an irreducible polynomial

$$P^* = P^*(z, w_1, \dots, w_l) \in \mathbf{C}[z, w_1, \dots, w_l], \qquad P^* \not\equiv 0. \tag{72}$$

In view of (71) and (72), the equation (70) becomes:

$$P^*(z, e^{\chi_1(z)}, \dots, e^{\chi_\nu(z)}, f_{\nu+1}(z), \dots, f_l(z)) = 0. \tag{73}$$

There are two possible cases.

Case (1) $\nu = l$. Then it follows from (73) that the functions $e^{\chi_1(z)}, \dots, e^{\chi_l(z)}$ are algebraically dependent over $\mathbf{C}(z)$; but this is impossible because of condition 2° of Theorem 6 and Lemma 10. Consequently, equation (73), along with (70), cannot hold. Thus, the set of functions (51) is algebraically independent, and the theorem is proved in case (1).

Case (2) $\nu < l$. We denote $\chi_i'(z) = -R_i(z)$, $i = 1, \dots, l$, and we consider the system of differential equations

$$\begin{aligned} w_i' + R_i w_i &= 0, && i = 1, \dots, \nu, \\ P_{1,i} w_i' + P_{0,i} w_i &= Q_i, && i = \nu + 1, \dots, l, \end{aligned} \tag{74}$$

and the associated differential operator

$$D = \frac{\partial}{\partial z} - \sum_{i=1}^{\nu} R_i w_i \frac{\partial}{\partial w_i} + \sum_{i=\nu+1}^{l} \frac{Q_i - P_{0,i} w_i}{P_{1,i}} \frac{\partial}{\partial w_i}, \tag{75}$$

$$D: \mathbf{C}(z)[w_1, \dots, w_l] \longrightarrow \mathbf{C}(z)[w_1, \dots, w_l].$$

Clearly, the functions

$$e^{\chi_1(z)}, \dots, e^{\chi_\nu(z)}, \; f_{\nu+1}(z), \dots, f_l(z) \tag{76}$$

form a solution of the system (74). Since the functions in (76) are connected by the relation (73) over $\mathbf{C}(z)$, it follows from the corollary of Lemma 4 that there exists a function $Q \in \mathbf{C}(z)$ such that

$$DP^* \equiv QP^*. \tag{77}$$

We now demonstrate some properties of a polynomial P^* satisfying the identity (77). Let $\deg P^* = s$. We write P^* in the following form:

$$P^* = \sum_{k_1 + \dots + k_l \le s} A_{k_1, \dots, k_l} w_1^{k_1} \cdots w_l^{k_l}, \tag{78}$$

$$A_{k_1,\dots,k_l} \in \mathbf{C}[z], \qquad k_i \in \mathbf{Z}^+, \qquad i = 1, \dots, l.$$

From (75), (77) and (78) we have

$$\sum_{k_1+\dots+k_l \le s} \left(\left(A'_{k_1,\dots,k_l} - \left(\sum_{i=1}^{\nu} k_i R_i + \sum_{i=\nu+1}^{l} k_i \frac{P_{0,i}}{P_{1,i}} \right) A_{k_1,\dots,k_l} \right) w_1^{k_1} \cdots w_l^{k_l} + \right.$$
$$\left. + \sum_{i=\nu+1}^{l} \left(k_i \frac{Q_i}{P_{1,i}} A_{k_1,\dots,k_l} w_i^{k_i - 1} \prod_{\substack{j=1 \\ j \ne i}}^{l} w_j^{k_j} \right) \right) \equiv$$
$$\equiv Q \sum_{k_1+\dots+k_l \le s} A_{k_1,\dots,k_l} w_1^{k_1} \cdots w_l^{k_l}. \tag{79}$$

We consider all of the nonzero terms in the polyomial P^* in (78) which depend on the variables $w_{\nu+1}, \dots, w_l$, and we choose any such term which has maximum total degree in $w_1, \dots, w_l$. Suppose that this term has the form

$$A_{r_1,\dots,r_l} w_1^{r_1} \cdots w_l^{r_l}, \qquad A_{r_1,\dots,r_l} \ne 0.$$

We set $r = r_1 + \dots + r_l$. It follows from (79) and the definition of r that every coefficient $A_{s_1,\dots,s_l}$ of P^* with $s_1 + \dots + s_l = r$ satisfies the differential equation

$$y' = \left(Q + \sum_{i=1}^{\nu} s_i R_i + \sum_{i=\nu+1}^{l} s_i \frac{P_{0,i}}{P_{1,i}} \right) y. \tag{80}$$

Let j_0, $\nu + 1 \le j_0 \le l$, be an index such that $r_{j_0} > 0$. We consider the coefficient $A_{k_1,\dots,k_l}$ of P^* having indices $k_i = r_i$ if $i \ne j_0$ and $k_{j_0} = r_{j_0} - 1$. Note that $A_{k_1,\dots,k_l}$ could be zero.

If we equate the coefficients of $w_1^{k_1} \cdots w_l^{k_l}$ in (79), we find that the polynomial $A_{k_1,\dots,k_l}$ satisfies the equation

$$y' - \left(Q + \sum_{i=1}^{\nu} k_i R_i + \sum_{i=\nu+1}^{l} k_i \frac{P_{0,i}}{P_{1,i}} \right) y +$$
$$+ \sum_{j=\nu+1}^{l} (k_j + 1) \frac{Q_j}{P_{1,j}} A_{k_1,\dots,k_j+1,\dots,k_l} = 0. \tag{81}$$

In (81) and below, for brevity we let $A_{k_1,\dots,k_j+1,\dots,k_l}$ denote the coefficient $A_{k_1,\dots,k_{j-1},k_j+1,k_{j+1},\dots,k_l}$. We now show that the same differential equation is also satisfied by the function

$$\psi(z) = -\sum_{j=\nu+1}^{l} (k_j + 1)A_{k_1,\dots,k_j+1,\dots,k_l} f_j(z). \tag{82}$$

To see this, we use the fact that, because $k_1 + \cdots + (k_j + 1) + \cdots + k_l = r$, the polynomial $A_{k_1,\dots,k_j+1,\dots,k_l}$ satisfies the differential equation (80) with parameters $k_1, \dots, k_j + 1, \dots, k_l$ in place of $s_1, \dots, s_l$. We obtain:

$$\psi'(z) - \left(Q + \sum_{i=1}^{\nu} k_i R_i + \sum_{i=\nu+1}^{l} k_i \frac{P_{0,i}}{P_{1,i}}\right)\psi(z) +$$

$$+ \sum_{j=\nu+1}^{l} (k_j + 1)\frac{Q_j}{P_{1,j}} A_{k_1,\dots,k_j+1,\dots,k_l} =$$

$$= -\sum_{j=\nu+1}^{l} (k_j + 1)\left(A'_{k_1,\dots,k_j+1,\dots,k_l} f_j(z) + \right.$$

$$\left. + \left(\frac{Q_j}{P_{1,j}} - \frac{P_{0,j}}{P_{1,j}} f_j(z)\right) A_{k_1,\dots,k_j+1,\dots,k_l} \right) +$$

$$+ \left(Q + \sum_{i=1}^{\nu} k_i R_i + \sum_{i=\nu+1}^{l} k_i \frac{P_{0,i}}{P_{1,i}}\right) \sum_{j=\nu+1}^{l} (k_j+1) A_{k_1,\dots,k_j+1,\dots,k_l} f_j(z) +$$

$$+ \sum_{j=\nu+1}^{l} (k_j + 1)\frac{Q_j}{P_{1,j}} A_{k_1,\dots,k_j+1,\dots,k_l} =$$

$$= \sum_{j=\nu+1}^{l} (k_j + 1) f_j(z) \left(\left(Q + \sum_{i=1}^{\nu} k_i R_i + \sum_{i=\nu+1}^{l} k_i \frac{P_{0,i}}{P_{1,i}} + \right.\right.$$

$$\left.\left. + \frac{P_{0,j}}{P_{1,j}} \right) A_{k_1,\dots,k_j+1,\dots,k_l} - A'_{k_1,\dots,k_j+1,\dots,k_l} \right) = 0.$$

Thus, the two functions $A_{k_1,\dots,k_l}$ and (82) satisfy the differential equation (81), and so the difference

$$Y_{k_1,\dots,k_l}(z) = A_{k_1,\dots,k_l}(z) - \psi(z)$$

is a solution of the differential equation (80) with parameters $k_1, \dots, k_l$.

We consider the function

$$\omega(z) = \frac{Y_{k_1,\dots,k_l}(z)}{A_{k_1,\dots,k_{j_0}+1,\dots,k_l}}. \tag{83}$$

Since the polynomial $A_{k_1,\dots,k_{j_0}+1,\dots,k_l}$ is a solution of the differential equation (80) with parameters $k_1, \dots, k_{j_0}+1, \dots, k_l$, and the function $Y_{k_1,\dots,k_l}(z)$ is a solution of the same differential equation, except with parameters $k_1, \dots, k_l$, it follows from the equality

$$\frac{\omega'(z)}{\omega(z)} = \frac{Y'_{k_1,\dots,k_l}(z)}{Y_{k_1,\dots,k_l}(z)} - \frac{A'_{k_1,\dots,k_{j_0}+1,\dots,k_l}}{A_{k_1,\dots,k_{j_0}+1,\dots,k_l}},$$

that the function $\omega(z)$ satisfies the differential equation

$$P_{1,j_0}y' + P_{0,j_0}y = 0. \tag{84}$$

Consequently, the function

$$\begin{aligned}\varphi(z) = f_{j_0}(z) - \frac{1}{k_{j_0}+1}\omega(z) &= \\ = -\frac{1}{(k_{j_0}+1)A_{k_1,\dots,k_{j_0}+1,\dots,k_l}}&\Bigg(A_{k_1,\dots,k_l} + \\ &+ \sum_{\substack{j=\nu+1\\ j\neq j_0}}^{l} (k_j+1)A_{k_1,\dots,k_j+1,\dots,k_l} f_j(z)\Bigg)\end{aligned} \tag{85}$$

satisfies the j_0-th differential equation of the system (50).

By the hypothesis of the theorem, all of the functions in (51) are transcendental, and the system of differential equations (50) satisfies the condition 1°. Hence, if we apply Lemma 12 to the system (50) and the solution

$$f_1(z), \dots, f_{j_0-1}(z), \varphi(z), f_{j_0+1}(z), \dots, f_m(z), \tag{86}$$

and if we take into account the linear equation (85) connecting the functions in (86) and 1 over $\mathbf{C}(z)$, we conclude that $\varphi(z)$ must be a rational function. But this cannot be the case, since the j_0-th differential equation in (50) does not have a rational function solution. This contradiction shows that the functions in (51) are algebraically independent over $\mathbf{C}(z)$ in case (2). This completes the proof of Theorem 7.

By the same token we have also proved Theorem 6. We now prove Theorems 4 and 5.

Proof of Theorem 4. We consider the set of $(n+1)m$ E-functions

$$\varphi_{\lambda_0}(\alpha_i z), \qquad \varphi_{\lambda_j}(\xi_i z), \qquad i = 1, \ldots, m, \qquad j = 1, \ldots, n,$$

$$\lambda_0 \in \mathbf{Z}^+, \qquad \lambda_j \in \mathbf{Q} \setminus \mathbf{Z}, \qquad (\lambda_{j_1} - \lambda_{j_2}) \notin \mathbf{Z}, \qquad 1 \le j_1 < j_2 < n.$$

From the differential equation (18) we find that these functions form a solution of the system of linear differential equations

$$\begin{aligned} y_i' &= \left(\alpha_i - \frac{\lambda_0}{z}\right) y_i + \frac{\lambda_0}{z}, \qquad i = 1, \ldots, m, \\ y_{i,j}' &= \left(\xi_i - \frac{\lambda_j}{z}\right) y_{i,j} + \frac{\lambda_j}{z}, \quad i = 1, \ldots, m, \quad j = 1, \ldots, m. \end{aligned} \tag{87}$$

The hypothesis of Theorem 4 obviously implies that condition 1° of Theorem 6 is fulfilled.

Integrating the first m equations in (87) (see (19)), we find that all of these equations have rational function solutions, since $\lambda_0 \in \mathbf{Z}^+$ in Theorem 4. The other mn equations of this system do not have solutions in $\mathbf{Q}(z)$, since otherwise there would be values of i and j, $1 \le i \le m$, $1 \le j \le n$, for which the transcendental entire function $\varphi_{\lambda_j}(\xi_i z)$ could be written in the form

$$\varphi_{\lambda_j}(\xi_i z) = c z^{-\lambda_j} e^{\xi_i z} + q, \qquad c \in \mathbf{C}, \qquad q \in \mathbf{C}(z),$$

and this is impossible because all of the $\lambda_1, \ldots, \lambda_n$ in Theorem 4 are fractional rational numbers.

From what was proved above, we know that $\chi_i(z) = \alpha_i z$, $i = 1, \ldots, m$, for the equations in (87) having rational solutions, and these polynomials are linearly independent over $\mathbf{Q}$, since in Theorem 4 the numbers $\alpha_1, \ldots, \alpha_m$ are assumed to be linearly independent over $\mathbf{Q}$. Thus, condition 2° of Theorem 6 is also fulfilled.

Using Theorem 6 with $z = 1$, we find that the numbers

$$\varphi_{\lambda_0}(\alpha_i), \qquad \varphi_{\lambda_j}(\xi_i), \qquad i = 1, \ldots, m, \qquad j = 1, \ldots, n,$$

are algebraically independent. Theorem 4 is proved.

To prove Theorem 5 we need one more lemma.

Lemma 13. *If a transcendental entire function of order 1 satisfies the first order linear differential equation (16), then*

$$\deg P_1(z) = \deg P_0(z).$$

This lemma is a special case of some general results on entire function solutions of linear differential equations with coefficients in $\mathbf{C}[z]$ which are contained in Valiron's monograph [81:1] and are proved by the Wiman–Valiron method. In particular, the following fact is proved there: if a transcendental entire function of order ρ satisfies the first order differential equation (16), $s_0 = \deg P_0(z)$, and $s_1 = \deg P_1(z)$, then $\rho \geq 1$ and $\rho = s_0 - s_1 + 1$. Lemma 13 follows from the latter equality with $\rho = 1$. We shall not prove the Valiron result in this book.

Proof of Theorem 5. From Lemma 13 it follows that the characteristic polynomial of the differential equation (16) has the form $\chi(z) = az$, where $a \in \mathbf{C}$, $a \neq 0$. We consider the m-tuple of E-functions $f(\xi_1 z), \ldots, f(\xi_m z)$, which satisfy the system of differential equations

$$\frac{1}{\xi_i} P_1(\xi_i z) y_i' + P_0(\xi_i z) y_i = Q(\xi_i z), \qquad i = 1, \ldots, m. \tag{88}$$

If we take the partial fraction decomposition of the difference $\xi_i P_0(\xi_i z)/ P_1(\xi_i z) - \xi_j P_0(\xi_j z)/P_1(\xi_j z)$, $i \neq j$, integrate, and use the fact that the numbers $\xi_1, \ldots, \xi_m$ are distinct, we find that condition $1°$ of Theorem 6 holds. For the system (88) we have $\chi_i(z) = a\xi_i z$, $i = 1, \ldots, m$, and it then obviously follows from the hypothesis of Theorem 5 that condition $2°$ of Theorem 6 is fulfilled. Hence, by Theorem 6 with $z = 1$, we conclude that the numbers (49) are algebraically independent. Theorem 5 is proved.

Lindemann's theorem and Theorem 2 follow quite easily from Theorem 6, since the verification of the conditions $1°$ and $2°$ is trivial in those cases.

Remarks

Lemmas 1 and 2 on hypergeometric functions can be found in Siegel's works [72:3,4].

In 1929 in [72:3], Siegel stated that the method in his paper could be used to prove Theorem 1. However, he did not give the proof. It is not possible to obtain Theorem 1 from Siegel's general theorem on algebraic independence of the values of E-functions in his 1949 monograph [72:4], since that theorem can be applied only to sets of E-functions which satisfy homogeneous linear differential equations, in view of the failure of solutions of non-homogeneous differential equations to satisfy the basic condition in the theorem.

Theorem 1 was first proved in 1954 in [70:1,9] as a corollary of a theorem that was more general than Siegel's. As mentioned above, the Second Fundamental Theorem, published in 1955 in [70:2,8], implies Theorem 1 without any additional arguments.

Theorem 2 is a consequence of the more general theorem published in 1959 in [70:10]. Theorem 4 was also proved in that paper. A special case of Theorem 4, concerning the values of a single function $\varphi_\lambda(z)$ at several linearly independent points, was proved in 1954 in [70:1,9]. Theorem 3 was also published at that time.

The transcendence of the integrals at the end of § 2 was asserted without proof in Siegel's 1929 paper [72:3]. More general facts concerning the algebraic independence of values of these integrals and the integrands follow from theorems which appeared in 1960–61 in [70:11,23].

Theorem 5 was published in 1961 in [70:12]. After it appeared, it became natural to combine into a general theorem all of the known results on algebraic independence of the values of E-functions which satisfy first order linear differential equations. This theorem (Theorem 6) was proved by Salikhov in 1973 in [66:1]. It contained no new results about the arithmetic properties of concrete E-functions, since all of the known E-functions which are solutions of first order linear differential equations had been studied earlier.

Chapter 6
Algebraic independence of the values of E-functions which satisfy second order linear differential equations

§ 1. A general theorem on algebraic independence of the values of an E-function and its derivative

We consider a transcendental E-function $f(z)$ which satisfies a linear differential equation of the form

$$P_2 y'' + P_1 y' + P_0 y = Q, \tag{1}$$

$$P_i = P_i(z) \in \mathbf{C}[z], \quad i = 0, 1, 2, \qquad Q = Q(z) \in \mathbf{C}[z].$$

If $f(z)$ does not satisfy any first order algebraic differential equation with coefficients in $\mathbf{C}[z]$, then, by Theorem 3 of Chapter 3, the numbers $f(\xi)$ and $f'(\xi)$ are algebraically independent for any $\xi \in \mathbf{A}$, $\xi P_2(\xi) \neq 0$. On the other hand, if $f(z)$ does satisfy such an algebraic differential equation, then, by Theorem 1 of Chapter 4, at least one of the numbers $f(\xi)$, $f'(\xi)$ is transcendental if $\xi \in \mathbf{A}$, $\xi P_2(\xi) \neq 0$. However, the other of the two numbers may turn out to be algebraic.

For example, the function $f(z) = (z-2)e^z$ satisfies the differential equation $y'' - 2y' + y = 0$, which has no singular points. The derivative is $f'(z) = (z-1)e^z$. At $\xi = 2$ we have $f(2) = 0$ and $f'(2) = e^2$, and at $\xi = 1$ we have $f'(1) = 0$ and $f(1) = -e$. In this example, $f(z)$ satisfies the first order linear differential equation $(z-2)y' + (z-1)y = 0$, which has a singular point at $z = 2$ and has $z = 1$ as the zero of the coefficient of y.

The following conjecture is very likely true: *If the transcendental E-function $f(z)$ satisfies the second order differential equation (1) and does not satisfy any first order linear differential equation with coefficients in* $\mathbf{C}[z]$, *then the numbers $f(\xi)$ and $f'(\xi)$ are transcendental for any $\xi \in$* $\mathbf{A}$ *with $\xi P_2(\xi) \neq 0$.*

The next lemma enables us to prove a general theorem about transcendence of the values at algebraic points of an E-function $f(z)$ which satisfies a differential equation of the form (1).

Lemma 1. *Suppose that the transcendental analytic function $f(z)$ is a solution of the second order linear differential equation (1) and also satisfies the first order differential equation*

$$P(z, f(z), f'(z)) = 0, \qquad P = P(z, y, y') \in \mathbf{C}[z][y, y'], \tag{2}$$

where P is an irreducible polynomial of degree k in y and y', $k \geq 1$. Then the polynomial P includes a term of the form

$$A(y')^k, \qquad A = A(z) \in \mathbf{C}[z], \qquad A \not\equiv 0. \tag{3}$$

Proof. Let

$$D = \frac{\partial}{\partial z} + y' \frac{\partial}{\partial y} - \left(\frac{P_1}{P_2} y' + \frac{P_0}{P_2} y - \frac{Q}{P_2}\right) \frac{\partial}{\partial y'} \tag{4}$$

be the differential operator associated to the differential equation (1). Then $DP \in \mathbf{C}(z)[y, y']$, and, by the corollary of Lemma 4 of Chapter 4, there exists a function $\omega \in \mathbf{C}(z)$ such that

$$DP \equiv \omega P. \tag{5}$$

From (5) it follows either that $DP \equiv 0$, or else that the polynomials P and DP have the same monomial terms in y and y' with nonzero coefficients. Let

$$A(y')^{k_2} y^{k_1}, \qquad A \in \mathbf{C}[z], \ A \not\equiv 0, \qquad k_2 \geq 0, \ k_1 \geq 0, \ k_2 + k_1 = k, \tag{6}$$

be the leading term in the lexicographical ordering with respect to powers of y' and y among all the homogeneous terms in P of maximum degree k. If $k_1 = 0$, then the lemma holds. If $k_1 > 0$, then it follows from the form of the differential operator (4) that the leading term in the polynomial DP among the homogeneous terms of degree k is

$$k_1 A(y')^{k_2+1} y^{k_1+1}, \tag{7}$$

which has higher order than the term in (6). Thus, $DP \not\equiv 0$.

However, we showed that the polynomials P and DP then have the same monomial terms in y and y'. Thus, the term (7) cannot occur in DP. This contradiction shows that we always have $k_1 = 0$, and the leading term in the polynomial has the form (3). The lemma is proved.

We note that Lemma 1 remains true if the differential equation (1) in the hypothesis has coefficients in a field $\mathcal{L}$ of analytic functions which is closed under differentiation, and if $P = P(y, y') \in \mathcal{L}[y, y']$.

Theorem 1. *Suppose that the transcendental E-function $f(z)$ is a solution of the second order linear differential equation (1), and also satisfies the first order algebraic differential equation (2), and let $\xi \in \mathbf{A}$, $\xi P_2(\xi)A(\xi) \neq 0$, where $A(z)$ is the coefficient of the leading term in the polynomial P in (2), which by Lemma 1 has the form (3). Then $f(\xi)$ is a transcendental number.*

Theorem 1 follows from Theorem 5 of Chapter 4 with $m = 2$ and $l = 1$ and from Lemma 1.

Once again, the simplest example of Theorem 1 is given by the functions $\sin z$ and $\cos z$.

When $\deg P > 1$, Theorem 1 leaves unresolved the question of the arithmetic properties of the numbers $f(\xi)$ and $f'(\xi)$ at points $\xi \in \mathbf{A}$, $\xi P_2(\xi) \neq 0$, for which $A(\xi) = 0$.

§ 2. The functions $K_\lambda(z)$ associated to Bessel functions

We consider the function

$$K_\lambda(z) = \sum_{n=0}^{\infty} \frac{(-1)^n}{n![\lambda+1,n]} \left(\frac{z}{2}\right)^{2n}, \qquad \lambda \neq -1, -2, \dots, \tag{8}$$

which satisfies the second order linear differential equation

$$y'' + \frac{2\lambda+1}{z} y' + y = 0. \tag{9}$$

This function differs only by a simple factor from the Bessel function J_λ with index λ:

$$J_\lambda(z) = \frac{1}{\Gamma(\lambda+1)} \left(\frac{z}{2}\right)^{\lambda} K_\lambda(z), \tag{10}$$

where $\Gamma(z)$ is the Euler gamma-function. If $\lambda \in \mathbf{Q}$, then by Lemma 2 of Chapter 5 the function (8) is a hypergeometric E-function.

Theorem 2 (C. Siegel). *If $\lambda \in \mathbf{Q}$, $\lambda \neq -k_1$, $k_1 \in \mathbf{N}$, $\lambda \neq (2k_2-1)/2$, $k_2 \in \mathbf{Z}$, and $\xi \in \mathbf{A} \setminus \{0\}$, then the numbers $K_\lambda(\xi)$ and $K'_\lambda(\xi)$ are algebraically independent.*

To prove this theorem we must show that the functions $K_\lambda(z)$ and $K'_\lambda(z)$ are algebraically independent over $\mathbf{C}(z)$.

Let $\mathcal{L}$ be a field of analytic functions which is closed under differentiation.

Lemma 2. *If the differential equation*

$$y'' + A_1 y' + A_0 y = 0, \qquad A_0, A_1 \in \mathcal{L}, \tag{11}$$

has a nontrivial solution y_0 such that the functions y_0 and y_0' are algebraically independent over the field $\mathcal{L}$, then this differential equation has a solution y_* whose logarithmic derivative y_*'/y_* is an algebraic function over $\mathcal{L}$.

Proof. If y_0 is an algebraic function over $\mathcal{L}$, then we can set $y_* = y_0$, since then y_*'/y_* will be an algebraic function over $\mathcal{L}$. Now suppose that y_0 is not an algebraic function over $\mathcal{L}$. Then there exists an irreducible polynomial

$$P = P(y, y') \in \mathcal{L}[y, y'], \qquad \deg_y P = k, \qquad k \geq 1, \tag{12}$$

such that

$$P(y_0, y_0') = 0. \tag{13}$$

We consider the differential operator

$$D = \frac{\partial}{\partial z} + y' \frac{\partial}{\partial y} - (A_1 y' + A_0 y) \frac{\partial}{\partial y'} \tag{14}$$

associated with the differential equation (11). Then $DP \in \mathcal{L}[y, y']$.

Since the transcendence degree of the two functions y_0 and y_0' over $\mathcal{L}$ is 1, by assumption, and these functions satisfy the differential equation (13), it follows by Lemma 4 of Chapter 5 that there exists a function $\omega \in \mathcal{L}$ such that

$$DP \equiv \omega P. \tag{15}$$

We let $Q = Q(y, y')$ denote the sum of all of the homogeneous terms of degree k in the polynomial (12). Since the differential equation (11) is homogeneous, it follows that, when the operator D in (14) is applied to P, each sum of homogeneous terms of P of a given degree goes either to the sum of homogeneous terms of the same degree in the polynomial DP or to zero. Thus, from (15) we find that

$$DQ \equiv \omega Q. \tag{16}$$

Now let y be an arbitrary solution of the differential equation (11). Then applying the operator D to the polynomial $Q(y, y')$ is equivalent to taking the total derivative with respect to z:

$$DQ(y, y') = \frac{d}{dz} Q(y, y'). \tag{17}$$

From (16) and (17) we obtain the following relation for any solution to the differential equation (11):

$$\frac{d}{dz} Q(y, y') = \omega\, Q(y, y'). \tag{18}$$

We consider the differential equation

$$v' = \omega v. \tag{19}$$

It has the general solution

$$v = c\, v_0, \qquad c \in \mathbf{C}, \qquad v_0 \not\equiv 0, \tag{20}$$

where $v_0 = v_0(z)$ is any particular solution.

From (18) it follows that the function $Q(y, y')$ is a solution of the differential equation (19) for any solution y of the differential equation (11). Let y_1 and y_2 be any two fixed linearly independent solutions of (11). Then

$$y = \gamma_1 y_1 + \gamma_2 y_2, \qquad \gamma_1, \gamma_2 \in \mathbf{C},$$

is also a solution of (11) for any γ_1 and γ_2. Hence,

$$Q(\gamma_1 y_1 + \gamma_2 y_2, \gamma_1 y_1' + \gamma_2 y_2') = c(\gamma_1, \gamma_2) v_0, \tag{21}$$

where $c(\gamma_1, \gamma_2)$ is a homogeneous polynomial in γ_1 and γ_2 with coefficients in $\mathbf{C}$, since the left side of (21) is a homogeneous polynomial in γ_1 and γ_2 of degree k. This implies that γ_1 and γ_2 can be chosen both nonzero in such a way that $c(\gamma_1, \gamma_2) = 0$.

We let $y_* = \gamma_1 y_1 + \gamma_2 y_2$ denote the solution corresponding to such a choice of γ_1 and γ_2. Then from (21) we find that $Q(y_*, y_*') = 0$, i.e.,

$$Q(1, y_*'/y_*) = 0. \tag{22}$$

From (22) it now follows that y_*'/y_* is an algebraic function over $\mathcal{L}$. The lemma is proved.

Lemma 3. *If $\lambda \in \mathbf{C}$, $\lambda \neq (2k-1)/2$, $k \in \mathbf{Z}$, and y is any nontrivial solution of the differential equation (9), then the functions y and y' are algebraically independent over $\mathbf{C}(z)$, i.e., y does not satisfy any first order algebraic differential equation with coefficients in $\mathbf{C}(z)$.*

Proof. We suppose the contrary, i.e., that y is a nontrivial solution of (9) such that y and y' are algebraically dependent over $\mathbf{C}(z)$. Then, by Lemma 2, the differential equation (9) has a solution y_* such that the function $u = y_*'/y_*$ is algebraic. The function u is an analytic function at all points $z \neq 0, \infty$. Hence its only possible branch points are $z = 0$ and $z = \infty$.

We prove that $z = \infty$ is not a branch point of u. Then $z = 0$ will not be a branch point either, since an algebraic function cannot have only one branch point. This will prove that u is a rational function.

Using the differential equation (9), we find that the function u satisfies the Riccati differential equation

$$u' + \frac{2\lambda + 1}{z} u + u^2 = -1. \tag{23}$$

Let

$$u = \sum_{k=0}^{\infty} c_k z^{r_k}, \qquad c_k \in \mathbf{C}, \qquad c_0 \neq 0, \qquad r_0 > r_1 > r_2 > \cdots, \tag{24}$$

be the power series expansion with rational exponents for a branch of u in a neighborhood of $z = \infty$. Substituting (24) in the differential equation (23), we obtain:

$$\sum_{k=0}^{\infty} (r_k + 2\lambda + 1) c_k z^{r_k - 1} + \sum_{k=0}^{\infty} \sum_{l=0}^{\infty} c_k c_l z^{r_k + r_l} = -1. \tag{25}$$

Comparing the coefficients of the terms with the highest power of z in (25), we find that $r_0 = 0$.

We next prove that all of the r_k with $k > 0$ are negative integers. Suppose this is not the case, and that r_{k_1} is the fractional exponent with minimal index $k_1 > 0$ for which the corresponding coefficient c_{k_1} in (24) is nonzero. In (25) we equate the coefficients of the power of z with the largest fractional exponent. All of the fractional exponents in the first sum in (25) are less than r_{k_1}, and when we add the two identical terms with exponent r_{k_1} in the double sum we obtain $2c_{k_1} c_0 z^{r_{k_1}}$. Because of the monotonic property of the exponents, the exponent is $< r_k$ in all of the other terms in the double sum having fractional exponent. Hence, $2c_{k_1} c_0 = 0$, and so $c_{k_1} = 0$. This contradiction shows that all of the r_k with $k > 0$ are negative integers, $z = \infty$ is not a branch point of u, and $u = y_*'/y_*$ is a rational function.

The function y_*'/y_* is not identically zero. In fact, otherwise we would have $y_* = c$ with $c \in \mathbf{C}$. Substituting this value for y_* in the differential equation (9), we obtain a contradiction. Thus, we may assume that

$$\frac{y_*'}{y_*} = -\frac{P_0}{P_1}, \qquad P_0 \not\equiv 0, \quad P_1 \not\equiv 0, \tag{26}$$

where P_0 and P_1 are relatively prime polynomials in $\mathbf{C}[z]$.

We consider the linear form in y and y'

$$R = P_0 y + P_1 y', \tag{27}$$

where y is an arbitrary solution of the differential equation (9).

Let

$$D = \frac{\partial}{\partial z} + y' \frac{\partial}{\partial y} - \left(\frac{2\lambda + 1}{z} y' + y \right) \frac{\partial}{\partial y'} \tag{28}$$

be the differential operator associated with the differential equation (9). Then

$$zDR = z\frac{d}{dz}R = z(P_0' - P_1)y + (zP_1' - (2\lambda+1)P_1 + zP_0)y'. \tag{29}$$

From (26), (27) and (29) it follows that the equations $R = 0$ and $zDR = 0$ hold when $y = y_*$. Hence, R and zDR are linearly dependent as linear forms in y and y' with coefficients in $\mathbf{C}[z]$. But the maximal degree of the coefficients of zDR may be at most 1 greater than the degree of the coefficients of R. Hence,

$$zDR \equiv (az + b)R, \qquad a, b \in \mathbf{C}. \tag{30}$$

If we equate the coefficients of y and y' on both sides of (30), making use of (27) and (29), we obtain:

$$\begin{aligned} &zP_1' - (2\lambda + 1 + az + b)P_1 + zP_0 = 0, \\ &zP_0' - zP_1 - (az + b)P_0 = 0. \end{aligned} \tag{31}$$

From (31) it follows that $a \neq 0$ and $\deg P_1 = \deg P_0$.

Let

$$P_1 = a_n z^n + a_{n-1} z^{n-1} + \cdots + a_0, \qquad a_n \neq 0, \qquad n \geq 0,$$

$$P_0 = b_n z^n + b_{n-1} z^{n-1} + \cdots + b_0, \qquad b_n \neq 0.$$

Comparing the coefficients of z^{n+1} in (31), we obtain

$$-a a_n + b_n = 0, \qquad a_n + a b_n = 0, \tag{32}$$

and hence

$$a^2 = -1. \tag{33}$$

Comparing the coefficients of z^n in (31), we find that

$$\begin{aligned} na_n - (2\lambda + 1 + b)a_n - aa_{n-1} + b_{n-1} &= 0, \\ nb_n - bb_n - a_{n-1} - ab_{n-1} &= 0. \end{aligned} \tag{34}$$

We multiply both sides of the second equality in (34) by a, and we then subtract the second equation from the first equation, using (33). As a result we obtain

$$n(a_n - ab_n) - (2\lambda + 1 + b)a_n + abb_n = 0. \tag{35}$$

From (32) we find that $b_n = aa_n$. We substitute this value for b_n in (35). If we use (33) and cancel a_n, we arrive at the relation

$$2n - (2\lambda + 1) - 2b = 0. \tag{36}$$

The polynomials P_0 and P_1 are nonzero and relatively prime. Consequently, they cannot both be divisible by z. It thus follows from (31) that either $2\lambda+1+b = 0$ or else $b = 0$. If $b = 0$, then from (36) we find that $2\lambda + 1 = 2n$. Hence,

$$\lambda = \frac{2n-1}{2}, \qquad n \geq 0. \tag{37}$$

If $2\lambda + 1 + b = 0$, then (36) gives us $2\lambda + 1 = -2n$, or

$$\lambda = -\frac{2n+1}{2}, \qquad n \geq 0. \tag{38}$$

Combining (37) and (38), we conclude that

$$\lambda = \frac{2k-1}{2}, \qquad k \in \mathbf{Z},$$

which contradicts the hypothesis of the lemma. This contradiction shows that, under the assumptions in the lemma, the functions y and y' are algebraically independent over $\mathbf{C}(z)$. The lemma is proved.

We note that the values $\lambda = -k$, $k \in \mathbf{N}$, are not excluded in Lemma 3.

Theorem 2 follows from Theorem 3 of Chapter 3 and Lemma 3.

We now show that in Theorem 2 the case when $\lambda = (2k-1)/2$, $k \in \mathbf{Z}$, is in fact exceptional, but that in that situation the numbers $K_\lambda(\xi)$ and $K'_\lambda(\xi)$ are transcendental.

Theorem 3. *If $\lambda = (2k-1)/2$, $k \in \mathbf{Z}$, and if $\xi \in \mathbf{A} \setminus \{0\}$, then $K_\lambda(\xi)$ and $K'_\lambda(\xi)$ are each transcendental numbers.*

To prove the theorem we need two lemmas.

Lemma 4. *If $\lambda = (2k-1)/2$, $k \in \mathbf{Z}$, then the functions $K_\lambda(z)$ and $K'_\lambda(z)$ are homogeneously algebraically independent (in particular, they are linearly independent) over $\mathbf{C}(z)$.*

Proof. Suppose that $K_\lambda(z)$ and $K'_\lambda(z)$ are homogeneously algebraically dependent over $\mathbf{C}(z)$. Then the logarithmic derivative $K'_\lambda(z)/K_\lambda(z)$ is an algebraic function. But this is a meromorphic function, and hence a rational function. Hence, the functions $K_\lambda(z)$ and $K'_\lambda(z)$ are linearly dependent over $\mathbf{C}(z)$, and so are connected by an equation

$$P_1(z)K'_\lambda(z) + P_0(z)K_\lambda(z) = 0, \qquad P_0(z), P_1(z) \in \mathbf{C}[z], \tag{39}$$

where $P_0(z)$ and $P_1(z)$ are relatively prime polynomials neither of which is identically zero.

In this case we can apply the argument used to prove Lemma 3, from which it follows that

$$\deg P_0(z) = \deg P_1(z). \tag{40}$$

If we replace z by $-z$ in (39) and use the fact that $K'_\lambda(z)$ and $K_\lambda(z)$ are an odd function and an even function, respectively, we obtain

$$-P_1(-z)K'_\lambda(z) + P_0(-z)K_\lambda(z) = 0. \tag{41}$$

The coefficients in equations (39) and (41) are proportional, and since $P_0(z)$ and $P_1(z)$ are relatively prime, it follows that those coefficients differ by a constant factor, which must be ± 1. This implies that one of the polynomials $P_0(z)$, $P_1(z)$ must be odd and other one even. Hence, $\deg P_0(z) \neq \deg P_1(z)$, contradicting (40). This contradiction proves the lemma.

Lemma 5. *If $\lambda = (2k-1)/2$, $k \in \mathbf{Z}$, then the functions $K_\lambda(z)$ and $K'_\lambda(z)$ are algebraically dependent over $\mathbf{C}(z)$.*

Proof. Let $\lambda \in \mathbf{C}$, $\lambda \neq -k$, $k \in \mathbf{Z}^+$, and $K_\lambda = K_\lambda(z)$. We first establish certain recursion relations between the functions K_λ and K'_λ and their derivatives.

It is easy to verify that

$$\frac{2\lambda}{z} K_{\lambda-1} = \left(\frac{z}{2}\right)^{-2\lambda} \frac{d}{dz}\left(\left(\frac{z}{2}\right)^{2\lambda} K_\lambda\right) = \frac{2\lambda}{z} K_\lambda + K'_\lambda,$$

or

$$2\lambda K_\lambda + zK'_\lambda = 2\lambda K_{\lambda-1}. \tag{42}$$

We differentiate (42) and then, using the differential equation (9), eliminate the function K''_λ. As a result we obtain

$$zK_\lambda = -2\lambda K'_{\lambda-1}. \tag{43}$$

From (42) and (43) we find that

$$z^2 K'_\lambda = 2\lambda z K_{\lambda-1} + 4\lambda^2 K'_{\lambda-1}. \tag{44}$$

Using induction, we now prove that for $k \in \mathbf{Z}^+$ the functions $K_{-\frac{1}{2}+k}$ and $K'_{-\frac{1}{2}+k}$ are algebraically dependent over $\mathbf{C}(z)$.

When $k = 0$, the defining equation (8) for $K_\lambda(z)$ shows that $K_{-\frac{1}{2}} = \cos z$ and $K'_{-\frac{1}{2}} = -\sin z$, and the assertion holds. We now suppose that the assertion is true for $k = l$, and we prove that it then holds for $k = l + 1$.

Setting $\lambda = k + 1/2$ in (43) and (44), we find that the functions $K_{k+\frac{1}{2}}$ and $K'_{k+\frac{1}{2}}$ depend algebraically over $\mathbf{C}(z)$ on the functions $K_{k-\frac{1}{2}}$ and $K'_{k-\frac{1}{2}}$; and, by the induction assumption, the latter functions are algebraically dependent over $\mathbf{C}(z)$. Thus, by induction, the claim holds for all $k \in \mathbf{Z}^+$.

Finally, using equations (42) and (43) in a similar way, we prove that for $k \in \mathbf{Z}^+$ the functions $K_{-\frac{1}{2}-k}$ and $K'_{-\frac{1}{2}-k}$ are algebraically dependent over $\mathbf{C}(z)$. This completes the proof of the lemma.

Proof of Theorem 3. Lemma 5 enables us to use the argument in the proof of Lemma 3. There exists a solution y_* of the differential equation (9) such that y'_*/y_* is a rational function. If we introduce the notation (26) and (27), where y is an arbitrary solution of (9), we arrive at the identity (30), in which $a^2 = -1$. Since $DR = \frac{d}{dz}R$, it follows that, after integrating the differential equation (30) and substituting $K_\lambda(z)$ for y in R, we obtain:

$$P_1(z)K'_\lambda(z) + P_0(z)K_\lambda(z) = cz^b e^{az}, \qquad c \in \mathbf{C}. \tag{45}$$

Since the left side of (45) is an entire function, we have $b \in \mathbf{Z}^+$, and $c \neq 0$ by Lemma 4.

According to Lemma 4 of Chapter 4, the coefficients of $P_0(z)$ and $P_1(z)$ and the number c can all be chosen in $\mathbf{A}$.

If we replace z by $-z$ in (45) and note that $K'_\lambda(z)$ and $K_\lambda(z)$ are odd and even functions, respectively, we obtain:

$$-P_1(-z)K'_\lambda(z) + P_0(-z)K_\lambda(z) = (-1)^b cz^b e^{-az}, \qquad b \in \mathbf{Z}^+, \quad c \neq 0. \tag{46}$$

Multiplying together (45) and (46), we arrive at the equation

$$-P_1(z)P_1(-z)(K'_\lambda(z))^2 + \big(P_1(z)P_0(-z) - P_1(-z)P_0(z)\big)K_\lambda(z)K'_\lambda(z)+ \\ + P_0(z)P_0(-z)K^2_\lambda(z) = (-1)^b c^2 z^{2b}, \qquad b \in \mathbf{Z}^+, \quad c \neq 0. \tag{47}$$

Let $\xi \in \mathbf{A}$, $\xi \neq 0$. By Theorem 1 of Chapter 4, at least one of the numbers $K'_\lambda(\xi)$, $K_\lambda(\xi)$ is transcendental. Hence, if any two of the coefficients of the products of $K'_\lambda(z)$ and $K_\lambda(z)$ on the left in (47) do not vanish at $z = \xi$, then it follows that both $K'_\lambda(\xi)$ and $K_\lambda(\xi)$ are transcendental numbers.

Since $c \neq 0$, it is impossible for all three of these coefficients to vanish at $z = \xi$. If the coefficients at both ends vanish at $z = \xi$, then (47) implies that both $K'_\lambda(\xi)$ and $K_\lambda(\xi)$ are transcendental.

Suppose that the first and middle coefficients vanish at $z = \xi$. Because $P_0(z)$ and $P_1(z)$ are relatively prime polynomials, we then have $P_1(\xi) = 0$ and $P_1(-\xi) = 0$, while $P_0(\xi) \neq 0$ and $P_0(-\xi) \neq 0$. By Theorem 6 of Chapter 2, it now follows from (45) and (46) that the number $K_\lambda(\xi)$ is transcendental; however, from (47) we know that this number is algebraic. Thus, this case is impossible.

We similarly prove that the case when the middle and last coefficients on the left in (47) vanish at $z = \xi$ is also impossible.

Thus, under the conditions of Theorem 3, both $K_\lambda(\xi)$ and $K'_\lambda(\xi)$ are transcendental numbers. The theorem is proved.

From Theorem 2 it follows that, under the conditions in that theorem, the number $K'_\lambda(\xi)/K_\lambda(\xi)$ is transcendental. This result can also be obtained from Theorem 2 of Chapter 3 and Lemma 3; and Lemma 4 enables one to extend the result to the case when $\lambda = (2k-1)/2$, $k \in \mathbf{Z}$.

Let $\lambda \in \mathbf{Q}$, $\lambda \neq -k$, $k \in \mathbf{N}$. From Theorems 2 and 3 we know that all of the zeros of the functions $K_\lambda(z)$ and $K'_\lambda(z)$, $z \neq 0$, are transcendental. Hence, from (10) we conclude that all of the (nonzero) zeros of the Bessel function are transcendental; and the equation

$$J'_\lambda(z) = \frac{1}{\Gamma(\lambda+1)}\left(\frac{z}{2}\right)^\lambda \left(\frac{\lambda}{z}K_\lambda(z) + K'_\lambda(z)\right)$$

tells us that all of the zeros of the derivative $J'_\lambda(z)$, $z \neq 0$, are also transcendental. Because of the relation

$$\frac{J'_\lambda(z)}{J_\lambda(z)} = \frac{\lambda}{z} + \frac{K'_\lambda(z)}{K_\lambda(z)} \tag{48}$$

we can further state that for any $\xi \in \mathbf{A}\setminus\{0\}$ the value of the logarithmic derivative $J'_\lambda(\xi)/J_\lambda(\xi)$ is transcendental.

We consider the function

$$v = f_\lambda(z) = \sum_{n=0}^{\infty} \frac{z^n}{n![\lambda+1, n]}.$$

Then

$$K_\lambda(z) = f_\lambda\left(-\frac{z^2}{4}\right).$$

The function $f_\lambda(z)$ satisfies the differential equation

$$zv'' + (\lambda+1)v' = v,$$

which, when we successively differentiate it, gives

$$zv^{(k)} + (\lambda + k - 1)v^{(k-1)} = v^{(k-2)}, \qquad k = 2, 3, \ldots.$$

From these differential equations we find that

$$\frac{v^(k-1)}{v^{(k)}} = \lambda + k + \frac{z}{\left(\frac{v^{(k)}}{v^{(k+1)}}\right)}, \qquad k = 1, 2, \ldots.$$

If we successively substitute the $(k+1)$-th equation in the k-th, we obtain the continued fraction expansion

$$\frac{v}{v'} = \lambda + 1 + \cfrac{z}{\lambda + 2 + \cfrac{z}{\lambda + 3 + \cfrac{z}{\lambda + 4 + \cdots}}}. \tag{49}$$

Since

$$\frac{v'}{v} = \frac{i}{\sqrt{z}} \frac{K'_\lambda(2i\sqrt{z})}{K_\lambda(2i\sqrt{z})},$$

it follows from Theorem 2 of Chapter 3 and Lemmas 2 and 3 that the value of the continued fraction (49) is transcendental for any $\lambda \in \mathbf{Q}$, $\lambda \neq -k$, $k \in \mathbf{N}$, and any $z \in \mathbf{A} \setminus \{0\}$. In particular, taking $\lambda = 0$ and $z = 1$, we find that

$$\alpha = 1 + \cfrac{1}{2 + \cfrac{1}{3 + \cfrac{1}{4 + \cdots}}}$$

is a transcendental number.

§ 3. The functions $K_\lambda(z)$ and e^z

We now prove a theorem which generalizes Theorem 2.

Theorem 4. *Let* $\lambda \in \mathbf{Q}$; $\lambda \neq -k_1$, $k_1 \in \mathbf{N}$; $\lambda \neq (2k_2-1)/2$, $k_2 \in \mathbf{Z}$; *and* $\xi, \eta \in \mathbf{A}$, $\xi\eta \neq 0$. *Then the numbers* $K_\lambda(\xi)$, $K'_\lambda(\xi)$ *and* e^η *are algebraically independent.*

We first prove some lemmas.

Lemma 6. *If* $\lambda \in \mathbf{C}$, $\lambda \neq (2k-1)/2$, $k \in \mathbf{Z}$,

$$P = P(z, y, y') \in \mathbf{C}[z, y, y'], \qquad P \not\equiv 0,$$

and D is the differential operator (28) associated to the differential equation (9), then the polynomial zDP is divisible by P as a polynomial in three independent variables if and only if P has the form

$$P = \sigma z^b, \qquad \sigma \in \mathbf{C}, \qquad \sigma \neq 0, \qquad b \in \mathbf{Z}^+. \tag{50}$$

Proof. Suppose that zDP is divisible by P. Since the total degree of zDP in y and y' is not greater than the degree of P, and since the degree of zDP in z is at most one greater than the degree of P, it follows that the quotient when zDP is divided by P is a polynomial in z of degree at most one. Thus,

$$zDP \equiv (az+b)P, \qquad a, b \in \mathbf{C}. \tag{51}$$

Now let y be an arbitrary solution of the differential equation (9). Then

$$zDP(y, y') = z\frac{d}{dz}P(y, y') = (az+b)P(y, y'). \tag{52}$$

If we set $y = \gamma_1 y_1 + \gamma_2 y_2$, where y_1 and y_2 are two linearly independent solutions of (9) and $\gamma_1, \gamma_2 \in \mathbf{C}$, and then integrate the differential equation (52), we obtain

$$P(z, y, y') = \sigma(\gamma_1, \gamma_2) z^b e^{az}. \tag{53}$$

We shall prove that the polynomial P does not depend on y or y'.

In fact, otherwise let Q be the sum of the homogeneous terms in y and y' of largest degree $n \geq 1$ which occur in P. Now when the operator zD is applied to

P, it takes any set of homogeneous terms of a given degree in y and y' either to a set of similar terms or to zero; thus, from (51) we have

$$zDQ = z\frac{d}{dz}Q = (az+b)Q,$$

and hence

$$Q(z,y,y') = \sigma_1(\gamma_1,\gamma_2)z^b e^{az}. \tag{54}$$

Since the left side of (54) is a homogeneous polynomial of positive degree in γ_1 and γ_2, it follows that $\sigma_1(\gamma_1,\gamma_2)$ is also a homogeneous polynomial of positive degree in γ_1 and γ_2. Hence, one can find γ_1 and γ_2, not both zero, such that $\sigma_1(\gamma_1,\gamma_2) = 0$. We then have $Q(z,y,y') = 0$ for the corresponding nontrivial solution of (9), and this is impossible, by Lemma 3. Thus, $n = 0$, and P does not depend on y or y'. It then follows from (53) that $a = 0$, $b \in \mathbf{Z}^+$, and (50) holds.

Conversely, if the polynomial P has the form (50), then zDP is obviously divisible by P. The lemma is proved.

Lemma 7. *If $\lambda \in \mathbf{C}$, $\lambda \neq (2k-1)/2$, $k \in \mathbf{Z}$; $\alpha \in \mathbf{C}$, $\alpha \neq 0$; and y is any nontrivial solution of the differential equation (9), then the functions y, y' and $e^{\alpha z}$ are algebraically independent over $\mathbf{C}(z)$.*

Proof. By Lemma 3, the functions y and y' are algebraically independent over $\mathbf{C}(z)$. We suppose that the lemma is false, i.e., that y, y' and $y_0 = e^{\alpha z}$ are algebraically dependent over $\mathbf{C}(z)$. Then there is an algebraic equation

$$P(z,y,y',e^{\alpha z}) = P_l e^{l\alpha z} + \cdots + P_1 e^{\alpha z} + P_0 = 0, \tag{55}$$

where $P = P(z,y,y',y_0)$ is an irreducible polynomial in $\mathbf{C}[z,y,y',y_0]$; $P_i = P_i(z,y,y') \in \mathbf{C}[z,y,y']$, $i = 0,1,\ldots,l$; and the set of P_i have no common factor.

Let

$$D_1 = D + \alpha\frac{d}{dy_0}, \tag{56}$$

where D is the differential operator (28). Since we have assumed that the set of functions $\{y, y', e^{\alpha z}\}$ has transcendence degree 2 over $\mathbf{C}(z)$, and the equation (55) holds, it follows by Lemma 4 of Chapter 5 that zD_1P is divisible by P as a polynomial in four independent variables. Arguing as in Lemma 4, we see that the ratio when zD_1P is divided by P is a polynomial in $\mathbf{C}[z]$ of degree at most 1. Thus,

$$zD_1P = (az+b)P, \qquad a,b \in \mathbf{C}. \tag{57}$$

From (55) and (56) we have:

$$zD_1P = (zDP_l + l\alpha P_l)e^{l\alpha z} + \dots + (zDP_1 + \alpha P_1)e^{\alpha z} + zDP_0 = 0, \tag{58}$$

and from (55) and (58), along with the identity (57), we find that:

$$zDP_k + k\alpha P_k \equiv (az + b)P_k, \qquad k = 0, 1, 2, \dots, l. \tag{59}$$

The identity (59) means that the polynomial zDP_k is divisible by the polynomial P_k for each $k = 0, 1, \dots, l$. Consequently, by Lemma 6,

$$P_k = \sigma_k z^{b_k}, \qquad \sigma_k \in \mathbf{C}, \qquad b_k \in \mathbf{Z}^k, \qquad k = 0, 1, \dots, l.$$

These equalities imply that P does not contain y or y'; but then the equation (55) is impossible, since $e^{\alpha z}$ is a transcendental function. This contradiction proves the lemma.

Proof of Theorem 4. The functions

$$u_1 = K_\lambda(z), \qquad u_2 = K'_\lambda(z), \qquad u_3 = e^{\frac{\eta}{\xi}z}$$

form a solution of the system of differential equations

$$u_1' = u_2, \qquad u_2' = -u_1 - \frac{2\lambda + 1}{z}u_2, \qquad u_3' = \frac{\eta}{\xi}u_3,$$

and so, by Lemma 7 with $\alpha = \eta/\xi$, they are algebraically independent over $\mathbf{C}(z)$. Thus, by the Second Fundamental Theorem in Chapter 3, the numbers $K_\lambda(\xi)$, $K'_\lambda(\xi)$ and e^η are algebraically independent.

§ 4. Kummer functions

We now consider the Kummer function

$$A_{\lambda,\nu}(z) = \sum_{n=0}^{\infty} \frac{[\nu, n]}{n![\lambda, n]} z^n, \qquad \lambda \neq 0, -1, -2, \dots, \tag{60}$$

which is a hypergeometric E-function for $\lambda, \nu \in \mathbf{Q}$. The function (60) satisfies the differential equation

$$y'' + \frac{\lambda - z}{z}y' - \frac{\nu}{z}y = 0. \tag{61}$$

Theorem 5 (V. A. Oleinikov). *Let $\lambda, \nu \in \mathbf{Q}$; $\lambda \neq -k$, $k \in \mathbf{Z}^+$; $\nu \notin \mathbf{Z}$; $\nu - \lambda \notin \mathbf{Z}^+$; and $\xi \in \mathbf{A} \setminus \{0\}$. Then the numbers $A_{\lambda,\nu}(\xi)$ and $A'_{\lambda,\nu}(\xi)$ are algebraically independent.*

To prove Theorem 5 we need two lemmas.

Lemma 8. *If $\lambda, \nu \in \mathbf{C}$, $\nu \notin \mathbf{Z}$, $\nu - \lambda \notin \mathbf{Z}$, and y is any nontrivial solution of the differential equation (61), then the functions y and y' are algebraically independent over $\mathbf{C}(z)$.*

Proof. We suppose the contrary, i.e., that y and y' are algebraically dependent over $\mathbf{C}(z)$. Then, by Lemma 2, there exists a solution y_* of the differential equation (61) such that $u = y'_*/y_*$ is an algebraic function. As in the proof of Lemma 3, the only possible branch points of u are $z = 0$ and $z = \infty$. We shall prove that $z = \infty$ cannot be a branch point of u; then, as in the proof of Lemma 3, we will be able to conclude that u is a rational function.

From (61) it follows that u satisfies the Riccati differential equation

$$u' + \frac{\lambda - z}{z} u + u^2 = \frac{\nu}{z}. \tag{62}$$

Let the right side of (24) be the power series expansion with decreasing rational exponents for some branch of u in a neighborhood of $z = \infty$. Substituting this series in (62), we obtain:

$$\sum_{k=0}^{\infty} (r_k + \lambda) c_k z^{r_k - 1} - \sum_{k=0}^{\infty} c_k z^{r_k} + \sum_{k=0}^{\infty} \sum_{l=0}^{\infty} c_k c_l z^{r_k + r_l} = \frac{\nu}{z}. \tag{63}$$

If we compare the coefficients of the terms in (63) having the highest degree in z, we see that either $r_0 = 0$ and $c_0 = 1$, or else $r_0 = -1$ and $c_0 = -\nu$.

We next prove that all of the exponents r_k for $k > 0$ are negative integers. In fact, suppose that this is not the case, and let r_{k_1} be the fractional exponent of smallest index among all exponents of powers of z in (24) for which $c_{k_1} \neq 0$. In both possible cases we compare the coefficients in (63) of the largest fractional power of z; we arrive at a contradiction unless $c_{k_1} = 0$. Consequently, the function u does not have a branch point at $z = \infty$. Then $z = 0$ cannot be a branch point of u either, and so u is a rational function.

We set

$$\frac{y'_*}{y_*} = -\frac{P_0}{P_1}, \qquad P_0, P_1 \in \mathbf{C}[z], \qquad P_1 \not\equiv 0. \tag{64}$$

If $y'_*/y_* = 0$, then $y'_* = 0$ and $y_* = c$. If we substitute this value for y_* in (61), we find that $\nu = 0$. In this case let $P_1 = 1$ and $P_0 = 0$. On the other hand, if $\nu \neq 0$, then we shall take P_0 and P_1 to be nonzero relatively prime polynomials.

We consider the linear form

$$R = P_1 y' + P_0 y, \tag{65}$$

where y is an arbitrary solution of the differential equation (61), and we apply to R the differential operator

$$D = \frac{\partial}{\partial z} + y' \frac{\partial}{\partial y} + \left(\frac{\nu}{z} y - \frac{\lambda - z}{z} y' \right) \frac{\partial}{\partial y'},$$

which is associated with the differential equation (61). Then

$$DR = \frac{d}{dz} R = \left(P_1' - \frac{\lambda - z}{z} P_1 + P_0 \right) y' + \left(P_0' + \frac{\nu}{z} P_1 \right) y. \tag{66}$$

From (64), (65) and (66) it follows that for $y = y_*$ we have the equations $R = 0$ and $zDR = 0$. Hence R and zDR are linearly dependent as linear forms in y and y' with coefficients in $\mathbf{C}[z]$.

As in the proof of Lemma 3, we obtain the identity

$$zDR \equiv (az + b)R, \qquad a, b \in \mathbf{C},$$

from which we obtain the equations

$$\begin{aligned} &P_1' - \left(\frac{\lambda + b}{z} + a - 1 \right) P_1 + P_0 = 0, \\ &P_0' - \left(\frac{b}{z} + a \right) P_0 + \frac{\nu}{z} P_1 = 0. \end{aligned} \tag{67}$$

From the first equation in (67) it follows that if $a \neq 1$, then the polynomials P_0 and P_1 have the same degree; then the second equation implies that $a = 0$. Thus, two cases are possible: $a = 0$ and $a = 1$.

1) $a = 0$. Let

$$\begin{aligned} P_1 &= a_n z^n + \cdots + a_0, \qquad a_n \neq 0, \qquad n \geq 0, \\ P_0 &= b_n z^n + \cdots + b_0, \qquad b_n \neq 0. \end{aligned}$$

If we compare the coefficients of the highest powers of z on the left in the equations in (67), we obtain:

$$a_n + b_n = 0, \qquad \nu a_n + (n - b) b_n = 0,$$

and hence

$$n - b - \nu = 0. \tag{68}$$

Because of the assumptions concerning P_0 and P_1 (that they are nonzero and relatively prime), the equations (67) imply that $P_0 \not\equiv 0$ and hence $\nu \neq 0$. Again two sub-cases are possible:

a) $\lambda + b = 0$. Substituting $b = -\lambda$ in place of b in the first equation in (67), we have

$$\nu - \lambda = n, \qquad n \in \mathbf{Z}^+. \tag{69}$$

b) $\lambda + b \neq 0$. From the first equation in (67) it follows that P_1 is divisible by z. This means that $n \geq 1$. But because P_0 and P_1 are relatively prime, it then follows from the second equation in (67) that $b = 0$. Substituting this value of b in (68), we find that

$$\nu = n, \qquad n \in \mathbf{N}. \tag{70}$$

2) $a = 1$. From (67) it follows that, if $P_0 \not\equiv 0$, then the degree of P_1 is 1 greater than the degree of P_0. On the other hand, if $P_0 = 0$, then $P_1 = 1$, as stipulated above. Let

$$P_1 = a_n z^n + \cdots + a_0, \qquad a_n \neq 0, \qquad n \geq 0,$$
$$P_0 = b_{n-1} z^{n-1} + \cdots + b_0,$$

where $n \geq 1$ if $P_0 \not\equiv 0$, and $n = 0$ if $P_0 \equiv 0$.

If we compare the coefficients of the highest powers of z on the left in (67), we obtain:

$$na_n - (\lambda + b)a_n + b_{n-1} = 0, \qquad \nu a_n - b_{n-1} = 0,$$

and hence

$$n + \nu - \lambda - b = 0. \tag{71}$$

Again there are two possible sub-cases.

a) $\lambda + b = 0$. Then from (71) we have

$$\nu = -n, \qquad n \in \mathbf{Z}^+. \tag{72}$$

b) $\lambda + b \neq 0$. As in the analogous sub-case above, from the first equation in (67) we find that P_1 is divisible by z. This means that $n \geq 1$. But then, since P_0 and P_1 are relatively prime, the second equation in (67) implies that $b = 0$. Substituting this value of b in (71), we obtain

$$\nu - \lambda = -n, \qquad n \in \mathbf{N}. \tag{73}$$

Combining (69), (70), (72) and (73), we conclude that either $\nu \in \mathbf{Z}$ or else $(\nu - \lambda) \in \mathbf{Z}$, and the lemma is proved.

We note that the values $\lambda = -k$, $k \in \mathbf{Z}^+$, are not excluded in Lemma 8.

Lemma 9. *If* $\lambda, \nu \in \mathbf{C}$, $\lambda \neq -k_1$, $k_1 \in \mathbf{Z}^+$, $\nu \notin \mathbf{Z}$, $\lambda - \nu = k_2$, $k_2 \in \mathbf{N}$, *then the functions* $A_{\lambda,\nu}(z)$ *and* $A'_{\lambda,\nu}(z)$ *are algebraically independent over* $\mathbf{C}(z)$.

Proof. We set $\lambda - \nu = k$, and denote $A_{\nu+k,\nu} = A_{\nu+k,\nu}(z)$. We use induction on k to prove that the functions $A_{\nu+k,\nu}$ and $A'_{\nu+k,\nu}$ are algebraically independent over $\mathbf{C}(z)$.

For $k = 1$,

$$A_{\nu+1,\nu} = \nu \sum_{n=0}^{\infty} \frac{z^{\nu}}{n!(\nu+n)} = \nu\omega(z), \tag{74}$$

where $\omega(z)$ is the function defined in (5.45) with λ replaced by ν. The following differential equation connects the function $A_{\nu+1,\nu}$ with its derivative and the function e^z:

$$A'_{\nu+1,\nu} = \frac{\nu}{z} e^z - \frac{\nu^2}{z} A_{\nu+1,\nu}. \tag{75}$$

By Lemma 7 of Chapter 5, the functions $\omega(z)$ and e^z are algebraically independent over $\mathbf{C}(z)$. Hence, it follows from (74) and (75) that $A_{\nu+1,\nu}$ and $A'_{\nu+1,\nu}$ are algebraically independent over $\mathbf{C}(z)$, since if that were not the case we would have an algebraic equation over $\mathbf{C}(z)$ relating the functions $\omega(z)$ and e^z.

Now suppose that our assertion is true for $k = l$, $l \in \mathbf{N}$. We must prove that it is then true for $k = l + 1$.

It is easy to verify the equality

$$\frac{\nu+k}{z} A_{\nu+k,\nu} = z^{-(\nu+k)} \left(\frac{d}{dz} z^{\nu+k} A_{\nu+k+1,\nu} \right) = \frac{\nu+k}{z} A_{\nu+k+1,\nu} + A'_{\nu+k+1,\nu},$$

from which we obtain

$$zA'_{\nu+k+1,\nu} + (\nu+k)A_{\nu+k+1,\nu} = (\nu+k)A_{\nu+k,\nu}. \tag{76}$$

We differentiate (76). Next, using the differential equation (61), we eliminate $A''_{\nu+k+1,\nu}$. As a result we obtain the equality

$$zA'_{\nu+k+1,\nu} + \nu A_{\nu+k+1,\nu} = (\nu+k)A'_{\nu+k,\nu}. \tag{77}$$

From (76) and (77) it follows that the functions $A_{\nu+k+1,\nu}$ and $A'_{\nu+k+1,\nu}$ and the functions $A_{\nu+k,\nu}$ and $A'_{\nu+k,\nu}$ are algebraically equivalent over $\mathbf{C}(z)$ (each pair of functions is algebraically dependent on the other pair). Hence they have the same transcendence degree over $\mathbf{C}(z)$. But by the induction assumption, $A_{\nu+k,\nu}$ and

$A'_{\nu+k,\nu}$ are algebraically independent over $\mathbf{C}(z)$. Thus, the functions $A_{\nu+k+1,\nu}$ and $A'_{\nu+k+1,\nu}$ are also algebraically independent over $\mathbf{C}(z)$. By induction, our assertion holds for any $k \in \mathbf{N}$, and the lemma is proved.

Theorem 5 now follows from Theorem 3 of Chapter 3 and Lemmas 8 and 9.

When $\lambda - \nu = k$, $k \in \mathbf{N}$, the differential equation (61) has a solution whose logarithmic derivative is a rational function. Nevertheless, by Lemma 8 the functions $A_{\lambda,\nu}(z)$ and $A'_{\lambda,\nu}(z)$ are algebraically independent over $\mathbf{C}(z)$, and Theorem 5 holds.

For example, when $\lambda - \nu = 1$, the differential equation (61) has the function $y = z^{-\lambda}$, which is algebraic for $\lambda \in \mathbf{Q}$, as a solution.

All of the other excluded values of λ and ν in Theorem 5 are really exceptional values. If $\nu = -k$, $k \in \mathbf{Z}^+$, then $A_{\lambda,\nu}(z)$ is obviously a polynomial. It is easy to verify that for $\nu = k$, $k \in \mathbf{N}$,

$$A_{\lambda,k}(z) = P_1\varphi_\lambda(z) + P_0, \qquad P_0, P_1 \in \mathbf{C}(z),$$

and for $\nu - \lambda = k$, $k \in \mathbf{Z}^+$,

$$A_{\lambda,\lambda+k}(z) = P_1 e^z + P_0, \qquad P_0, P_1 \in \mathbf{C}(z).$$

In these cases the functions $A_{\lambda,\nu}(z)$, $A'_{\lambda,\nu}(z)$ and 1 are linearly dependent over $\mathbf{C}(z)$.

By Theorem 1 of Chapter 4, if $\nu = k$, $k \in \mathbf{N}$, or $\nu - \lambda = k$, $k \in \mathbf{Z}^+$, and if $\xi \in \mathbf{A}$, $\xi \neq 0$, then one of the numbers $A_{\lambda,\nu}(\xi)$, $A'_{\lambda,\nu}(\xi)$ is transcendental. The linear equations relating $A_{\lambda,\nu}(z)$ and $A'_{\lambda,\nu}(z)$ determine the points $\xi \in \mathbf{A}$ at which one of the numbers $A_{\lambda,\nu}(\xi)$ or $A'_{\lambda,\nu}(\xi)$ is algebraic.

For example, if $\nu - \lambda = 1$, then we have

$$A_{\lambda,\lambda+1}(z) = \sum_{n=1}^{\infty} \frac{\lambda + n}{\lambda n!} z^n = e^z \left(1 + \frac{z}{\lambda}\right).$$

The numbers $A_{\lambda,\lambda+1}(-\lambda)$ and $A'_{\lambda,\lambda+1}(-\lambda - 1)$ are algebraic.

If $\nu = 2$, then

$$A_{\lambda,2}(z) = (z - \lambda + 2)\varphi_{\lambda-1}(z) + \lambda - 1,$$

and $A_{\lambda,2}(\lambda - 2)$ is an algebraic number.

These functions are examples of E-functions which satisfy second order linear differential equations and which take algebraic values at certain algebraic points which are not zero or critical points of the differential equation.

§ 5. Solutions of non-homogeneous linear differential equations

We consider functions which satisfy non-homogeneous second order linear differential equations:

$$K_{\lambda,\mu}(z) = \sum_{n=0}^{\infty} \frac{(-1)^n}{[\lambda+1,n][\mu+1,n]} \left(\frac{z}{2}\right)^{2n}, \qquad \lambda,\mu \neq -1,-2,\ldots, \tag{78}$$

is a solution of the equation

$$y'' + \frac{2\lambda+2\mu+1}{z} y' + \left(1 + \frac{4\lambda\mu}{z^2}\right) y = \frac{4\lambda\mu}{z^2}; \tag{79}$$

and

$$A_{\lambda,\mu,\nu}(z) = \sum_{n=0}^{\infty} \frac{[\nu+1,n]}{[\lambda+1,n][\mu+1,n]} z^n, \qquad \lambda,\mu \neq -1,-2,\ldots, \tag{80}$$

is a solution of the equation

$$y'' + \left(\frac{\lambda+\mu+1}{z} - 1\right) y' + \frac{\lambda\mu - (\nu+1)z}{z^2} y = \frac{\lambda\mu}{z^2}. \tag{81}$$

When the parameters are rational numbers they are hypergeometric E-functions. We prove the following theorems about these functions.

Theorem 6. *If $\lambda,\mu \in \mathbf{Q}$; $\lambda,\mu \neq -k_1$, $k_1 \in \mathbf{N}$; $\lambda - \mu \neq (2k_2-1)/2$, $k_2 \in \mathbf{Z}$; and $\xi \in \mathbf{A}\setminus\{0\}$, then the numbers $K_{\lambda,\mu}(\xi)$ aned $K'_{\lambda,\mu}(\xi)$ are algebraically independent.*

Theorem 7. *If $\lambda,\mu,\nu \in \mathbf{Q}$; $\lambda,\mu,\nu \neq -k$, $k \in \mathbf{N}$; $\nu - \lambda \notin \mathbf{Z}$; $\nu - \mu \notin \mathbf{Z}$; and $\xi \in \mathbf{A} \setminus \{0\}$, then the numbers $A_{\lambda,\mu,\nu}(\xi)$ and $A'_{\lambda,\mu,\nu}(\xi)$ are algebraically independent.*

The lemma below reduces the proof of algebraic independence over $\mathbf{C}(z)$ of the functions in Theorems 6 and 7 to the analogous fact for the solutions of the corresponding homogeneous linear differential equations; the latter problem was treated in § 3 and 4.

Let $\mathcal{L}$ be some field of analytic functions which is closed under differentiation.

Lemma 10. *Suppose that the system of m, $m \geq 2$, linear (homogeneous or non-homogeneous) differential equations*

$$y'_k = Q_{k,0} + \sum_{i=1}^{m} Q_{k,i} y_i, \qquad k = 1,\ldots,m, \qquad Q_{k,i} \in \mathcal{L}, \tag{82}$$

has a solution $y_1^0, \dots, y_m^0$ *such that the functions* $y_1^0, \dots, y_{m-1}^0$ *are algebraically independent over the field* $\mathcal{L}$ *but*

$$P(y_1^0, \dots, y_m^0) = 0, \qquad P = P(y_1, \dots, y_m) \in \mathcal{L}[y_1, \dots, y_m], \tag{83}$$

where P *is an irreducible polynomial. Furthermore, let* $P_s = P_s(y_1, \dots, y_m)$ *be the homogeneous part of* P *of greatest total degree* s, $s \geq 1$, *in* $y_1, \dots, y_m$. *Then there exists a nontrivial solution* $y_1^*, \dots, y_m^*$ *of the system of homogeneous linear differential equations*

$$y_k' = \sum_{i=1}^{m} Q_{k,i} y_i, \qquad k = 1, \dots, m, \qquad Q_{k,i} \in \mathcal{L}, \tag{84}$$

corresponding to (82) which has the property that

$$P_s(y_1^*, \dots, y_m^*) = 0. \tag{85}$$

Proof. We let

$$D_0 = \frac{\partial}{\partial z} + \sum_{k=1}^{m} \left(\sum_{i=1}^{m} Q_{k,i} y_i \right) \frac{\partial}{\partial y_k} \tag{86}$$

denote the differential operator associated with the system (84), and we let

$$D = D_0 + \sum_{k=1}^{m} Q_{k,0} \frac{\partial}{\partial y_k} \tag{87}$$

denote the differential operator associated with the system (82).

First let $y_1, \dots, y_m$ be independent variables. The conditions of the lemma enable us to use Lemma 4 of Chapter 5 when we apply D to the polynomial P in (83); hence, there exists a function $\omega \in \mathcal{L}$ such that

$$DP \equiv \omega P. \tag{88}$$

We let P_s^* denote the homogeneous part of DP of total degree s in $y_1, \dots, y_m$. Then from (88) we obtain

$$P_s^* \equiv \omega P_s. \tag{89}$$

From (86) and (87) it follows that when one goes from the polynomial P to the polynomial DP or $D_0 P$ the homogeneous part of highest degree s remains the same. Thus, $D_0 P_s = P_s^*$, and so from (89) we have

$$D_0 P_s \equiv \omega P_s. \tag{90}$$

Now let $y_1, \dots, y_m$ be any solution of the system of differential equations (84). From (90) it follows that

$$D_0 P_s(y_1, \dots, y_m) = \frac{d}{dz} P_s(y_1, \dots, y_m) \equiv \omega P_s(y_1, \dots, y_m). \tag{91}$$

We consider the differential equation

$$v' = \omega v, \qquad \omega \in \mathcal{L}. \tag{92}$$

Its general solution is $v = cv_0$, where c is a constant and v_0 is a particular solution, $v_0 \not\equiv 0$. From (91) we then find that the function $P_s(y_1, \dots, y_m)$ is a solution of the differential equation (92) for any solution $y_1, \dots, y_m$ of (84). Let $y_{1,1}, \dots, y_{m,1}$ and $y_{1,2}, \dots, y_{m,2}$ be any two fixed linearly independent solutions of (84). Then $y_k = \gamma_1 y_{k,1} + \gamma_2 y_{k,2}$, $k = 1, \dots, m$, is also a solution of the system for arbitrary constants γ_1 and γ_2. Hence,

$$P_s(\gamma_1 y_{1,1} + \gamma_2 y_{1,2}, \dots, \gamma_1 y_{m,1} + \gamma_2 y_{m,2}) = c(\gamma_1, \gamma_2) v_0, \tag{93}$$

where $c(\gamma_1, \gamma_2)$ is a homogeneous polynomial of degree s with coefficients in $\mathbf{C}$. We can then find γ_1 and γ_2, not both zero, for which $c(\gamma_1, \gamma_2) = 0$. If we set $y_k^* = \gamma_1 y_{k,1} + \gamma_2 y_{k,2}$, $k = 1, \dots, m$, for this choice of γ_1 and γ_2, we obtain (85) from (93). The lemma is proved.

Lemma 11. *Suppose that the (homogeneous or non-homogeneous) differential equation*

$$y^{(m)} + Q_{m-1} y^{(m-1)} + \dots + Q_1 y' + Q_0 y = R, \qquad m \geq 2, \tag{94}$$

$$R, Q_0, Q_1, \dots, Q_{m-1} \in \mathcal{L},$$

has solution y_0 which does not satisfy any algebraic differential equation of order less than $m-1$ with coefficients in $\mathcal{L}$ (if $m = 2$ and (94) is non-homogeneous, y_0 is not an algebraic function over $\mathcal{L}$), but which does satisfy the equation

$$P(y_0, y_0', \dots, y_0^{(m-1)}) = 0, \qquad P = P(y, y_1, \dots, y_{m-1}) \in \mathcal{L}[y, y_1, \dots, y_{m-1}],$$

where P is an irreducible polynomial. Let $P_s(y, y_1, \dots, y_{m-1})$ be the homogeneous part of P of highest total degree in $y, y_1, \dots, y_{m-1}$. Then there exists a solution y_* of the homogeneous linear differential equation corresponding to (94) such that

$$P_s(y_*, y_*', \dots, y_*^{(m-1)}) = 0.$$

In particular, for $m = 2$ the solution y_* has the property that its logarithmic derivative is an algebraic function over $\mathcal{L}$.

Proof. If $m = 2$ in the homogeneous case and y_0 is an algebraic function, then the lemma is obvious. Otherwise the lemma follows from Lemma 10.

Proof of Theorem 6. We consider the function $y_0 = z^{2\mu} K_{\lambda,\mu}(z)$, which satisfies the differential equation

$$y'' + \frac{2(\lambda - \mu) + 1}{z} y' + y = 4\lambda\mu z^{\mu-2}. \tag{95}$$

Since $K_{\lambda,\mu}(z)$ is entire, it follows that y_0 is not an algebraic function. Suppose that y_0 satisfies a first order algebraic differential equation; then, by Lemma 11 with $\mathcal{L} = \mathbf{C}(z)$, we find that the homogeneous differential equation corresponding to (95) has a solution y_* whose logarithmic derivative is an algebraic function. But this contradicts Lemma 3. Consequently, y_0 and y_0' are algebraically independent over $\mathbf{C}(z)$. Then, by Lemma 8 of Chapter 5, the functions $K_{\lambda,\mu}(z)$ and $K'_{\lambda,\mu}(z)$ are also algebraically independent over $\mathbf{C}(z)$.

Applying Theorem 3 of Chapter 3, we obtain Theorem 6.

Proof of Theorem 7. If we consider the function $y_0 = z^{-\mu} A_{\lambda,\mu,\nu}(z)$, we see that it satisfies the differential equation

$$y'' + \left(\frac{\lambda - \mu + 1}{z} - 1\right) y' - \frac{\nu - \mu + 1}{z} y = \lambda\mu z^{\mu-2}, \tag{96}$$

and it is not an algebraic function if $\lambda \neq -k$, $k \in \mathbf{N}$. The rest of the proof is the same as for Theorem 6, except that we use the equation (96) and Lemma 8.

We note that Theorem 7 also holds when $\nu - \lambda = -k$, $k \in \mathbf{N}$, and when $\nu - \mu = -k$, $k \in \mathbf{N}$. These cases require a separate argument, as in Theorem 5. We shall omit the proof, except to say that it is similar to the analogous case of Theorem 5 and uses Theorem 2 of [70:10].

The following theorem, which is definitive and more general than Theorem 6, was proved as a special case of more general theorems by I. I. Belogrivov [3:3] and V. A. Oleinikov [56:7], working independently and by different methods.

Theorem 8. *Suppose that $\lambda, \mu \in \mathbf{Q}$; $\lambda, \mu \neq -k_1$, $k_1 \in \mathbf{N}$; λ and μ do not have the form of a pair k_2 and $k_3 + 1/2$ with $k_2 \in \mathbf{N}$, $k_3 \in \mathbf{Z}$; and $\xi \in \mathbf{A} \setminus \{0\}$. Then the numbers $K_{\lambda,\mu}(\xi)$ and $K'_{\lambda,\mu}(\xi)$ are algebraically independent.*

This theorem is a special case of Theorem 5 of Chapter 10.

Remarks

Lemma 1 and Theorem 1 were proved in 1960 in [70:11]. It would be of interest to know whether Lemma 1 can be generalized to the case of an E-function which satisfies a linear differential equation of order $m > 2$ with coefficients in $\mathbf{C}[z]$ and also an algebraic differential equation of order $m-1$ with coefficients in $\mathbf{C}[z]$ but does not satisfy an algebraic differential equation of the same type having order less than $m-1$. The attempted proof of this assertion in [70:11] contains an error.

Theorem 2, along with Lemmas 2 and 3, were proved in Siegel's paper [72:3] and are included in his monograph [72:4]. The second part of our proof of Lemma 3 differs from Siegel's proof. Theorem 3 follows from well-known properties of Bessel functions and from Lindemann's Theorem. We give a version of the proof which is conceptually close to the proof of Lemma 3.

In [72:3], Siegel stated a theorem without proof which is more general than Theorem 4. Theorem 4 with $\eta = \xi$ is also included in a more general theorem published in 1959 in [70:10].

Theorem 5 was proved in 1962 by Oleinikov [56:1]. In the book we give a proof of Lemma 8 which is similar to that of Lemma 3.

Theorem 6 was first proved in 1954 in [70:1,9]. That proof was rather long and complicated. Then in 1966, Lemma 10 was proved in [70:16]; Theorems 6 and 7 follow rather easily from that lemma. A proof of Lemma 11 is contained in the book [70:35].

We note that I. Yakabe studied the Kummer functions (60) in 1959 in [90:1]. Using Siegel's work, he showed that, if $\lambda, \nu \in \mathbf{Q}$, $\nu \notin \mathbf{Z}$, $\lambda - \nu \notin \mathbf{Z}$, and $\xi \in \mathbf{A} \setminus \{0\}$, then at least one of the numbers $A_{\lambda,\nu}(\xi)$ or $A'_{\lambda,\nu}(\xi)$ is transcendental.

Chapter 7
Solutions of certain linear differential equations of arbitrary order

§ 1. Solutions of non-homogeneous differential equations

In this chapter we use the Second Fundamental Theorem of Chapter 3 to obtain several algebraic independence results for the values of certain sets of E-functions. In order to prove that our functions are algebraically independent over $\mathbf{C}(z)$, we use a method based on the differential properties of these functions and the divisibility properties of polynomials in several variables. This is a further development of the method used in Chapters 5 and 6.

We set

$$\begin{aligned} \omega_0(z) &= e^z = \sum_{n=0}^{\infty} \frac{z^n}{n!}, \\ \omega_k(z) &= \int_0^z \frac{\omega_{k-1}(t) - 1}{t}\, dt + 1 = 1 + \sum_{n=1}^{\infty} \frac{z^n}{n! n^k}, \qquad k = 1, \dots, m. \end{aligned} \tag{1}$$

The function $\omega_0(z) = e^z$ is an E-function. For $k \geq 1$ we represent the function $\omega_k(z)$ as follows:

$$\omega_k(z) = 1 + z \sum_{n=0}^{\infty} \frac{[1, n]^k}{[2, n]^{k+1}}\, z^n.$$

The sum on the right is a hypergeometric function of the form (5.1). According to Lemmas 1 and 2 of Chapter 5 and the properties of E-functions, we conclude that $\omega_k(z)$ for any $k \geq 1$ is an E-function which satisfies a non-homogeneous linear differential equation of order $k + 1$ with coefficients in $\mathbf{Q}(z)$ and with the one singular point $z = 0$.

The set of functions in (1) form a solution of the system of linear differential equations

$$y_0' = y_0, \qquad y_k' = \frac{1}{z} y_{k-1} - \frac{1}{z}, \qquad k = 1, \dots, m. \tag{2}$$

We prove a theorem about the functions in (1) which is a generalization of Lindemann's Theorem.

Theorem 1. *For any $m \in \mathbf{Z}^+$ and $n \in \mathbf{N}$, if the algebraic numbers $\xi_1, \dots, \xi_n$ are linearly independent over $\mathbf{Q}$, then:* 1) *the numbers $\omega_k(\xi_i)$, $i = 1, \dots, n$, $k = 0, 1, \dots, m$, are algebraically independent;* 2) *the numbers $\omega_m^{(l)}(\xi_i)$, $i = 1, \dots, n$, $l = 0, 1, \dots, m$, are algebraically independent.*

To prove this we need some lemmas. Let $\mathcal{L}$ be a field of analytic functions containing $\mathbf{C}$ and closed under differentiation.

Lemma 1. *If $\varphi(z)$ is an algebraic function over $\mathcal{L}$, and if $\varphi'(z) \in \mathcal{L}$, then also $\varphi(z) \in \mathcal{L}$.*

Proof. Let

$$P(y) = y^n + a_{n-1}y^{n-1} + \cdots + a_1 y + a_0, \qquad P(y) \in \mathcal{L}[y], \qquad n \geq 1,$$

be an irreducible polynomial such that $P(\varphi(z)) = 0$. If $n = 1$, then the lemma holds. If $n > 1$, then we differentiate the equation $P(\varphi(z)) = 0$, obtaining

$$(n\varphi'(z) + a'_{n-1})\varphi^{n-1}(z) + \cdots + a'_0 = 0. \tag{3}$$

Since $\varphi(z)$ is an algebraic function of degree n over $\mathcal{L}$, and since $(n\varphi'(z) + a'_{n-1}) \in \mathcal{L}$ by the hypothesis of the lemma, it follows from (3) that $n\varphi'(z) + a'_{n-1} = 0$. Integrating, we find that $n\varphi(z) + a_{n-1} = c \in \mathbf{C} \subset \mathcal{L}$. This implies that $\varphi(z) \in \mathcal{L}$, and the lemma is proved.

We denote

$$y_{k,i} = y_{k,i}(z) = \omega_k(\xi_i z), \qquad \xi_i \in \mathbf{C}, \qquad i = 1, \dots, n, \quad k = 1, \dots, m. \tag{4}$$

From (2) it follows that the functions (4) form a solution of the system of differential equations

$$y'_{0,i} = \xi_i y_{0,i}, \quad y'_{k,i} = \frac{1}{z} y_{k-1,i} - \frac{1}{z}, \quad i = 1, \dots, n, \ k = 1, \dots, m. \tag{5}$$

Lemma 2. *For any $m \in \mathbf{Z}^+$ and $n \in \mathbf{N}$, if the complex numbers $\xi_1, \dots, \xi_n$ are linearly independent over $\mathbf{Q}$, then the functions in (4) are algebraically independent over $\mathbf{C}(z)$.*

Proof. We arrange the functions (4) in the following order:

$$y_{0,1}, \dots, y_{0,n};\ y_{1,1}, \dots, y_{1,n};\ \dots\ ;\ y_{m,1}, \dots, y_{m,n}. \tag{6}$$

We then number them in increasing order:

$$y_s = y_{k,i}, \quad s = 1, 2, \dots, (m+1)n, \quad s = kn + i, \quad 0 \le k \le m, \; 1 \le i \le n.$$

Then the system of differential equations (5) takes the form

$$\begin{aligned} y_s' &= \xi_s y_s, \qquad s = 1, \dots, n, \\ y_s' &= \frac{1}{z} y_{s-n} - \frac{1}{z}, \qquad s = n+1, \dots, (m+1)n. \end{aligned} \tag{7}$$

The functions $y_1, \dots, y_n$ are algebraically independent over $\mathbf{C}(z)$ by Lemma 21 of Chapter 3. We suppose that $y_1, \dots, y_s$, where $n \le s < (m+1)n$, are algebraically independent over $\mathbf{C}(z)$ and then prove that $y_1, \dots, y_s, y_{s+1}$ are algebraically independent over $\mathbf{C}(z)$. By induction, this will prove the lemma for any $m \in \mathbf{Z}^+$ and $n \in \mathbf{N}$.

Suppose the contrary, i.e., that there exists an irreducible polynomial

$$P = P(y_{s+1}, y_s, \dots, y_1, z) \in \mathbf{C}[y_{s+1}, y_s, \dots, y_1, z]$$

in which y_{s+1} occurs and such that

$$P(y_{s+1}, y_s, \dots, y_1, z) = 0. \tag{8}$$

We consider the field of analytic functions $\mathcal{L} = \mathbf{C}(y_s, \dots, y_1, z)$. From the differential equations (7) it follows that $\mathcal{L}$ is closed under differentiation. The equation (8) implies that y_{s+1} is an algebraic function over $\mathcal{L}$. And from the corresponding differential equation in (7) we can conclude that $y_{s+1}' \in \mathcal{L}$. Thus, by Lemma 1, $y_{s+1} \in \mathcal{L}$. Consequently, the equation (8) has the form

$$P = P_1 y_{s+1} + P_0 = 0, \qquad P_0, P_1 \in \mathbf{C}[y_s, \dots, y_1, z], \qquad P_0 P_1 \not\equiv 0, \tag{9}$$

where P_0 and P_1 are relatively prime polynomials.

We differentiate both sides of the equation $y_{s+1} + P_0/P_1 = 0$ and then use the differential equations (7). As a result we obtain

$$\frac{1}{z}(y_{s+1-n} - 1) + \frac{P_0'}{P_1} - \frac{P_0 P_1'}{P_1^2} = 0, \tag{10}$$

where P_0' and P_1' denote the total derivative with respect to z of the functions P_0 and P_1, in which the derivatives $y_1', \dots, y_s'$ that appear after differentiation are replaced by the right sides of the corresponding differential equations in (7). Then $zP_0', zP_1' \in \mathbf{C}[y_s, \dots, y_1, z]$.

The left side of (10) is a rational function in $y_s, \dots, y_1$ over $\mathbf{C}(z)$. However, the functions $y_s, \dots, y_1$ are algebraically independent over $\mathbf{C}(z)$. Consequently, the equality (10) holds as an identity in $y_s, \dots, y_1$ and z. Multiplying both sides of this identity by zP_1, we obtain

$$zP_0' - P_0 \frac{zP_1'}{P_1} + (y_{s+1-n} - 1)P_1 \equiv 0. \tag{11}$$

P_0 and P_1 are relatively prime polynomials in $\mathbf{C}[y_s, \dots, y_1, z]$. It hence follows from (11) that zP_1' is divisible by P_1, and

$$\frac{zP_1'}{P_1} \equiv az + b, \qquad a, b \in \mathbf{C}, \tag{12}$$

since zP_1' and P_1 have the same degree in $y_s, \dots, y_1$, and the degree of zP_1' in z may be 1 greater than that of P_1.

We arrange the terms in the polynomial P_1 in lexicographic order according to powers of $y_s, \dots, y_1$ with coefficients in $\mathbf{C}[z]$. Let

$$Qy_s^{k_s} \cdots y_{n+1}^{k_{n+1}} y_n^{k_n} \cdots y_1^{k_1}, \qquad Q \in \mathbf{C}[z],$$

be the leading term of P_1. Then

$$z(Q' + (k_1\xi_1 + \dots + k_n\xi_n)Q)y_s^{k_s} \cdots y_{n+1}^{k_{n+1}} y_n^{k_n} \cdots y_1^{k_1}$$

is the leading term of the polynomial zP_1'. Hence,

$$\frac{zP_1'}{P_1} = \frac{zQ'}{Q} + (k_1\xi_1 + \dots + k_n\xi_n)z \equiv az + b,$$

from which we obtain

$$a = k_1\xi_1 + \dots + k_n\xi_n, \qquad b = \deg Q \in \mathbf{Z}^+. \tag{13}$$

Integrating (12), we have

$$P_1 = cz^b e^{az} = cz^b y_n^{k_n} \cdots y_1^{k_1}, \qquad c \in \mathbf{C}, \qquad c \neq 0. \tag{14}$$

Thus, the polynomial P_1 does not depend on $y_s, \dots, y_{n+1}$. Substituting the expression we have found for P_1 in the identity (11), we find that

$$zP_0' - (az + b)P_0 + c(y_{s+1-n} - 1)z^b e^{az} \equiv 0, \qquad c \neq 0. \tag{15}$$

Two cases are possible:

A. $s+1-n=i\le n$, i.e., $y_{s+1-n}=y_i=e^{\xi_i z}$. From (15) and (14) it follows that y_i appears in P_0. Then

$$\begin{aligned}&P_0=Q_l y_i^l+\cdots+Q_1 y_i+Q_0, \qquad Q_l\not\equiv 0, \qquad l\ge 1,\\ &Q_0,Q_1,\dots,Q_l\in \mathbf{C}[y_s,\dots,y_{i+1},y_{i-1},\dots,y_1,z],\\ &P_0'=(Q_l'+l\xi_i Q_l)y_i^l+\cdots+Q_0'.\end{aligned} \tag{16}$$

In (15) we compare the terms which do not depend on y_i. We consider two sub-cases.

1) We have $k_i=0$ in (14). Then from (15) it follows that $Q_0\not\equiv 0$ and

$$zQ_0'-(az+b)Q_0-cz^b e^{az}=0, \qquad c\ne 0.$$

Integrating this differential equation, we obtain

$$Q_0=c_0 z^b e^{az}+cz^b e^{az}\ln z, \qquad c_0\in\mathbf{C}, \qquad c\ne 0,$$

which is impossible, since Q_0 and $z^b e^{az}$ are entire functions and $\ln z$ has a logarithmic branch point.

2) We have $k_i>0$ in (14). Then this equality implies that $Q_0\not\equiv 0$, since P_0 and P_1 are relatively prime polynomials. Consequently,

$$zQ_0'-(az+b)Q_0=0,$$

and hence

$$Q_0=c_0 z^b e^{az}, \qquad c_0\in\mathbf{C}\setminus\{0\}, \qquad e^{az}=y_n^{k_n}\cdots y_1^{k_1}.$$

But this last equality is impossible, because Q_0 does not depend on y_i, and, by the induction assumption, the functions $y_s,\dots,y_1$ are algebraically independent over $\mathbf{C}(z)$.

B. $s+1-n=i>n$. Again let P_0 be written as in (16). Differentiating, we have

$$\begin{aligned}zP_0'=zQ_l'y_i^l+\big(l(y_{i-n}-1)Q_l+zQ_{l-1}'\big)y_i^{l-1}+\cdots\\ \cdots+(y_{i-n}-1)Q_1+Q_0'.\end{aligned} \tag{17}$$

We consider three sub-cases.

1) $l>2$. Using (16) and (17) and comparing the coefficients of y_i^l and y_i^{l-1} in (15), we obtain

$$\begin{aligned}zQ_l'-(az+b)Q_l=0,\\ l(y_{i-n}-1)Q_l+zQ_{l-1}'-(az+b)Q_{l-1}=0.\end{aligned}$$

From this we obtain:

$$\begin{aligned} &Q_l = c_l z^b e^{az}, \qquad c_l \in \mathbf{C} \setminus \{0\}, \\ &zQ'_{l-1} - (az+b)Q_{l-1} + lc_l(y_{i-n} - 1)z^b e^{az} = 0. \end{aligned} \tag{18}$$

If we integrate the linear differential equation (18), we find that

$$Q_{l-1} = c_{l-1} z^b e^{az} + lc_l z^b e^{az} y_i, \qquad c_{l-1} \in \mathbf{C} \setminus \{0\}.$$

But this is impossible, because $Q_{l-1} \in \mathbf{C}[y_s, \ldots, y_{i+1}, y_{i-1}, \ldots \ldots, y_1, z]$, and $y_s, \ldots, y_1$ are algebraically independent over $\mathbf{C}(z)$.

2) $l = 2$. We similarly find that

$$\begin{aligned} &Q_2 = c_2 z^b e^{az}, \qquad c_2 \in \mathbf{C} \setminus \{0\}, \\ &zQ'_1 - (az+b)Q_1 + 2c_2 z^b (y_{i-n} - 1)e^{az} + cz^b e^{az} = 0. \end{aligned}$$

Integrating, we obtain

$$Q_1 = c_1 z^b e^{az} + 2c_2 z^b e^{az} y_i + cz^b e^{az} \ln z, \qquad c_1 \in \mathbf{C}, \qquad cc_2 \neq 0. \tag{19}$$

But this last equation is impossible, since all of the terms on both sides are entire functions except for the last term on the right, which has a logarithmic branch point.

3) $l = 1$. In a similar way we arrive at the equation

$$zQ'_1 - (az+b)Q_1 + cz^b e^{az} = 0;$$

integrating this, we obtain

$$Q_1 = c_1 z^b e^{az} + cz^b e^{az} \ln z, \qquad c_1 \in \mathbf{C}, \qquad cc_1 \neq 0.$$

But this equation is impossible for the same reason as (19).

Thus, the identity (15) always leads to a contradiction, and the lemma is proved by induction.

Let $\mathbf{V}$ be a field, and let $\mathbf{W}$ be a commutative ring or field which contains $\mathbf{V}$.

Lemma 3. *Suppose that two m-tuples of elements of* $\mathbf{W}$

$$w_1, \ldots, w_m, \tag{20}$$

$$w_1^*, \ldots, w_m^* \tag{21}$$

are connected by linear equations

$$w_k^* = v_{k,0} + \sum_{i=1}^{m} v_{k,i} w_i, \qquad k = 1, \ldots, m, \qquad v_{k,i} \in \mathbf{V}, \tag{22}$$

such that

$$|v_{k,i}|_{k,i=1,\ldots,m} \neq 0. \tag{23}$$

Then

$$\deg \operatorname{tr}_{\mathbf{V}}\{w_1, \ldots, w_m\} = \deg \operatorname{tr}_{\mathbf{V}}\{w_1^*, \ldots, w_m^*\}.$$

In particular, if the elements of one of the m-tuples (20) or (21) are algebraically independent over **V***, then so are the elements of the other m-tuple.*

Proof. From (22) it follows that each of the elements in (21) depends algebraically over **V** on the elements in (20). Because of the condition (23), the equations (22) can be used to express each of the elements in (20) linearly in terms of the elements in (21) and 1 with coefficients in **V**. Hence, each of the elements in (20) depends algebraically over **V** on the elements in (21). Consequently, the sets (20) and (21) are algebraically equivalent over **V**, and it follows from the properties of algebraic independence that they have the same transcendence degree over **V**.

Proof of Theorem 1. Part 1) follows from the Second Fundamental Theorem of Chapter 3 and Lemma 2, since the functions in (4) form a solution of the system of differential equations (5).

Part 2). We first consider the case $n = 1$. We differentiate the equation

$$\omega_m'(\xi_1 z) = \frac{1}{z}\omega_m(\xi_1 z) - \frac{1}{z}$$

$m - 1$ times. After each differentiation we replace each derivative $\omega_k'(\xi_1 z)$ which occurs ($k = m - 1, \ldots, 1, 0$) by the right side of the corresponding differential equation in (5). As a result we find that each of the functions $\omega_m(\xi_1 z), \omega_m'(\xi_1 z), \ldots, \omega_m^{(m)}(\xi_1 z)$ can be expressed linearly in terms of the functions $\omega_m(\xi_1 z), \ldots, \omega_1(\xi_1 z), \omega_0(\xi_1 z)$ and 1 with coefficients in $\mathbf{Q}\left[\frac{1}{z}\right]$. The matrix of coefficients of the functions $\omega_m(\xi_1 z), \ldots, \omega_1(\xi_1 z)$, $\omega_0(\xi_1 z)$ will be a triangular matrix whose diagonal entries are all nonzero when $z \neq 0$. Thus, the determinant of this matrix is nonzero for $z \neq 0$.

If we set $z = 1$ in these linear equations, apply Lemma 3 with $\mathbf{V} = \mathbf{A}$ and $\mathbf{W} = \mathbf{C}$, and use part 1) of the theorem, we obtain part 2) of the theorem for $n = 1$.

The case of arbitrary n is treated similarly. One merely has to arrange the functions (4) in the following order:

$$\omega_m(\xi_1 z), \dots, \omega_1(\xi_1 z), \omega_0(\xi_1 z), \dots, \omega_m(\xi_n z), \dots, \omega_1(\xi_n z), \omega_0(\xi_n z).$$

§ 2. Solutions of homogeneous differential equations

Let $\lambda_k \in \mathbf{Q}$, $\lambda_k \neq 0, -1, -2, \dots$, $k = 1, \dots, m$. We set

$$\chi_0(z) = e^z, \quad \chi_k(z) = \sum_{n=0}^{\infty} \frac{z^n}{n!(\lambda_1 + n) \cdots (\lambda_k + n)}, \qquad k = 1, \dots, m. \tag{24}$$

Then

$$\chi_k(z) = \frac{1}{\lambda_1 \cdots \lambda_k} \sum_{n=0}^{\infty} \frac{[\lambda_1, n] \cdots [\lambda_k, n]}{[1, n][\lambda_1 + 1, n] \cdots [\lambda_k + 1, n]} z^n.$$

The sum on the right is a hypergeometric E-function. Using the properties of E-functions and Lemmas 1 and 2 of Chapter 5, we see that $\chi_k(z)$ for any $k \geq 0$ is an E-function which satisfies a homogeneous linear differential equation of order $k + 1$ with coefficients in $\mathbf{Q}(z)$ and with its only singular point at $z = 0$.

It is easy to verify that (recall the symbol $\int^z$ from § 2 of Chapter 5)

$$\chi_k(z) = z^{-\lambda_k} \int^z t^{\lambda_k - 1} \chi_{k-1}(t) dt, \qquad k = 1, \dots, m. \tag{25}$$

Hence, taking the derivative, we find that the set of E-functions in (24) form a solution of the system of homogeneous linear differential equations

$$y_0' = y_0, \qquad y_k' = -\frac{\lambda_k}{z} y_k + \frac{1}{z} y_{k-1}, \qquad k = 1, \dots, m. \tag{26}$$

Theorem 2. *Suppose that $\lambda_k \in \mathbf{Q} \setminus \mathbf{Z}$, $k = 1, \dots, m$, $m \in \mathbf{Z}^+$; the differences $\lambda_i - \lambda_k$ do not belong to $\mathbf{Z} \setminus \{0\}$ for $i \neq k$, $i, k = 1, \dots, m$; and $\xi \in \mathbf{A} \setminus \{0\}$. Then* 1) *the $m + 1$ numbers*

$$e^{\xi}, \quad \chi_1(\xi), \dots, \chi_m(\xi) \tag{27}$$

are algebraically independent, and 2) *the $m + 1$ numbers $\chi_m^{(l)}(\xi)$, $l = 0, 1, \dots, m$, are algebraically independent.*

The proof of Theorem 2 requires the following function theoretic lemma.

Lemma 4. *Under the conditions of Theorem 2, the functions in (24) are algebraically independent over* $\mathbf{C}(z)$.

Proof. We use induction on m. If $m = 0$, then the lemma holds, because e^z is a transcendental function. Suppose that the lemma holds for $m = n - 1$, $n \in \mathbf{N}$; we shall prove that it then holds for $m = n$.

Suppose the contrary, i.e., that there exists an irreducible polynomial

$$P = P(y_n, \dots, y_1, y_0, z) \in \mathbf{C}[y_n, \dots, y_1, y_0, z]$$

in which y_n appears such that

$$P(\chi_n(z), \dots, \chi_1(z), \chi_0(z), z) = 0. \tag{28}$$

Let

$$D = \frac{\partial}{\partial z} + y_0 \frac{\partial}{\partial y_0} - \sum_{i=1}^{n-1} \left(\frac{\lambda_i}{z} y_i - \frac{1}{z} y_{i-1} \right) \frac{\partial}{\partial y_i}$$

and

$$D_1 = D - \left(\frac{\lambda_n}{z} y_n - \frac{1}{z} y_{n-1} \right) \frac{\partial}{\partial y_n}$$

be the differential operators associated with the system of differential equations (26). Then $zD_1P \in \mathbf{C}[y_n, \dots, y_1, y_0, z]$.

By the induction assumption, the functions (24) have transcendence degree n over $\mathbf{C}(z)$, and they are connected by the algebraic equation (28). Hence, by Lemma 4 of Chapter 5, zD_1P is divisible by P as a polynomial in $n + 2$ independent variables, and there exists $\omega \in \mathbf{C}(z)$ such that $zD_1P \equiv \omega P$. If we compare the degrees in z and in $y_n, \dots, y_1, y_0$ of zD_1P and P, we see that $\omega = az + b$, $a, b \in \mathbf{C}$. Thus,

$$zD_1P \equiv (az + b)P. \tag{29}$$

The polynomial P is irreducible. It therefore contains terms which do not depend on y_0. Let T denote the sum of all such terms. Then

$$P = Sy_0 + T, \qquad S \in \mathbf{C}[y_n, \dots, y_1, y_0, z],$$

$$T \in \mathbf{C}[y_n, \dots, y_1, z], \qquad T \not\equiv 0.$$

When D_1 is applied to P, the terms containing y_0 go to terms of the same type. The polynomial T does not contain terms with y_0, but such terms can appear in

zD_1T. In (29) the leading term of T in the variables $y_n, \dots, y_1$ can only cancel with the leading term of the same order in the polynomial zD_1T.

Let

$$Qy_n^{s_n} \cdots y_1^{s_1}, \qquad Q \in \mathbf{C}[z], \qquad Q \not\equiv 0,$$

be the leading term of T. Then

$$\left(zQ' - \left(\sum_{i=1}^{n} \lambda_i s_i\right) Q\right) y_n^{s_n} \cdots y_1^{s_1}$$

is the leading term of zD_1T. If we compare the coefficients of the leading terms of zD_1T and T in (29) and divide both sides of the resulting identity by Q, we obtain

$$\frac{zQ'}{Q} - \sum_{i=1}^{n} \lambda_i s_i \equiv az + b, \qquad \frac{zQ'}{Q} = s, \qquad s = \deg Q.$$

From this identity we find that $a = 0$ and $b \in \mathbf{Q}$. Then (29) takes the form

$$zD_1P \equiv bP. \tag{30}$$

We write the polynomial P as follows:

$$\begin{aligned} &P = P_l y_n^l + P_{l-1} y_n^{l-1} + \cdots + P_0, \qquad P_l P_0 \not\equiv 0, \qquad l \geq 1, \\ &P_i \in \mathbf{C}[y_{n-1}, \dots, y_1, y_0, z], \qquad i = 0, 1, \dots, l. \end{aligned} \tag{31}$$

We apply the operator D_1 to P. Then

$$\begin{aligned} zD_1P = (zDP_l - l\lambda_n P_l) y_n^l + (l y_{n-1} P_l + zDP_{l-1} - \\ - (l-1)\lambda_n P_{l-1}) y_n^{l-1} + \cdots + (y_{n-1} P_1 + zDP_0). \end{aligned} \tag{32}$$

Comparing the coefficients of y_n^l and y_n^{l-1} in (30) and using (31) and (32), we obtain

$$\begin{aligned} zDP_l - (b + l\lambda_n) P_l \equiv 0, \\ l y_{n-1} P_l + zDP_{l-1} - (b + (l-1)\lambda_n) P_{l-1} \equiv 0. \end{aligned} \tag{33}$$

We now set $y_i = \chi_i(z)$, $i = 0, 1, \dots, n$. Then $DP_l = \frac{d}{dz} P_l$. Integrating the first identity in (33), we find that

$$P_l = c_l z^{b + l\lambda_n}, \qquad c_l \in \mathbf{C} \setminus \{0\}.$$

Since P_l is an entire function, we have $(b + l\lambda_n) \in \mathbf{Z}^+$. If we substitute this expression for P_l in the second identity in (33), we have

$$lc_l z^{b+l\lambda_n} y_{n-1} + zDP_{l-1} - (b + (l-1)\lambda_n)P_{l-1} \equiv 0, \qquad lc_l \neq 0. \tag{34}$$

Since $DP_{l-1} = \frac{d}{dz}P_{l-1}$, if we integrate the differential equation (34) and use the corresponding equation in (25) we obtain

$$P_{l-1} = -lP_l y_n + c_{l-1} z^{b+(l-1)\lambda_n}, \qquad c_{l-1} \in \mathbf{C}, \qquad c_l \neq 0. \tag{35}$$

From (35) it follows that $c_{l-1} = 0$, since P_{l-1} and $P_l y_n$ are entire functions, $\lambda_n \in \mathbf{Q} \setminus \mathbf{Z}$ and $(b + l\lambda_n) \in \mathbf{Z}^+$, and it also follows that $l = 1$ and $P = P_1 y_n + P_0$.

We set $b + \lambda_n = p$, $p \in \mathbf{Z}^+$. Then the identity (34) takes the form

$$z^p y_{n-1} + zDP_0 - (p - \lambda_n)P_0 \equiv 0. \tag{36}$$

When the operator D is applied to the polynomial P_0, each homogeneous part of given total degree in $y_{n-1}, \ldots, y_1, y_0$ goes to a similar homogeneous part or to zero. From (36) it follows that P_0 has a linear homogeneous term in $y_{n-1}, \ldots, y_1, y_0$, which we shall denote R, $R \not\equiv 0$. Then from (36) we obtain

$$z^p y_{n-1} + zDR - (p - \lambda_n)R \equiv 0, \tag{37}$$

from which it follows that y_{n-1} occurs in R.

We prove that (37) is impossible. There are two possible cases.

A. $n = 1$. In this case $R = Q_0 y_0$, $Q_0 \in \mathbf{C}[z]$, $Q_0 \not\equiv 0$, and $zDR = z(Q_0' + Q_0)y_0$. Comparing the coefficients of y_0 in (37), we obtain

$$z^p + zQ_0' + (z + \lambda_1 - p)Q_0 \equiv 0, \qquad p \in \mathbf{Z}^+. \tag{38}$$

The polynomial zQ_0 has degree 1 greater than that of zQ_0'. Consequently, $p = \deg Q_0 + 1$, and so $p \geq 1$. But $\lambda_1 \in \mathbf{Q} \setminus \mathbf{Z}$. Hence, $\lambda_1 - p \neq 0$, and (38) implies that Q_0 is divisible by z.

Let $Q_0 = z^s Q$, where Q is a polynomial which is not divisible by z and $s \in \mathbf{N}$. Then if we cancel z^s in (38), the identity takes the form

$$z^{p-s} + zQ' + (z + \lambda_1 + 1 + s - p)Q \equiv 0, \qquad p - s \geq 1,$$

from which it follows that we must have $\lambda_1 + s - p = 0$. But this cannot be, since $(\lambda_1 + s - p) \in \mathbf{Q} \setminus \mathbf{Z}$.

B. $n > 1$. In this case

$$R = Q_{n-1}y_{n-1} + Q_{n-2}y_{n-2} + \cdots + Q_0 y_0,$$
$$Q_{n-1} \not\equiv 0, \qquad Q_i \in \mathbf{C}[z], \qquad i = 0, 1, \ldots, n-1,$$
$$zDR = (zQ'_{n-1} - \lambda_{n-1}Q_{n-1})y_{n-1} + (Q_{n-1} + zQ'_{n-2} - \lambda_{n-2}Q_{n-2})y_{n-2} + \cdots + (Q_1 + zQ'_0 + zQ_0)y_0.$$

Comparing the coefficients of $y_{n-2}, \ldots, y_1, y_0$ in (37), we obtain the system of identities

$$\begin{aligned} z^p + zQ'_{n-1} - (p + \lambda_{n-1} - \lambda_n)Q_{n-1} &\equiv 0, \\ Q_{n-1} + zQ'_{n-2} - (p + \lambda_{n-2} - \lambda_n)Q_{n-2} &\equiv 0, \\ \cdots\cdots\cdots\cdots\cdots \\ Q_2 + zQ'_1 - (p + \lambda_1 - \lambda_n)Q_1 &\equiv 0, \\ Q_1 + zQ'_0 - (p - z - \lambda_n)Q_0 &\equiv 0. \end{aligned} \tag{39}$$

There are two possible sub-cases.

1) $\lambda_i - \lambda_n \neq 0$, $i = 1, \ldots, n-1$.

Integrating the first differential equation in (39), we obtain

$$Q_{n-1} = \frac{1}{\lambda_{n-1} - \lambda_n} z^p + c_{n-1} z^{p+\lambda_{n-1}-\lambda_n}, \qquad c_{n-1} \in \mathbf{C}.$$

But $p + \lambda_{n-1} - \lambda_n \notin \mathbf{Z}$. Hence $c_{n-1} = 0$ and

$$Q_{n-1} = \gamma_{n-1} z^p, \qquad \gamma_{n-1} = \frac{1}{\lambda_{n-1} - \lambda_n} \neq 0.$$

If we substitute this expression for Q_{n-1} in the second identity in (39), we similarly obtain

$$Q_{n-2} = \gamma_{n-2} z^p, \qquad \gamma_{n-2} = \frac{1}{(\lambda_{n-2} - \lambda_{n-1})(\lambda_{n-1} - \lambda_n)}.$$

Continuing this process, we arrive at the equality

$$Q_1 = \gamma_1 z^p, \qquad \gamma_1 \neq 0,$$

and the last identity in (39) takes the form

$$\gamma_1 z^p + zQ'_0 + (z + \lambda_n - p)Q_0 = 0, \qquad \gamma_1 \neq 0. \tag{40}$$

But (40) is an identity of the same type as (38), and so cannot hold.

2) r is the largest index such that $\lambda_r - \lambda_n = 0$, $1 \leq r \leq n-1$.

We consider the first $n-r$ identities in (39). If $r = n-1$, then there is only one identity

$$z^p + zQ'_{n-1} - pQ_{n-1} \equiv 0, \tag{41}$$

which, when we integrate, gives

$$Q_{n-1} = -z^p \ln z + c_{n-1} z^p, \qquad c_{n-1} \in \mathbf{C},$$

which is impossible, since $Q_{n-1} \in \mathbf{C}[z]$.

On the other hand, if $r < n-1$, then, since $\lambda_i - \lambda_n \neq 0$ for $r+1 \leq i \leq n-1$, as in the previous sub-case we obtain

$$Q_i = \gamma_i z^p, \qquad \gamma_i \in \mathbf{C} \setminus \{0\}, \qquad i = r+1, \ldots, n-1.$$

Then the last in our subset of identities from (39) takes the form

$$\gamma_{r+1} z^p + zQ'_r - pQ_r \equiv 0, \qquad \gamma_{r+1} \neq 0,$$

which, as in the case of (41), we see is impossible.

Thus, the identity (37) always leads to a contradiction, and the lemma holds for $m = n$. By induction, the lemma holds for all m.

Since the E-functions in (24) form a solution of the system of differential equations (26), part 1) of Theorem 2 follows from the Second Fundamental Theorem of Chapter 3 and from Lemma 4. Part 2) is proved in the same way as was part 2) of Theorem 1.

§ 3. Corollaries of Theorems 1 and 2

Now let $\lambda_k = \lambda$, $k = 1, \ldots, m$, $\lambda \neq 0, -1, -2, \ldots$. We set

$$\omega_{\lambda,0}(z) = e^z, \qquad \omega_{\lambda,k}(z) = \sum_{n=0}^{\infty} \frac{z^n}{n!(\lambda+n)^k}, \qquad k = 1, \ldots, m. \tag{42}$$

Then Theorem 2 implies

Theorem 3. *If $\lambda \in \mathbf{Q} \setminus \mathbf{Z}$ and $\xi \in \mathbf{A} \setminus \{0\}$, then for any $m \in \mathbf{N}$ the numbers e^ξ, $\omega_{\lambda,k}(\xi)$, $k = 1, \ldots, m$, are algebraically independent.*

We note that the function $\omega_{\lambda,1}(z)$ was considered in § 2 of Chapter 5, and that Theorem 3 with $m = 1$ follows from Theorem 3 of Chapter 5.

From (42) we have

$$\frac{\partial^l}{\partial\lambda^l}\omega_{\lambda,l}(z) = (-1)^l l!\, \omega_{\lambda,l+1}(z), \qquad l = 1,\dots,m. \tag{43}$$

Hence, Theorem 3 is equivalent to the following result.

Theorem 4. *If $\lambda \in \mathbf{Q}\setminus\mathbf{Z}$ and $\xi \in \mathbf{A}\setminus\{0\}$, then for any $m \in \mathbf{N}$ the numbers e^{ξ}, $\frac{\partial^l}{\partial\lambda^l}\omega_{\lambda,l}(\xi)$, $l = 0,1,\dots,m$, are algebraically independent.*

Let

$$\varphi_\lambda(z) = \sum_{n=0}^{\infty} \frac{z^n}{(\lambda+1)\cdots(\lambda+n)}, \qquad \lambda \neq -1,-2,\dots, \tag{44}$$

be the function that was introduced in § 2 of Chapter 5. Then

$$\varphi_\lambda(z) = \lambda z^{-\lambda} e^z \int_0^z t^{\lambda-1} e^{-t}\, dt, \qquad \lambda \neq 0,-1,-2,\dots. \tag{45}$$

From (44) and (45) we obtain

$$\varphi_{\lambda-1}(z) = 1 + \frac{z}{\lambda}\varphi_\lambda(z) = 1 + z^{1-\lambda} e^z \int_0^z t^{\lambda-1} e^{-t}\, dt,$$

and, after replacing λ by $\lambda+1$,

$$\begin{aligned}\varphi_\lambda(z) &= 1 + z^{-\lambda} e^z \int_0^z t^{\lambda} e^{-t}\, dt \\ &= 1 + z e^z \omega_{\lambda+1,1}(-z), \qquad \lambda \neq -1,-2,\dots.\end{aligned} \tag{46}$$

If we differentiate (46) with respect to λ, we find that

$$\frac{\partial^l}{\partial\lambda^l}\varphi_\lambda(z) = z e^z \frac{\partial^l}{\partial\lambda^l}\omega_{\lambda+1,1}(-z), \;\; \lambda \neq -1,-2,\dots, \;\; l = 1,\dots,m. \tag{47}$$

For $l = 1$ the function in (47) has the form

$$\frac{\partial}{\partial\lambda}\varphi_\lambda(z) = -\sum_{n=1}^{\infty}\left(\frac{1}{\lambda+1} + \cdots + \frac{1}{\lambda+n}\right)\frac{z^n}{(\lambda+1)\cdots(\lambda+n)}.$$

If we use Theorem 3 with ξ replaced by $-\xi$, Theorem 4, and (43) and (47), we obtain the following theorem.

Theorem 5 (K. Mahler). *If* $\lambda \in \mathbf{Q} \setminus \mathbf{Z}$ *and* $\xi \in \mathbf{A} \setminus \{0\}$, *then for any* $m \in \mathbf{N}$ *the numbers* $e^\xi, \dfrac{\partial^l}{\partial \lambda^l}\varphi_\lambda(\xi)$, $l = 0, 1, \ldots, m$, *are algebraically independent.*

The functions in (45) and (47) form a solution of the system of linear differential equations

$$\begin{aligned}
y_0' &= \left(1 - \frac{\lambda}{z}\right) y_0 + \frac{1}{z}, \\
y_1' &= \left(1 - \frac{\lambda}{z}\right) y_1 - \frac{1}{z} y_0 + \frac{1}{z}, \\
y_2' &= \left(1 - \frac{\lambda}{z}\right) y_2 - \frac{2}{z} y_1, \\
&\dots\dots\dots\dots\dots \\
y_m' &= \left(1 - \frac{\lambda}{z}\right) y_m - \frac{m}{z} y_{m-1}.
\end{aligned}$$

Hence, the method used to prove Theorem 2 will also give the following result.

Theorem 6 (K. Mahler). *If* $\lambda \in \mathbf{Q}$, $\lambda \neq -1, -2, \cdots$, *and* $\xi \in \mathbf{A} \setminus \{0\}$, *then for any* $m \in \mathbf{N}$ *the numbers* $\dfrac{\partial^l}{\partial \lambda^l}\varphi_\lambda(\xi)$, $l = 0, 1, \ldots, m$, *are algebraically independent.*

We shall not give the proof of Theorem 6. Mahler's book [42:7] contains a proof based on a different idea.

We show that Theorem 6 with $\lambda = 0$ follows from Theorem 1.

Theorem 7. *If* $\xi \in \mathbf{A} \setminus \{0\}$, *then for any* $m \in \mathbf{N}$ *the numbers* $e^\xi, \dfrac{\partial^l}{\partial \lambda^l}\varphi_\lambda(\xi)|_{\lambda=0}$, $l = 1, \ldots, m$, *are algebraically independent.*

Proof. It is easy to verify that

$$\begin{aligned}
\omega_{1,1}(z) &= \frac{\omega_0(z) - 1}{z} = \frac{e^z - 1}{z}, \\
\omega_{1,k}(z) &= \frac{\omega_{k-1}(z) - 1}{z}, \qquad k = 2, 3, \ldots,
\end{aligned} \tag{48}$$

where the functions $\omega_k(z)$, $k = 0, 1, 2, \ldots$, are defined by (1). Using (46), (47), (43) and (48), we find that

$$\begin{aligned}
&\varphi_0(z) = e^z, \\
&\frac{\partial^l}{\partial \lambda^l}\varphi_\lambda(z)|_{\lambda=0} = z e^z \frac{\partial^l}{\partial \lambda^l}\omega_{\lambda+1,1}(-z)|_{\lambda=0} = \\
&\qquad = z e^z (-1)^l l!\, \omega_{1,l+1}(-z) = e^z (-1)^{l+1} l! (\omega_l(-z) - 1).
\end{aligned} \tag{49}$$

From (49) and Theorem 1 with ξ replaced by $-\xi$ we obtain Theorem 7.

We note that

$$\frac{\partial}{\partial\lambda}\varphi_\lambda(z)|_{\lambda=0} = -\sum_{n=1}^{\infty}\left(1+\frac{1}{2}+\cdots+\frac{1}{n}\right)\frac{z^n}{n!}.$$

Since

$$\omega_{\lambda,1}(z) = z^{-\lambda}\int_0^z t^{\lambda-1}e^t dt = \int_0^1 u^{\lambda-1}e^{zu}du,$$

it follows that differentiating with respect to the parameter gives

$$\frac{\partial^k}{\partial\lambda^k}\omega_{\lambda,1}(z) = \int_0^1 u^{\lambda-1}\ln^k u\, e^{zu}\, du.$$

Hence, we obtain the following result from Theorem 4.

Theorem 8. *If* $\lambda \in \mathbf{Q}\setminus\mathbf{Z}$, $\lambda > 0$, *and* $\xi \in \mathbf{A}\setminus\{0\}$, *then for any* $m \in \mathbf{N}$ *the numbers*

$$e^\xi, \quad \int_0^1 t^{\lambda-1}\ln^k t\, e^{\xi t}\, dt, \qquad k = 0, 1, \ldots, m,$$

are algebraically independent.

Theorem 8 with only the integrals (without e^ξ) was proved by Mahler in 1968 in [42:5,7].

Remarks

Lemma 1 is due to K. Mahler [42:5]. Theorem 1 for $m = 1$ was proved in 1954 in [70:1,9]. Theorems 1 and 2 are contained in a more general theorem concerning the values of $\varphi_\lambda(z)$ and the functions

$$\sum_{n=0}^{\infty}\frac{z^n}{[\lambda, n](\mu_1+n)\cdots(\mu_l+n)}, \qquad l = 1, \ldots, m,$$

with different values of the parameters λ and μ_i and different values of the argument, subject to certain natural restrictions. The latter theorem was proved in 1959 in [70:10]. Theorems 1 and 3 were proved by a different method in 1961 in [70:13]. Those proofs are given in Chapter 8. The first result on the arithmetic properties of the values of the functions $\omega_{\lambda,1}(z)$ and $\omega_{\lambda,2}(z)$ are contained in [70:1,9].

In [70:10] general theorems were also proved about the values of the functions

$$K_\lambda(z) = \sum_{n=0}^{\infty} \frac{(-1)^n}{n![\lambda, n]} \left(\frac{z}{2}\right)^{2n},$$

$$\sum_{n=0}^{\infty} \frac{(-1)^n}{n![\lambda, n](\mu_1 + 1)\cdots(\mu_l + 1)} \left(\frac{z}{2}\right)^{2n}, \qquad l = 1, \ldots, m,$$

and the functions

$$H_{\lambda,l}(z) = z^{-\mu_l} e^z \int^z t^{\mu_l - 1} e^{-t} H_{l-1}(t)dt, \qquad l = 1, \ldots, m.$$

Theorems of somewhat less generality but of a similar type were proved in 1954 in [70:1,9].

The arithmetic properties of values of the derivatives of hypergeometric E-functions with respect to the parameter were first investigated in 1967 by I. I. Belogrivov [3,1]. He proved theorems on algebraic independence of the values at different algebraic points of the functions e^z, $\varphi_\lambda(z)$, $\frac{\partial}{\partial\lambda}\varphi_\lambda(z)|_{\lambda=0}$ and $\frac{\partial}{\partial\lambda}\varphi_\lambda(z)$ and the derivatives of the last two functions with respect to z. The most basic of these results are contained in Theorem 5 of [70:1,9] and Theorem 2 of [70:10]. In the same paper [3:1], Belogrivov proved algebraic independence theorems for the values at algebraic points of the functions $K_\lambda(z)$, $K_\lambda'(z)$, $\frac{\partial}{\partial\lambda}K_\lambda(z)$, and the first and second derivatives with respect to z of the third function.

In 1968, Mahler [42:5] generalized Belogrivov's results to the functions $\frac{\partial^l}{\partial\lambda^l}K_\lambda(z)$, $l = 0, 1, 2, 3$, and $\frac{\partial^l}{\partial\lambda^l}\varphi_\lambda(z)$, $l = 0, 1, 2, \ldots$.

In 1971, Belogrivov [3:5] proved several theorems on algebraic independence of the values at algebraic points of the derivatives of $K_\lambda(z)$ of arbitrary order with respect to the parameter and of certain derivatives with respect to z.

In 1972–73, K. Väänänen [79:1,3] also proved algebraic independence of the values at algebraic points of the functions $\frac{\partial^l}{\partial\lambda^l}K_\lambda(z)$, $l = 0, 1, 2, \ldots$, and of their derivatives with respect to z; and he proved some other theorems about the arithmetic properties of the values of the derivatives of $K_\lambda(z)$ with respect to the parameter. In 1975, Väänänen [79:4,5] proved a theorem of the same type for the derivatives with respect to the parameters of the Kummer functions (5.60) which we studied in Chapter 6.

The results of Mahler given in § 3 are contained in his book [42:7].

Chapter 8
Arithmetic methods applied to solutions of linear differential equations of arbitrary order

§ 1. Statement of the theorems

In this chapter we prove several theorems about algebraic independence of the values of E-functions. In order to prove that our functions are algebraically independent over $\mathbf{C}(z)$, we use a method based on the arithmetic properties of the power series coefficients of the functions. This method was used by W. Maier [43:1] and C. Siegel [72:3] for similar purposes. All of the results proved in § 1-4 were published in [70:13], and the ones in § 5 were published in [70:6,23].

We set

$$\omega_0(z) = e^z, \qquad \omega_s(z) = 1 + \int_0^z \frac{\omega_{s-1}(t) - 1}{t}\, dt, \qquad s = 1, \ldots, m, \qquad m \geq 1; \tag{1}$$

$$\gamma_0(z) = \cos z, \qquad \gamma_s(z) = 1 + \int_0^z \frac{\gamma_{s-1}(t) - 1}{t}\, dt, \qquad s = 1, \ldots, m, \qquad m \geq 1. \tag{2}$$

Then

$$\omega_s(z) = 1 + \sum_{n=1}^{\infty} \frac{z^n}{n!\, n^s}, \qquad s = 0, 1, \ldots, m, \qquad m \geq 1; \tag{3}$$

$$\gamma_s(z) = 1 + \sum_{n=1}^{\infty} \frac{(-1)^n z^{2n}}{(2n)!(2n)^s}, \qquad s = 0, 1, \ldots, m, \qquad m \geq 1. \tag{4}$$

The first theorem that follows is actually contained in Theorem 1 of Chapter 7. We give it here as a simple example of how the method used in this chapter works.

Theorem 1. *If* $\xi \in \mathbf{A}\setminus\{0\}$, *then:* 1) *the numbers* $\omega_s(\xi)$, $s = 0, 1, \ldots, m$, *are algebraically independent;* 2) *the numbers* $\omega_m^{(l)}(\xi)$, $l = 0, 1, \ldots, m$, *are algebraically independent.*

Theorem 2. *If* $\xi \in \mathbf{A}\setminus\{0\}$, *then:* 1) *the numbers* $\gamma_s(\xi)$, $s = 0, 1, \ldots, m$, *are algebraically independent;* 2) *the numbers* $\gamma_m^{(l)}(\xi)$, $l = 0, 1, \ldots, m$, *are algebraically independent.*

Let $\mu \in \mathbf{Q}\setminus\mathbf{Z}$. Using the symbol $\int^z$ that was introduced in § 2 of Chapter 5, we define the notation

$$\begin{aligned} &\omega_{\mu,0}(z) = e^z, \\ &\omega_{\mu,s}(z) = z^{-\mu}\int^z t^{\mu-1}\omega_{\mu,s-1}(t)dt, \qquad s = 1, \ldots, m, \qquad m \geq 1; \end{aligned} \tag{5}$$

$$\begin{aligned} &\gamma_{\mu,0}(z) = \cos z, \\ &\gamma_{\mu,s}(z) = z^{-\mu}\int^z t^{\mu-1}\gamma_{\mu,s-1}(t)dt, \qquad s = 1, \ldots, m, \qquad m \geq 1; \end{aligned} \tag{6}$$

$$\begin{aligned} &\delta_{\mu,0}(z) = \sin z, \\ &\delta_{\mu,s}(z) = z^{-\mu}\int^z t^{\mu-1}\delta_{\mu,s-1}(t)dt, \qquad s = 1, \ldots, m, \qquad m \geq 1. \end{aligned} \tag{7}$$

Then

$$\omega_{\mu,s}(z) = \sum_{n=0}^{\infty}\frac{z^n}{n!(n+\mu)^s}, \qquad s = 0, 1, \ldots, m, \qquad m \geq 1; \tag{8}$$

$$\gamma_{\mu,s}(z) = \sum_{n=0}^{\infty}\frac{(-1)^n z^{2n}}{(2n)!(2n+\mu)^s}, \qquad s = 0, 1, \ldots, m, \qquad m \geq 1; \tag{9}$$

$$\delta_{\mu,s}(z) = \sum_{n=0}^{\infty}\frac{(-1)^n z^{2n-1}}{(2n-1)!(2n-1+\mu)^s}, \qquad s = 0, 1, \ldots, m, \qquad m \geq 1. \tag{10}$$

We shall allow the index $\mu = 0$ in the case of the functions $\delta_{\mu,s}(z)$.

It is not hard to verify that, if $f(z)$ is an E-function and $k = \operatorname{ord} f(z)$, then the function

$$z^{-\mu}\int^z t^{\mu-1}f(t)dt, \qquad \mu \in \mathbf{Q}, \qquad \mu \neq -k, -(k+1), -(k+2), \ldots,$$

is also an E-function. Hence, the functions in (3), (4), (8)-(10) are all E-functions. Using the relations (1), (2), (5), (6) and (7), we see that the functions (3) and (5) satisfy linear differential equations of order $s+1$ with coefficients in $\mathbf{Q}[z]$, and the functions (4), (6) and (7) satisfy similar differential equations of order $s+2$.

Theorem 3. *If $\mu \in \mathbf{Q} \setminus \mathbf{Z}$ and $\xi \in \mathbf{A} \setminus \{0\}$, then: 1) the numbers $\omega_{\mu,s}(\xi)$, $s = 0, 1, \ldots, m$, are algebraically independent; 2) the numbers $\omega^{(l)}_{\mu,m}(\xi)$, $l = 0, 1, \ldots, m$, are algebraically independent.*

Theorem 4. *If $\mu \in \mathbf{Q} \setminus \mathbf{Z}$ and $\xi \in \mathbf{A} \setminus \{0\}$, then: 1) the numbers $\gamma_{\mu,s}(\xi)$, $s = 0, 1, \ldots, m$, are algebraically independent; 2) the numbers $\gamma^{(l)}_{\mu,m}(\xi)$, $l = 0, 1, \ldots, m$, are algebraically independent.*

Theorem 5. *If $\mu \in \mathbf{Q} \setminus \mathbf{Z}$ or $\mu = 0$, and if $\xi \in \mathbf{A} \setminus \{0\}$, then: 1) the numbers $\delta_{\mu,s}(\xi)$, $s = 0, 1, \ldots, m$, are algebraically independent; 2) the numbers $\delta^{(l)}_{\mu,m}(\xi)$, $l = 0, 1, \ldots, m$, are algebraically independent.*

If we denote

$$\operatorname{si} z = \int_0^z \frac{\sin t}{t}\, dt, \qquad \operatorname{ci} z = \int_0^z \frac{1 - \cos t}{t}\, dt,$$

then it follows from Theorems 2 and 5 that for any $\xi \in \mathbf{A} \setminus \{0\}$ the numbers $\operatorname{si}\xi$ and $\sin\xi$ and also the numbers $\operatorname{ci}\xi$ and $\cos\xi$ are algebraically independent.

Let $\lambda \in \mathbf{Q}$, $\lambda > 0$. By means of a change of variable we can conclude from Theorems 3–5 that each of the functions of z

$$\int_0^z e^{-t^\lambda}\, dt, \qquad \int_0^z \sin t^\lambda\, dt, \qquad \int_0^z \cos t^\lambda\, dt \tag{11}$$

takes a transcendental value for any $z \in \mathbf{A} \setminus \{0\}$. If in addition $\lambda \neq 1/k$ for $k \in \mathbf{N}$, then for any $z \in \mathbf{A} \setminus \{0\}$ the value of each of the functions (11) is algebraically independent of the value of the corresponding function e^{z^λ}, $\sin z^\lambda$, $\cos z^\lambda$. In particular, this holds for the "incomplete" probability integral and Fresnel integrals

$$\int_0^z e^{-t^2}\, dt, \qquad \int_0^z \sin t^2\, dt, \qquad \int_0^z \cos t^2\, dt \tag{12}$$

and the corresponding functions e^{z^2}, $\sin z^2$, $\cos z^2$.

We note that these results were proved for the first integrals in (11) and (12) in § 2 of Chapter 5 as a consequence of Theorem 3 of that chapter. Theorem 3 above implies a similar result for the "incomplete" gamma-function and e^z, which was also proved in § 2 of Chapter 5.

We set

$$\psi_k(z) = \sum_{n=0}^{\infty} \frac{1}{(n!)^k} \left(\frac{z}{k}\right)^{kn}, \qquad k \geq 1, \tag{13}$$

and

$$\psi_{k,s}(z) = 1 + \sum_{n=1}^{\infty} \frac{1}{((n-1)!)^k n^s} \left(\frac{z}{k}\right)^{kn}, \tag{14}$$

$$s = 1, \ldots, m_k, \qquad m_k \geq k; \qquad k = 1, \ldots, r, \qquad r \geq 1.$$

We have

$$\begin{aligned} &\psi_{k,k}(z) = \psi_k(z), \qquad k = 1, \ldots, r; \\ &\psi_{1,s}(z) = \omega_{s-1}(z). \end{aligned} \tag{15}$$

The functions (13) and (14) are hypergeometric E-functions. By Lemma 2 of Chapter 5, the function in (13) satisfies a k-th order linear differential equation with coefficients in $\mathbf{C}[z]$, and the function $\psi_{k,m_k}(z)$ satisfies an m_k-th order differential equation of the same type.

Theorem 6. *If $\xi \in \mathbf{A}\backslash\{0\}$, then for any k and for $m_k \geq k$: 1) the numbers $\psi_{k,s}(\xi)$, $s = 0, 1, \ldots, m_k$, are algebraically independent; 2) the numbers $\psi^{(l)}_{k,m_k}(\xi)$, $l = 0, 1, \ldots, m_k - 1$, are algebraically independent. In particular, the k numbers $\psi^{(l)}_k(\xi)$, $l = 0, 1, \ldots, k-1$, are algebraically independent.*

Theorem 7. *If $\xi \in \mathbf{A} \setminus \{0\}$, then for any r and for $m_1 \geq 1, \ldots, m_r \geq r$: 1) the $m_1 + \cdots + m_r$ numbers $\psi_{k,s}(\xi)$, $s = 1, \ldots, m_k$, $k = 1, \ldots, r$, are algebraically independent; 2) the $m_1 + \cdots + m_r$ numbers $\psi^{(l)}_{k,m_k}(\xi)$, $l = 0, 1, \ldots, m_k - 1$, $k = 1, \ldots, r$, are algebraically independent. In particular the $r(r+1)/2$ numbers $\psi^{(l)}_k(\xi)$, $l = 0, 1, \ldots, k-1$, $k = 1, \ldots, r$, are algebraically independent.*

§ 2. Auxiliary lemmas

The basic idea of the method used in this chapter to prove algebraic independence of our functions over $\mathbf{C}(z)$ is contained in the following lemma.

Lemma 1. *Let*

$$\varphi_\nu(z) = \sum_{n=0}^{\infty} a_{n,\nu} z^n, \qquad \nu = 0, 1, \ldots, u, \qquad u \geq 0, \tag{16}$$

$$f_\nu(z) = \sum_{n=0}^{\infty} b_{n,\nu} z^n, \qquad \nu = 1, \ldots, v, \qquad v \geq 1, \tag{17}$$

be power series with coefficients in an algebraic number field $\mathbf{K}$ *which satisfy the following conditions:*

1°. *The series in (16) are algebraically independent over* $\mathbf{C}$.

2°. *For every* $t \in \mathbf{N}$ *there exists* $l \in \mathbf{N}$, $l > t$, *such that the exact denominators of the numbers* $b_{l,1}, \ldots, b_{l,v}$ *are divisible by a certain prime ideal* $\mathcal{P}_l$ *to the exact powers* $k_1, \ldots, k_v$, *respectively, where*

$$1 \leq k_1 < k_2 < \cdots < k_v, \tag{18}$$

and the denominators of the numbers

$$\begin{aligned} &a_{n,\nu}, \qquad n = 0, 1, \ldots, l+t, \qquad \nu = 0, 1, \ldots, u, \\ &b_{n,\nu}, \qquad n = 0, 1, \ldots, l-1, \qquad \nu = 1, \ldots, v, \end{aligned} \tag{19}$$

are not divisible by $\mathcal{P}_l$.

Then the $u + v + 1$ *power series in (16) and (17) are algebraically independent over* $\mathbf{C}$.

Proof. Suppose that the series in (16) and (17) are algebraically dependent over $\mathbf{C}(z)$. We may suppose that the series in (16) along with $f_\nu(z)$, $\nu = 1, \ldots, k-1$, $1 \leq k \leq v$, are algebraically independent, but that the series in (16) along with $f_\nu(z)$, $\nu = 1, \ldots, k$, are algebraically dependent over $\mathbf{C}$. Then there exists an algebraic equation

$$P = P(f_1(z), \ldots, f_k(z); \varphi_0(z), \ldots, \varphi_u(z)) = 0, \tag{20}$$

where P is an irreducible polynomial in $u + k + 1$ variables with coefficients in $\mathbf{C}$, in which $f_k(z)$ appears. By Lemma 4 of Chapter 4, P can be chosen so that its coefficients are in $\mathbf{Z_K}$.

Without loss of generality we may assume that each of the variables $f_1(z), \ldots$ $\ldots, f_k(z)$ occurs in P, since otherwise instead of the functions $f_1(z), \ldots, f_v(z)$ we could consider only the ones which occur in P and index them so that condition 2° of the lemma is fulfilled.

Then each of the partial derivatives

$$\frac{\partial P}{\partial f_\nu(z)}, \qquad \nu = 1, \dots, k,$$

is a polynomial in the same $u+k+1$ variables as P which is not identically zero. In fact, if we had, for example,

$$\frac{\partial P}{\partial f_s(z)} \equiv 0, \tag{21}$$

then the polynomial $\partial P/\partial f_s(z)$ would have to be divisible by the irreducible polynomial P as a polynomial in $u+k+1$ variables. Otherwise, by eliminating $f_k(z)$ from the two equation (20) and (21), we could obtain an algebraic equation connecting the functions in (16) and $f_\nu(z)$, $\nu = 1, \dots, k-1$, over $\mathbf{K}$, and this is impossible, by assumption. However, the polynomial $\partial P/\partial f_s(z)$ cannot be divisible by P, since it is not identically zero in the variables (16) and $f_\nu(z)$, $\nu = 1, \dots, k$, and it has lower degree in $f_s(z)$ than P.

Let

$$\frac{\partial P(f_1(z), \dots, f_k(z); \varphi_0(z), \dots, \varphi_u(z))}{\partial f_\nu(z)} = \gamma_\nu z^{t_\nu} + \cdots, \tag{22}$$

$$\gamma_\nu \in \mathbf{K}, \qquad \gamma_\nu \neq 0, \qquad t_\nu \geq 0, \qquad \nu = 1, \dots, k,$$

be the power series representation for the function on the left. We let t, $t \geq 0$, be the smallest of the exponents $t_1, \dots, t_k$, and we let s, $1 \leq s \leq k$, be the largest value of ν for which $t_\nu = t$, so that

$$\begin{aligned} t_\nu &\geq t, \qquad \nu = 1, \dots, s-1, \\ t_s &= t, \\ t_\nu &> t, \qquad \nu = s+1, \dots, k. \end{aligned} \tag{23}$$

We set

$$\begin{aligned} g_\nu(z) &= \sum_{n=0}^{l} b_{n,\nu} z^\nu, \qquad \nu = 1, \dots, s-1, \\ g_\nu(z) &= \sum_{n=0}^{l-1} b_{n,\nu} z^\nu, \qquad \nu = s, \dots, k, \end{aligned} \tag{24}$$

where l is a natural number which will be chosen later.

By Taylor's formula we have

$$\begin{aligned}
&P(f_1(z),\dots,f_k(z);\varphi_0(z),\dots,\varphi_u(z)) = \\
&\quad = P(g_1(z),\dots,g_k(z);\varphi_0(z),\dots,\varphi_u(z)) + \\
&\quad + \sum_{\nu=1}^{k} \frac{\partial P(g_1(z),\dots,g_k(z);\varphi_0(z),\dots,\varphi_u(z))}{\partial g_\nu(z)}(f_\nu(z)-g_\nu(z)) + \\
&\quad + \frac{1}{2}\sum_{\nu=1}^{k}\sum_{\rho=1}^{k} \frac{\partial^2 P(g_1(z),\dots,g_k(z);\varphi_0(z),\dots,\varphi_u(z))}{\partial g_\nu(z)\partial g_\rho(z)} \cdot \\
&\qquad \cdot (f_\nu(z)-g_\nu(z))(f_\rho(z)-g_\rho(z)) + \cdots .
\end{aligned}$$

Hence, in view of (20), we obtain

$$\begin{aligned}
&\sum_{\nu=1}^{k} \frac{\partial P(g_1(z),\dots,g_k(z);\varphi_0(z),\dots,\varphi_u(z))}{\partial g_\nu(z)}(f_\nu(z)-g_\nu(z)) = \\
&\quad = -P(g_1(z),\dots,g_k(z);\varphi_0(z),\dots,\varphi_u(z)) - \\
&\quad - \frac{1}{2}\sum_{\nu=1}^{k}\sum_{\rho=1}^{k} \frac{\partial^2 P(g_1(z),\dots,g_k(z);\varphi_0(z),\dots,\varphi_u(z))}{\partial g_\nu(z)\partial g_\rho(z)} \cdot \\
&\qquad \cdot (f_\nu(z)-g_\nu(z))(f_\rho(z)-g_\rho(z)) - \cdots .
\end{aligned} \tag{25}$$

We let T denote the largest of the exponents $t_1,\dots,t_k$ in (22), and we choose $l > T$ (hence $l > t$) in such a way that condition 2° of the lemma is fulfilled. It then follows from (17) and (24) that for each $\nu = 1,\dots,k$, the terms in the series $f_\nu(z)$ and $g_\nu(z)$ are equal until the z^T-term. Hence, by (22), we also have

$$\frac{\partial P(g_1(z),\dots,g_k(z);\varphi_0(z),\dots,\varphi_u(z))}{\partial g_\nu(z)} = \gamma_\nu z^{t_\nu} + \cdots, \quad \nu = 1,\dots,k. \tag{26}$$

Using (16), (17), (24), (26), (23) and the inequality $l > t$, we equate the coefficients of z^{l+t} on both sides of (25). As a result we obtain the equality

$$\gamma_s b_{l,s} = A_l, \qquad \gamma_s \neq 0, \tag{27}$$

where A_l is a polynomial in the variables (19) and $b_{l,1},\dots,b_{l,s-1}$ with coefficients in $\mathbf{Z_K}$.

Since the polynomial A_l is the coefficient of z^{l+t} on the right in (25), it follows that in each of its terms if we multiply the first subscript of each variable (19) and $b_{l,1},\dots,b_{l,s-1}$ with the power to which the variable occurs in the term and then add the result for all of the variables, we always obtain $l+t$. But $l > t$. Thus,

A_l is a linear form in $b_{l,1}, \dots, b_{l,s-1}$ and 1 whose coefficients are polynomials in the variables (19) having coefficients in $\mathbf{Z}_{\mathbf{K}}$.

The prime ideal $\mathcal{P}_l$ does not divide the denominators of the numbers (19), and for $s > 1$ it divides the denominators of $b_{l,1}, \dots, b_{l,s-1}$ exactly to the powers $k_1, \dots, k_{s-1}$, respectively, where (18) holds. Hence, the exact denominator of $A_l \in \mathbf{Q}$ can be divisible by $\mathcal{P}_l$ only if $s > 1$ and only to a power no greater than k_{s-1}. It then follows from (27) that the exact denominator of $\gamma_s b_{l,s}$ can be divisible by $\mathcal{P}_l$ only if $s > 1$ and only to a power no greater than k_{s-1}. Here the exact denominator of $b_{l,s}$ is divisible by $\mathcal{P}_l$ to the power $k_s \geq 1$, where $k_s > k_{s-1}$ for $s > 1$. Hence, the numerator of γ_s is divisible by $\mathcal{P}_l$.

By condition 2° of the lemma, the last divisibility assertion holds for an infinite set of values of l; and the prime ideals $\mathcal{P}_l$ corresponding to different l are distinct. But the numerator of γ_s cannot be divisible by infinitely many distinct prime ideals $\mathcal{P}_l$. Thus, (27) is impossible, and the lemma is proved.

Lemma 1 is not the most general statement about algebraic independence of power series which can be proved by this method. However, Lemma 1 has a form suitable for proving various theorems in this chapter. If one wants to use this method to prove algebraic independence of different types of power series, one can modify the conditions in Lemma 1. For example, in condition 2° one can allow $\mathcal{P}_l$ to divide the exact denominators of some of the coefficients (19) but not to very high powers.

The next lemma is a generalization of Lemma 1 which is also suitable for proving some of the theorems of Chapter 8. It can be strengthened if needed.

Lemma 2. *Suppose that all of the coefficients in the power series (16) and (17) are in the algebraic number field* $\mathbf{K}$ *and satisfy condition 1° of Lemma 1 and the following condition:*

2°. *For any* $t \in \mathbf{N}$ *there exists an* $l \in \mathbf{N}$, $l > t$, *such that the exact denominators of* $b_{l,1}, \dots, b_{l,v}$ *are divisible by some prime ideal* $\mathcal{P}_l$ *of* $\mathbf{K}$ *to the respective powers* $k_1, \dots, k_v$, *which satisfy (18), and the denominators of the numbers*

$$
\begin{aligned}
&a_{n,\nu}, \qquad n = 0, 1, \dots, \left[\frac{l}{2}\right] - 1, \qquad \nu = 0, 1, \dots, u, \\
&b_{n,\nu}, \qquad n = 0, 1, \dots, \left[\frac{l}{2}\right] - 1, \qquad \nu = 1, \dots, v,
\end{aligned}
\tag{28}
$$

are not divisible by $\mathcal{P}_l$.

Furthermore, $\mathcal{P}_l$ *may divide the exact denominators of the numbers*

$$
\begin{aligned}
&a_{n,\nu}, \qquad n = \left[\frac{l}{2}\right], \dots, l-1, \qquad \nu = 0, 1, \dots, u, \\
&b_{n,\nu}, \qquad n = \left[\frac{l}{2}\right], \dots, l-1, \qquad \nu = 1, \dots, v,
\end{aligned}
\tag{29}
$$

to powers less than $k_1/2$, and it may divide the exact denominators of the numbers

$$a_{n,\nu}, \qquad n = l, \dots, l+t, \qquad \nu = 0, 1, \dots, u, \tag{30}$$

to powers less than k_1.

Then the $u+v+1$ power series in (16) and (17) are algebraically independent over $\mathbf{C}$.

Proof. We first repeat all of the arguments in the proof of Lemma 1 until we reach the equality (27). From condition 2° it follows that for every $t \in \mathbf{N}$ there exists an infinite set of integers $l > t$ and corresponding prime ideals $\mathcal{P}_l$ which satisfy all of the requirements in the condition. Hence, we may choose l so as to satisfy condition 2° and also the inequalities $l \geq t$, $l \geq 2t+4$. For later use we observe that such a choice of l can be made in infinitely many ways, with distinct corresponding prime ideals $\mathcal{P}_l$. Then $l+t \leq \frac{3}{2}l - 2$.

In the proof of Lemma 1 we showed that right side A_l of (27) is a polynomial in the variables (19) and $b_{l,1}, \dots, b_{l,s-1}$ with coefficients in $\mathbf{Z_K}$, and it has the property that in any of its monomial terms, if we multiply the first subscripts of the variables by the corresponding exponents and add the result, we always obtain $l+t$. We regard A_l as a polynomial in the variables (29), (30), and $b_{l,1}, \dots, b_{l,s-1}$ whose coefficients are polynomials in the variables (28). The first subscript of each of the variables (29), (30) and $b_{l,1}, \dots, b_{l,s-1}$ is at least $[l/2] \geq (l-1)/2$. Hence, A_l has total degree in the variables (29), (30) and $b_{l,1}, \dots, b_{l,s-1}$ no greater than two. In fact, otherwise the sum of the product of the first subscript with the degree over the variables which occur in a term of total degree greater than two, would be at least $(3l-3)/2$, which is $> l+t$, since $l+t \leq 3l/2 - 2$.

We similarly see that the variables (30) and $b_{l,1}, \dots, b_{l,s-1}$ can only appear in terms of A_l of total degree 1 in the variables (29), (30) and $b_{l,1}, \dots, b_{l,s-1}$, because the first subscripts are at least l. In fact, if these variables appeared in a second degree term of A_l, then taking the sum of the product of the first subscript with the degree would give a result at least $(3l-1)/2 > 3l/2 - 2$, as before.

From what was proved above and condition 2° it follows that the exact denominators of the terms of A_l of degree two are divisible by a power of $\mathcal{P}_l$ less than k_1, and the exact denominators of the terms of degree 1 are divisible by $\mathcal{P}_l$ only if $s > 1$, and then to a power at most k_{s-1}. This means that the exact denominator of the number A_l is divisible by $\mathcal{P}_l$ to a power $< k_1$ for $s = 1$ and to a power $\leq k_{s-1}$ for $s > 1$.

We then argue as at the end of the proof of Lemma 1, obtaining a contradiction which proves the lemma.

We note that, if the power series in (16) and (17) are assumed to have rational coefficients in Lemmas 1 and 2, then the prime ideals $\mathcal{P}_l$ can be replaced by prime numbers p_l in the lemmas. In Theorems 1–8 below, Lemmas 1 and 2 will be applied only to function whose power series have rational coefficients. The only exception to this is Theorem 9.

§ 3. Proof of Theorems 1–5

In Theorems 1–7, as well as Theorems 8 and 9 given below, Part 2) will be a consequence of Part 1), as we see using Lemma 3 of Chapter 7, as in the proof of Theorem 1 of Chapter 7. Thus, in what follows we shall only give the proof of Part 1).

Proof of Theorem 1. From (1) we find that the set of E-functions $\omega_s(z)$, $s = 0, 1, \ldots, m$, forms a solution of the system of linear differential equations

$$\omega_0' = \omega_0, \qquad \omega_s' = \frac{1}{z}\omega_{s-1} - \frac{1}{z}, \qquad s = 1, \ldots, m. \tag{31}$$

In Lemma 1 we set $u = 0$, $v = m + 1$, and

$$\varphi_0(z) = z, \qquad f_\nu(z) = \omega_{\nu-1}(z), \qquad \nu = 1, \ldots, m+1. \tag{32}$$

The function $\varphi_0(z)$ satisfies condition 1° of Lemma 1.

Let t be any natural number. We set $l = p$, where p is any prime $p > t$, and we denote $p_l = p$. Then (3) shows that the exact denominators of the coefficients $b_{l,1}, \ldots, b_{l,m+1}$ are divisible by p to the first, second, ..., $(m + 1)$-th powers, respectively; and the exact denominators of the coefficients in (19) are not divisible by p. Consequently, the functions (32) satisfy condition 2° of Lemma 1.

According to Lemma 1, the functions in (32) are algebraically independent over C, and this means that the E-functions $\omega_s(z)$, $s = 0, 1, \ldots, m$, are algebraically independent over $\mathbf{C}(z)$. If we apply the Second Fundamental Theorem of Chapter 3 to these functions, we obtain Part 1) of Theorem 1.

Proof of Theorem 2. From (2) we see that the $m + 2$ E-functions $\gamma_s(z)$, $s = 0, 1, \ldots, m$, and $\gamma(z) = \sin z$ form a solution of the system of linear differential equations

$$\gamma_0' = -\gamma, \qquad \gamma' = \gamma_0, \qquad \gamma_s' = \frac{1}{z}\gamma_{s-1} - \frac{1}{z}, \quad s = 1, \ldots, m. \tag{33}$$

In Lemma 2 we set $u = 1$, $v = m$, and

$$\varphi_0(z) = z, \quad \varphi_1(z) = \gamma_0(z) = \cos z, \quad f_\nu(z) = \gamma_\nu(z), \quad \nu = 1, \ldots, m. \tag{34}$$

The function $\cos z$ is transcendental. Hence $\varphi_0(z)$ and $\varphi_1(z)$ satisfy condition 1° of Lemmas 1 and 2.

Let t be any natural number. We choose a prime $p > t$ and set $l = 2p$. We denote $p_l = p$. Thus, $l + t < 3p_l$. From (4) we then find that p_l divides the exact denominators of the coefficients $b_{l,1}, \dots, b_{l,m}$ to the respective powers $3, \dots, m+2$, and it does not divide the exact denominators of the coefficients

$$\begin{aligned} &a_{n,0}, && n = 0, 1, \dots, l + t; \\ &a_{n,1}, && n = 0, 1, \dots, \frac{l}{2} - 1; \\ &b_{n,\nu}, && n = 0, 1, \dots, \frac{l}{2} - 1, \qquad \nu = 1, \dots, m. \end{aligned}$$

Furthermore, p_l divides the exact denominators of the coefficients

$$\begin{aligned} &a_{n,1}, && n = \frac{l}{2}, \dots, l - 1; \\ &b_{n,\nu}, && n = \frac{l}{2}, \dots, l - 1, \qquad \nu = 1, \dots, m, \end{aligned} \tag{35}$$

with even first subscript to the first power; it divides the exact denominators of the coefficients

$$a_{n,1}, \qquad n = l, \dots, l + t, \tag{36}$$

with even first subscript to the second power; and it does not divide the exact denominators of the coefficients (35) and (36) with odd first subscript. Since $k_1 = 3$ in this case, it follows that the functions (34) satisfy condition 2° of Lemma 2.

By Lemma 2, the functions (34) are algebraically independent, and this means that the $(m + 1)$-tuple of functions $\gamma_s(z)$, $s = 0, 1, \dots, m$, is algebraically independent over $\mathbf{C}(z)$. Then the E-functions $\gamma_s(z)$, $s = 0, 1, \dots, m$, and $\gamma(z) = \sin z$ are connected only by the one algebraic equation over $\mathbf{C}(z)$

$$\gamma_0^2(z) + \gamma^2(z) = 1,$$

which has constant coefficients. Now if we apply either Theorem 6 or Theorem 8 of Chapter 4 to this set of E-functions and use (33), we obtain Theorem 2.

Proof of Theorem 3. From (5) we find that the E-functions $\omega_{\mu,s}$, $s = 0, 1, \dots, m$, form a solution of the system of linear differential equations

$$\begin{aligned} \omega'_{\mu,0} &= \omega_{\mu,0}, \\ \omega'_{\mu,s} &= -\frac{\mu}{z}\,\omega_{\mu,s} + \frac{1}{z}\,\omega_{\mu,s-1}(z), \qquad s = 1, \dots, m. \end{aligned} \tag{37}$$

In Lemma 1 we set $u = 1$, $v = m$, and

$$\varphi_0(z) = z, \quad \varphi_1(z) = \omega_{\mu,0}(z) = e^z, \qquad f_\nu(z) = \omega_{\mu,\nu}(z), \quad \nu = 1, \ldots, m. \tag{38}$$

Condition 1° of Lemma 1 holds because e^z is a transcendental function. The number μ is a rational fraction, so we set

$$\mu = \frac{a}{b}, \qquad (a, b) = 1, \qquad a, b \in \mathbf{Z}, \qquad b \geq 2. \tag{39}$$

Then from (8) we have

$$b_{n,\nu} = \frac{b^\nu}{n!(bn + a)^\nu}, \qquad \nu = 1, \ldots, m, \quad n = 0, 1, \ldots. \tag{40}$$

Let t be any natural number. Since $(a, b) = 1$, it follows from Dirichlet's theorem on primes in an arithmetic progression that there are infinitely many primes among the numbers $bn + 1$, $n = 1, 2, \ldots$. We choose $l \in \mathbf{N}$ so that

$$l > 2t, \qquad bl + a \geq \frac{3}{2}l, \qquad bl + a = p, \tag{41}$$

where p is a prime. We denote $p_l = p$. From (40) we see that p_l divides the exact denominators of the coefficients $b_{l,1}, \ldots, b_{l,m}$ to the powers $1, \ldots, m$, respectively. From (41), together with the inequality $l+t < \frac{3}{2}l$, it follows that p_l does not divide the exact denominators of the coefficients (19). Thus, the functions in (38) satisfy condition 2° of Lemma 1.

The rest of the proof of Part 1) of Theorem 3 is like the analogous part of the proof of Theorem 1, except that we use the equations (37).

Proof of Theorem 4. Using (6), we see that the $(m + 2)$-tuple of E-functions $\gamma_{\mu,s}(z)$, $s = 0, 1, \ldots, m$, and $\gamma(z) = \sin z$ form a solution of the system of linear differential equations

$$\begin{aligned} \gamma'_{\mu,0} &= -\gamma, \qquad \gamma' = \gamma_{\mu,0}, \\ \gamma'_{\mu,s} &= -\frac{\mu}{z}\gamma_{\mu,s} + \frac{1}{z}\gamma_{\mu,s-1}, \qquad s = 1, \ldots, m. \end{aligned} \tag{42}$$

In Lemma 2 we set $u = 1$, $v = m$, and

$$\begin{aligned} &\varphi_0(z) = z, \qquad \varphi_1(z) = \gamma_{\mu,0}(z) = \cos z, \\ &f_\nu(z) = \gamma_{\mu,\nu}(z), \qquad \nu = 1, \ldots, m. \end{aligned} \tag{43}$$

It suffices to show that the functions in (43) satisfy condition 2° of Lemma 2; then the rest of the proof of Theorem 4 will go through as in the case of Theorem 2, since the functions (43) satisfy the differential equations (42).

We write μ as in (39). Then from (9) we have

$$b_{2n-1,\nu} = 0, \qquad b_{2n,\nu} = \frac{b^\nu}{(2n)!(2bn+a)^\nu}, \tag{44}$$

$$\nu = 1, \dots, m, \qquad n = 0, 1, \dots.$$

Let t be any natural number. There are two possible cases.

1) a is an odd number. Then $(2b, a) = 1$, and, by Dirichlet's theorem on primes in an arithmetic progression, there are infinitely many primes among the numbers $2bn+a$, $n = 1, 2, \dots$. We choose an even number l so that the conditions (41) hold, where p is a prime. We denote $p_l = p$. Using (44), as in the case of Theorem 3 we see that the functions in (43) satisfy condition 2° of Lemma 2.

2) a is an even number. Then b is odd, since $(a, b) = 1$. Hence, $b \geq 3$. We set $a = 2c$. We have

$$2bn + a = 2(bn + c), \qquad (b, c) = 1, \qquad b \geq 3.$$

Hence, the progression $bn + c$, $n = 1, 2, \dots$, contains infinitely many primes.

We choose an even number $l = 2l_1$ for which

$$l > 3t, \qquad bl_1 + c \geq \frac{4}{3} l, \qquad bl_1 + c = p,$$

where p is a prime. We denote $p_l = p$. Then $l + t < \frac{4}{3} l$. Arguing exactly as in case 1), we conclude that the functions (43) satisfy condition 2° of Lemma 2.

Proof of Theorem 5. Using (7), we see that the $(m+2)$-tuple of E-functions $\delta_{\mu,s}(z)$, $s = 0, 1, \dots, m$, and $\delta(z) = \cos z$ form a solution of the system of differential equations

$$\begin{aligned} &\delta'_{\mu,0} = \delta, \qquad \delta' = -\delta_{\mu,0}, \\ &\delta'_{\mu,s} = -\frac{\mu}{z}\,\delta_{\mu,s} + \frac{1}{z}\,\delta_{\mu,s-1}, \qquad s = 1, \dots, m. \end{aligned} \tag{45}$$

If $\mu = 0$, then in Lemma 1 we set $u = 0$, $v = m$, and

$$\varphi_0(z) = z, \qquad f_\nu(z) = \delta_{0,\nu-1}(z), \qquad \nu = 1, \dots, m+1. \tag{46}$$

For any $t \in \mathbf{N}$ we choose an odd prime $l > t$ and set $p_l = l$. Arguing exactly as in the proof of Theorem 1, we find that the functions (46) satisfy all of the conditions of Lemma 1. Then the rest of the proof goes through just as in the case of Theorem 2 if we use the system of differential equations (45).

Now suppose that μ is a rational fraction. In Lemma 1 we set $u = 1$, $v = m$, and

$$\varphi_0(z) = z, \qquad \varphi_1(z) = \delta_{\mu,0}(z) = \sin z,$$
$$f_\nu(z) = \delta_{\mu,\nu}(z), \qquad \nu = 1, \ldots, m,$$

and we write μ as in (39). Then

$$b_{2n,\nu} = 0, \qquad b_{2n-1,\nu} = \frac{b^\nu}{(2n-1)!(2bn+a-b)^\nu},$$

$$\nu = 1, \ldots, m, \qquad n = 0, 1, \ldots.$$

The proof of the theorem goes through as in the case of Theorem 4 if we consider the two cases:

1) One of the numbers a, b is even and the other is odd. Then since $(a, b) = 1$, we have $(2b, a-b) = 1$.

2) Both a and b are odd. Then $a - b = 2c$, and

$$2nb + a - b = 2(nb + c), \qquad (b, c) = 1, \qquad b \geq 3.$$

§ 4. Proof of Theorems 6 and 7

Proof of Theorem 6. The set of E-functions $\psi_{k,s}(z)$, $s = 1, \ldots, m_k$, defined by the series (14) forms a solution of the system of differential equations

$$\begin{aligned} \psi'_{k,s} &= \frac{1}{z}\psi_{k,s-1}, \qquad s = 2, 3, \ldots, m_k, \\ \psi'_{k,1} &= z^{k-1}\psi_{k,k}. \end{aligned} \tag{47}$$

In Lemma 1 we set $u = 0$, $v = m$, and

$$\varphi_0(z) = z, \qquad f_\nu(z) = \psi_{k,\nu}(z), \qquad \nu = 1, \ldots, m_k. \tag{48}$$

Let t be any natural number. We choose $l = kp$, $l > t$, where p is a prime. We set $p_l = p$. Then the proof that the functions (48) are algebraically independent over **C** and the rest of the proof of Theorem 6 go through just as in the case of Theorem 1, except that we use the differential equations (47).

Proof of Theorem 7. We consider the set of E-functions:

$$\psi_{k,s}(z), \quad s = 1, \ldots, m_k, \quad k = q, \ldots, r, \quad \varphi_0(z) = z, \tag{49}$$

and we prove that for any q, $1 \leq q \leq r$, the functions (46) are algebraically independent over $\mathbf{C}$.

If $q = r$, our claim of algebraic independence follows from the proof of Theorem 6. We suppose that the claim is true for $q+1$, where $1 \leq q \leq r-1$, and we prove that it then holds for q. In this way the claim will be proved for any q, $1 \leq q \leq r$.

In Lemma 1 we set $u = m_{q+1} + \cdots + m_r + 1$, $v = m_q$, and $\varphi_0(z) = z$, and we let $\varphi_1(z), \ldots, \varphi_u(z)$ denote the functions

$$\psi_{k,s}(z), \qquad s = 1, \ldots, m_k, \quad k = q+1, \ldots, r, \tag{50}$$

numbered in any order. We further denote

$$f_\nu(z) = \psi_{q,\nu}(z), \qquad \nu = 1, \ldots, m_q. \tag{51}$$

Let t be any natural number. We set $l = qp$, $p > t$, where p is any prime. We set $p_l = p$. Then the exact denominators of the coefficients $b_{l,1}, \ldots, b_{l,m_q}$ are divisible by the prime p_l to the powers $1, \ldots, m_q$, respectively.

We note that the first power series coefficient of the function $\psi_{k,s}(z)$ whose exact denominator is divisible by p_l has index kp_l. Hence, p_l does not divide the exact denominators of the power series coefficients of any of the functions in (50) with indices less than $(q+1)p$ or the exact denominators of the power series coefficients of any of the functions in (51) with indices less than $l = qp_l$. Since $l + t < qp_l + p_l = (q+1)p_l$, this means that the prime p_l does not divide the exact denominators of any of the coefficients (19).

Thus, the functions

$$\psi_{k,s}(z), \qquad s = 1, \ldots, m_k, \quad k = 1, \ldots, r, \tag{52}$$

and z are algebraically independent over $\mathbf{C}$, and this means that the functions (52) are algebraically independent over $\mathbf{C}(z)$. The latter functions form a solution of the system of differential equations consisting of all of the equations in (47) for $k = 1, \ldots, r$. Hence, by the First Fundamental Theorem of Chapter 3, the numbers $\psi_{k,s}(\xi)$, $s = 1, \ldots, m_k$, $k = 1, \ldots, r$, are algebraically independent for any $\xi \in \mathbf{A} \setminus \{0\}$.

§ 5. Further results

We consider the functions

$$A_{\lambda,k}(z) = \sum_{n=0}^{\infty} \frac{1}{((\lambda+1)\cdots(\lambda+n))^k} \left(\frac{z}{k}\right)^{kn}, \tag{53}$$

$$A_{\lambda,k,s}(z) = 1 + \sum_{n=1}^{\infty} \frac{1}{((\lambda+1)\cdots(\lambda+n-1))^k(\lambda+n)^s} \left(\frac{z}{k}\right)^{kn}, \tag{54}$$
$$\lambda \neq -1,-2,\ldots, \quad k \geq 1, \quad m_k \geq k, \quad s = 1,\ldots,m_k.$$

Note that $A_{\lambda,k,k}(z) = A_{\lambda,k}(z)$. When $\lambda \in \mathbf{Q}$, the function (53) is a hypergeometric E-function which satisfies a k-th order linear differential equation with coefficients in $\mathbf{C}[z]$. As in the case of the functions (7.1) in § 1 of Chapter 7, we see that for $\lambda \in \mathbf{Q}$ the functions in (54) are E-functions, and the function $A_{\lambda,k,m_k}(z)$ satisfies an m_k-th order linear differential equation.

Theorem 8. *Let $k, m_k \in \mathbf{N}$, $m_k \geq k$; $\lambda \in \mathbf{Q}$, $\lambda \neq -1,-2,\ldots$; $\xi \in \mathbf{A} \setminus \{0\}$. Then:*

1) *the numbers $A_{\lambda,k,s}(\xi)$, $s = 1,\ldots,m_k$, are algebraically independent;*

2) *the numbers $A^{(l)}_{\lambda,k,m_k}(\xi)$, $l = 0,1,\ldots,m_k-1$, are algebraically independent. In particular, the numbers $A^{(l)}_{\lambda,k}(\xi)$, $l = 0,1,\ldots,k-1$, are algebraically independent.*

Proof. We apply Lemma 1, setting $u = 0$, $v = m$,

$$\varphi_0(z) = z, \qquad f_\nu(z) = A_{\lambda,k,\nu}(z), \qquad \nu = 1,\ldots,m_k. \tag{55}$$

The function $\varphi_0(z)$ obviously satisfies condition 1° of Lemma 1. Let

$$\lambda = \frac{a}{b}, \qquad a,b \in \mathbf{Z}, \qquad (a,b) = 1, \qquad b \geq 1.$$

Then, in the notation of Lemma 1, the functions (55) satisfy:

$$\begin{aligned} &a_{0,0} = 0, \qquad a_{1,0} = 1, \qquad a_{n,0} = 0, \qquad n = 2,3,\ldots, \\ &b_{n,\nu} = 0, \qquad \text{if } n \not\equiv 0 \ (mod\ k), \qquad b_{0,\nu} = 1, \\ &b_{kn,\nu} = \frac{b^{k(n-1)+\nu}}{k^{kn}((b+a)(2b+a)\cdots((n-1)b+a))^k(nb+a)^\nu}, \\ &n = 1,2,\ldots, \quad \nu = 1,\ldots,m_k. \end{aligned} \tag{56}$$

Let t be any natural number. The arithmetic progression $bn + a$ contains infinitely many primes, so we can choose $l = kn_1$, $l > t$, so that $bn_1 + a = p_l$, where p_l is a prime and $p_l > \max(k, b)$. It then follows from (56) that the exact denominators of $b_{l,1}, \dots, b_{l,m}$ are divisible by the prime p_l to the powers $1, \dots, m_k$, respectively, and the exact denominators of all of the numbers $a_{n,0}$, $n = 0, 1, \dots, l+t$, and $b_{n,\nu}$, $\nu = 1, \dots, m$, $n = 0, 1, \dots, l-1$, are not divisible by p_l. Thus, the functions (55) satisfy condition 2° of Lemma 1.

By Lemma 1, the functions in (55) are algebraically independent over $\mathbf{C}$, and this means that the functions in (54) are algebraically independent over $\mathbf{C}(z)$.

The E-functions in (54) form a solution of the system of linear differential equations

$$\begin{aligned} y_1' &= -\frac{k}{z} y_1 + \left(\frac{z}{k}\right)^{k-1} y_k + \frac{k\lambda}{z}, \\ y_s' &= \frac{k}{z} y_{s-1} - \frac{k\lambda}{z} y_s + \frac{(k\lambda - 1)}{z}, \qquad s = 2, 3, \dots, m_k. \end{aligned} \tag{57}$$

which has only one singular point at $z = 0$; and these functions are algebraically independent over $\mathbf{C}(z)$. Thus, by the Second Fundamental Theorem of Chapter 3, the numbers $A_{\lambda,k,s}(\xi)$, $s = 1, \dots, m_k$, are algebraically independent.

Suppose we are given an s-tuple of rational numbers

$$\frac{a_1}{b_1}, \dots, \frac{a_s}{b_s}, \quad a_i \in \mathbf{Z}, \quad b_i \in \mathbf{N}, \quad (a_i, b_i) = 1, \quad i = 1, \dots, s, \quad s \geq 2. \tag{58}$$

We shall say that (58) is a *regular* set if for any indices i and j with $1 \leq i < j \leq s$, the difference of the numerators $a_i - a_j$ is not divisible by the greatest common divisor of the denominators (b_i, b_j).

Now let $\lambda_i \in \mathbf{Q}$, $k_i \in \mathbf{N}$, $i = 1, \dots, r$. We set

$$\lambda_i = \frac{a_i}{b_i}, \quad a_i \in \mathbf{Z}, \quad b_i \in \mathbf{N}, \quad (a_i, b_i) = 1, \quad i = 1, \dots, r. \tag{59}$$

We divide the numbers (59) into equivalence classes in each of which the ratio b_i/k_i has the same value.

We say that the r-tuple (59) is *normal* relative to $k_1, \dots, k_r$ if the set of elements in each equivalence class either consists of only one element or else is a regular set.

Theorem 9. *Suppose that* $k_i, m_i \in \mathbf{N}$, $m_i \geq k_i$, $i = 1, \dots, r$; $\lambda_i \in \mathbf{Q}$, $\lambda_i \neq -1, -2, \dots$, $i = 1, \dots, r$; *the numbers* $\lambda_1, \dots, \lambda_r$ *form a normal set relative to* $k_1, \dots, k_r$; *and* $\xi_i \in \mathbf{A} \setminus \{0\}$, $i = 1, \dots, r$. *Then:*

1) *the* $m_1 + \dots + m_r$ *numbers* $A_{\lambda_i,k_i,s}(\xi_i)$, $s = 1, \dots, m_i$, $i = 1, \dots, r$, *are algebraically independent;*

2) *the* $m_1+\cdots+m_r$ *numbers* $A^{(l)}_{\lambda_i,k_i,m_i}(\xi_i)$, $l=0,1,\ldots,m_i-1$, $i=1,\ldots,r$, *are algebraically independent. In particular, the* $k_1+\cdots+k_r$ *numbers* $A^{(l)}_{\lambda_i,k_i}(\xi_i)$, $l=0,1,\ldots,k_i-1$, $i=1,\ldots,r$, *are algebraically independent.*

To prove the theorem we first need the following lemma.

Lemma 3. *Under the conditions of Theorem 9, the functions*

$$A_{\lambda_i,k_i,s}(\xi_i z),\qquad s=1,\ldots,m_i,\quad i=1,\ldots,r, \tag{60}$$

are algebraically independent over $\mathbf{C}(z)$.

Proof. We let $\mathbf{K}$ denote an algebraic number field containing all of the numbers $\xi_1,\ldots,\xi_r$. Let $q\in\mathbf{N}$ be such that $q\xi_1,\ldots,q\xi_r\in\mathbf{Z_K}$. We set

$$\lambda_i=\frac{a_i}{b_i},\quad a_i\in\mathbf{Z},\quad b_i\in\mathbf{N},\quad (a_i,b_i)=1,\quad i=1,\ldots,r,$$

and we partition the numbers $\lambda_1,\ldots,\lambda_r$ into equivalence classes according to the value of b_i/k_i, as above. Since we can index these numbers as we choose, we order them according to increasing b_i/k_i, and within each equivalence class with more than one element we order them arbitrarily. Then for any λ_i and λ_{i+1}, $1\le i\le r-1$, we have $\frac{b_{i+1}}{k_{i+1}}>\frac{b_i}{k_i}$ if they belong to different classes, and $\frac{b_{i+1}}{k_{i+1}}=\frac{b_i}{k_i}$ if they belong to the same class.

To prove the lemma it is sufficient to show that the functions (60) and z are algebraically independent over $\mathbf{C}$. We prove this by induction on r. For $r=1$ the claim follows from Lemma 1. We suppose that it holds for $r=j$, $j\ge 1$, and we prove the claim for $r=j+1$.

To do this we apply Lemma 1 with $u=m_1+\cdots+m_j$, $v=m_{j+1}$, $\varphi_0(z)=z$; with $\varphi_1(z),\ldots,\varphi_u(z)$ equal to $A_{\lambda_1,k_1,1}(\xi_1 z),\ldots,A_{\lambda_1,k_1,m_1}(\xi_1 z),\ldots\ldots,A_{\lambda_j,k_j,1}(\xi_j z),\ldots,A_{\lambda_j,k_j,m_j}(\xi_j z)$; and with $f_\nu(z)=A_{\lambda_{j+1},k_{j+1},\nu}(\xi_{j+1}z)$, $\nu=1,\ldots,v$.

Condition 1° of Lemma 1 is obviously fulfilled. We now show that condition 2° also holds.

We consider the coefficients of $\varphi_\nu(z)$ and $f_\nu(z)$. For $n=0,1,2,\ldots$ and $\nu=1,\ldots,m_{j+1}$, the coefficients $a_{n,0}$ and $b_{n,\nu}$ have the same form as in (56), except that a, b, k and m are replaced by a_{j+1}, b_{j+1}, k_{j+1} and m_{j+1}, and the coefficients $b_{k_{j+1}n,\nu}$ have an additional factor $\xi_{j+1}^{k_{j+1}n}$ in the numerator. The coefficients $a_{n,\nu}$, $n=0,1,2,\ldots$, $\nu=1,\ldots,u$, are zero for $n\not\equiv 0\ (mod\ k_i)$; they are equal to 1 for

$n = 0$; and $a_{k_i n,\nu}$ has the form

$$\frac{\xi_i^{k_i n} b_i^{k_i(n-1)+s}}{k_i^{k_i n}((b_i + a_i)(2b_i + a_i)\cdots((n-1)b_i + a_i))^{k_i}(nb_i + a_i)^s}, \tag{61}$$

$$n = 1, 2, \ldots,$$

where $1 \leq i \leq j$ and $1 \leq s \leq m$ depend on which indices correspond to $\varphi_\nu = A_{\lambda_i, k_i, s}(\xi_i z)$.

We denote

$$g = \max_{1 \leq i \leq r} (b_i, |a_i|, k_i), \qquad g \geq 1, \tag{62}$$

and we let t be any natural number. The progression $b_{j+1}n + a_{j+1}$ contains infinitely many prime numbers, so we can set $l = k_{j+1}n_1$ and choose n_1 so that $b_{j+1}n_1 + a_{j+1} = p_l$, where p_l is a prime which does not divide the discriminant of $\mathbf{K}$ and is not divisible by any of the finitely many prime ideals which are divisors of the numbers $q, q\xi_1, \ldots, q\xi_r$, and where the following inequalities also hold:

$$n_1 > g^2(t+2), \qquad p_l > g(t+2). \tag{63}$$

From (63) it follows that $l > t$.

We now choose a prime ideal $\mathcal{P}_l$ which divides the prime number p_l. Since p_l does not divide the discriminant of $\mathbf{K}$, it follows that p_l is not divisible by $\mathcal{P}_l^2$. Because the prime ideal factors of different prime numbers are different, if we use the equalities of the type (56), the notation (62) and the inequalities (63), we see that the exact denominators of $b_{l,1}, \ldots, b_{l,m_{j+1}}$ are divisible by the prime ideal $\mathcal{P}_l$ to the powers $1, \ldots, m_{j+1}$, respectively, and the exact denominators of all of the numbers $b_{n,\nu}$, $n = 0, 1, \ldots, l-1$, $\nu = 1, \ldots, m_{j+1}$, and $a_{n,0}$, $n = 0, 1, \ldots, l+t$, are not divisible by $\mathcal{P}_l$.

We prove that $\mathcal{P}_l$ does not divide the exact denominator of any of the numbers

$$a_{n,\nu}, \qquad n = 0, 1, \ldots, l+t, \qquad \nu = 1, \ldots, u. \tag{64}$$

From (61) it follows that we need only show that the prime p_l does not divide any of the numbers in the arithmetic progressions $b_i n + a_i$, $i = 1, \ldots, j$, for which n satisfies the inequalities

$$0 \leq k_i n \leq l + t = k_{j+1}n_1 + t, \tag{65}$$

or

$$0 \leq n \leq \frac{k_{j+1}}{k_i} n_1 + \frac{t}{k_i}. \tag{66}$$

Let i be any index in the range $1, \dots, j$. There are two possible cases.

Case 1. $\frac{b_i}{k_i} < \frac{b_{j+1}}{k_{j+1}}$. Then the relation $b_i < \frac{k_i}{k_{j+1}} b_{j+1}$ implies that

$$b_i \le \frac{k_i b_{j+1} - 1}{k_{j+1}}. \tag{67}$$

Using (63), (65) and (66) and the notation (62), we find that

$$\begin{aligned}
|b_i n + a_i| &\le b_i n + |a_i| \le \\
&\le \frac{k_i b_{j+1} - 1}{k_{j+1}} \left(\frac{k_{j+1}}{k_i} n_1 + \frac{t}{k_i} \right) + g = \\
&= b_{j+1} n_1 + a_{j+1} - \frac{n_1}{k_i} + \frac{b_{j+1} t}{k_{j+1}} - \frac{t}{k_i k_{j+1}} + g - a_{j+1} < \\
&< p_l - \frac{n_1}{g} + gt + 2g < p_l.
\end{aligned}$$

The last inequality shows that in case 1 the numbers $b_i n + a$ are not divisible by p_l for the values of n being considered.

Case 2. $\frac{b_i}{k_i} = \frac{b_{j+1}}{k_{j+1}}$. Then if we use the inequalities in (63) and (65), the notation (62), and the relation $b_i = \frac{k_i}{k_{j+1}} b_{j+1}$, we obtain the following inequalities for values of n satisfying (65):

$$\begin{aligned}
|b_i n + a_i| &\le b_i n + |a_i| \le \frac{k_i}{k_{j+1}} b_{j+1} \left(\frac{k_{j+1}}{k_j} n_1 + \frac{t}{k_i} \right) + g = \\
&= b_{j+1} n_1 + a_{j+1} + \frac{b_{j+1}}{k_{j+1}} t + g - a_{j+1} \le p_l + gt + 2g < 2p_l.
\end{aligned} \tag{68}$$

If $b_i n + a_i$ is divisible by p_l, then, in view of the second condition in (63) and the definition of g, it follows from (68) that $b_i n + a_i = p_l = b_{j+1} n_1 + a_{j+1}$. This equality implies that the greatest common divisor (b_i, b_{j+1}) divides the difference $a_{j+1} - a_i$, and this is impossible, since the r-tuple $\lambda_1, \dots, \lambda_r$ is normal relative to $k_1, \dots, k_r$, and λ_i and λ_{j+1} belong to the same equivalence class. Thus, in case 2 as well the numbers $b_i n + a_i$ with n satisfying (65) cannot be divisible by p_l.

We have shown that $\mathcal{P}_l$ does not divide the exact denominators of any of the numbers in (64), thereby proving that condition 2° of Lemma 1 is fulfilled. By Lemma 1, then, the functions (60) and z are algebraically independent over **C**. Lemma 3 is proved.

Proof of Theorem 9. The E-functions in (60) form a solution of the system of $m_1 + \cdots + m_r$ first order linear differential equations which consists of the r subsystems obtained from (57) by replacing λ, k and m_k by λ_i, k_i and m_i,

$i = 1, \dots, r$, replacing z by $\xi_i z$, and renumbering the functions y_s with two subscripts as $y_{s,i}$, $s = 1, \dots, m_i$, $i = 1, \dots, r$. This system has only one singular point at $z = 0$. Hence, by the Second Fundamental Theorem of Chapter 3 and Lemma 3, the numbers $A_{\lambda_i, k_i, s}(\xi_i)$, $s = 1, \dots, m_i$, $i = 1, \dots, r$, are algebraically independent.

We give some obvious conditions under which Theorem 9 holds.

1. The numbers $\lambda_1, \dots, \lambda_r$ form a normal set relative to $k_1, \dots, k_r$ if $b_i/k_i \neq b_j/k_j$ for $i \neq j$, i.e., if each equivalence class has only one element.

2. The numbers $\lambda_1, \dots, \lambda_r$ form a normal set relative to $k_1, \dots, k_r$ if each equivalence class with more than one element consists of numbers having the same denominator and if $\lambda_i \not\equiv \lambda_j \ (mod\ 1)$ for $i \neq j$.

In particular, we see that Theorem 9 holds in the following cases:

a) $\lambda_i = \lambda$, $i = 1, \dots, r$, and $k_i \neq k_j$ for $i \neq j$;

b) $\lambda_i = a_i/b$, $i = 1, \dots, r$, where $r = \varphi(b)$, $a_1, \dots, a_r$ are a reduced set of residues modulo b, and $k_i = k$, $i = 1, \dots, r$;

c) $\lambda_i \not\equiv \lambda_j \ (mod\ 1)$ for $i \neq j$, and $k_i = k$, $i = 1, \dots, r$.

In [70:23] we describe other cases when $\lambda_1, \dots, \lambda_r$ is a normal r-tuple relative to $k_1, \dots, k_r$.

Theorems 6, 7 and 8 follow from Theorem 9. If we are given a concrete set of numbers $\lambda_1, \dots, \lambda_r$, then it is usually not hard to determine directly whether it is normal relative to $k_1, \dots, k_r$.

We note that the conditions in Theorem 9 are only sufficient conditions.

In 1967, Belogrivov [3:2,3] developed the method of this chapter in order to investigate certain classes of hypergeometric functions. In particular, he proved the following theorem about the functions

$$A_{m,2}(z) = \sum_{n=0}^{\infty} \frac{1}{[\lambda + 1, n]^{m_1} [\mu + 1, n]^{m_2}} \left(\frac{z}{m}\right)^{mn},$$

$$m_1 + m_2 = m, \qquad \lambda, \mu \neq -1, -2, \dots.$$

Theorem 10. *Let $\lambda, \mu \in \mathbf{Q}$, where if one of these two numbers is an integer then the other one is not half of an odd integer, and let $\xi \in \mathbf{A} \setminus \{0\}$. Then the numbers $A_{m,2}(\xi), A'_{m,2}(\xi), \dots, A_{m,2}^{(m-1)}(\xi)$ are algebraically independent.*

Theorem 10 with $m_1 = m_2 = 1$ contains Theorem 6 of Chapter 6.

The same papers of Belogrivov prove several theorems about algebraic independence of the values of the functions

$$A_{m,s}(z) = \sum_{n=0}^{\infty} \frac{1}{[\lambda_1 + 1, n]^{m_1} [\lambda_2 + 1, n]^{m_2} \cdots [\lambda_s + 1, n]^{m_s}} \left(\frac{z}{m}\right)^{mn},$$

$$m_1 + m_2 + \cdots + m_s = m, \qquad \lambda_1, \lambda_2, \ldots, \lambda_s \neq -1, -2, \ldots,$$

of their successive derivatives, and of certain associated functions with specific restrictions on the values of the parameters. We state the simplest result of this type.

Theorem 11. *Let* $\lambda_1, \ldots, \lambda_s \in \mathbf{Q}$, *where the fractional parts of these numbers which are nonzero are either all greater than* $1/2$ *or else all less than* $1/2$, *and let* $\xi \in \mathbf{A} \setminus \{0\}$. *Then the numbers* $A_{m,s}(\xi), A'_{m,s}(\xi), \ldots, A^{(m-1)}_{m,s}(\xi)$ *are algebraically independent.*

Chapter 9
Siegel's Theorem

§ 1. Statement of the theorem and some basic auxiliary results

In § 2 of Chapter 6 we proved Theorem 2 on algebraic independence of the values at algebraic points of the function

$$K_\lambda(z) = \sum_{n=0}^{\infty} \frac{(-1)^n}{n!(\lambda+1)\cdots(\lambda+n)} \left(\frac{z}{2}\right)^{2n}, \qquad \lambda \neq -1, -2, \ldots, \tag{1}$$

and its derivative. This theorem was proved by Siegel in 1929 in [72:3]. An important part of his proof was a method which makes it possible to prove algebraic independence over $\mathbf{C}(z)$ of the solutions and the derivatives of the solutions of a second order homogeneous linear differential equation with coefficients in $\mathbf{C}(z)$.

In the same paper Siegel developed a method for proving algebraic independence over $\mathbf{C}(z)$ of the solutions and the derivatives of the solutions of a set of concrete second order homogeneous linear differential equations with coefficients in $\mathbf{C}(z)$. This enabled him to prove the following theorem.

Siegel's Theorem. *Suppose that the numbers $\lambda_1, \ldots, \lambda_m \in \mathbf{Q}$, $m \geq 1$, satisfy the conditions*

$$\lambda_i \neq \frac{2k-1}{2}, \qquad k \in \mathbf{Z}, \qquad -\lambda_i \notin \mathbf{N}, \qquad i = 1, \ldots, m;$$

$$(\lambda_{i_1} \pm \lambda_{i_2}) \notin \mathbf{Z}, \qquad i_1, i_2 = 1, \ldots, m, \qquad i_1 \neq i_2,$$

and $\xi_1, \ldots, \xi_n$, $n \geq 1$, are algebraic numbers whose squares are distinct and nonzero. Then the $2mn$ numbers

$$K_{\lambda_i}(\xi_j), \; K'_{\lambda_i}(\xi_j), \qquad i = 1, \ldots, m, \qquad j = 1, \ldots, n,$$

are algebraically independent.

Siegel's proof of this theorem was complicated. After Siegel's method was generalized and the Second Fundamental Theorem was published (see Chapter 3), the proof was shortened somewhat. It became no longer necessary to verify the Siegel normality condition for products of powers of the functions under consideration.

The function (1) is a solution of the differential equation (6.9), and $K_{\lambda_i}(\xi_j z)$ is a solution of the corresponding differential equation in (2):

$$y''_{i,j} + \frac{2\lambda_i + 1}{z} y'_{i,j} + \xi_j^2 y_{i,j} = 0, \qquad i = 1, \ldots, m, \qquad j = 1, \ldots, n. \quad (2)$$

To prove Siegel's Theorem we first prove a lemma.

Lemma 1. *Suppose that* $\lambda_i \in \mathbf{Q}$, $\lambda_i \neq \frac{2k-1}{2}$, $k \in \mathbf{Z}$, $i = 1, \ldots, m$; $(\lambda_{i_1} \pm \lambda_{i_2}) \notin \mathbf{Z}$, $i_1, i_2 = 1, \ldots, m$, $i_1 \neq i_2$; $\xi_1, \ldots, \xi_n \in \mathbf{C} \setminus \{0\}$, $\xi_{j_1}^2 \neq \xi_{j_2}^2$, $j_1, j_2 = 1, \ldots, n$, $j_1 \neq j_2$; *and* $y_{i,j}$ *is any nontrivial solution of the differential equation in (2) corresponding to* i, j. *Then the functions* $y_{i,j}, y'_{i,j}$, $i = 1, \ldots, m$, $j = 1, \ldots, n$, *are algebraically independent over* $\mathbf{C}(z)$.

We denote

$$y_{i,j,1} = K_{\lambda_i}(\xi_j z), \qquad y_{i,j,2} = K'_{\lambda_i}(\xi_j z), \qquad i = 1, \ldots, m, \qquad j = 1, \ldots, n.$$

This set of functions is a solution of a system of $2mn$ homogeneous linear differential equations which splits into mn subsystems

$$y'_{i,j,1} = y_{i,j,2}, \qquad y'_{i,j,2} = -\frac{2\lambda_i + 1}{z} y_{i,j,2} - \xi_i^2 y_{i,j,1},$$

$$i = 1, \ldots, m, \qquad j = 1, \ldots, n.$$

Hence, Siegel's Theorem follows from the Second Fundamental Theorem of Chapter 3 and Lemma 1.

In 1971, I. I. Belogrivov [3:4] somewhat simplified Siegel's demonstration of the algebraic independence of the functions in Siegel's Theorem, and he generalized the argument to the Kummer functions (6.60) and some related functions. This resulted in a simplification of the proof of Siegel's Theorem.

In 1968, E. R. Kolchin [32:2] proved a general theorem about the properties of solutions of a set of homogeneous linear differential equations of arbitrary order m. From this theorem with $m = 2$ he obtained a result for a set of Bessel functions from which Lemma 1 follows.

In Chapter 9 we shall prove Siegel's Theorem and some related theorems. We shall give a simpler version of the proof, which was prepared by Yu. V. Nesterenko at the author's request.

Lemma 1 will be proved using the following general theorem.

Theorem 1. *Suppose that the differential equations*

$$y'' + a_i y' + b_i y = 0, \qquad a_i, b_i \in \mathbf{C}(z), \qquad i = 1, \ldots, n, \qquad n \geq 1, \quad (3)$$

satisfy the following conditions:

1°. *There exist functions* $v_i \in \mathbf{C}(z)$, $v_i \not\equiv 0$, $i = 1, \ldots, n$, *such that*

$$v_i' + a_i v_i = 0, \qquad i = 1, \ldots, n.$$

2°. *Any nontrivial solution of any of the equations in (3) is algebraically independent of its derivative over* $\mathbf{C}(z)$.

Further suppose that $f_1, \ldots, f_n$ *is a set of nontrivial solutions of the differential equations (3). Then either the* $2n$ *functions* $f_1, f_1', \ldots, f_n, f_n'$ *are algebraically independent over* $\mathbf{C}(z)$, *or else there exist nontrivial solutions* φ_i, φ_j, $i \neq j$, *of the corresponding differential equations in (3) and also functions* $u_1, u_2 \in \mathbf{C}(z)$ *and* ω, $\omega^2 \in \mathbf{C}(z)$, *such that*

$$\varphi_i = \omega(u_1 \varphi_j + u_2 \varphi_j').$$

Theorem 1 is the special case $m = 2$ of the theorem of Kolchin mentioned above, which he proved by methods of differential algebra and the theory of algebraic groups. Theorem 1 was proved in a somewhat weaker form (with $\omega \in \mathbf{A}(z)$) by Belogrivov [3:4] by means of a generalization of Siegel's argument in [72:3]. Nesterenko's proof largely follows that of Siegel and Belogrivov. We note that there is no known application to concrete functions of Kolchin's general theorem with $m \geq 3$.

In order to prove Theorem 1, in § 2 and 3 we shall establish some lemmas.

§ 2. Some lemmas

Let $\mathcal{L}$ denote a field of analytic functions of z which contains $\mathbf{C}$ and is closed under differentiation. In particular, $\mathcal{L}$ could be $\mathcal{C}(z)$.

We shall state without proof some facts about algebraic field extensions which will be needed later. They are proved, for example, in Chevalley's book [11:1].

Let

$$\varphi_1 = \varphi_1(z), \ldots, \varphi_m = \varphi_m(z) \tag{4}$$

be functions which are algebraic over $\mathcal{L}$, and let $\mathcal{K} = \mathcal{L}(\varphi_1, \ldots, \varphi_m)$ be the field obtained by adjoining the functions in (4) to $\mathcal{L}$. There exists an algebraic function $\varphi = \varphi(z)$ over $\mathcal{L}$ such that $\mathcal{K} = \mathcal{L}(\varphi)$. Consequently, there exist functions $R_i(x) \in \mathcal{L}(x)$ for which

$$\varphi_i(z) = R_i(\varphi(z)), \qquad i = 1, \ldots, m.$$

The function $\varphi(z)$ is called a *primitive element of the field* $\mathcal{K}$.

If

$$P(x) = x^n + a_{n-1}x^{n-1} + \cdots + a_0, \qquad a_k \in \mathcal{L}, \qquad k = 0, 1, \ldots, n-1,$$

is the *minimal polynomial* of $\varphi(z)$ over $\mathcal{L}$ (which is irreducible over $\mathcal{L}$, and hence of minimal degree), and if $\psi(z)$ is an algebraic function over $\mathcal{L}$ which satisfies the equation $P(\psi(z)) = 0$, then the functions

$$\psi_i(z) = R_i(\psi(z)), \qquad i = 1, \ldots, m,$$

are called the *conjugates of the functions (4) over the field* $\mathcal{L}$. There is a finite set of m-tuples of functions which are conjugate over $\mathcal{L}$ to the functions (4). This set does not depend on the choice of primitive element $\varphi(z)$ of $\mathcal{K}$. If $Q(x_1, \ldots, x_m) \in \mathcal{L}(x_1, \ldots, x_m)$ and $Q(\varphi_1(z), \ldots, \varphi_m(z)) = 0$, then we also have $Q(\psi_1(z), \ldots, \psi_m(z)) = 0$ for any m-tuple of functions $\psi_1(z), \ldots, \psi_m(z)$ which are conjugate over $\mathcal{L}$ to the functions (4).

Lemma 2. *Suppose that the functions (4) are algebraic over $\mathcal{L}$ and form a solution of the system of linear differential equations*

$$y_k' = Q_{k,0} + \sum_{i=1}^{m} Q_{k,i} y_i, \quad k = 1, \ldots, m, \quad Q_{k,i} \in \mathcal{L}, \quad m \geq 1. \tag{5}$$

If

$$\psi_1 = \psi_1(z), \ldots, \psi_m = \psi_m(z) \tag{6}$$

is an m-tuple of functions conjugate over $\mathcal{L}$ to the functions (4), then these functions (6) also form a solution of the system of differential equations (5).

Proof. Let $P_k(x) \in \mathcal{L}[x]$ be the minimal polynomial over $\mathcal{L}$ of the algebraic function $\varphi_k(z)$. Then

$$P_k(\varphi_k(z)) = 0, \tag{7}$$

and, since $\psi_k(z)$ is conjugate to $\varphi_k(z)$ over $\mathcal{L}$, it follows that

$$P_k(\psi_k(z)) = 0. \tag{8}$$

Differentiating (7) and (8), we obtain

$$\frac{\partial P_k(\varphi_k)}{\partial z} + \frac{\partial P_k(\varphi_k)}{\partial x} \varphi_k' = 0, \tag{9}$$

$$\frac{\partial P_k(\psi_k)}{\partial z} + \frac{\partial P_k(\psi_k)}{\partial x}\psi_k' = 0. \tag{10}$$

If we use the fact that the functions (4) satisfy the system of differential equations (5), then from (9) we find that

$$\frac{\partial P_k(\varphi_k)}{\partial z} + \frac{\partial P_k(\varphi_k)}{\partial x}\left(Q_{k,0} + \sum_{i=1}^{m} Q_{k,i}\varphi_i\right) = 0. \tag{11}$$

The functions (6) are conjugate over $\mathcal{L}$ to the functions (4). Hence, it follows from (11) that

$$\frac{\partial P_k(\psi_k)}{\partial z} + \frac{\partial P_k(\psi_k)}{\partial x}\left(Q_{k,0} + \sum_{i=1}^{m} Q_{k,i}\psi_i\right) = 0. \tag{12}$$

Since $P_k(x)$ is the minimal polynomial of the algebraic function $\varphi_k(z)$ over $\mathcal{L}$, it does not have multiple roots. It then follows from (8) that $\frac{\partial P_k(\psi_k)}{\partial x} \neq 0$. If we now compare (10) and (12), we obtain

$$\psi_k' = Q_{k,0} + \sum_{i=1}^{m} Q_{k,i}\psi_i.$$

The lemma is proved.

Corollary. *Suppose that the function $\varphi(z)$ is algebraic over $\mathcal{L}$ and satisfies the linear differential equation*

$$\begin{gathered} y^{(m)} + A_{m-1}y^{(m-1)} + \cdots + A_1 y' + A_0 y = B, \qquad m \geq 1, \\ A_0, A_1, \ldots, A_{m-1}, B \in \mathcal{L}. \end{gathered} \tag{13}$$

Then if $\psi(z)$ is a conjugate of $\varphi(z)$ over $\mathcal{L}$, it follows that $\psi(z)$ also satisfies the differential equation (13).

Proof. The functions

$$\varphi_1(z) = \varphi(z), \ \varphi_2(z) = \varphi'(z), \ldots, \ \varphi_m(z) = \varphi^{(m-1)}(z)$$

are all in the field $\mathcal{K} = \mathcal{L}(\varphi(z))$ and, by assumption, form a solution of the system of linear differential equations

$$\begin{aligned} y_k' &= y_{k+1}, \qquad k = 1, \ldots, m-1, \\ y_m' &= B - A_0 y_1 - A_1 y_2 - \cdots - A_{m-1} y_m. \end{aligned} \tag{14}$$

If $\varphi_i(z) = R_i(\varphi(z))$, $R_i(x) \in \mathcal{L}(x)$, $i = 1, \ldots, m$, then, by Lemma 2, the functions $\psi_i(z) = R_i(\psi(z))$, $i = 1, \ldots, m$, form a solution of the system of linear differential equations (14). Since $\psi_1(z) = \psi(z)$, it follows from (14) that $\psi(z)$ is a solution of the differential equation (13). The corollary is proved.

We consider the ring $\mathcal{L}[x_1, \ldots, x_m]$ of polynomials in $x_1, \ldots, x_m$ with coefficients in $\mathcal{L}$. We state without proof some facts that will be needed later about ideals of the ring $\mathcal{L}[x_1, \ldots, x_m]$.

Let $\mathcal{I}$ be an ideal of the ring $\mathcal{L}[x_1, \ldots, x_m]$, i.e., a subset of $\mathcal{L}[x_1, \ldots, x_m]$ which is closed under addition and under multiplication by elements of $\mathcal{L}[x_1, \ldots, x_m]$. According to the Hilbert basis theorem, for every ideal $\mathcal{I} \subset \mathcal{L}[x_1, \ldots, x_m]$ there exists a finite set of polynomials $P_1, \ldots, P_s$ in $\mathcal{L}[x_1, \ldots, x_m]$ such that $\mathcal{I}$ is the set of all possible linear combinations of $P_1, \ldots, P_s$ with coefficients in $\mathcal{L}[x_1, \ldots, x_m]$. The latter fact is denoted $\mathcal{I} = (P_1, \ldots, P_s)$. The set of polynomials $P_1, \ldots, P_s$ is called a *basis* of the ideal $\mathcal{I}$. If the basis polynomials are homogeneous, then $\mathcal{I}$ is called a *homogeneous ideal.* If the basis polynomials are homogeneous in some subset of the variables, then $\mathcal{I}$ is said to be *homogeneous with respect to those variables.* If $\mathcal{I}$ has a basis consisting of a single polynomial P, i.e., if $\mathcal{I} = (P)$, then we say that $\mathcal{I}$ is a *principal ideal.* The principal ideal (1) is the entire ring $\mathcal{L}[x_1, \ldots, x_m]$.

An m-tuple of functions $(g_1(z), \ldots, g_m(z))$ is said to be a *zero of the ideal* $\mathcal{I}$ if $P(g_1(z), \ldots, g_m(z)) = 0$ for every polynomial $P \in \mathcal{I}$. According to the Hilbert Nullstellensatz, every ideal $\mathcal{I} \subset \mathcal{L}[x_1, \ldots, x_m]$ which is not the whole ring has a zero whose components can be chosen to be algebraic functions over $\mathcal{L}$. If $P \in \mathcal{L}[x_1, \ldots, x_m]$, and if every zero of $\mathcal{I}$ is also a zero of P, then there exists $M \in \mathbf{N}$ such that $P^M \in \mathcal{I}$. In particular, every homogeneous ideal has the *trivial zero* $(0, \ldots, 0)$; and if $\mathcal{I}$ has no other zeros, then there exists $M \in \mathbf{N}$ such that $x_i^M \in \mathcal{I}$, $i = 1, \ldots, m$.

We consider the differential operator associated to the system of differential equations (5)

$$D = \frac{\partial}{\partial z} + \sum_{k=1}^{m} \left(Q_{k,0} + \sum_{i=1}^{m} Q_{k,i} x_i \right) \frac{\partial}{\partial x_k},$$

which acts on the ring $\mathcal{L}[x_1, \ldots, x_m]$. If $f_1(z), \ldots, f_m(z)$ is a solution of the system (5), then for every $P \in \mathcal{L}[x_1, \ldots, x_m]$ we have

$$DP(f_1(z), \ldots, f_m(z)) = \frac{d}{dz} P(f_1(z), \ldots, f_m(z)). \tag{15}$$

We suppose that the functions $f_1(z), \ldots, f_m(z)$ are algebraically independent over $\mathcal{L}$. Then the set of all polynomials $P \in \mathcal{L}[x_1, \ldots, x_m]$ for which $P(f_1(z), \ldots, f_m(z)) = 0$ is obviously an ideal $\mathcal{I}$ of the ring $\mathcal{L}[x_1, \ldots, x_m]$. From

(15) it follows that $\mathcal{I}$ *is closed under the operator* D, i.e., $D\mathcal{I} \subset \mathcal{I}$. The converse is also true.

Lemma 3. *Let* $\mathcal{I}$ *be an ideal of* $\mathcal{L}[x_1, \ldots, x_m]$ *which is not the whole ring and is closed under the operator* D. *Then:*

1°. *There exists a solution* $f_1(z), \ldots, f_m(z)$ *of the system of differential equations (5) such that for any polynomial* $P \in \mathcal{I}$ *one has*

$$P(f_1(z), \ldots, f_m(z)) = 0. \tag{16}$$

2°. *If the system (5) is homogeneous, and if the ideal* $\mathcal{I}$ *is also homogeneous and has a nontrivial zero, then (5) has a nontrivial solution* $f_1(z), \ldots, f_m(z)$ *which satisfies (16) for any* $P \in \mathcal{I}$.

3°. *If (5) splits into several subsystems of differential equations, and if the ideal is homogeneous relative to each of the corresponding sets of variables and has a zero which is nontrivial in each set of variables, then the above solution of (5) can be chosen so as also to be nontrivial in each subset of components corresponding to the subsystems.*

Proof. Since $\mathcal{I} \neq \mathcal{L}[x_1, \ldots, x_m]$, by Hilbert's Nullstellensatz there exist algebraic functions $g_1(z), \ldots, g_m(z)$ over $\mathcal{L}$ such that $P(g_1(z), \ldots, g_m(z)) = 0$ for any $P \in \mathcal{I}$.

We let $P_1, \ldots, P_s$ denote a basis of the ideal $\mathcal{I}$. Let $\alpha \in \mathbf{C}$ have the property that all of the coefficients in (5), the functions $g_1(z), \ldots, g_m(z)$, and all of the coefficients of the polynomials $P_1, \ldots, P_s$ are analytic at α. We consider a solution $f_1(z), \ldots, f_m(z)$ of (5) for which $f_k(\alpha) = g_k(\alpha)$, $k = 1, \ldots, m$. The functions $f_1(z), \ldots, f_m(z)$ are analytic at α. Let P be one of the polynomials $P_1, \ldots, P_s$. We denote $\Phi(z) = P(f_1(z), \ldots, f_m(z))$. By our choice of α, the function $\Phi(z)$ is analytic at α.

Using (15), we obtain

$$\frac{d^k\Phi(z)}{dz^k} = D^k P(f_1(z), \ldots, f_m(z)), \qquad k = 0, 1, 2, \ldots,$$

and

$$\frac{d^k\Phi(\alpha)}{dz^k} = D^k P(f_1(\alpha), \ldots, f_m(\alpha)) = D^k P(g_1(\alpha), \ldots, g_m(\alpha)) = 0,$$

since $D^k P \in \mathcal{I}$, $k = 0, 1, 2, \ldots$. Consequently, $\Phi(z) \equiv 0$, i.e., $P_i(f_1(z), \ldots, f_m(z)) = 0$, $i = 1, \ldots, s$; hence, $P(f_1(z), \ldots, f_m(z)) = 0$ for any $P \in \mathcal{I}$. This proves part 1°.

Part 2° follows from the observation that, if $g_1(z), \dots, g_m(z)$ is a nontrivial zero of $\mathcal{I}$, then the point α can be chosen in such a way that at least one of the numbers $g_i(\alpha)$ is nonzero.

Part 3° holds, because in that case α can be chosen so that in each subset of variables at least one of the numbers $g_i(\alpha)$ is nonzero.

§ 3. Some properties of solutions of second order homogeneous linear differential equations

We consider the homogeneous linear differential equation

$$y'' + ay' + by = 0, \qquad a, b \in \mathcal{L}. \tag{17}$$

Let φ_1 and φ_2 be solutions of (17) which are linearly independent over $\mathbf{C}$. We set $\mathcal{K} = \mathcal{L}(\varphi_1, \varphi_1', \varphi_2, \varphi_2')$. The field $\mathcal{K}$ obviously does not depend on the choice of the fundamental pair of solutions φ_1, φ_2 of (17).

Lemma 4. *Suppose that φ and φ' are algebraically independent over $\mathcal{L}$ for any nontrivial solution φ of (17), and there exists $v \in \mathcal{L}$, $v \not\equiv 0$, such that $v' + av = 0$. Then* $\deg\operatorname{tr}_{\mathcal{L}}\mathcal{K} = 3$.

Proof. Let φ_1, φ_2 be a fundamental pair of solutions of (17). By assumption, φ_1 and φ_1' are algebraically independent over $\mathcal{L}$. Hence, $\deg\operatorname{tr}_{\mathcal{L}}\mathcal{K} \geq 2$.

From the equality

$$(\varphi_2'\varphi_1 - \varphi_1'\varphi_2)' = \varphi_2''\varphi_1 - \varphi_1''\varphi_2 = -a(\varphi_2'\varphi_1 - \varphi_1'\varphi_2)$$

it follows that there exists a constant $c \in \mathbf{C}$ such that

$$\varphi_2'\varphi_1 - \varphi_1'\varphi_2 = cv \in \mathcal{L}. \tag{18}$$

Consequently, $\deg\operatorname{tr}_{\mathcal{L}}\mathcal{K} \leq 3$.

We now prove that $\deg\operatorname{tr}_{\mathcal{L}}\mathcal{K} = 3$. Suppose that $\deg\operatorname{tr}_{\mathcal{L}}\mathcal{K} = 2$. We set $\mathcal{L}_1 = \mathcal{L}(\varphi_1, \varphi_1')$. Since φ_1 is a solution of (17), it follows that $\mathcal{L}_1$ is closed under differentiation. By assumption, φ_2, and hence also φ_2/φ_1, is an algebraic function over $\mathcal{L}_1$. We have

$$\left(\frac{\varphi_2}{\varphi_1}\right)' = \frac{\varphi_2'\varphi_1 - \varphi_1'\varphi_2}{\varphi_1^2} = \frac{cv}{\varphi_1^2} \in \mathcal{L}_1. \tag{19}$$

Because of (19), we can apply Lemma 1 of Chapter 7 to the function φ_2/φ_1 and the field $\mathcal{L}_1$; we conclude that $\varphi_2/\varphi_1 \in \mathcal{L}_1$, and so $\varphi_2 \in \mathcal{L}_1$. But then also $\varphi_2' \in \mathcal{L}_1$.

Let

$$\varphi_2 = \frac{A_1(\varphi_1, \varphi_1')}{A_0(\varphi_1, \varphi_1')}, \qquad \varphi_2' = \frac{A_2(\varphi_1, \varphi_1')}{A_0(\varphi_1, \varphi_1')}, \tag{20}$$

where $A_k = A_k(x_1, x_2)$, $k = 0, 1, 2$, are polynomials in the ring $\mathcal{L}[x_1, x_2]$ which do not have a nontrivial common factor. We let

$$D = \frac{\partial}{\partial z} + x_2 \frac{\partial}{\partial x_1} - (ax_2 + bx_1)\frac{\partial}{\partial x_2}$$

be the differential operator associated to the differential equation (17), which acts on $\mathcal{L}[x_1, x_2]$. We have

$$\varphi_2'' = \frac{d}{dz}\left(\frac{A_2(\varphi_1, \varphi_1')}{A_0(\varphi_1, \varphi_1')}\right) = \left.\frac{(DA_2)A_0 - (DA_0)A_2}{A_0^2}\right|_{\substack{x_1=\varphi_1 \\ x_2=\varphi_1'}}. \tag{21}$$

If we substitute the right sides of the equations in (20) and (21) in place of φ_2, φ_2', φ_2'' in the differential equation (17) and use the algebraic independence of φ_1 and φ_1' over $\mathcal{L}$, we obtain the following equation in $\mathcal{L}[x_1, x_2]$:

$$\frac{(DA_2)A_0 - (DA_0)A_2}{A_0^2} + a\frac{A_2}{A_0} + b\frac{A_1}{A_0} = 0,$$

from which it follows that A_0 divides the polynomial $A_2 DA_0$. In addition, from (20) we have

$$\frac{A_2}{A_0} = \frac{A_0 DA_1 - A_1 DA_0}{A_0^2},$$

from which we conclude that A_0 divides the polynomial $A_1 DA_0$. But the polynomials A_0, A_1 and A_2 have no nontrivial common factor; so if A_0 divides $A_2 DA_0$ and $A_1 DA_0$, it must divide DA_0. However, $\deg_{\mathbf{x}}(DA_0) \leq \deg_{\mathbf{x}}(A_0)$. Consequently,

$$DA_0 \equiv \omega_0 A_0, \qquad \omega_0 \in \mathcal{L}. \tag{22}$$

We prove that $A_0 \in \mathcal{L}$. In fact, otherwise consider the principal ideal $\mathcal{I} = (A_0)$. From (22) it follows that $\mathcal{I}$ is closed under the operator D; and the polynomial A_0 obviously has a nontrivial zero. Applying Lemma 3 to the ideal $\mathcal{I}$, we find that there exists a nontrivial solution ψ of (17) for which $A_0(\psi, \psi') = 0$. But this contradicts the hypothesis of the lemma. Consequently, $A_0 \in \mathcal{L}$, and we may take $A_0 = 1$, $\varphi_2 = A_1(\varphi_1, \varphi_1')$, $\varphi_2' = A_2(\varphi_1, \varphi_1')$.

We consider the polynomial

$$B(x_1, x_2) = x_1 A_2(x_1, x_2) - x_2 A_1(x_1, x_2) - cv, \qquad B(x_1, x_2) \in \mathcal{L}[x_1, x_2],$$

where cv is given by (18). Since the solutions φ_1, φ_2 are linearly independent over C, it follows from (18) that $c \neq 0$; and since $B(0,0) = -cv \neq 0$, it follows that $B(x_1, x_2) \not\equiv 0$. We have

$$B(\varphi_1, \varphi_1') = \varphi_2'\varphi_1 - \varphi_1'\varphi_2 - cv = 0.$$

But this equation contradicts the algebraic independence of the functions φ_1 and φ_1' over $\mathcal{L}$. This contradiction enables us to conclude that $\deg\operatorname{tr}_{\mathcal{L}}\mathcal{K} = 3$, and the proof is complete.

Lemma 5. *Suppose that each of the two differential equations*

$$y'' + a_1 y' + b_1 y = 0, \qquad y'' + a_2 y' + b_2 y = 0, \tag{23}$$

satisfies the conditions of Lemma 4; ψ_1 and ψ_2 are nontrivial solutions of (23); and the functions ψ_1, ψ_1', ψ_2, ψ_2' are algebraically independent over $\mathcal{L}$. Then there exist nontrivial solutions φ_1 and φ_2 of (23) such that φ_2 is algebraic over $\mathcal{L}_1 = \mathcal{L}(\varphi_1, \varphi_1')$.

Proof. Suppose that the lemma is false. Then the functions φ_1, φ_1' and φ_2 are algebraically independent over $\mathcal{L}$ for any nontrivial solutions φ_1 and φ_2 of (23). In particular, the functions ψ_1, ψ_1' and ψ_2 are algebraically independent over $\mathcal{L}$. There exists an irreducible polynomial $P = P(x_{1,1}, x_{1,2}, x_{2,1}, x_{2,2}) \in \mathcal{L}[x_{1,1}, x_{1,2}, x_{2,1}, x_{2,2}] = G$ such that

$$P(\psi_1, \psi_1', \psi_2, \psi_2') = 0. \tag{24}$$

We let

$$D = \frac{\partial}{\partial z} + x_{1,2}\frac{\partial}{\partial x_{1,1}} - (a_1 x_{1,2} + b_1 x_{1,1})\frac{\partial}{\partial x_{1,2}} + \\ + x_{2,2}\frac{\partial}{\partial x_{2,1}} - (a_2 x_{2,2} + b_2 x_{2,1})\frac{\partial}{\partial x_{2,2}}$$

denote the differential operator which is associated to the differential equations (23) and acts on the ring G.

Since $\deg\operatorname{tr}_{\mathcal{L}}[\psi_1, \psi_1', \psi_2, \psi_2'] = 3$ and (24) holds, it follows from Lemma 4 of Chapter 5 that the polynomial DP is divisible by the irreducible polynomial P, and that the quotient ω is in $\mathcal{L}$, i.e.,

$$DP \equiv \omega P. \tag{25}$$

Let P_1 denote the sum of the terms of P having maximum total degree in $x_{1,1}$ and $x_{1,2}$, and let P_2 denote the sum of the terms of P_1 having maximum total degree in $x_{2,1}$ and $x_{2,2}$. The polynomial P_2 is homogeneous in each of the sets of variables $x_{1,1}, x_{1,2}$ and $x_{2,1}, x_{2,2}$. Since D takes a sum of terms of P homogeneous relative to $x_{1,1}, x_{1,2}$ (or relative to $x_{2,1}, x_{2,2}$) to a similar sum of terms of DP or to zero, it follows from (23) that

$$DP_2 \equiv \omega P_2, \qquad \omega \in \mathcal{L}. \tag{26}$$

Suppose that $Q = Q(\mathbf{x}) = Q(x_{1,1}, x_{1,2}, x_{2,1}, x_{2,2})$ is an irreducible factor of P_2, $\deg_{\mathbf{x}} Q \geq 1$, $P_2 = Q^n Q_1$, where Q and Q_1 are relatively prime polynomials. Using (26), we obtain

$$DP_2 = nQ^{n-1}(DQ)Q_1 + Q^n DQ_1 \equiv \omega Q^n Q_1,$$

from which it follows that DQ is divisible by the polynomial Q. Since $\deg_{\mathbf{x}} Q \geq \deg_{\mathbf{x}} DQ$, we have

$$DQ \equiv \omega_1 Q, \qquad \omega_1 \in \mathcal{L}. \tag{27}$$

We note that Q is a homogeneous polynomial in each of the sets of variables $x_{1,1}, x_{1,2}$ and $x_{2,1}, x_{2,2}$.

We consider the principal ideal $\mathcal{I} = (Q)$. The polynomial Q is irreducible, and, by the hypothesis of the lemma, all four variables $x_{i,j}$ occur in it. Hence it clearly has a zero which is nontrivial in both sets of variables $x_{1,1}, x_{1,2}$ and $x_{2,1}, x_{2,2}$. Thus, the ideal $\mathcal{I}$ has a zero which is nontrivial in both of these sets of variables. From (27) it follows that $\mathcal{I}$ is closed under the operator D. If we apply Lemma 3 to $\mathcal{I}$, which is a homogeneous ideal in each set of variables $x_{1,1}, x_{1,2}$ and $x_{2,1}, x_{2,2}$, we find that there exist nontrivial solutions φ_1 and φ_2 of the differential equations (23) such that

$$Q(\varphi_1, \varphi_1', \varphi_2, \varphi_2') = 0. \tag{28}$$

Differentiating (27) with respect to $x_{1,1}$, we obtain

$$D\left(\frac{\partial Q}{\partial x_{1,1}}\right) - b_1 \frac{\partial Q}{\partial x_{1,2}} = \omega_1 \frac{\partial Q}{\partial x_{1,1}},$$

or

$$D\left(\frac{\partial Q}{\partial x_{1,1}}\right) = \omega_1 \frac{\partial Q}{\partial x_{1,1}} + b_1 \frac{\partial Q}{\partial x_{1,2}}.$$

According to the conditions in the lemma, there exist functions $v_1, v_2 \in \mathcal{L}$, $v_1 v_2 \neq 0$, such that $v_1' + a_1 v_1 = 0$ and $v_2' + a_2 v_2 = 0$. We set

$$R_1(\mathbf{x}) = R_1(x_{1,1}, x_{1,2}, x_{2,1}, x_{2,2}) = \frac{v_1}{x_{1,2}} \frac{\partial Q}{\partial x_{1,1}},$$

$$R_1(\mathbf{x}) \in \mathcal{L}(x_{1,1}, x_{1,2}, x_{2,1}, x_{2,2}), \qquad R_1(\mathbf{x}) \neq 0,$$

and we set $F_1 = R_1(\boldsymbol{\varphi}) = R_1(\varphi_1, \varphi_1', \varphi_2, \varphi_2')$. Since $\deg_{x_{1,1}} \frac{\partial Q}{\partial x_{1,1}} < \deg_{x_{1,1}} Q$, it follows that $F_1 \not\equiv 0$. We have

$$\begin{aligned} F_1' &= \frac{v_1'}{\varphi_1'} \frac{\partial Q(\boldsymbol{\varphi})}{\partial x_{1,1}} + \frac{v_1}{(\varphi_1')^2}(a_1\varphi_1' + b_1\varphi_1)\frac{\partial Q(\boldsymbol{\varphi})}{\partial x_{1,1}} + \frac{v_1}{\varphi_1'} D\left(\frac{\partial Q(\mathbf{x})}{\partial x_{1,1}}\right)\Bigg|_{\mathbf{x}=\boldsymbol{\varphi}} = \\ &= \frac{b_1 v_1 \varphi_1}{(\varphi_1')^2} \frac{\partial Q(\boldsymbol{\varphi})}{\partial x_{1,1}} + \frac{v_1 \omega_1}{\varphi_1'} \frac{\partial Q(\boldsymbol{\varphi})}{\partial x_{1,1}} + \frac{v_1 b_1}{\varphi_1'} \frac{\partial Q(\boldsymbol{\varphi})}{\partial x_{1,2}} = \\ &= \frac{b_1 v_1}{(\varphi_1')^2}\left(\varphi_1 \frac{\partial Q(\boldsymbol{\varphi})}{\partial x_{1,1}} + \varphi_1' \frac{\partial Q(\boldsymbol{\varphi})}{\partial x_{1,2}}\right) + \omega_1 \frac{v_1}{\varphi_1'} \frac{\partial Q(\boldsymbol{\varphi})}{\partial x_{1,1}}. \end{aligned} \tag{29}$$

Since Q is a homogeneous polynomial in the variables $x_{1,1}$ and $x_{1,2}$, it follows from Euler's theorem on homogeneous functions that the polynomial

$$x_{1,1}\frac{\partial Q}{\partial x_{1,1}} + x_{1,2}\frac{\partial Q}{\partial x_{1,2}}$$

is divisible by Q, and hence

$$\varphi_1 \frac{\partial Q(\boldsymbol{\varphi})}{\partial x_{1,1}} + \varphi_1' \frac{\partial Q(\boldsymbol{\varphi})}{\partial x_{1,2}} = 0.$$

From (29) we now have

$$F_1' = \omega_1 F_1, \qquad \omega_1 \in \mathcal{L}. \tag{30}$$

If we denote

$$R_2(\mathbf{x}) = \frac{v_2}{x_{2,2}} \frac{\partial Q}{\partial x_{2,1}}, \qquad F_2 = R_2(\boldsymbol{\varphi}),$$

then we similarly obtain

$$F_2' = \omega_1 F_2. \tag{31}$$

Equations (30) and (31) mean that for some constant $\sigma \in \mathbf{C}$ we have the equality

$$F_2 - \sigma F_1 = 0. \tag{32}$$

If we set

$$A = A(x_{1,1}, x_{1,2}, x_{2,1}, x_{2,2}) = v_2 x_{1,2} \frac{\partial Q}{\partial x_{2,1}} - \sigma v_1 x_{2,2} \frac{\partial Q}{\partial x_{1,1}},$$

$$A \in \mathcal{L}[x_{1,1}, x_{1,2}, x_{2,1}, x_{2,2}],$$

then from (32) we find that

$$A(\varphi_1, \varphi_1', \varphi_2, \varphi_2') = 0. \tag{33}$$

Since $\deg_{\mathbf{x}} \frac{\partial Q}{\partial x_{2,1}} < \deg_{\mathbf{x}} Q$, it follows that $\frac{\partial Q(\boldsymbol{\varphi})}{\partial x_{2,1}} \not\equiv 0$. Consequently, $F_2 \not\equiv 0$ and $\sigma \neq 0$. The polynomials $v_2 x_{1,2} \frac{\partial Q}{\partial x_{2,1}}$ and $\sigma v_1 x_{2,2} \frac{\partial Q}{\partial x_{1,1}}$ have different total degrees in $x_{1,1}, x_{1,2}$. Hence, $A \not\equiv 0$.

By the assumption at the beginning of the proof, $\deg \operatorname{tr}_{\mathcal{L}}[\varphi_1, \varphi_1', \varphi_2] = 3$. Hence, from (28) and (33) it follows that A is divisible by the irreducible polynomial Q. But $\deg_{\mathbf{x}} A = \deg_{\mathbf{x}} Q$. Consequently,

$$A \equiv \omega_2 Q, \qquad \omega_2 \in \mathcal{L}. \tag{34}$$

But (34) is impossible, since Q is a homogeneous polynomial in the variables $x_{1,1}, x_{1,2}$, and A does not have this property. This contradiction proves Lemma 5.

§ 4. Algebraic independence of solutions of a set of second order homogeneous linear differential equations

Proof of Theorem 1. The proof is by induction on the number n of differential equations in (3). If $n = 1$, the result holds, because f and f' are algebraically independent over $\mathbf{C}(z)$, by condition 1°.

We now suppose that the theorem has been proved for $n = k - 1$, $k \geq 2$, and that there do not exist functions φ_i and φ_j, $1 \leq i < j \leq k$, with the properties in the theorem. We shall prove that f_i and f_i', $i = 1, \ldots, k$, are algebraically independent over $\mathbf{C}(z)$. This will prove the theorem for $n = k$, and so for all n, by induction.

We denote $\mathcal{L} = \mathbf{C}(z)$ if $k = 2$ and $\mathcal{L} = \mathbf{C}(z, f_3, f_3', \dots, f_k, f_k')$ if $k \geq 3$. From the induction assumption it follows that $\deg \operatorname{tr}_{\mathbf{C}(z)} \mathcal{L} = 2(k-2)$. Hence it suffices to prove that the functions f_1, f_1', f_2, f_2' are algebraically independent over $\mathcal{L}$.

Suppose that this is not the case. Then by Lemma 5 (whose hypothesis is satisfied, by the induction assumption), there exist nontrivial solutions ψ_1 and ψ_2 of the first two differential equations in (3) such that ψ_1 is algebraic over the field $\mathcal{L}_1 = \mathcal{L}(\psi_2, \psi_2')$. Let ψ_1^* be a conjugate of ψ_1 over $\mathcal{L}_1$. By the corollary of Lemma 2, the function ψ_1^*, like ψ_1, satisfies the first differential equation in (3). If $\psi_1^*/\psi_1 \notin \mathbf{C}$, then ψ_1 and ψ_1^* form a fundamental pair of solutions of the first differential equation in (3). Since ψ_1 and ψ_1^* are algebraic over $\mathcal{L}_1$, it follows that the field $\mathcal{K} = \mathcal{L}_1(\psi_1, \psi_1', \psi_1^*, {\psi_1^*}')$ is algebraic over $\mathcal{L}_1$, which means that $\deg \operatorname{tr}_{\mathcal{L}} \mathcal{K} \leq 2$. But this contradicts Lemma 4. Hence we have $\psi_1^* = c^* \psi_1$ with $c^* \in \mathbf{C}$.

Let $\chi(x) = x^r + u_{r-1} x^{r-1} + \dots + u_0$ be the minimal polynomial of ψ_1 over $\mathcal{L}_1$. If we make the substitution $x = \psi_1 t$, we see that all of the roots of the equation

$$t^r + \frac{u_{r-1}}{\psi_1} t^{r-1} + \dots + \frac{u_0}{\psi_1^r} = 0$$

are complex numbers. But then $u_j = c_j \psi_1^{r-j}$, $j = 0, 1, \dots, r-1$, $c_j \in \mathbf{C}$. Since $\deg_{\mathcal{L}_1} \psi_1 = r$, we find that $u_j = 0$, $j = 1, \dots, r-1$, and $\psi_1^r = u \in \mathcal{L}_1$, where $u = -u_0$. This equality implies that $\psi_1'/\psi_1 \in \mathcal{L}_1$, and so there exist relatively prime polynomials $A, B \in \mathbf{C}(z)[x_{2,1}, x_{2,2}, \dots, x_{k,1}, x_{k,2}]$ such that

$$A(\boldsymbol{\psi}_2, \mathbf{f}_3, \dots, \mathbf{f}_k)\psi_1 - B(\boldsymbol{\psi}_2, \mathbf{f}_3, \dots, \mathbf{f}_k)\psi_1' = 0, \tag{35}$$

where $\boldsymbol{\psi}_2$ and the $\mathbf{f}_i$ denote pairs consisting of a function and its derivative. If we differentiate (35) and use the fact that ψ_1 satisfies the first differential equation in (3), we obtain the equation

$$(DA + b_1 B)\psi_1 - (DB - A - a_1 B)\psi_1' = 0, \tag{36}$$

where

$$D = \frac{\partial}{\partial z} + \sum_{i=1}^{k} \left(x_{i,2} \frac{\partial}{\partial x_{i,1}} - (a_i x_{i,2} + b_i x_{i,1}) \frac{\partial}{\partial x_{i,2}} \right)$$

is the differential operator which is associated to the system of differential equations (3) and acts on the ring $\mathbf{C}(z)[x_{1,1}, x_{1,2}, \dots, x_{k,1}, x_{k,2}]$. (For brevity we have omitted $\psi_2, \psi_2', f_3, f_3', \dots, f_k, f_k'$ in the notation in the coefficients of ψ_1

and ψ_1'.) Since $\psi_1 \neq 0$ and the functions $\psi_2, \psi_2', f_3, f_3', \ldots, f_k, f_k'$ are algebraically independent over $\mathbf{C}(z)$, it follows from (35) and (36) that the polynomial $DP = (DA + b_1 B)x_{1,1} - (DB - A - a_1 B)x_{1,2}$ is divisible by the irreducible polynomial $P = Ax_{1,1} - Bx_{1,2}$. If we compare the degrees of P and DP, we obtain the identity

$$DP = \omega_0 P, \qquad \omega_0 \in \mathbf{C}(z). \tag{37}$$

We let P_2 denote the sum of all of the homogeneous terms in P of maximum total degree in $x_{2,1}, x_{2,2}$. We then let P_3 denote the sum of all of the homogeneous terms in P_2 of maximum total degree in $x_{3,1}, x_{3,2}$. We continue this process until we get to the polynomial P_k. We set $Q = P_k$. Then

$$Q = A_1 x_{1,1} - B_1 x_{1,2}, \qquad A_1, B_1 \in \mathbf{C}(z)[x_{2,1}, x_{2,2}, \ldots, x_{k,1}, x_{k,2}],$$

and Q is a homogeneous polynomial relative to each pair of variables $x_{i,1}, x_{i,2}$, $i = 1, \ldots, k$.

Equating similar terms in (37), we obtain

$$DQ = \omega_0 Q. \tag{38}$$

From (38) we find that

$$DA_1 = -b_1 B_1 + \omega_0 A_1, \qquad DB_1 = A_1 + (\omega_0 + a_1)B_1,$$

from which it follows that the ideal $\mathcal{I} = (A_1, B_1) \subset \mathbf{C}(z)[x_{2,1}, x_{2,2}, \ldots, x_{k,1}, x_{k,2}]$ is closed under the operator D, i.e., $D\mathcal{I} \subset \mathcal{I}$. The ideal $\mathcal{I}$ is homogeneous relative to each pair of variables $x_{i,1}, x_{i,2}$, $i = 2, \ldots, k$, and if the polynomials A_1 and B_1 depend on at least two of these pairs of variables, then there exists a zero $\xi_{2,1}, \xi_{2,2}, \ldots, \xi_{k,1}, \xi_{k,2}$ of the ideal $\mathcal{I}$ such that each pair $\xi_{i,1}, \xi_{i,2}$ has a nonzero component. If we apply Lemma 3 to the ideal $\mathcal{I}$, we find that there exist nontrivial solutions $g_2, \ldots, g_k$ of the differential equations in (3) with indices $i = 2, \ldots, k$, such that the $(2k-2)$-tuple $g_2, g_2', \ldots, g_k, g_k'$ is a zero of the ideal $\mathcal{I}$, and so these functions are algebraically dependent over $\mathbf{C}(z)$. But this is impossible, by the induction assumption. Hence, the polynomials A_1 and B_1 depend only on a single pair of variables $x_{i,1}, x_{i,2}$, let us say $x_{2,1}, x_{2,2}$, to be definite.

We consider the principal ideal $\mathcal{I}_1 = (Q)$. From (38) it follows that $D\mathcal{I}_1 \subset \mathcal{I}_1$. Since $\mathcal{I}_1$ has a nontrivial zero, Lemma 3 implies that there exist nontrivial solutions φ_1 and φ_2 of the first two differential equations in (3) such that

$$Q(\varphi_1, \varphi_1', \varphi_2, \varphi_2') = 0. \tag{39}$$

Since A_1 and B_1 are homogeneous polynomials of the same degree, it follows from (39) that the function φ_2'/φ_2 satisfies an equation

$$L_s x^s + L_{s-1}x^{s-1} + \cdots + L_0 = 0, \tag{40}$$

where $s = \deg A_1 = \deg B_1$, and the L_i are linear forms in the functions φ_1 and φ_1' with coefficients in $\mathbf{C}(z)$.

Since the functions $\varphi_1, \varphi_1', \varphi_2, \varphi_2'$ are algebraically dependent over $\mathbf{C}(z)$, if we repeat the argument used at the beginning of the proof of the theorem we find that

$$\frac{\varphi_2'}{\varphi_2} = \frac{M(\varphi_1, \varphi_1')}{N(\varphi_1, \varphi_1')}, \tag{41}$$

where $M = M(x_{1,1}, x_{1,2})$ and $N = N(x_{1,1}, x_{1,2})$ are relatively prime polynomials in $\mathbf{C}(z)[x_{1,1}, x_{1,2}]$. If M and N were not homogeneous polynomials of the same degree, then it would follow from (41) and the relation

$$A_1(1, \varphi_2'/\varphi_2)\varphi_1 - B_1(1, \varphi_2'/\varphi_2)\varphi_1' = 0$$

that the functions φ_1 and φ_1' are algebraic over the field $\mathbf{C}(z, \varphi_2'/\varphi_2)$, i.e., they are algebraically dependent over $\mathbf{C}(z)$. But that is impossible, by condition 2° of the theorem. Hence, M and N are homogeneous polynomials of the same degree.

If we substitute the right side of (41) in place of x in (40) and again use the algebraic independence of φ_1 and φ_1' over $\mathbf{C}(z)$, we find that M divides L_0 and N divides L_s. This means that M and N are linear forms. Consequently, the polynomial Q has degree 2 in the variables $x_{1,1}, x_{1,2}, x_{2,1}, x_{2,2}$.

If we apply the argument in the proof of Lemma 5 to the polynomial Q, which is homogeneous in each pair of variables $x_{1,1}, x_{1,2}$ and $x_{2,1}, x_{2,2}$, we obtain a polynomial

$$A = A(x_{1,1}, x_{1,2}, x_{2,1}, x_{2,2}) = v_2 x_{1,2}\frac{\partial Q}{\partial x_{2,1}} - \sigma v_1 x_{2,2}\frac{\partial Q}{\partial x_{1,1}},$$

$$A \in \mathcal{L}[x_{1,1}, x_{1,2}, x_{2,1}, x_{2,2}],$$

such that $A(\varphi_1, \varphi_1', \varphi_2, \varphi_2') = 0$. Since $\deg Q = 2$, it follows from the last equation and (41) that

$$\varphi_2^2 = \frac{S(\varphi_1, \varphi_1')}{T(\varphi_1, \varphi_1')},$$

where S and T are relatively prime homogeneous polynomials with $\deg S - \deg T = 2$. Differentiating this equality and using (41), we obtain

$$2\frac{S}{T}\frac{M}{N} = \frac{TDS - SDT}{T^2},$$

from which it follows that T divides NDT and S divides NDS. If $T = N^q T_1$, where $(N, T_1) = 1$, $q \in \mathbf{Z}^+$, then the equation $NDT = qN^q T_1 DN + N^{q+1} DT_1$ and the fact that T divides NDT imply that T_1 divides DT_1. Consequently, we have $DT_1 = \omega_1 T_1$, where $\omega_1 \in \mathbf{C}(z)$. Hence, the principal ideal $\mathcal{I}_2 = (T_1)$ satisfies the condition $D\mathcal{I}_2 \subset \mathcal{I}_2$, and since it has a nontrivial zero, it follows by Lemma 3 that there exists a nontrivial solution of the first differential equation in (3) which depends algebraically on its derivative. But this contradicts the hypothesis of the theorem. Thus, $T_1 \in \mathbf{C}(z)$. We similarly prove that $S = N^p S_1$, $p \in \mathbf{N}$, where $S_1 \in \mathbf{C}(z)$. Since $\deg N = 1$ and $\deg S - \deg T = 2$ with $(S, T) = 1$, it follows that $q = 0$ and $p = 2$. Hence, $\varphi_2^2 = \omega_2 N^2(\varphi_1, \varphi_1')$, $\omega_2 \in \mathbf{C}(z)$, and so

$$\varphi_2 = \omega N(\varphi_1, \varphi_1'), \qquad \omega^2 \in \mathbf{C}(z).$$

This equality contradicts the supposition at the beginning of the proof, and thereby proves the theorem.

§ 5. Proof of Siegel's Theorem

We make use of Theorem 1 to prove Lemma 1. But Theorem 1 cannot be applied directly to the solutions of the set of differential equations (2), since that system does not satisfy condition 1° of the theorem. The function $K_\lambda(z)$ is connected with the Bessel function $J_\lambda(z)$ by the equality $J_\lambda(z) = \frac{1}{\Gamma(\lambda+1)}\left(\frac{z}{2}\right)^\lambda K_\lambda(z)$. The function $J_\lambda(z)$ satisfies the Bessel differential equation

$$y'' + \frac{1}{z}y' + \left(1 - \frac{\lambda^2}{z^2}\right)y = 0, \tag{42}$$

and the function $J_{\lambda_i}(\xi_j z)$ satifies the differential equation

$$y'' + \frac{1}{z}y' + \left(\xi_j^2 - \frac{\lambda_i^2}{z^2}\right)y = 0, \qquad i = 1, \dots, m, \quad j = 1, \dots, n. \tag{43}$$

Condition 1° of Theorem 1 is fulfilled for the differential equations (43). Hence, we shall prove a lemma for the solutions of these equations (proved by Siegel in [72:3]) which is analogous to Lemma 1.

Lemma 6. *Suppose that $\lambda_i \in \mathbf{C}$, $\lambda_i \neq \frac{2k-1}{2}$, $k \in \mathbf{Z}$, $i = 1, \ldots, m$; $(\lambda_{i_1} \pm \lambda_{i_2}) \notin \mathbf{Z}$, $i_1, i_2 = 1, \ldots, m$, $i_1 \neq i_2$; $\xi_1, \ldots, \xi_n \in \mathbf{C} \setminus \{0\}$, $\xi_{j_1}^2 \neq \xi_{j_2}^2$, $j_1, j_2 = 1, \ldots, n$, $j_1 \neq j_2$; and $v_{i,j}$ is a nontrivial solution of the (i,j)-th differential equation in (43). Then the functions $v_{i,j}$, $v'_{i,j}$, $i = 1, \ldots, m$, $j = 1, \ldots, n$, are algebraically independent over $\mathbf{C}(z)$.*

Since $v_{i,j} = z^{\lambda_i} y_{i,j}$, where $y_{i,j}$ is a nontrivial solution of the corresponding differential equation in (2), it follows from Lemma 8 of Chapter 5 that Lemma 1 is a consequence of Lemma 6. Thus, to prove Siegel's Theorem it suffices to prove Lemma 6. But we first prove one more auxiliary lemma.

Lemma 7. *Suppose that $a, b, \mu, \nu \in \mathbf{C}$, $ab \neq 0$, $\mu, \nu \neq \frac{2k-1}{2}$, $k \in \mathbf{Z}$, and f and φ are nontrivial solutions of the differential equations*

$$\begin{aligned} y'' + \frac{1}{z} y' + \left(a^2 - \frac{\mu^2}{z^2}\right) y &= 0, \\ y'' + \frac{1}{z} y' + \left(b^2 - \frac{\nu^2}{z^2}\right) y &= 0. \end{aligned} \tag{44}$$

If

$$f = \omega(u_1 \varphi + u_2 \varphi'), \qquad \omega^2, u_1, u_2 \in \mathbf{C}(z), \tag{45}$$

then $a^2 = b^2$, and either $\mu + \nu \in \mathbf{Z}$ or else $\mu - \nu \in \mathbf{Z}$.

Proof. If we differentiate (45) and use the fact that φ satisfies the second differential equation in (44), we find that the function f' can also be written as a linear combination of φ and φ' analogous to (45). Thus, there exist polynomials $B, A_{1,1}, A_{1,2}, A_{2,1}, A_{2,2} \in \mathbf{C}[z]$ such that the following matrix equation holds:

$$\begin{pmatrix} f \\ f' \end{pmatrix} = \frac{\omega}{B} \begin{pmatrix} A_{1,1} & A_{1,2} \\ A_{2,1} & A_{2,2} \end{pmatrix} \begin{pmatrix} \varphi \\ \varphi' \end{pmatrix}, \tag{46}$$

where we may suppose that $(B, A_{1,1}, \ldots, A_{2,2}) = 1$, $\omega^2 = A \in \mathbf{C}[z]$, the polynomial A does not have multiple roots, and the polynomials A and B have leading coefficient 1.

The analytic functions $f, \varphi, \varphi', u_1, u_2$ do not have branch points at $z \in \mathbf{C}$ if $z \neq 0, \infty$. Hence, ω can only have branch points at $z = 0$ or $z = \infty$. Consequently, we have $A = z^h$ with $h = 0$ or 1.

We differentiate (46) and use the fact that $\omega' = \frac{h}{2z}\omega$ and

$$\begin{pmatrix} f' \\ f'' \end{pmatrix} = U_{a,\mu} \begin{pmatrix} f \\ f' \end{pmatrix}, \qquad \begin{pmatrix} \varphi' \\ \varphi'' \end{pmatrix} = U_{b,\nu} \begin{pmatrix} \varphi \\ \varphi' \end{pmatrix},$$

where

$$U_{a,\mu} = \begin{pmatrix} 0 & 1 \\ \mu^2 z^{-2} - a^2 & -z^{-1} \end{pmatrix},$$

and $U_{b,\nu}$ is obtained from $U_{a,\mu}$ by replacing a by b and μ by ν. As a result we obtain the relation

$$\begin{aligned} U_{a,\mu}\begin{pmatrix} f \\ f' \end{pmatrix} =\omega\Bigg(&\left(\frac{h}{2zB} - \frac{B'}{B^2}\right)\begin{pmatrix} A_{1,1} & A_{1,2} \\ A_{2,1} & A_{2,2} \end{pmatrix} + \\ &+ \frac{1}{B}\begin{pmatrix} A'_{1,1} & A'_{1,2} \\ A'_{2,1} & A'_{2,2} \end{pmatrix} + \frac{1}{B}\begin{pmatrix} A_{1,1} & A_{1,2} \\ A_{2,1} & A_{2,2} \end{pmatrix} U_{b,\nu}\Bigg)\begin{pmatrix} \varphi \\ \varphi' \end{pmatrix}. \end{aligned} \tag{47}$$

Now if we compare (46) and (47) and use the fact that φ and φ' are linearly independent over $\mathbf{C}(z)$, by Lemma 3 of Chapter 6, then we obtain the equality

$$\begin{aligned} \begin{pmatrix} A'_{1,1} & A'_{1,2} \\ A'_{2,1} & A'_{2,2} \end{pmatrix} = U_{a,\mu}&\begin{pmatrix} A_{1,1} & A_{1,2} \\ A_{2,1} & A_{2,2} \end{pmatrix} - \begin{pmatrix} A_{1,1} & A_{1,2} \\ A_{2,1} & A_{2,2} \end{pmatrix} U_{b,\nu} + \\ &+ \left(\frac{B'}{B} - \frac{h}{2z}\right)\begin{pmatrix} A_{1,1} & A_{1,2} \\ A_{2,1} & A_{2,2} \end{pmatrix}. \end{aligned} \tag{48}$$

From (48) and the condition $(B, A_{1,1}, \dots, A_{2,2}) = 1$ it follows that z^2B' is divisible by the polynomial B. But this is possible only if $B = z^k$, $k \in \mathbf{Z}$. Hence, (48) can be rewritten in the following form:

$$\begin{aligned} \begin{pmatrix} A'_{1,1} & A'_{1,2} \\ A'_{2,1} & A'_{2,2} \end{pmatrix} = U_{a,\mu}&\begin{pmatrix} A_{1,1} & A_{1,2} \\ A_{2,1} & A_{2,2} \end{pmatrix} - \begin{pmatrix} A_{1,1} & A_{1,2} \\ A_{2,1} & A_{2,2} \end{pmatrix} U_{b,\nu} + \\ &+ \frac{2k-h}{2z}\begin{pmatrix} A_{1,1} & A_{1,2} \\ A_{2,1} & A_{2,2} \end{pmatrix}, \end{aligned} \tag{49}$$

or

$$\begin{aligned} A'_{1,1} &= \left(k - \frac{h}{2}\right)\frac{1}{z}A_{1,1} + \left(b^2 - \frac{\nu^2}{z^2}\right)A_{1,2} + A_{2,1}, \\ A'_{1,2} &= -A_{1,1} + \left(k - \frac{h}{2} + 1\right)\frac{1}{z}A_{1,2} + A_{2,2}, \\ A'_{2,1} &= -\left(a^2 - \frac{\mu^2}{z^2}\right)A_{1,1} + \left(k - \frac{h}{2} - 1\right)\frac{1}{z}A_{2,1} + \left(b^2 - \frac{\nu^2}{z^2}\right)A_{2,2}, \\ A'_{2,2} &= -\left(a^2 - \frac{\mu^2}{z^2}\right)A_{1,2} - A_{2,1} + \left(k - \frac{h}{2}\right)\frac{1}{z}A_{2,2}. \end{aligned} \tag{50}$$

Let $s = \max \deg A_{i,j}$ and

$$A_{i,j} = a_{i,j} z^s + b_{i,j} z^{s-1} + \cdots + d_{i,j}.$$

Comparing the coefficients of z^s on both sides of (50), we obtain

$$\begin{aligned} b^2 a_{1,2} + a_{2,1} &= 0, \\ -a_{1,1} + a_{2,2} &= 0, \\ -a^2 a_{1,1} + b^2 a_{2,2} &= 0, \\ a^2 a_{1,2} + a_{2,1} &= 0. \end{aligned} \tag{51}$$

If $a^2 \neq b^2$, then it follows from (51) that $a_{1,2} = a_{2,1} = 0$ and $a_{1,1} = a_{2,2} = 0$. But this is impossible; hence, $a^2 = b^2$.

Comparing the coefficients of z^{s-1} in (50), we obtain

$$\begin{aligned} s a_{1,1} &= \left(k - \frac{h}{2}\right) a_{1,1} + b^2 b_{1,2} + b_{2,1}, \\ s a_{1,2} &= -b_{1,1} + \left(k - \frac{h}{2} + 1\right) a_{1,2} + b_{2,2}, \\ s a_{2,1} &= -a^2 b_{1,1} + \left(k - \frac{h}{2} - 1\right) a_{2,1} + b^2 b_{2,2}, \\ s a_{2,2} &= -a^2 b_{1,2} - b_{2,1} + \left(k - \frac{h}{2}\right) a_{2,2}. \end{aligned} \tag{52}$$

Suppose that $h = 1$. If we add the first and fourth equations in (52) and use the relations $a^2 = b^2$ and $a_{2,2} = a_{1,1}$ that were established above, we find that $a_{2,2} = a_{1,1} = 0$. If we then subtract the third equation in (52) from the second one multiplied by a^2, and if we observe that $a^2 a_{1,2} = -a_{2,1}$ and $a^2 \neq 0$, we obtain $a_{1,2} = a_{2,1} = 0$. Since the numbers $a_{i,j}$ cannot all be zero, we conclude that $h \neq 1$; hence $h = 0$.

From (49) it follows that when we substitute $f = z^{-k}(A_{1,1}\varphi + A_{1,2}\varphi')$ in the first equation in (44) and substitute the corresponding expression for f' as a linear combination of φ and φ', we obtain an equality which holds identically in z, φ and φ'. Hence, if φ_1 is any solution of the second differential equation in (44), and if f_1 is defined by setting

$$f_1(z) = z^{-k}(A_{1,1}\varphi_1 + A_{1,2}\varphi_1'), \tag{53}$$

then $f_1(z)$ is a solution of the first differential equation in (44).

We set

$$\varphi_1(z) = J_\nu(bz) = \left(\frac{bz}{2}\right)^\nu \frac{1}{\Gamma(\nu+1)} \sum_{n=0}^{\infty} \frac{(-1)^n \left(\frac{bz}{2}\right)^{2n}}{n![\nu+1,n]},$$

where $J_\nu(z)$ is the Bessel function. If we replace φ_1 and φ_1' by their power series in (53), we find that

$$f_1(z) = z^{\nu+m} \sum_{n=0}^{\infty} a_n z^n, \qquad a_j \in \mathbf{C}, \qquad a_0 \neq 0, \qquad m \in \mathbf{Z}. \tag{54}$$

We now substitute the expansion of $f_1(z)$ in (54) in the first differential equation in (44). We then compare the coefficients of $z^{\nu+m-2}$ in this differential equation, obtaining

$$((\nu+m)(\nu+m-1) + (\nu+m) - \mu^2)a_0 = 0,$$

from which we find that, since $a_0 \neq 0$, we have $(\nu+m)^2 = \mu^2$, i.e., $\nu+m = \pm\mu$. The lemma is proved.

Proof of Lemma 6. Since $v_{i,j} = z^{\lambda_i} y_{i,j}$, it follows from Lemma 3 of Chapter 6 and Lemma 8 of Chapter 5 that, under the conditions of the lemma, each pair $v_{i,j}, v_{i,j}'$ is algebraically independent. Hence, Lemma 6 follows from Theorem 1 and Lemma 7.

The proof of Lemma 6, which implies Lemma 1, completes the proof of Siegel's Theorem.

§ 6. Solutions of non-homogeneous linear differential equations

Neither the method in § 2–5, nor various other methods of proof of algebraic independence of the solutions of linear differential equations, enable us to investigate directly the algebraic independence of the solutions of non-homogeneous differential equations.

In Chapter 6 (Theorems 6 and 7) we showed that, when our functions which satisfy a non-homogeneous linear differential equation or system of differential equations are connected by at most a single algebraic equation over $\mathrm{C}(z)$, the question of algebraic independence of these functions reduces to the analogous question for the solutions of the corresponding homogeneous differential equation or system of homogeneous differential equations. In this way we were able to apply those theorems to specific examples of E-functions.

In § 6 we shall show that the transition from non-homogeneous to homogeneous differential equations can be carried out when our functions are connected by an arbitrary number of algebraic equations over $\mathbf{C}(z)$. This transition step was carried out by Nesterenko in 1969 in [52:1]. All of the theorems proved in § 7 are contained in that paper.

As before, $\mathcal{L}$ will denote a field of analytic functions of z which contains $\mathbf{C}$ and is closed under differentiation.

Theorem 2. *Suppose that the analytic functions*

$$f_1(z), \ldots, f_m(z), \qquad m \geq 1, \tag{55}$$

form a solution of the system of linear differential equations

$$y_k' = Q_{k,0} + \sum_{i=1}^{m} Q_{k,i} y_i, \qquad k = 1, \ldots, m, \qquad Q_{k,i} \in \mathcal{L}, \tag{56}$$

and that any nontrivial solution of the corresponding homogeneous system of differential equations

$$y_k' = \sum_{i=1}^{m} Q_{k,i} y_i, \qquad k = 1, \ldots, m, \tag{57}$$

consists of functions which are algebraically independent over $\mathcal{L}$. Then either the functions (55) are algebraically independent over $\mathcal{L}$, or else they all belong to $\mathcal{L}$.

Proof. Suppose that the functions (55) are algebraically dependent over $\mathcal{L}$. For any polynomial $P \in \mathcal{L}[x_1, \ldots, x_m]$ we let P^0 denote the sum of the terms of P of maximum total degree. Let $\mathcal{I}$ be the ideal of $\mathcal{L}[x_1, \ldots, x_m]$ consisting of all polynomials $P = P(x_1, \ldots, x_m)$ such that $P(f_1(z), \ldots, f_m(z)) = 0$, and let $\mathcal{I}^0$ be the ideal generated by all of the polynomials P^0 for $P \in \mathcal{I}$. Let

$$D = \frac{\partial}{\partial z} + \sum_{k=1}^{m} \left(Q_{k,0} + \sum_{i=1}^{m} Q_{k,i} x_i \right) \frac{\partial}{\partial x_k}$$

and

$$D^0 = \frac{\partial}{\partial z} + \sum_{k=1}^{m} \left(\sum_{i=1}^{m} Q_{k,i} x_i \right) \frac{\partial}{\partial x_k}$$

be the differential operators which are associated with the systems of differential equations (56) and (57) and which act on the ring $\mathcal{L}[x_1, \ldots, x_m]$.

From (15) it follows that $D\mathcal{I} \subset \mathcal{I}$. Let $P \in \mathcal{I}$. Since $D\mathcal{I} \subset \mathcal{I}$, it follows that $(DP)^0 \in \mathcal{I}^0$; and since D^0P^0 is either zero or else $(DP)^0$, we have $D^0P^0 \in \mathcal{I}^0$. This implies that $D^0\mathcal{I}^0 \subset \mathcal{I}^0$. Suppose that $\mathcal{I}^0$ has a nontrivial zero. Then by part 2° of Lemma 3, we have a contradiction with the hypothesis of the theorem. Consequently, the only zero of $\mathcal{I}^0$ is $(0,\dots,0)$. It then follows from Hilbert's Nullstellensatz that for some M we have $x_i^M \in \mathcal{I}^0$ for all i, $1 \le i \le m$.

Suppose that $P_1,\dots,P_s \in \mathcal{I}$ are such that $P_1^0,\dots,P_s^0$ form a basis of the ideal $\mathcal{I}^0$. Such polynomials can be found, because $\mathcal{I}^0$ is generated by the polynomials P^0 with $P \in \mathcal{I}$. Then there exist homogeneous polynomials $H_{k,i} \in \mathcal{L}[x_1,\dots,x_m]$ for which

$$x_k^M = \sum_{i=1}^{s} H_{k,i}P_i^0, \qquad k = 1,\dots,m.$$

We set

$$G_k = \sum_{i=1}^{s} H_{k,i}P_i, \qquad k = 1,\dots,m.$$

Since $P_i \in \mathcal{I}$, we have $G_k \in \mathcal{I}$, and so $G_k(f_1(z),\dots,f_m(z)) = 0$. Since $G_k^0 = x_k^M$, it follows that f_k^M is a linear combination of monomial expressions

$$f_1^{k_1}(z)\cdots f_m^{k_m}(z), \tag{58}$$

with exponents satisfying the inequalities $k_1 + \cdots + k_m < M$, $k_i \ge 0$, and with coefficients in $\mathcal{L}$. But this implies that the vector space over $\mathcal{L}$ spanned by all possible monomials (58) with $k_i \ge 0$, $i = 1,\dots,m$, is finite dimensional, and hence all of the functions (55) are algebraic over $\mathcal{L}$.

Let $\psi_1(z),\dots,\psi_m(z)$ be an m-tuple of conjugates over $\mathcal{L}$ of the functions (55). By Lemma 2, the m-tuple $\psi_1(z),\dots,\psi_m(z)$ also form a solution of the system of differential equations (56). Then the functions $\varphi_i(z) = f_i(z) - \psi_i(z)$, $i = 1,\dots,m$, form a solution of the homogeneous system (57) which consists of functions algebraic over $\mathcal{L}$. By the hypothesis of the theorem, this is possible only if $\varphi_i(z) \equiv 0$, $i = 1,\dots,m$. Thus, any set of conjugates over $\mathcal{L}$ of the functions (55) must coincide with (55). This means that $f_i(z) \in \mathcal{L}$, $i = 1,\dots,m$. The theorem is proved.

We note that either of the situations in the conclusion of Theorem 2 may occur. For example, the equation $y' = y - z + 1$ has both a rational solution $y_1 = z$ and a transcendental solution $y_2 = e^z + z$.

We now prove a more general theorem on the solutions of a set of non-homogeneous systems of differential equations.

We consider a set of r systems of linear differential equations

$$Q_i : \qquad y_k' = Q_{k,0}^{(i)} + \sum_{j=1}^{m_i} Q_{k,j}^{(i)} y_j, \tag{59}$$

$$k = 1, \ldots, m_i, \qquad Q_{k,j}^{(i)} \in \mathcal{L}, \qquad m_i \geq 1, \qquad i = 1, \ldots, r,$$

and the corresponding set of homogeneous systems

$$Q_i^0 : \qquad y_k' = \sum_{j=1}^{m_i} Q_{k,j}^{(i)} y_j, \qquad k = 1, \ldots, m_i, \quad i = 1, \ldots, r.$$

Theorem 3. *Suppose that the $(m_1 + \cdots + m_r)$-tuple of functions in any set of nontrivial solutions of the systems Q_i^0 is algebraically independent over $\mathcal{L}$, and let*

$$f_{1,i}, \ldots, f_{m_i,i}, \qquad i = 1, \ldots, r, \tag{60}$$

be a set of solutions of the systems (59) such that for each i at least one of the functions in (60) is not in $\mathcal{L}$. Then the functions (60) are algebraically independent over $\mathcal{L}$.

Proof. We use induction on the number r of systems in (59). If $r = 1$, then Theorem 3 follows from Theorem 2. Suppose that the theorem holds if there are fewer than r systems in (59), where $r \geq 2$. We prove that the theorem then holds for the r systems in (59).

Let l denote the number of non-homogeneous systems among the systems in (59). If $l = 0$, then the theorem holds. Suppose that the theorem has been proved for all sets of r systems of which fewer than l are non-homogeneous systems. If $l < r$, then without loss of generality we may assume that Q_1 is a homogeneous system.

We let $\mathcal{L}_1 = \mathcal{L}(f_{1,1}, \ldots, f_{m_1,1})$ denote the field obtained by adjoining the functions $f_{i,1}$ to $\mathcal{L}$. This field is closed under differentiation. Any set of nontrivial solutions of the systems $Q_2^0, \ldots, Q_r^0$ consists of functions which are algebraically independent over $\mathcal{L}_1$. If $l < r$ this follows from our choice of Q_1 and from the hypothesis of the theorem. If $l = r \geq 2$, then it follows from the supposition about sets with fewer than l non-homogeneous systems.

Suppose that there exists i, say $i = 2$, such that all of the functions which make up the corresponding solution belong to $\mathcal{L}_1$:

$$f_{j,2} \in \mathcal{L}_1, \qquad j = 1, \ldots, m_2.$$

We let D_1 and D_1^0 denote the differential operators corresponding to the systems Q_1 and Q_1^0. In what follows, for brevity we shall let $Q_{k,j}$ rather than $Q_{k,j}^{(2)}$ denote the coefficients in the system Q_2. Thus, there exist rational functions $R_k = R_k(x_1, \dots, x_{m_1})$, $k = 1, \dots, m_2$, such that

$$f_{k,2} = R_k(f_{1,1}, \dots, f_{m_1,1}), \qquad k = 1, \dots, m_2.$$

The action of the operator D_1 can be extended from the ring $\mathcal{L}[x_1, \dots, x_{m_1}]$ to the field of rational functions $\mathcal{L}(x_1, \dots, x_{m_1})$ by means of the rule

$$D_1\left(\frac{A}{B}\right) = \frac{BD_1A - AD_1B}{B^2}. \tag{61}$$

Then for $R \in \mathcal{L}(x_1, \dots, x_{m_1})$ we have

$$\frac{d}{dz} R(f_{1,1}, \dots, f_{m_1,1}) = D_1 R(f_{1,1}, \dots, f_{m_1,1}).$$

In particular,

$$\begin{aligned} D_1 R_k(f_{1,1}, \dots, f_{m_1,1}) &= \frac{d}{dz} R_k(f_{1,1}, \dots, f_{m_1,1}) = f'_{k,2} = \\ &= \sum_{i=1}^{m_2} Q_{k,i} f_{i,2} + Q_{k,0} = \\ &= \sum_{i=1}^{m_2} Q_{k,i} R_i(f_{1,1}, \dots, f_{m_1,1}) + Q_{k,0}. \end{aligned}$$

Since the functions $f_{1,1}, \dots, f_{m_1,1}$ are algebraically independent over $\mathcal{L}$, by Theorem 2, it follows that

$$D_1 R_k = Q_{k,0} + \sum_{i=1}^{m_2} Q_{k,i} R_i, \qquad k = 1, \dots, m_2. \tag{62}$$

If $R = A/B$ is a rational function, then we define the degree of R to be the difference of the degrees of A and B.

Let N be the maximum of the degrees of the functions R_k, $k = 1, \dots, m_2$. Set $R_k^0 = 0$ if R_k has degree less than N; if R_k has degree N, let R_k^0 be the ratio of the homogeneous parts of maximum degree in the numerator and denominator of R_k.

Suppose that $N \neq 0$. Then if we compare the terms of degree N on both sides of (62), we obtain:

$$D_1^0 R_k^0 = \sum_{i=1}^{m_2} Q_{k,i} R_i^0, \qquad k = 1, \dots, m_2. \tag{63}$$

Let $f_{1,1}^0, \dots, f_{m_1,1}^0$ be a nontrivial solution of the system Q_1^0. We set

$$f_{k,2}^0 = R_k^0 \left(f_{1,1}^0, \dots, f_{m_1,1}^0 \right). \tag{64}$$

At least one of the functions $f_{1,2}^0, \dots, f_{m_2,2}^0$ is nonzero, since the R_k^0 are not all zero and $f_{1,1}^0, \dots, f_{m_1,1}^0$ are algebraically independent over $\mathcal{L}$. From (63) it follows that $f_{1,2}^0, \dots, f_{m_2,2}^0$ is a solution of Q_2^0; but then (64) contradicts the hypothesis of the theorem. Consequently, $N = 0$, and from (62) it follows that

$$D_1^0 R_k^0 = Q_{k,0} + \sum_{i=1}^{m_2} Q_{k,i} R_i^0, \qquad k = 1, \dots, m_2, \tag{65}$$

where we have $\deg R_k^0 = 0$, $k = 1, \dots, m_2$.

Let $R_k^0 = A_k/B$, where B is a least common denominator for all of the rational functions R_k^0. We claim that $B \in \mathcal{L}$. This is obviously the case if $m_1 = 1$. Suppose that $m_1 > 1$. Then, if we substitute $R_k^0 = A_k/B$ in (65) and use (61), we obtain the relation

$$A_k D_1^0 B = B \left(D_1^0 A_k - \sum_{i=1}^{m_2} Q_{k,i} A_i - Q_{k,0} B \right).$$

Since B is a least common denominator, we conclude that B divides $D_1^0 B$, and so the homogeneous principal ideal (B) is closed under the operator D_1^0. If we suppose that $\deg B > 0$, then, since $m_1 > 1$, the ideal (B) must have a nontrivial zero. Using Lemma 3, we arrive at a contradiction with the hypothesis of the theorem. Thus, $\deg B = 0$, i.e., $B \in \mathcal{L}$. Moreover, since each R_k^0 has degree 0, it follows that all of the R_k^0 are also in $\mathcal{L}$. This implies that $\frac{dR_k^0}{dz} = D_1^0 R_k^0$, and then it follows from (65) that R_k^0, $k = 1, \dots, m_2$, is a solution of Q_2.

We denote $\varphi_k = f_{k,2} - R_k^0$. Since $f_{k,2} \in \mathcal{L}_1$ and $R_k^0 \in \mathcal{L} \subset \mathcal{L}_1$, it follows that $\varphi_k \in \mathcal{L}_1$. The m_2-tuple $(\varphi_1, \dots, \varphi_{m_2})$ is a solution of Q_2^0, because $(f_{1,2}, \dots, f_{m_2,2})$ and $(R_1^0, \dots, R_{m_2}^0)$ are solutions of Q_2. But by what was proved above, the nontrivial solutions of the systems Q_k^0, $k = 2, \dots, r$, are algebraically independent over $\mathcal{L}_1$. Hence, $\varphi_k = 0$, $k = 1, \ldots, m_2$, and then $f_{k,2} = R_k^0 \in \mathcal{L}$, $k = 1, \ldots, m_2$.

This result contradicts the assumptions of the theorem and shows that for each i, $i = 2, \ldots, r$, not all of the functions $f_{1,i}, \ldots, f_{m_i,i}$ can belong to $\mathcal{L}_1$. Using the induction assumption, we conclude that the $m_2 + \cdots + m_r$ functions $f_{1,i}, \ldots, f_{m_i,i}$, $i = 2, \ldots, r$, are algebraically independent over $\mathcal{L}$. This proves the theorem, since the functions $f_{1,1}, \ldots, f_{m_1,1}$ are algebraically independent over $\mathcal{L}$ by Theorem 2.

A version of Theorem 3 can be given for solutions of several linear differential equations.

Theorem 4. *Suppose that the differential equations*

$$y^{(m_k)} + A_{m_k-1,k}\, y^{(m_k-1)} + \cdots + A_{0,k}\, y = B_k, \tag{66}$$

$$B_k, A_{i,k} \in \mathcal{L}, \qquad m_k \geq 1, \qquad k = 1, \ldots, r, \qquad r \geq 1,$$

have the property that, if $\varphi_1, \ldots, \varphi_r$ are any nontrivial solutions of the homogeneous differential equations corresponding to (66), then the functions

$$\varphi_k, \varphi_k', \ldots, \varphi_k^{(m_k-1)}, \qquad k = 1, \ldots, r,$$

are algebraically independent over $\mathcal{L}$. Let $f_1, \ldots, f_r$ be solutions of (66) such that $f_k \notin \mathcal{L}$, $k = 1, \ldots, r$. Then the functions

$$f_k, f_k', \ldots, f_k^{(m_k-1)}, \qquad k = 1, \ldots, r,$$

are algebraically independent over $\mathcal{L}$.

We apply these results to some specific functions. Consider the hypergeometric function

$$K_{\lambda,\mu}(z) = \sum_{n=0}^{\infty} \frac{(-1)^n}{[\lambda+1, n][\mu+1, n]} \left(\frac{z}{2}\right)^{2n}, \qquad \lambda, \mu \neq -1, -2, \ldots,$$

which satisfies the non-homogeneous linear differential equation

$$y'' + \frac{2\lambda + 2\mu + 1}{z} y' + \left(1 + \frac{4\lambda\mu}{z^2}\right) y = \frac{4\lambda\mu}{z^2}.$$

Theorem 5. *Suppose that $\lambda_i, \mu_i \in \mathbf{Q}$, $i = 1, \ldots, m$, $m \geq 1$, satisfy the conditions*

$$\lambda_i, \mu_i \neq -k_1, \quad k_1 \in \mathbf{N}, \quad \lambda_i - \mu_i \neq \frac{2k_2 - 1}{2}, \quad k_2 \in \mathbf{Z}^+, \quad i = 1, \ldots, m;$$

$$(\lambda_{i_1} - \mu_{i_1}) \pm (\lambda_{i_2} - \mu_{i_2}) \neq k_3, \quad k_3 \in \mathbf{Z}, \quad i_1, i_2 = 1, \dots, m, \quad i_1 \neq i_2;$$

and let $\xi_1, \dots, \xi_n$, $n \geq 1$, *be algebraic numbers such that* $\xi_1^2, \dots, \xi_n^2$ *are distinct from one another and nonzero. Then the* $2mn$ *numbers*

$$K_{\lambda_i,\mu_i}(\xi_j), \quad K'_{\lambda_i,\mu_i}(\xi_j), \qquad i = 1, \dots, m, \quad j = 1, \dots, n, \tag{67}$$

are algebraically independent.

To prove Theorem 5 we need the following lemma.

Lemma 8. *Suppose that* λ_i, μ_i, $i = 1, \dots, m$, *satisfy the conditions in Theorem 5, and* $\xi_1, \dots, \xi_n$ *are complex numbers such that* $\xi_1^2, \dots, \xi_n^2$ *are distinct from one another and nonzero. Then the* $2mn$ *functions*

$$K_{\lambda_i,\mu_i}(\xi_j z), \quad K'_{\lambda_i,\mu_i}(\xi_j z), \qquad i = 1, \dots, m, \quad j = 1, \dots, n, \tag{68}$$

are algebraically independent over $\mathbf{C}(z)$.

Proof. Let $\mathcal{L}$ be the field of algebraic functions of z. It is obviously sufficient to show that the functions (68) are algebraically independent over $\mathcal{L}$.

We consider the set of functions

$$v_{i,j}(z) = z^{\lambda_i+\mu_i} K_{\lambda_i,\mu_i}(\xi_j z), \qquad i = 1, \dots, m, \quad j = 1, \dots, n. \tag{69}$$

These are transcendental functions, and they satisfy the differential equations

$$v''_{i,j} + \frac{1}{z} v'_{i,j} + \left(\xi_j^2 - \frac{(\lambda_i - \mu_i)^2}{z^2} \right) v_{i,j} = 4\lambda_i \mu_i z^{\lambda_i+\mu_i+2}, \tag{70}$$

$$i = 1, \dots, m, \qquad j = 1, \dots, n.$$

The coefficients in the differential equations (70) belong to $\mathcal{L}$, and the corresponding homogeneous differential equations are of the same type as the differential equations (43) obtained from the Bessel equation. These homogeneous differential equations satisfy the conditions in Lemma 6. It then follows from Lemma 6 that all of the conditions in Theorem 4 hold for the differential equations (70), and so the functions (69) along with the functions $v'_{i,j}$ are algebraically independent over $\mathcal{L}$. Since $z^{\lambda_i+\mu_i} \in \mathcal{L}$, it follows that the functions (68) are also algebraically independent over $\mathcal{L}$. The lemma is proved.

Proof of Theorem 5. The $(2mn)$-tuple of E-functions in (68) is a solution of the system of linear differential equations

$$v'_{i,j,1} = v_{i,j,2},$$
$$v'_{i,j,2} = \frac{4\lambda_i\mu_i}{z^2} - \frac{2\lambda_i + 2\mu_i + 1}{z} v_{i,j,2} - \left(\xi_j + \frac{4\lambda_i\mu_i}{z^2}\right) v_{i,j,1},$$
$$i = 1, \ldots, m, \qquad j = 1, \ldots, n,$$

and, by Lemma 8, this $(2mn)$-tuple is algebraically independent over $\mathbf{C}(z)$. Hence, Theorem 5 follows from the Second Fundamental Theorem in Chapter 3.

We note that Theorem 5 was first proved in 1969 by A. A. Shmelev [71:1]. His proof that the functions involved are algebraically independent was rather complicated. It used the method of [70:9].

In 1971 in [3:4], Belogrivov studied the Kummer functions

$$A_{\lambda,\nu}(z) = \sum_{n=0}^{\infty} \frac{[\nu, n]}{n![\lambda, n]} z^n, \qquad \lambda \neq 0, -1, -2, \ldots,$$

(see § 4 of Chapter 6). He proved the following analogue of Siegel's Theorem.

Theorem 6. *Suppose that $\lambda_i, \nu_i \in \mathbf{Q}$, $i = 1, \ldots, m$, $m \geq 1$, satisfy the conditions*

$$-\lambda_i \notin \mathbf{Z}^+, \quad \nu_i \notin \mathbf{Z}, \quad (\nu_i - \lambda_i) \notin \mathbf{Z}, \qquad i = 1, \ldots, m;$$

$$(2\nu_{i_1} - \lambda_{i_1}) - (2\nu_{i_2} - \lambda_{i_2}) \pm (\lambda_{i_1} \pm \lambda_{i_2}) \neq 2k,$$

$$k \in \mathbf{Z}, \qquad i_1, i_2 = 1, \ldots, m, \qquad i_1 \neq i_2;$$

and let $\xi_1, \ldots, \xi_n$, $n \geq 1$, be algebraic numbers which are linearly independent over $\mathbf{Q}$. Then the $2mn$ numbers

$$A_{\lambda_i,\nu_i}(\xi_j), \quad A'_{\lambda_i,\nu_i}(\xi_j), \qquad i = 1, \ldots, m, \quad j = 1, \ldots, n,$$

are algebraically independent.

Theorem 6 can be proved in the same way as Siegel's Theorem, using the auxiliary results in § 2–6.

We consider the hypergeometric E-function

$$A_{\lambda,\mu,\nu}(z) = \sum_{n=0}^{\infty} \frac{[\nu, n]}{[\lambda, n][\mu, n]} z^n, \qquad \lambda, \mu, \nu \neq 0, -1, -2, \ldots,$$

which satisfies the differential equation

$$y'' + \left(\frac{\lambda+\mu-1}{z} - 1\right) y' + \frac{(\lambda-1)(\mu-1) - \nu z}{z^2} y = \frac{(\lambda-1)(\mu-1)}{z^2}.$$

If we use the algebraic independence over $\mathbf{C}(z)$ of functions which form a solution of the set of differential equations considered in Theorem 6, then we can use Theorem 4 to prove a theorem similar to Theorem 5.

Theorem 7. *Suppose that $\lambda_i, \mu_i, \nu_i \in \mathbf{Q}$, $i = 1, \ldots, m$, $m \geq 1$, satisfy the conditions:*

$$\lambda_i, \mu_i, \nu_i \neq -k_1, \quad k_1 \in \mathbf{Z}^+, \quad (\nu_i - \lambda_i) \notin \mathbf{Z}, \quad (\nu_i - \mu_i) \notin \mathbf{Z}, \quad i = 1, \ldots, m;$$

$$(2\nu_{i_1} - \mu_{i_1} - \lambda_{i_1}) - (2\nu_{i_2} - \mu_{i_2} - \lambda_{i_2}) \pm ((\lambda_{i_1} \pm \lambda_{i_2}) - (\mu_{i_1} \pm \mu_{i_2})) \neq 2k_2,$$

$$k_2 \in \mathbf{Z}, \qquad i_1, i_2 = 1, \ldots, m, \qquad i_1 \neq i_2;$$

and let $\xi_1, \ldots, \xi_n$, $n \geq 1$, be algebraic numbers which are linearly independent over $\mathbf{Q}$. Then the $2mn$ numbers

$$A_{\lambda_i,\mu_i,\nu_i}(\xi_j), \quad A'_{\lambda_i,\mu_i,\nu_i}(\xi_j), \qquad i = 1, \ldots, m, \quad j = 1, \ldots, n,$$

are algebraically independent.

§ 7. Generalizations of Siegel's Theorem

If $\lambda \in \mathbf{Q} \setminus \mathbf{Z}$, $\lambda \neq \frac{2k-1}{2}$, $k \in \mathbf{Z}$, then Siegel's Theorem cannot be used to conclude that the four numbers $K_\lambda(\xi)$, $K'_\lambda(\xi)$, $K_{-\lambda}(\xi)$, and $K'_{-\lambda}(\xi)$, where $\xi \in \mathbf{A} \setminus \{0\}$, are algebraically independent. In this case the corresponding functions turn out to be algebraically dependent over $\mathbf{C}(z)$.

In fact, we consider the Bessel functions $J_\lambda(z)$ and $J_{-\lambda}(z)$, which are linearly independent solutions of the Bessel differential equation (42). We have the relations

$$\begin{aligned} J_\lambda(z) &= \frac{1}{\Gamma(1+\lambda)} \left(\frac{z}{2}\right)^{\lambda} K_\lambda(z), \\ J_{-\lambda}(z) &= \frac{1}{\Gamma(1-\lambda)} \left(\frac{z}{2}\right)^{-\lambda} K_{-\lambda}(z). \end{aligned} \tag{71}$$

According to Liouville's formula for the Bessel function,

$$J_\lambda(z)J'_{-\lambda}(z) - J'_\lambda(z)J_{-\lambda}(z) = \frac{c}{z}, \qquad c \in \mathbf{C}. \tag{72}$$

In (72), if we replace $J_\lambda(z)$ and $J_{-\lambda}(z)$ and their derivatives by the right sides of equations (71) and their derivatives, and if we compute the value of c, we obtain the equation

$$K_\lambda(z)K'_{-\lambda}(z) - K'_\lambda(z)K_{-\lambda}(z) - \frac{2\lambda}{z}K_\lambda(z)K_{-\lambda}(z) + \frac{2\lambda}{z} = 0. \tag{73}$$

We now prove an algebraic independence result for the values of these functions.

Theorem 8. *If* $\lambda \in \mathbf{Q}\setminus\mathbf{Z}$, $\lambda \neq \frac{2k-1}{2}$, $k \in \mathbf{Z}$, *and* $\xi \in \mathbf{A}\setminus\{0\}$, *then any three of the four numbers* $K_\lambda(\xi)$, $K'_\lambda(\xi)$, $K_{-\lambda}(\xi)$ *and* $K'_{-\lambda}(\xi)$ *are algebraically independent.*

Proof. From Lemma 3 of Chapter 6 and Lemma 8 of Chapter 5 it follows that, if λ is as in the theorem, the Bessel differential equation satisfies the conditions of Lemma 4 with $\mathcal{L} = \mathbf{C}(z)$. Lemma 4 then says that

$$\deg \operatorname{tr}_{\mathbf{C}(z)}\{\varphi_1, \varphi_1', \varphi_2, \varphi_2'\} \geq 3$$

for any two linearly independent solutions φ_1, φ_2 of (42). In view of the relation (73), we then see that any three of the functions $K_\lambda(z)$, $K'_\lambda(z)$, $K_{-\lambda}(z)$ and $K'_{-\lambda}(z)$ are algebraically independent over $\mathbf{C}(z)$. These four functions make up a solution of the system of differential equations

$$y_1' = y_2, \qquad y_2' = -\frac{2\lambda+1}{z}y_2 - y_1,$$
$$y_3' = y_4, \qquad y_4' = \frac{2\lambda-1}{z}y_4 - y_3.$$

Applying the Third Fundamental Theorem of Chapter 4, we find that

$$\deg \operatorname{tr}\{K_\lambda(\xi), K'_\lambda(\xi), K_{-\lambda}(\xi), K'_{-\lambda}(\xi)\} = 3.$$

Because of the relation (73) with $z = \xi$, which shows that each of the four numbers is algebraically dependent on the other three, the theorem follows from the transcendence degree result.

We now show that the restrictions on the rational parameters $\lambda_1, \ldots, \lambda_n$ in Siegel's Theorem are essential. In what follows let $\xi \in \mathbf{A}\setminus\{0\}$.

If $\lambda = (2k-1)/2$, $k \in \mathbf{Z}$, then in § 2 of Chapter 6 it was shown that $K_\lambda(z)$ and $K'_\lambda(z)$ are algebraically dependent over $\mathbf{C}(z)$, and $\deg \operatorname{tr}\{K_\lambda(\xi), K'_\lambda(\xi)\} = 1$.

Note that the recursive equations (6.42) and (6.43), in which $\lambda \neq -k_1$, $k_1 \in \mathbf{Z}^+$, enable us to conclude that for any $k \in \mathbf{N}$ and any λ the functions $K_{\lambda\pm k}(z)$ and $K'_{\lambda\pm k}(z)$ can be expressed as a linear combination of $K_\lambda(z)$ and $K'_\lambda(z)$ with coefficients in $\mathbf{Q}(z)$. Hence, the numbers $K_{\lambda\pm k}(\xi)$ and $K'_{\lambda\pm k}(\xi)$ can be expressed as a linear combination of $K_\lambda(\xi)$ and $K'_\lambda(\xi)$ with coefficients in $\mathbf{A}$.

From this observation it follows that the case when $(\lambda_{i_1} - \lambda_{i_2}) \in \mathbf{Z}$ must be excluded in Siegel's Theorem, since in that case there are at most two algebraically independent numbers among

$$K_{\lambda_{i_1}}(\xi), \quad K'_{\lambda_{i_1}}(\xi), \quad K_{\lambda_{i_2}}(\xi), \quad K'_{\lambda_{i_2}}(\xi) \tag{74}$$

(there are two if $\lambda_{i_1} \neq (2k-1)/2$, and there is one if $\lambda_{i_1} = (2k-1)/2$ for $k \in \mathbf{Z}$).

Similarly, in the case when $(\lambda_{i_1} - \lambda_{i_2}) \in \mathbf{Z}$, if we use (73) and Theorem 8, we see that there are at most three algebraically independent numbers among the four in (74) (there are three if $\lambda_{i_1} \neq (2k-1)/2$, and there is one if $\lambda_{i_1} = (2k-1)/2$ for $k \in \mathbf{Z}$).

In the case $\lambda = \mu_0$, where $\mu_0 = (2k-1)/2$, $k \in \mathbf{Z}$, it follows from the equalities $K_{1/2}(z) = \cos z$, $K_{-1/2}(z) = -\sin z$ and the recursive relations (6.42) and (6.43) that the numbers $K_{\mu_0}(\xi)$ and $K'_{\mu_0}(\xi)$ can be expressed as a linear combination of $e^{\xi i}$ and $e^{-\xi i}$ with coefficients in $\mathbf{A}$.

We shall prove the following generalization of Siegel's Theorem.

Theorem 9. *Suppose that λ_i and μ_i, $i = 0, 1, \ldots, m$, $m \geq 1$, satisfy the conditions*

$$\lambda_0 \in \mathbf{Z}^+; \quad \lambda_i \in \mathbf{Q} \setminus \mathbf{Z}, \quad \lambda_i \neq \frac{2k-1}{2}, \quad k \in \mathbf{Z}, \quad i = 1, \ldots, m;$$

$$(\lambda_{i_1} - \lambda_{i_2}) \notin \mathbf{Z}, \quad i_1, i_2 = 1, \ldots, m, \quad i_1 \neq i_2;$$

$$\mu_0 = \frac{2k_1 - 1}{2}, \quad k_1 \in \mathbf{Z}; \quad (\lambda_i + \mu_i) \in \mathbf{Z}, \quad i = 1, \ldots, m.$$

Further suppose that $\xi_1, \ldots, \xi_n \in \mathbf{A} \setminus \{0\}$, $n \geq 1$, $\xi_{i_1}^2 \neq \xi_{i_2}^2$, $1 \leq i_1 < i_2 \leq n$; and $\eta_1, \ldots, \eta_n$ are numbers in $\mathbf{A}$ which are linearly independent over $\mathbf{Q}$. Then the $3n(m+1)$ numbers $K_{\lambda_0}(\xi_j)$, $K'_{\lambda_0}(\xi_j)$, $K_{\lambda_i}(\xi_j)$, $K'_{\lambda_i}(\xi_j)$, $K_{\mu_0}(\eta_j)$, $K_{\mu_j}(\eta_j)$, $i = 1, \ldots, m$, $j = 1, \ldots, n$, are algebraically independent.

From the argument given above we see that it suffices to prove Theorem 9 in the case when $\mu_i = -\lambda_i$, $i = 1, \ldots, m$, and the numbers $K_{\mu_0}(\eta_j)$ are replaced by $e^{i\eta_j}$. We first prove two lemmas that we shall need for the proof of the theorem. Let the field $\mathcal{L}$ be defined as in § § 2,3,4,6.

Lemma 9. *Suppose that the differential equations*

$$y'' + a_i y' + b_i y = 0, \qquad a_i, b_i \in \mathcal{L}, \qquad i = 1, \dots, n, \ \ n \geq 1, \tag{75}$$

satisfy the conditions:

1°. *There exist functions $v_i \in \mathcal{L}$, $v_i \neq 0$, $i = 1, \dots, n$, such that $v_i' + a_i v_i = 0$, $i = 1, \dots, n$.*

2°. *For any nontrivial solutions $\varphi_1, \dots, \varphi_n$ of the corresponding differential equations in (75), the functions φ_i, φ_i', $i = 1, \dots, n$, are algebraically independent over $\mathcal{L}$.*

Further suppose that $\psi_1, \dots, \psi_n$ are solutions of the corresponding differential equations in (75) such that ψ_i is linearly independent of φ_i over $\mathbf{C}$, $i = 1, \dots, n$. Then the functions $\varphi_i, \varphi_i', \psi_i$, $i = 1, \dots, n$, are algebraically independent over $\mathcal{L}$.

Proof. We consider the field $\mathcal{L}_1 = \mathcal{L}(\varphi_1, \varphi_1', \dots, \varphi_n, \varphi_n')$. To prove the lemma it suffices to show that the functions $\psi_1, \dots, \psi_n$ are algebraically independent over $\mathcal{L}_1$. We prove this claim by induction on the number of functions $\psi_1, \dots, \psi_k$, $1 \leq k \leq n$.

If $k = 1$, we apply Lemma 4 with $\mathcal{L}$ replaced by $\mathcal{L}_2 = \mathcal{L}(\varphi_2, \varphi_2', \dots, \varphi_n, \varphi_n')$ to the first equation in (75). By assumption, the functions φ_1 and φ_1' are algebraically independent over $\mathcal{L}_2$. Hence, Lemma 4 implies that

$$\deg \operatorname{tr}_{\mathcal{L}_2}\{\varphi_1, \varphi_1', \psi_1, \psi_1'\} = 3.$$

This shows that ψ_1 is a transcendental function over the field $\mathcal{L}_1$.

Suppose that our claim has been proved for the functions $\psi_1, \dots, \psi_k$, $1 \leq k < n$. We show that it then holds for $\psi_1, \dots, \psi_{k+1}$ as well. We apply Lemma 4 to the $(k+1)$-st equation in (75) with $\mathcal{L}$ replaced by the field

$$\mathcal{L}_{k+1} = \mathcal{L}(\varphi_1, \varphi_1', \dots, \varphi_k, \varphi_k', \varphi_{k+2}, \varphi_{k+2}', \dots, \varphi_n, \varphi_n', \psi_1, \dots, \psi_k),$$

and we use the induction assumption. Just as in the case $k = 1$, we find that the functions $\psi_1, \dots, \psi_{k+1}$ are algebraically independent over $\mathcal{L}$. The lemma is proved.

Lemma 10. *Suppose that the numbers $\lambda_1, \dots, \lambda_m$ and $\xi_1, \dots, \xi_n$ satisfy the conditions in Lemma 6; the numbers $\eta_1, \dots, \eta_n \in \mathbf{C}$ are linearly independent over $\mathbf{Q}$; $v_{i,j,1}$ and $v_{i,j,2}$ are arbitrary linearly independent solutions of the corresponding differential equation in (43); and $y_{i,j,1}$ and $y_{i,j,2}$ are arbitrary linearly independent solutions of the corresponding differential equation in (2). Then:*

1°. *The functions $v_{i,j,1}$, $v_{i,j,1}'$ and $v_{i,j,2}$, $i = 1, \dots, m$, $j = 1, \dots, n$, are algebraically independent over $\mathbf{C}(z)$.*

2°. *The functions $v_{i,j,1}$, $v_{i,j,1}'$, $v_{i,j,2}$, $e^{\eta_j z}$, $i = 1, \dots, m$, $j = 1, \dots, n$, are algebraically independent over $\mathbf{C}(z)$.*

3°. *The functions* $y_{i,j,1}$, $y'_{i,j,1}$, $y_{i,j,2}$, $e^{\eta_j z}$, $i = 1, \ldots, m$, $j = 1, \ldots, n$, *are algebraically independent over* $\mathbf{C}(z)$.

Proof. 1°. By Lemma 6, the functions $v_{i,j,1}$, $v'_{i,j,1}$, $i = 1, \ldots, m$, $j = 1, \ldots, n$, are algebraically independent over $\mathbf{C}(z)$. Hence, the conditions in Lemma 9 hold for the differential equations (43), and so the functions $v_{i,j,1}$, $v'_{i,j,1}$, $v_{i,j,2}$ are algebraically independent over $\mathbf{C}(z)$.

2°. Let $\gamma_1, \ldots, \gamma_k$ be distinct nonzero complex numbers, $k \geq 1$. We shall prove that the functions $1, e^{\gamma_1 z}, \ldots, e^{\gamma_k z}$ are linearly independent over the field

$$\mathcal{L}_0 = \mathbf{C}(z, v_{1,1,1}, v'_{1,1,1}, v_{1,1,2}, \ldots, v_{m,n,1}, v'_{m,n,1}, v_{m,n,2}).$$

This will prove part 2° of the lemma.

Suppose that this is not the case. Then

$$P_0 + P_1 e^{\gamma_1 z} + \cdots + P_k e^{\gamma_k z} = 0, \qquad P_i \not\equiv 0, \qquad k \geq 1, \tag{76}$$

$$P_i \in \mathbf{C}[z, v_{1,1,1}, v'_{1,1,1}, v_{1,1,2}, \ldots, v_{m,n,1}, v'_{m,n,1}, v_{m,n,2}],$$

where $P_0, P_1, \ldots, P_k$ have no polynomial factor in common, and k is the smallest integer for which such an equation holds. Differentiating (76), we obtain

$$\frac{dP_0}{dz} + \left(\frac{dP_1}{dz} + \gamma_1 P_1\right) e^{\gamma_1 z} + \cdots + \left(\frac{dP_k}{dz} + \gamma_k P_k\right) e^{\gamma_k z} = 0. \tag{77}$$

The differential equations (43), along with Liouville's formulas for these functions

$$v_{i,j,1} v'_{i,j,2} - v'_{i,j,1} v_{i,j,2} = \frac{c_{i,j}}{z}, \qquad c_{i,j} \in \mathbf{C},$$

imply that the field $\mathcal{L}_0$ is closed under differentiation. Then, because of the minimality of k, we find that the coefficients of $e^{\gamma_1 z}, \ldots, e^{\gamma_k z}$ in equations (76) and (77) must be proportional, i.e.

$$\frac{d \ln P_0}{dz} = \frac{d \ln P_1}{dz} + \gamma_1 = \cdots = \frac{d \ln P_k}{dz} + \gamma_k.$$

Integrating this, we obtain

$$P_0 = c P_1 e^{\gamma_1 z}, \qquad c \in \mathbf{C}. \tag{78}$$

This means that $k = 1$ and $c = -1$.

We consider both sides of (78) as functions of z, and we replace z by lz, where $l \in \mathbf{N}$ and

$$\min_{1 \le i \le m} |\xi_i| l > \max_{1 \le i \le m} |\xi_i|.$$

Then the numbers $\xi_1^2, \ldots, \xi_n^2, (l\xi_1)^2, \ldots, (l\xi_n)^2$ are all distinct and nonzero. Let $P_{0,l}$ and $P_{1,l}$ be the polynomials P_0 and P_1 after our change of variables. If we eliminate the function $e^{\gamma_1 z}$ from the two equations $P_{0,l} = -P_{1,l}e^{l\gamma_1 z}$ and $P_0 = -P_1 e^{\gamma_1 z}$, we obtain the relation $P_{0,l}P_1^l = (-1)^{l-1} P_{1,l} P_0^l$, which contradicts what was proved in part 1° if n is replaced by $2n$ there. This contradiction proves part 2°.

3°. Since $v_{i,j,1} = z^{\lambda_i} y_{i,j,1}$ and $v_{i,j,2} = z^{\lambda_i} y_{i,j,2}$, we can use Lemma 8 of Chapter 5 to conclude that part 3° of Lemma 10 is a consequence of part 2°.

Proof of Theorem 9. Using Lemma 9, the Third Fundamental Theorem of Chapter 4, and equations of the type (73) with $\lambda = \lambda_1, \ldots, \lambda_m$, and arguing as in the proof of Theorem 8, we prove that the numbers $K_{\lambda_0}(\xi_j)$, $K'_{\lambda_0}(\xi_j)$, $K_{\lambda_i}(\xi_j)$, $K'_{\lambda_i}(\xi_j)$, $K_{-\lambda_i}(\xi_j)$, e^{η_j} are algebraically independent. As noted before, this implies the theorem.

Theorems 8 and 9 were proved in 1962 and 1969 in [70:14,24].

In 1971, Belogrivov [3:4] proved the following theorem, which generalizes Siegel's Theorem to the case of a set of values of the functions $K_\lambda(z)$, $A_{\mu,\nu}(z)$ and e^z with different values of the parameters and arguments.

Theorem 10. *Suppose that* λ_i, μ_i, ν_i, $i = 1, \ldots, m$, $m \ge 1$, *are rational numbers which satisfy the conditions:*

$$\lambda_i \ne \frac{2k-1}{2}, \qquad k \in \mathbf{Z}, \qquad -\lambda_i \notin \mathbf{N}, \qquad i = 1, \ldots, m;$$

$$(\lambda_{i_1} \pm \lambda_{i_2}) \notin \mathbf{Z}, \qquad i_1, i_2 = 1, \ldots, m, \qquad i_1 \ne i_2;$$

$$-\mu_i \notin \mathbf{Z}^+, \qquad \nu_i \notin \mathbf{Z}, \qquad (\nu_i - \mu_i) \notin \mathbf{Z}, \qquad i = 1, \ldots, m;$$

$$(2\nu_{i_1} - \mu_{i_1}) - (2\nu_{i_2} - \mu_{i_2}) \pm (\mu_{i_1} \pm \mu_{i_2}) \ne 2k_1, \qquad k_1 \in \mathbf{Z},$$

$$i_1, i_2 = 1, \ldots, m, \qquad i_1 \ne i_2;$$

$\xi_1, \ldots, \xi_n$, $n \ge 1$, *are algebraic numbers whose squares are distinct and nonzero; and* $\eta_1, \ldots, \eta_n$ *and* $\gamma_1, \ldots, \gamma_n$ *are algebraic numbers which are linearly independent over* $\mathbf{Q}$ *and satisfy the condition* $\eta_i^2 \ne -4\xi_i^2$, $i = 1, \ldots, n$. *Then the* $(4m+1)n$

numbers

$$K_{\lambda_i}(\xi_j),\ K'_{\lambda_i}(\xi_j),\ A_{\mu_i,\nu_i}(\eta_j),\ A'_{\mu_i,\nu_i}(\eta_j),\ e^{\gamma_j},$$

$$i = 1,\dots,m, \qquad j = 1,\dots,n,$$

are algebraically independent.

In 1981, V. A. Gorelov [25:1] proved an algebraic independence theorem for the values of the functions $A_{\lambda,\mu,\nu}(z)$ and for the values of the functions $K_{\lambda,\mu}(z)$ and $A_{\lambda,\mu,\nu}(z)$ and their derivatives with different rational values of the parameters and different algebraic values of the arguments, subject to natural restrictions.

Chapter 10
Solutions of linear differential equations of prime order p

§ 1. Statement of the basic results

In Chapters 5, 6 and 9 we considered the problem of algebraic independence over $\mathbf{C}(z)$ of a set of analytic functions which satisfy first and second order linear differential equations with coefficients in $\mathbf{C}[z]$. We could then use the general theorems in Chapters 3 and 4 on transcendence and algebraic independence of the values of E-functions in order to prove transcendence and algebraic independence of the values at algebraic points of many specific sets of E-functions which satisfy first and second order linear differential equations. In Chapters 7 and 8 we considered the analogous problem for functions which satisfy certain special classes of higher order differential equations. When one studies algebraic independence of the solutions of more general differential equations of order greater than two, some essential difficulties arise which have not yet been completely overcome.

In 1966–69, V. A. Oleinikov [56:3–6] studied algebraic independence of the solutions of third order linear differential equations. He proposed a general approach to such problems, which was a refinement and generalization of Siegel's argument [72:3,4] (see Chapters 5 and 9). We shall give a brief exposition of the basic idea in Oleinikov's method.

Let $\mathcal{L}$ be a field of analytic functions which contains $\mathbf{C}(z)$. We say that the algebraic differential equation

$$P = P(z, y, y', \ldots, y^{(m)}) = 0, \qquad P \in \mathbf{C}[z, z_0, z_1, \ldots, z_m], \tag{1}$$

is *reducible over* $\mathcal{L}$ if it has a solution $y(z) \not\equiv 0$ which satisfies an algebraic differential equation with coefficients in $\mathcal{L}$ of order less than m (which is a nonzero algebraic function over $\mathcal{L}$, if $m = 1$). If this is not the case, we say that the differential equation (1) is *irreducible over* $\mathcal{L}$.

The algebraic differential equation (1) is said to be *linearly (homogeneously linearly) reducible over* $\mathcal{L}$ if it has a solution $y(z) \not\equiv 0$ which satisfies a linear (a homogeneous linear) differential equation with coefficients in $\mathcal{L}$ of order less than m. If this is not the case, we say that the differential equation (1) is *linearly (homogeneously linearly) irreducible over* $\mathcal{L}$.

For each $k \in \mathbf{N}$, $1 \le k \le m$, we consider the operator G_k which is defined as follows on the set of analytic functions $f = f(z, z_1, \ldots, z_{k-1})$ of k complex variables which are defined in some region of $\mathbf{C}^k$:

$$G_k f = \frac{\partial f}{\partial z_{k-1}} f + \frac{\partial f}{\partial z_{k-2}} z_{k-1} + \cdots + \frac{\partial f}{\partial z_1} z_2 + \frac{\partial f}{\partial z}, \quad 2 \le k \le m; \quad G_1 f = f'(z).$$

One of the theorems which Oleinikov proves which is basic for his method is

Theorem 1. *The algebraic differential equation (1) is irreducible over $\mathcal{L}$ if and only if, for any $k \in \mathbf{N}$, $k \le m$, there does not exist a function $f = f(z, z_1, \ldots, z_{k-1})$ which is algebraic over $\mathcal{L}(z, z_1, \ldots, z_{k-1})$ and satisfies the equation*

$$P(z, z_1, \ldots, z_{k-1}, f, G_k f, \ldots, G_k^{m-k} f) = 0. \tag{2}$$

Thus, the problem of proving algebraic independence of every solution of (1) with its derivatives through the $(m-1)$-st order inclusive reduces to the study of certain partial differential equations and the proof that they have no algebraic solutions.

For $k = 1$, i.e., for functions of one variable, in this situation Siegel used the expansion of the algebraic function $f(z)$ in rational powers of z in a neighborhood of the singular point $z = 0$ or $z = \infty$. If one substitutes these expansions in the corresponding differential equations and compares coefficients, one obtains a series of relations between the powers and coefficients in the expansions, and these relations lead to the desired conclusions (see § 2 of Chapter 6).

In his investigation of the equations (2), Oleinikov uses the same device with algebraic functions of several variables. Such a function can be represented by a series

$$f(z, z_1, \ldots, z_{k-1}) = h_0(z_1, \ldots, z_{k-1}) z^{\epsilon_0} + h_1(z_1, \ldots, z_{k-1}) z^{\epsilon_1} + \cdots, \tag{3}$$

where $h_0, h_1, \ldots$ are algebraic functions of $z_1, \ldots, z_{k-1}$, and the fractional exponents $\epsilon_0, \epsilon_1, \ldots$ all have the same denominator. Substituting the expansions (3) in the equation (2) and comparing coefficients of powers of z, one arrives at partial differential equations similar to (2) for the functions $h_0(z_1, \ldots, z_{k-1})$, $h_1(z_1, \ldots, z_{k-1})$, ... These are functions of a smaller number of variables. The equations for those functions can then be studied in an analogous fashion.

The realization of this scheme in concrete situations is complicated. Oleinikov was able to do this only for certain third order differential equations; he applied the results to prove algebraic independence of the values at algebraic points of certain hypergeometric E-functions which satisfy third order linear differential equations.

In 1969, Nesterenko [52:1] used the theory of polynomial ideals to prove that his functions are algebraically independent. He proved that, if a (homogeneous or non-homogeneous) linear differential equation is reducible, then the corresponding homogeneous differential equation has a solution which satisfies an algebraic homogeneous differential equation of lower order (see Theorem 2 of Chapter 9 and Lemma 1 of § 2).

In 1977, V. Kh. Salikhov [66:2] used Nesterenko's result to show that, with certain restrictions on the coefficients of the linear equation (1), the functions $h_i(z_1, \dots, z_{k-1})$ in (3) are linear forms in $z_1, \dots, z_{k-1}$. He proved the following general theorem.

Theorem 2. *Let p be a prime number, and let $r_0, r_1, \dots, r_{p-1} \in \mathbf{C}(z)$, where the functions $r_1, \dots, r_{p-1}$ vanish at $z = \infty$ and r_0 takes a finite nonzero value at $z = \infty$. Then:*

1°. *The differential equation*

$$y^{(p)} + r_{p-1}y^{(p-1)} + \cdots + r_1 y' + r_0 y = 0 \tag{4}$$

is reducible over $\mathbf{C}(z)$ if and only if it is homogeneously linearly reducible over the field of algebraic functions of z.

2°. *If, in addition, the functions $r_0, r_1, \dots, r_{p-1}$ have no poles in the region $z \neq 0$, then the differential equation (4) is reducible over $\mathbf{C}(z)$ if and only if it is homogeneously linearly reducible over the field $\mathbf{C}(z)$.*

A key element of the proof of this theorem is that the p-th roots of unity $\zeta_1, \dots, \zeta_p$ are connected by a unique linear equation $\zeta_1 + \cdots + \zeta_p = 0$ with coefficients in $\mathbf{Z}$ (up to a constant multiple). This is why Theorem 2 cannot be extended to differential equations (4) whose order is a composite number.

Theorem 2 reduces the proof that our differential equation is irreducible to a proof of homogeneous linear irreducibility, which is significantly simpler.

We note that essentially it was when Oleinikov examined the case $p = 3$ in [56:5] that he discovered that the functions h_i in (3) are linear forms.

Salikhov applied Theorem 2 to the hypergeometric functions

$$\varphi_{\lambda_1, \dots, \lambda_p}(z) = \sum_{n=0}^{\infty} \frac{1}{[\lambda_1 + 1, n] \cdots [\lambda_p + 1, n]} \left(\frac{z}{p}\right)^{pn},$$

$$\lambda_i \neq -1, -2, \dots, \qquad i = 1, \dots, p.$$

By Lemma 1 of Chapter 5, these functions satisfy the following p-th order linear differential equation with coefficients in $\mathbf{C}(z)$:

$$\left(\prod_{i=1}^{p}(\delta + p\lambda_i) - z^p\right) y = \prod_{i=1}^{p}(p\lambda_i), \qquad \delta = z\frac{d}{dz}. \tag{5}$$

By proving a certain sufficient condition for the homogeneous equation corresponding to (5) to be homogeneously linearly irreducible, Salikhov obtained the following theorem.

Theorem 3. *Let p be a prime number, and let $\lambda_1, \dots, \lambda_p$ be complex numbers which are not negative integers. Suppose that the following condition holds for all $k \in \mathbf{N}$, $1 \le k \le p-1$, and for any k-tuple $\{i_1, \dots, i_k\} \subset \{1, \dots, p\}$:*

$$\sum_{j=1}^{k} \left(p\lambda_{i_j} - \sum_{s=1}^{p} \lambda_s \right) + \frac{k(k-p)}{2} \neq 0, 1, 2, \dots.$$

Then the homogeneous differential equation corresponding to (5) is irreducible over $\mathbf{C}(z)$, and the functions

$$\varphi_{\lambda_1,\dots,\lambda_p}(z),\ \varphi'_{\lambda_1,\dots,\lambda_p}(z),\ \dots,\ \varphi^{(p-1)}_{\lambda_1,\dots,\lambda_p}(z) \tag{6}$$

are algebraically independent over $\mathbf{C}(z)$.

This theorem, along with Theorem 3 of Chapter 3, implies an algebraic independence result for the values of the functions (6).

Theorem 4 (V. Kh. Salikhov). *If p is a prime number and $\lambda_1, \dots, \lambda_p$ are rational numbers satisfying the condition in Theorem 3, and if $\xi \in \mathbf{A} \setminus \{0\}$, then the p numbers*

$$\varphi_{\lambda_1,\dots,\lambda_p}(\xi),\ \varphi'_{\lambda_1,\dots,\lambda_p}(\xi),\ \dots,\ \varphi^{(p-1)}_{\lambda_1,\dots,\lambda_p}(\xi) \tag{7}$$

are algebraically independent.

The conditions in Theorem 3 are only sufficient conditions for the functions (6) to be algebraically independent over $\mathbf{C}(z)$. If one wants to refine the conditions in Theorems 3 and 4, one must investigate homogeneous linear irreducibility of the homogeneous linear differential equation corresponding to (5). One such investigation was carried out by Oleinikov in [56:8]. Using a result of that paper along with Theorem 2, he proved a theorem which gives another sufficient condition for the values of the functions (6) to be algebraically independent.

Oleinikov then made the following conjecture: *the functions (6) are algebraically dependent over $\mathbf{C}(z)$ if and only if $p\lambda_i \in \mathbf{Z}$, $i = 1, \dots, p$, and the set $\{p\lambda_1, \dots, p\lambda_p\}$ is a complete system of residues modulo p.* This conjecture is true for $p = 2$ (see Theorem 8 of Chapter 6).

In 1985, using different methods, V. V. Kazakov and V. Kh. Salikhov simultaneously and independently carried out further studies of linear irreducibility connected with the differential equation (5).

Kazakov proved a sufficient condition for homogeneous linear irreducibility of the homogeneous differential equation corresponding to (5) which enabled him to refine Theorem 4. The condition on the parameters $\lambda_1, \dots, \lambda_p$ in Theorem 4 were replaced by the following: *the numbers* $p(\lambda_i - \lambda_1)$, $i = 2, \dots, p$, *are not integers forming a reduced system of residues modulo p*. It can be shown that when (5) is a homogeneous differential equation, the latter condition is also necessary.

Salikhov proved a necessary and sufficient condition for homogeneous linear irreducibility of the homogeneous differential equation corresponding to (5), proved the conjecture of Oleinikov mentioned above concerning algebraic independence of the functions (6) over $\mathbf{C}(z)$, and thereby obtained a necessary and sufficient condition for the conclusion of Theorem 4 to hold. He also found a necessary and sufficient condition for homogeneous linear irreducibility of the homogeneous differential equation corresponding to (5.5) in the case of the general hypergeometric function (5.1).

The results of Kazakov and Salikhov are not yet in print. However, the author of this book obtained their permission to use the proofs. In this chapter we prove Salikhov's theorem. We shall use an argument of Kazakov to establish a sufficient condition for homogeneous linear irreducibility of the corresponding differential equation.

We set $\psi_1(z) = \varphi_{\lambda_1,\dots,\lambda_p}(z)$ and

$$\psi_i(z) = \sum_{n=0}^{\infty} \frac{(n+\lambda_1)\cdots(n+\lambda_{i-1})}{[\lambda_1+1,n]\cdots[\lambda_p+1,n]} \left(\frac{z}{p}\right)^{pn-i+1}, \qquad i = 2, \dots, p.$$

It is easy to verify that the functions $\psi_1(z), \dots, \psi_p(z)$ form a solution of the system of linear differential equations

$$\begin{aligned}
y_1' &= -\frac{\beta_1}{z} y_1 + y_2, \\
&\cdots\cdots\cdots \\
y_{p-1}' &= -\frac{\beta_{p-1}}{z} y_{p-1} + y_p, \\
y_p' &= -\frac{\beta_p}{z} y_p + y_1 + \lambda_1 \cdots \lambda_p \left(\frac{z}{p}\right)^{-p}, \\
\beta_i &= p\lambda_i + i - 1, \qquad i = 1, \dots, p.
\end{aligned}$$

Let $c, \lambda_1, \dots, \lambda_m, \mu_1, \dots, \mu_m \in \mathbf{C}$, $m \in \mathbf{N}$. We denote

$$\{\lambda\} = \{\lambda_1, \dots, \lambda_m\}, \qquad \{\lambda + c\} = \{\lambda_1 + c, \dots, \lambda_m + c\}.$$

If we have $(\lambda_j - \mu_{i_j}) \in \mathbf{Z}$, $j = 1, \dots, m$, for some permutation $i_1, \dots, i_m$ of the numbers $1, \dots, m$, then we introduce the notation

$$\{\lambda\} \equiv \{\mu\} \ (mod\ \mathbf{Z}).$$

Theorem 5 (V. Kh. Salikhov). *Let p be a prime number, let $\lambda_1, \dots, \lambda_p$ be rational numbers which are not negative integers, and let $\xi \in \mathbf{A} \setminus \{0\}$. Then:*

1°. *The p numbers $\varphi_{\lambda_1,\dots,\lambda_p}(\xi), \varphi'_{\lambda_1,\dots,\lambda_p}(\xi), \dots, \varphi^{(p-1)}_{\lambda_1,\dots,\lambda_p}(\xi)$ are algebraically independent if and only if the following conditions holds:*

$$\{\lambda\} = \{\lambda_1, \dots, \lambda_p\} \not\equiv \left\{0, \frac{1}{p}, \dots, \frac{p-1}{p}\right\} \quad (mod\ \mathbf{Z}). \tag{8}$$

2°. *The p numbers $\psi_1(\xi), \dots, \psi_p(\xi)$ are algebraically independent if and only if the condition (8) holds.*

Condition (8) means that the numbers $p\lambda_1, \dots, p\lambda_p$ are not integers which form a complete system of residues modulo p.

To prove Theorem 5 we shall need the following theorem.

Theorem 6 (V. Kh. Salikhov). *Let p be a prime number, and let $\lambda_1, \dots, \lambda_p$ be complex numbers, $-\lambda_i \notin \mathbf{N}$, $i = 1, \dots, p$. Then:*

1°. *The p functions $\varphi_{\lambda_1,\dots,\lambda_p}(z), \varphi'_{\lambda_1,\dots,\lambda_p}(z), \dots, \varphi^{(p-1)}_{\lambda_1,\dots,\lambda_p}(z)$ are algebraically independent over $\mathbf{C}(z)$ if and only if condition (8) holds.*

2°. *The p functions $\psi_1(z), \dots, \psi_p(z)$ are algebraically independent over $\mathbf{C}(z)$ if and only if condition (8) holds.*

Again let $\mathcal{L}$ denote a field of analytic functions which contains $\mathbf{C}(z)$.

A system of first order linear differential equations with coefficients in $\mathbf{C}(z)$ is said to be *reducible* (*linearly reducible*) (*homogeneously linearly reducible*) over $\mathcal{L}$ if it has a solution whose components are algebraically dependent (are linearly dependent along with 1) (are linearly dependent) over $\mathcal{L}$ and are not all zero. If this is not the case, then the system is said to be *irreducible* (*linearly irreducible*) (*homogeneously linearly irreducible*) over $\mathcal{L}$.

We shall subsequently prove a theorem which generalizes Theorem 2 to solutions of a system of first order linear differential equations.

Theorem 7. *Suppose that*

$$y'_k = \sum_{i=1}^{m} Q_{k,i} y_i, \qquad k = 1, \dots, m, \qquad Q_{k,i} \in \mathbf{C}(z), \tag{9}$$

where each of the functions $Q_{k,i}$ takes a finite value $a_{k,i}$ at the point $z = \infty$, and the roots $\alpha_1, \dots, \alpha_m$ of the characteristic polynomial $A = \|a_{k,i}\|$, $k, i = 1, \dots, m$, have the property that the two-dimensional vectors $(1, \alpha_i)$, $i = 1, \dots, m$, are linearly independent over $\mathbf{Q}$. Then:

1°. *If the system (9) is reducible over $\mathbf{C}(z)$, then it is homogeneously linearly reducible over the field of algebraic functions over $\mathbf{C}(z)$.*

2°. *If, in addition, none of the functions* $Q_{k,i}$ *has a pole in the region* $z \neq 0$, *then the system (9) is homogeneously linearly reducible over* $\mathbf{C}(z)$.

This theorem was proved by Nesterenko in 1983, but his proof has not yet been published. The exposition given here was prepared by Nesterenko at the author's request. The proof of Theorem 7 uses some fundamental ideas from Salikhov's proof of Theorem 2. At the same time it is characterized by some technical improvements and simplifications.

Theorem 2 is a rather simple consequence of Theorem 7. Theorem 6 will be proved using Theorems 7 and 2. Theorem 5 is deduced from Theorem 6 and the Second Fundamental Theorem of Chapter 3.

§ 2. The homogeneous ideal $\mathcal{I}$

The proof of Theorem 7 is divided into several parts. The first part is based on Lemma 1 below. Before proving Lemma 1, we recall some concepts from the theory of polynomial ideals.

An *ideal* $\mathcal{I}$ *of the ring* $\mathbf{C}(z)[x_1, \dots, x_m]$ is called a *homogeneous ideal* if it has a basis consisting of homogeneous polynomials. If P is a polynomial in a homogeneous ideal $\mathcal{I}$, then every sum of all of the terms of P of fixed total degree is also in $\mathcal{I}$.

An ideal $\mathcal{I}$ of $\mathbf{C}(z)[x_1, \dots, x_m]$ is called a *prime ideal* if for any polynomials $P, Q \in \mathbf{C}(z)[x_1, \dots, x_m]$, the relations $PQ \in \mathcal{I}$ and $P \notin \mathcal{I}$ imply that $Q \in \mathcal{I}$. We say that the prime ideal $\mathcal{I}$ has *dimension* l if l is the largest integer for which there exist distinct indices $i_1, \dots, i_l$ with $\mathcal{I} \bigcap \mathbf{C}(z)[x_{i_1}, \dots, x_{i_l}] = (0)$. We let $\dim \mathcal{I}$ denote the dimension of $\mathcal{I}$. If two prime ideals have the same dimension, and one is contained in the other, then they are equal.

Lemma 1. *If the system of differential equations (9) has a solution whose components are algebraically dependent over* $\mathbf{C}(z)$, *then either it has a nontrivial solution all of whose components are algebraic over* $\mathbf{C}(z)$, *or else it has a nontrivial solution* $\psi_1(z), \dots, \psi_m(z)$ *such that the set of all polynomials in* $\mathbf{C}(z)[x_1, \dots, x_m]$ *which vanish when* $\psi_1(z), \dots, \psi_m(z)$ *are substituted for* $x_1, \dots, x_m$ *is a homogeneous prime ideal.*

Proof. Let $f_1(z), \dots, f_m(z)$ be a nonzero solution of (9) having the minimal number of functions algebraically independent over $\mathbf{C}(z)$. The set of all polynomials in $\mathbf{C}(z)[x_1, \dots, x_m]$ which vanish when $f_1(z), \dots, f_m(z)$ are substituted for $x_1, \dots, x_m$ is obviously a prime ideal, which we denote $\mathcal{I}$.

We let

$$D = \frac{\partial}{\partial z} + \sum_{k=1}^{m} \left(\sum_{i=1}^{m} Q_{k,i} x_i \right) \frac{\partial}{\partial x_k}$$

be the differential operator which is associated to the system of differential equations (9) and acts on the ring $\mathbf{C}(z)[x_1, \ldots, x_m]$. Then $D\mathcal{I} \subset \mathcal{I}$. We define the homogeneous ideal $\mathcal{I}^0$ just as in the proof of Theorem 2 of § 7, Chapter 9, where we also proved that $D\mathcal{I}^0 \subset \mathcal{I}^0$, and that, if the ideal $\mathcal{I}^0$ has only the trivial zero $(0, 0, \ldots, 0)$, then the functions $f_1(z), \ldots, f_m(z)$ are algebraic over $\mathbf{C}(z)$. Since the lemma holds in that case, in what follows we shall suppose that $\mathcal{I}^0$ has a nontrivial zero. By Lemma 3 of Chapter 9, the system (9) has a solution $\psi_1(z), \ldots, \psi_m(z)$ which is a zero of the ideal $\mathcal{I}^0$.

Let $\mathcal{J}$ be the prime ideal consisting of all polynomials in $\mathbf{C}(z)[x_1, \ldots, x_m]$ which vanish when $\psi_1(z), \ldots, \psi_m(z)$ are substituted for $x_1, \ldots, x_m$. We let $\mathcal{J}_1$ denote the maximal homogeneous ideal containing $\mathcal{I}^0$ and contained in $\mathcal{J}$. Then $\mathcal{I}^0 \subset \mathcal{J}_1 \subset \mathcal{J}$.

We prove that $\mathcal{J}_1$ is a prime ideal. Suppose the contrary, i.e., that there exist polynomials $P, Q \in \mathbf{C}(z)[x_1, \ldots, x_m]$ such that $PQ \in \mathcal{J}_1$, $P \notin \mathcal{J}_1$, and $Q \notin \mathcal{J}_1$. Let P_i and Q_j be the sum of the homogeneous terms of P of degree i and of Q of degree j, respectively, and let r and s be the smallest integers such that $P_r \notin \mathcal{J}_1$, $Q_s \notin \mathcal{J}_1$. Since $\mathcal{J}_1$ is a homogeneous ideal, we have

$$\sum_{i+j=r+s} P_i Q_j \in \mathcal{J}_1.$$

By the minimality of r and s, this relation implies that $P_r Q_s \in \mathcal{J}_1 \subset \mathcal{J}$. But at the same time, since $\mathcal{J}_1$ is the maximal homogeneous ideal contained in $\mathcal{J}$, it follows from $P_r \notin \mathcal{J}_1$ and $Q_s \notin \mathcal{J}_1$ that $P_r \notin \mathcal{J}$ and $Q_s \notin \mathcal{J}$. This is impossible, since $\mathcal{J}$ is a prime ideal. This contradiction proves that $\mathcal{J}_1$ is a prime ideal.

The assumption at the beginning of the proof concerning the functions $f_1(z)$, $\ldots, f_m(z)$ implies that

$$\dim \mathcal{I} \leq \dim \mathcal{J} \leq \dim \mathcal{J}_1.$$

On the other hand, if we have $\mathcal{J}_1 \bigcap \mathbf{C}(z)[x_{i_1}, \ldots, x_{i_l}] = (0)$ for some indices $i_1, \ldots, i_l$, then it follows that $\mathcal{I}^0 \bigcap \mathbf{C}(z)[x_{i_1}, \ldots$ $\ldots, x_{i_l}] = (0)$, and hence $\mathcal{I} \bigcap \mathbf{C}(z)[x_{i_1}, \ldots, x_{i_l}] = (0)$. This implies that $\dim \mathcal{J}_1 \leq \dim \mathcal{I}$. Thus,

$$\dim \mathcal{I} = \dim \mathcal{J} = \dim \mathcal{J}_1,$$

and, since $\mathcal{J}$ and $\mathcal{J}_1$ are prime ideals, from the last equality it follows that $\mathcal{J}_1 = \mathcal{J}$, i.e., $\mathcal{J}$ is a homogeneous ideal. The lemma is proved.

If the system (9) has a nontrivial solution all of whose components are algebraic over $\mathbf{C}(z)$, then part 1° of Theorem 7 obviously holds. If this is not the case, then, by Lemma 1, the system (9) has a solution such that the ideal of all algebraic equations over $\mathbf{C}(z)$ satisfied by its components is a homogeneous ideal. Without loss of generality we may assume that the m-tuple $f_1(z), \dots, f_m(z)$ is such a solution, and that $\mathcal{I}$ is the homogeneous prime ideal consisting of all polynomials in $\mathbf{C}(z)[x_1, \dots, x_m]$ which vanish when $f_1(z), \dots, f_m(z)$ are substituted for $x_1, \dots, x_m$.

We set $k = \dim \mathcal{I} \geq 1$, and we let

$$\mathcal{R} = \mathbf{C}(z)[x_1, \dots, x_m]/\mathcal{I}$$

be the quotient ring of $\mathbf{C}(z)[x_1, \dots, x_m]$ modulo the ideal $\mathcal{I}$. If $g_1, \dots, g_m$ are the images of $x_1, \dots, x_m$ in $\mathcal{R}$, then $\mathcal{R} = \mathbf{C}(z)[g_1, \dots, g_m]$, and there are k and no more than k of the elements $g_1, \dots, g_m$ which are algebraically independent over $\mathbf{C}(z)$.

§ 3. Algebraic functions of several variables

Let $\mathbf{U}$ denote the ring of rational functions of z which are regular at infinity.

In the second part of the proof we show that it suffices to prove Theorem 7 under the assumption that the homogeneous ideal $\mathcal{I}$ has the property that the elements $g_i \in \mathcal{R}$ are algebraic and integral over the ring $\mathbf{U}[g_1, \dots, g_k]$, i.e., each g_i satisfies an algebraic equation of the form

$$x^r + \varphi_{r-1} x^{r-1} + \cdots + \varphi_1 x + \varphi_0 = 0,$$

$$\varphi_j \in \mathbf{U}[g_1, \dots, g_k], \qquad j = 0, 1, \dots, r-1.$$

We let v denote the valuation at infinity of the field $\mathbf{C}(z)$, i.e., the function $v : \mathbf{C}(z) \to \mathbf{Z}$ such that $v(A/B) = \deg B - \deg A$ for $A, B \in \mathbf{C}[z]$. Then $v(u) \geq 0$ for all $u \in \mathbf{U}$.

Lemma 2. *Suppose that $\xi_1, \dots, \xi_d$ are elements of the ring $\mathcal{R}$, $\deg \operatorname{tr}_{\mathbf{C}(z)} \mathbf{C}(z, \xi_1, \dots, \xi_d) = k$, and the ideal of all algebraic equations over $\mathbf{C}(z)$ satisfied by $\xi_1, \dots, \xi_d$ is a homogeneous ideal. Then there exist linear forms $u_1, \dots, u_k$ in $\xi_1, \dots, \xi_d$ with coefficients in $\mathbf{C}$ such that $\mathbf{U}[\xi_1, \dots, \xi_d] \supset \mathbf{U}[u_1, \dots, u_k]$ is an integral ring extension.*

We recall that an extension of rings $\mathcal{R}_1 \supset \mathcal{R}_2$ is called an *integral extension* if every element of $\mathcal{R}_1$ is a root of an algebraic equation with coefficients in $\mathcal{R}_2$ and with leading coefficient 1. If $\mathcal{R}_1 \supset \mathcal{R}_2$ and $\mathcal{R}_2 \supset \mathcal{R}_3$ are integral ring extensions, then $\mathcal{R}_1 \supset \mathcal{R}_3$ is also an integral ring extension.

Proof of Lemma 2. We use induction on d. If $d = k$, then we set $u_i = \xi_i$, $i = 1, \dots, k$, and the lemma is obviously true.

Now suppose that $d > k$, and that the lemma holds for rings generated by $d-1$ elements. There exists a homogeneous polynomial $P \in \mathbf{U}[x_1, \dots, x_d]$ such that $P(\xi_1, \dots, \xi_d) = 0$. We may assume that P is irreducible, that $\partial P/\partial x_d \not\equiv 0$, and that $v(P) = 0$, where $v(P)$ is the minimum value of v on the coefficients of P. We let Q denote a polynomial in $\mathbf{C}[x_1, \dots, x_d]$ for which $v(P - Q) \geq 1$. Since P is a homogeneous polynomial, Q is also homogeneous, and hence $Q(x_1, \dots, x_{d-1}, 1) \not\equiv 0$. Let $a_1, \dots, a_{d-1} \in \mathbf{C}$ be such that $Q(a_1, \dots, a_{d-1}, 1) \neq 0$.

We set $\eta_i = \xi_i - a_i\xi_d$, $i = 1, \dots, d-1$, and $\eta_d = \xi_d$. Obviously, $\mathbf{U}[\xi_1, \dots, \xi_d] = \mathbf{U}[\eta_1, \dots, \eta_d]$, and the ideal of algebraic equations satisfied by $\eta_1, \dots, \eta_d$ over $\mathbf{C}(z)$ is a homogeneous ideal. We now prove that η_d is an integral element over the ring $\mathbf{U}[\eta_1, \dots, \eta_{d-1}]$.

We denote

$$S = S(x_1, \dots, x_d) = P(a_1, \dots, a_{d-1}, 1)^{-1} P(x_1 + a_1 x_d, \dots, x_{d-1} + a_{d-1}x_d, x_d).$$

Since $v(P - Q) \geq 1$ and $Q(a_1, \dots, a_{d-1}, 1) \neq 0$, it follows that $v(P(a_1, \dots$ $\dots, a_{d-1}, 1)) = 0$. Hence, $S \in \mathbf{U}[x_1, \dots, x_d]$. We set $T(x) = S(\eta_1, \dots, \eta_{d-1}, x)$. The coefficients of $T(x)$ are in the ring $\mathbf{U}[\eta_1, \dots, \eta_{d-1}]$, and the coefficient of the highest power of x is

$$S(0, \dots, 0, 1) = P(a_1, \dots, a_{d-1}, 1)^{-1} P(a_1, \dots, a_{d-1}, 1) = 1.$$

Since

$$T(\eta_d) = S(\eta_1, \dots, \eta_{d-1}, \eta_d) = P(a_1, \dots, a_{d-1}, 1)^{-1} P(\xi_1, \dots, \xi_d) = 0,$$

it follows that η_d is an integral element over the ring $\mathbf{U}[\eta_1, \dots, \eta_{d-1}]$, and so $\mathbf{U}[\xi_1, \dots, \xi_d] \supset \mathbf{U}[\eta_1, \dots, \eta_{d-1}]$ is an integral ring extension.

According to the induction assumption, there exist linear forms $u_1, \dots, u_k$ in $\eta_1, \dots, \eta_{d-1}$ — and hence in $\xi_1, \dots, \xi_d$ — with coefficients in $\mathbf{C}$ such that $\mathbf{U}[\eta_1, \dots, \eta_{d-1}] \supset \mathbf{U}[u_1, \dots, u_k]$ is an integral ring extension. This gives us the conclusion of the lemma.

By Lemma 2, there exist numbers $c_{i,j} \in \mathbf{C}$ such that for

$$u_i = \sum_{j=1}^{m} c_{i,j} g_j, \qquad i = 1, \dots, k, \tag{10}$$

the ring $\mathbf{U}[g_1, \dots, g_m]$ is integral over $\mathbf{U}[u_1, \dots, u_k]$. We let

$$C = \|c_{i,j}\|, \qquad i, j = 1, \dots, m,$$

denote any nonsingular matrix whose first k rows are given by (10), and we consider the system of differential equations

$$y_k' = \sum_{i=1}^{m} Q_{k,i}^* y_i, \qquad k = 1, \dots, m, \qquad Q_{k,i}^* \in \mathbf{C}(z), \tag{11}$$

whose coefficient matrix Q^* is given by the relation

$$Q^* = \|Q_{k,i}^*\| = CQC^{-1},$$

where Q is the coefficient matrix of the system of differential equations (9).

If $\mathbf{f}$ is the column-vector of functions $f_1(z), \dots, f_m(z)$, then the column-vector $\boldsymbol{\psi} = C\mathbf{f}$ consists of functions $\psi_1(z), \dots, \psi_m(z)$ satisfying the system of differential equations (11). In fact,

$$\boldsymbol{\psi}' = C\mathbf{f}' = CQ\mathbf{f} = CQC^{-1}\boldsymbol{\psi} = Q^*\boldsymbol{\psi}.$$

The functions $\psi_1(z), \dots, \psi_m(z)$ are algebraically dependent over $\mathbf{C}(z)$, and the ideal $\mathcal{J}_0 \subset \mathbf{C}(z)[x_1, \dots, x_m]$ consisting of the algebraic equations satisfied by these functions over $\mathbf{C}(z)$ is obtained from $\mathcal{I}$ by replacing the vector $\mathbf{x}$ by the vector $C^{-1}\mathbf{x}$ in the polynomials $P \in \mathcal{I}$. $\mathcal{J}_0$ is obviously a homogeneous ideal, and if $\mathbf{u} = C\mathbf{g}$, where $\mathbf{g}$ is the column-vector with entries $g_1, \dots, g_m$, then we have

$$\mathbf{C}(z)[x_1, \dots, x_m]/\mathcal{J}_0 = \mathbf{C}(z)[u_1, \dots, u_m],$$

and the ring $\mathbf{U}[u_1, \dots, u_m] = \mathbf{U}[g_1, \dots, g_m]$ is an integral extension of $\mathbf{U}[u_1, \dots$ $\dots, u_k]$.

If Theorem 7 holds for the system of differential equations (11), then it obviously also holds for the system (9). Hence, in what follows we may assume that, under the conditions of Theorem 7, $\mathcal{I}$ is a homogeneous ideal, $\dim \mathcal{I} = k$, and $\mathbf{U}[g_1, \dots, g_m] \supset \mathbf{U}[g_1, \dots, g_k]$ is an integral ring extension.
Since $\deg \operatorname{tr}_{\mathbf{C}(z)} \mathbf{C}(z, g_1, \dots, g_m) = k$, and since $\mathbf{U}[g_1, \dots, g_m] \supset \mathbf{U}[g_1, \dots, g_k]$ is an algebraic extension, it follows that the elements $g_1, \dots, g_k$ are algebraically independent over $\mathbf{C}(z)$. We let $\mathcal{K}$ denote the algebraic closure of the field $\mathbf{C}(g_1, \dots$ $\dots, g_k)$.

The field $\mathcal{K}(z, g_1, \dots, g_m)$ is an algebraic extension of $\mathcal{K}(z)$. Every element of $\mathcal{K}(z)$ can be expanded as a series of decreasing powers of z with coefficients in $\mathcal{K}$. Similarly, the elements of $\mathcal{K}(z, g_1, \dots, g_m)$ can be expanded as a series of decreasing fractional powers of z having the same denominator e in the exponents. The coefficients of this expansion are also in $\mathcal{K}$.

We note that these are the same expansions (3) used by Oleinikov. He proved that the series converge. In the proof below we shall operate with them as formal power series. We state the result we need in the form of a lemma.

Lemma 3. *There exist a number $e \in \mathbf{N}$ and an imbedding of $\mathcal{K}(z, g_1, \dots, g_m)$ in the field $\mathcal{L}$ of formal power series in t with coefficients in $\mathcal{K}$ which is the identity on $\mathcal{K}$ and has the property that $t^e = z^{-1}$.*

For the proof of this lemma, see Chapter IV, § 6 of Chevalley's book [11:1].

In what follows we shall regard $\mathcal{K}(z, g_1, \dots, g_m)$ as a subfield of $\mathcal{L}$.

We define a function $V : \mathcal{L} \to \mathbf{Z}$ by setting $V(\alpha) = s$ for every $\alpha \in \mathcal{L}$, where α is of the form

$$\alpha = \sum_{l \geq s} a_l t^l, \qquad a_l \in \mathcal{K}, \qquad a_s \neq 0.$$

We call this function a *valuation* on the field $\mathcal{L}$. It has the following properties.

1°. If $\alpha, \beta \in \mathcal{L}$, then

$$V(\alpha\beta) = V(\alpha) + V(\beta).$$

2°. If $\alpha, \beta \in \mathcal{L}$, then

$$V(\alpha + \beta) \geq \min(V(\alpha), V(\beta)).$$

Here we take $V(0) = +\infty$ with the usual conventions for operations with the symbol $+\infty$.

3°. $V(t) = 1$.

4°. If $\alpha = P(z)/Q(z)$ with $P, Q \in \mathcal{K}[z]$, then $V(\alpha) = e(\deg_z Q - \deg_z P)$.

Since $V(g_1) = \dots = V(g_k) = 0$, and since $\mathbf{U}$ is the ring of rational functions of z which are regular at infinity, it follows that the valuation V takes nonnegative values on the ring $\mathbf{U}[g_1, \dots, g_k]$. If $\alpha \in \mathbf{U}[g_1, \dots, g_m]$, then α is algebraic and integral over $\mathbf{U}[g_1, \dots, g_k]$. This means that for some $\varphi_0, \varphi_1, \dots, \varphi_{r-1} \in \mathbf{U}[g_1, \dots, g_k]$ it satisfies the equation

$$\alpha^r + \varphi_{r-1}\alpha^{r-1} + \dots + \varphi_1\alpha + \varphi_0 = 0.$$

But then

$$r\,V(\alpha) \geq \min_{0 \leq i \leq r-1} (V(\varphi_i) + iV(\alpha)) \geq \min_{0 \leq i \leq r-1} (iV(\alpha)) = \min(0, (r-1)V(\alpha)).$$

This inequality is possible only if $V(\alpha) \geq 0$. We have thus shown that V takes nonnegative values on the ring $\mathbf{U}[g_1, \dots, g_m]$.

§ 4. The differential operator G

Let $\mathbf{V}$ be a field. We call a map ∂ from $\mathbf{V}$ to itself a *differential operator* if the following relations hold for any two elements $\alpha, \beta \in \mathbf{V}$:

$$\partial(\alpha+\beta) = \partial\alpha + \partial\beta, \qquad \partial(\alpha\beta) = \alpha\,\partial\beta + \beta\,\partial\alpha.$$

The third part of the proof of Theorem 7 involves defining a certain differential operator G and determining its properties.

Since the elements $t, g_1, \ldots, g_k$ are algebraically independent over $\mathbf{C}$, it follows that the partial derivatives $\frac{\partial}{\partial t}, \frac{\partial}{\partial g_1}, \ldots, \frac{\partial}{\partial g_k}$ can be defined on the field $\mathbf{C}(t, g_1, \ldots, g_k)$. The action of these partial derivatives extends in a natural way to the field $\mathcal{K}(t, g_1, \ldots, g_m)$. For example, if $\alpha \in \mathcal{K}(t, g_1, \ldots, g_m)$, and if $P(x, t, x_1, \ldots, x_k)$ is an irreducible polynomial with coefficients in $\mathbf{C}$ such that $P(\alpha, t, g_1, \ldots, g_k) = 0$, then the partial derivative $\frac{\partial\alpha}{\partial t}$ is determined from the equation

$$\frac{\partial P(\alpha, t, g_1, \ldots, g_k)}{\partial x}\frac{\partial\alpha}{\partial t} + \frac{\partial P(\alpha, t, g_1, \ldots, g_k)}{\partial t} = 0.$$

We now set

$$G = -\frac{1}{e}t^{e+1}\frac{\partial}{\partial t} + \sum_{i=1}^{k}\left(\sum_{j=1}^{m} Q_{i,j}g_j\right)\frac{\partial}{\partial g_i}. \tag{12}$$

The differential operator G acts on the field $\mathcal{K}(t, g_1, \ldots, g_m)$.

Lemma 4. *We have the relations* $Gz = 1$ *and*

$$Gg_i = \sum_{j=1}^{m} Q_{i,j}g_j, \qquad i = 1, \ldots, m. \tag{13}$$

Proof. If we apply G to the equality $t^e = z^{-1}$, we obtain:

$$-\frac{1}{e}t^{e+1}et^{e-1} = -z^{-2}Gz,$$

from which it follows that $Gz = 1$.

For $i = 1, \ldots, k$, the relation (13) for g_i follows from the definition of the operator G.

Now suppose that $i > k$, and that $P = P(z, x_1, \ldots, x_k, x_i)$ is an irreducible polynomial with coefficients in $\mathbf{C}$ such that

$$P(z, g_1, \ldots, g_k, g_i) = 0. \tag{14}$$

We apply G to (14), as a result of which we obtain:

$$\begin{aligned}&\frac{\partial P(z,g_1,\dots,g_k,g_i)}{\partial z}+\\&+\sum_{j=1}^{k}\frac{\partial P(z,g_1,\dots,g_k,g_i)}{\partial x_j}\sum_{i=1}^{m}Q_{j,l}g_l+\frac{\partial P(z,g_1,\dots,g_k,g_i)}{\partial x_i}Gg_i=0.\end{aligned}\tag{15}$$

Since $P\in\mathcal{I}$, it follows that $DP\in\mathcal{I}$. Hence,

$$\begin{aligned}DP(z,g_1,\dots,g_k,g_i)=&\frac{\partial P(z,g_1,\dots,g_k,g_i)}{\partial z}+\\&+\sum_{j=1}^{k}\frac{\partial P(z,g_1,\dots,g_k,g_i)}{\partial x_j}\sum_{i=1}^{m}Q_{j,l}g_l+\\&+\frac{\partial P(z,g_1,\dots,g_k,g_i)}{\partial x_i}\sum_{l=1}^{m}Q_{i,l}g_l\\=&0.\end{aligned}\tag{16}$$

Since $\frac{\partial P(z,g_1,\dots,g_k,g_i)}{\partial x_i}\neq 0$, we obtain (13) from (15) and (16).

Lemma 5. *Let ∂ be a differential operator which acts on the field $\mathcal{K}(t,g_1,\dots,g_m)$ and maps $\mathcal{K}$ to itself. Then there exists an integer c such that for all $\alpha\in\mathcal{K}$ $(t,g_1,\dots,g_m)$*

$$V(\partial\alpha)\geq V(\alpha)-c.$$

For the proof, see Chapter VI, § 4, Lemma 4 of Chevalley's book [11:1].

If we apply Lemma 5 to the partial derivatives $\frac{\partial}{\partial t},\frac{\partial}{\partial g_1},\dots,\frac{\partial}{\partial g_k}$, we find that there exists an integer c_0 such that for all $\alpha\in\mathcal{K}(t,g_1,\dots,g_m)$ we have

$$\min\left(V\left(\frac{\partial\alpha}{\partial t}\right),V\left(\frac{\partial\alpha}{\partial g_1}\right),\dots,V\left(\frac{\partial\alpha}{\partial g_k}\right)\right)\geq V(\alpha)-c_0.\tag{17}$$

Since $Q_{i,j}\in\mathbf{U}$ in Theorem 7 by assumption, and since V takes nonnegative values on the ring $\mathbf{U}[g_1,\dots,g_m]$, as proved above, it follows from (12) and (17) that for any $\alpha\in\mathcal{K}(t,g_1,\dots,g_m)$ we have

$$V(G\alpha)\geq V(\alpha)-c_0.\tag{18}$$

We now define the operator G on the field $\mathcal{L}$. Suppose that $\alpha\in\mathcal{L}$, and $\alpha_n\in\mathcal{K}(t)$ is a sequence of elements such that $\lim_{n\to\infty}V(\alpha-\alpha_n)=+\infty$. From (18) it follows that $\{G\alpha_n\}$ is a Cauchy sequence relative to the metric determined by V on $\mathcal{L}$. Hence, there is a limit $\lim_{n\to\infty} G\alpha_n$ in $\mathcal{L}$, which we denote $G\alpha$. This limit does not depend on the choice of sequence $\{\alpha_n\}$, and the operator G on $\mathcal{L}$ defined in this way is a differential operator. (For a proof of these facts, see Chapter VI, § 4, Lemma 5 of Chevalley's book [11:1].)

Lemma 6. *The differential operator G acting on the field $\mathcal{K}(t, g_1, \dots, g_m)$ can be extended to the field $\mathcal{L}$, and for any $\alpha \in \mathcal{L}$ one has the inequality*

$$V(G\alpha) \geq V(\alpha). \tag{19}$$

Proof. It remains for us to prove the relation (19). Suppose that $\alpha \in \mathcal{L}$, and

$$\alpha = \sum_{l \geq s} a_l t^l, \qquad a_l \in \mathcal{K}, \qquad a_s \neq 0.$$

We choose n large enough so that the element

$$\alpha_n = \sum_{i=s}^{n} a_i t^i$$

satisfies the inequality

$$V(G\alpha - G\alpha_n) > V(G\alpha).$$

This inequality implies that

$$V(G\alpha) = V(G\alpha_n) = V\left(\sum_{i=s}^{n} G(a_i t^i)\right) \geq \min_{s \leq i \leq n} V(G(a_i t^i)). \tag{20}$$

Since $a_i \in \mathcal{K}$, we have $\frac{\partial a_i}{\partial t} = 0$, $\frac{\partial a_i}{\partial g_j} \in \mathcal{K}$, $j = 1, \dots, k$, and hence

$$V(Ga_i) = V\left(\sum_{j=1}^{k} \left(\sum_{l=1}^{m} Q_{j,l} g_l\right) \frac{\partial a_i}{\partial g_j}\right) \geq 0.$$

Consequently, for all $i = s, \dots, n$,

$$V(G(a_i t^i)) = V\left((Ga_i)t^i - a_i \frac{i}{e} t^{i+e}\right) \geq s. \tag{21}$$

From (20) and (21) we conclude that (19) holds for all $\alpha \in \mathcal{L}$.

§ 5. The differential operators S and δ

In the fourth part of the proof of Theorem 7, we construct a certain prime ideal $\mathcal{P}$ of the ring $\mathbf{C}[x_1, \dots, x_m]$ and study its properties.

Since $V(g_j) \geq 0$, $j = 1, \dots, m$, it follows that there exist elements $h_1, \dots$ $\dots, h_m \in \mathcal{K}$ such that

$$V(g_j - h_j) \geq 1, \qquad j = 1, \dots, m. \tag{22}$$

Here we have $h_j = g_j$ for $j \leq k$. Hence, at least k of the $h_1, \dots, h_m$ are algebraically independent over $\mathbf{C}$. In particular, they are not all zero.

Lemma 7. *Let $\mathcal{P}$ be the prime ideal of $\mathbf{C}[x_1, \dots, x_m]$ consisting of all polynomials which vanish when $h_1, \dots, h_m$ are substituted for $x_1, \dots, x_m$, and consider the differential operator*

$$S = \sum_{i=1}^{m} \left(\sum_{j=1}^{m} a_{i,j} x_j \right) \frac{\partial}{\partial x_j}$$

acting on the ring $\mathbf{C}[x_1, \dots, x_m]$ (where the $a_{i,j}$ are given in the statement of the theorem). Then $\mathcal{P}$ is a homogeneous ideal, $\dim \mathcal{P} = \dim \mathcal{I} = k$, and $S\mathcal{P} \subset \mathcal{P}$.

Proof. We let $\mathcal{I}_0$ denote the set of polynomials P in $\mathbf{C}[x_1, \dots, x_m]$ which have the property that there exists $Q \in \mathcal{I}$ with $v(P - Q) \geq 1$ (the valuation v was defined right before Lemma 2). Recall that the value of v at an element of $\mathbf{C}(z)[x_1, \dots, x_m]$ was determined in the proof of Lemma 2. We prove that $\mathcal{I}_0$ is a homogeneous ideal of $\mathbf{C}[x_1, \dots, x_m]$, $\mathcal{I}_0 \subset \mathcal{P}$, and $S\mathcal{I}_0 \subset \mathcal{I}_0$.

If $P_1, P_2 \in \mathcal{I}_0$, then for some $Q_1, Q_2 \in \mathcal{I}$ we have $v(P_i - Q_i) \geq 1$. If we set $P = P_1 + P_2$ and $Q = Q_1 + Q_2 \in \mathcal{I}$, then we find that $v(P - Q) \geq 1$, i.e., $P \in \mathcal{I}_0$. If $P \in \mathcal{I}_0$ and $T \in \mathbf{C}[x_1, \dots, x_m]$, then for some $Q \in \mathcal{I}$ we have

$$v(P - Q) \geq 1, \qquad v(PT - QT) \geq 1 + v(T) = 1.$$

Since $TQ \in \mathcal{I}$, it follows that $PT \in \mathcal{I}_0$. Hence, $\mathcal{I}_0$ is an ideal.

Suppose that $P \in \mathcal{I}_0$, $P = \sum P_j$, where P_j is the homogeneous component of P of degree j. If $Q \in \mathcal{I}$ with $v(P - Q) \geq 1$, and if $Q = \sum Q_j$, where Q_j is the homogeneous component of Q of degree j, then $Q_j \in \mathcal{I}$, since $\mathcal{I}$ is a homogeneous ideal. Since $v\left(\sum(P_j - Q_j)\right) \geq 1$ and the $P_j - Q_j$ are homogeneous polynomials of different degrees, we have $v(P_j - Q_j) \geq 1$, and so $P_j \in \mathcal{I}_0$. Thus, $\mathcal{I}_0$ is a homogeneous ideal.

Let $P \in \mathcal{I}_0$, and let $Q \in \mathcal{I}$ be such that $v(P - Q) \geq 1$. It is easy to see that $v(DP - DQ) \geq 1$ and $v(DP - SP) \geq 1$. These inequalities imply that

$v(SP - DQ) \geq 1$. Since $DQ \in \mathcal{I}$ and $SP \in \mathbf{C}[x_1, \ldots, x_m]$, it follows that $SP \in \mathcal{I}_0$. Hence, $S\mathcal{I}_0 \subset \mathcal{I}_0$.

Furthermore, since $Q(g_1, \ldots, g_m) = 0$, it follows that $V(P(g_1, \ldots, g_m)) \geq 1$, and hence $V(P(h_1, \ldots, h_m)) \geq 1$. Because $P(h_1, \ldots, h_m) \in \mathcal{K}$, we find that $P(h_1, \ldots, h_m) = 0$, i.e., the m-tuple $h_1, \ldots, h_m$ is a zero of the ideal $\mathcal{I}_0$. Thus, $\mathcal{I}_0 \subset \mathcal{P}$.

Exactly as in the proof of Lemma 1, from this we show that $\mathcal{P}$ is a homogeneous ideal, $k = \dim \mathcal{I} = \dim \mathcal{P}$, and, moreover, $\mathcal{P}$ is minimal in the set of prime ideals containing $\mathcal{I}_0$.

We now recall some facts which we need to complete the proof of the lemma.

For every ideal $\mathcal{I}_0$ of the ring $\mathbf{C}[x_1, \ldots, x_m]$ there is a uniquely determined set of prime ideals $\mathcal{P}_1, \ldots, \mathcal{P}_s$, called the associated prime ideals of $\mathcal{I}_0$, such that the intersection $\mathcal{P}_1 \bigcap \cdots \bigcap \mathcal{P}_s$ is precisely the set of polynomials $P \in \mathbf{C}[x_1, \ldots, x_m]$ for which $P^n \in \mathcal{I}_0$ for some $n \in \mathbf{N}$. Any prime ideal which is minimal in the set of prime ideals containing $\mathcal{I}_0$ must be one of the ideals $\mathcal{P}_1, \ldots, \mathcal{P}_s$. (See § 117-120 of Van der Waerden's book [83:1].)

We now resume the proof of the lemma. Let $\mathcal{P} = \mathcal{P}_1, \mathcal{P}_2, \ldots, \mathcal{P}_s$ be the associated prime ideals of $\mathcal{I}_0$. Since $\mathcal{P}$ is minimal in the set of prime ideals containing $\mathcal{I}_0$, it follows that $\mathcal{P}_j \neq \mathcal{P}$ for $j \geq 2$, and so there exist polynomials $Q_2, \ldots, Q_s$ such that $Q_j \in \mathcal{P}_j$, $Q_j \notin \mathcal{P}$, $j \geq 2$. Let $P \in \mathcal{P}$. Then $PQ_2 \cdots Q_s \in \mathcal{P} \bigcap \mathcal{P}_2 \bigcap \cdots \bigcap \mathcal{P}_s$, and hence we have $P^n Q \in \mathcal{I}_0$ for some $n \in \mathbf{N}$, where $Q = Q_2^n \cdots Q_s^n \notin \mathcal{P}$. Since $S\mathcal{I}_0 \subset \mathcal{I}_0$, it follows that $S^n(P^n Q) \in \mathcal{I}_0 \subset \mathcal{P}$. Since $P \in \mathcal{P}$, we have $(SP)^n Q \in \mathcal{P}$, and since $Q \notin \mathcal{P}$ and $\mathcal{P}$ is a prime ideal, we have $SP \in \mathcal{P}$. This shows that $S\mathcal{P} \subset \mathcal{P}$, and completes the proof of the lemma.

As in the case of Lemma 4, the fact that $S\mathcal{P} \subset \mathcal{P}$ implies that one can define a differential operator δ on the field $\mathbf{C}(h_1, \ldots, h_m)$ in such a way that

$$\delta h_i = \sum_{j=1}^{m} a_{i,j} h_j, \qquad i = 1, \ldots, m.$$

Since $\mathcal{K} \supset \mathbf{C}(h_1, \ldots, h_m)$ is an algebraic extension, the operator δ (just like G before) can be uniquely extended to the field $\mathcal{K}$.

Lemma 8. *Let $\xi_1, \ldots, \xi_d$ be nonzero elements of $\mathcal{K}$ which are connected by a homogeneous algebraic equation with coefficients in $\mathbf{C}$, and suppose that one has*

$$\delta \xi_i = \beta_i \xi_i, \qquad \beta_i \in \mathbf{C}, \qquad i = 1, \ldots, d.$$

Then there exist integers $s_1, \ldots, s_d$, not all zero, such that $s_1 + \cdots + s_d = 0$ and $s_1\beta_1 + \cdots + s_d\beta_d = 0$.

Proof. Let

$$\sum_{l_1+\cdots+l_d=\nu} b_{l_1,\ldots,l_d}\xi_1^{l_1}\cdots\xi_d^{l_d}=0, \qquad b_{l_1,\ldots,l_d}\in\mathbf{C},$$

be the homogeneous equation satisfied by $\xi_1,\ldots,\xi_d$. If we apply the differential operator δ to this equation μ times, we obtain:

$$\sum_{l_1+\cdots+l_d=\nu}(l_1\beta_1+\cdots+l_d\beta_d)^{\mu}b_{l_1,\ldots,l_d}\xi_1^{l_1}\cdots\xi_d^{l_d}=0. \tag{23}$$

We regard (23) for $\mu=0,1,\ldots,\binom{\nu+d-1}{d-1}-1$ as a system of homogeneous linear equations relative to the terms $b_{l_1,\ldots,l_d}\xi_1^{l_1}\cdots\xi_d^{l_d}$. Since not all of these terms are zero, it follows that the determinant of the coefficient matrix of the system must vanish. But this is a Vandermonde determinant formed from the numbers $l_1\beta_1+\cdots+l_d\beta_d$ with $l_1+\cdots+l_d=\nu$. Hence, one can find two different d-tuples of nonnegative integers l_i', l_i'', $i=1,\ldots,d$, such that

$$l_1'+\cdots+l_d'=\nu, \qquad l_1''+\cdots+l_d''=\nu,$$

$$l_1'\beta_1+\cdots+l_d'\beta_d=l_1''\beta_1+\cdots+l_d''\beta_d.$$

This implies the lemma.

Lemma 9. *The ideal $\mathcal{P}$ in Lemma 7 has a basis consisting of $m-k$ linear forms in the variables $x_1,\ldots,x_m$. The roots of the characteristic polynomial of the matrix A in Theorem 7 can be numbered in such a way that $\mathcal{P}=(l_1,\ldots,l_{m-k})$ with the linear forms l_i satisfying the equalities*

$$Sl_i=\alpha_i l_i, \qquad i=1,\ldots,m-k.$$

Proof. Here and later we shall write $\mathbf{h}$ and $\mathbf{g}$ for the vectors with coordinates $(h_1,\ldots,h_m)$ and $(g_1,\ldots,g_m)$. The differential operator S may be regarded as a linear transformation on the space of linear forms in $x_1,\ldots,x_m$ with coefficients in $\mathbf{C}$. This transformation has matrix A in the coordinates $x_1,\ldots,x_m$, and, by the hypothesis of Theorem 7, this matrix has distinct eigenvalues $\alpha_1,\ldots,\alpha_m$. Hence, there exist linear forms $l_1,\ldots,l_m$ such that

$$Sl_i=\alpha_i l_i, \qquad i=1,\ldots,m.$$

We also know that the forms $l_1,\ldots,l_m$ are linearly independent over $\mathbf{C}$ (since $\alpha_i\neq\alpha_j$).

Consequently, the maximum number of algebraically independent elements over **C** among the $\xi_i = l_i(\mathbf{h})$, $i = 1, \dots, m$, is equal to $\deg \operatorname{tr}_{\mathbf{C}} \mathbf{C}(h_1, \dots, h_m) = k$, and the ideal of algebraic equations with coefficients in **C** satisfied by $\xi_1, \dots, \xi_m$ is a homogeneous ideal (see Lemma 7).

We renumber the roots of the characteristic polynomial of A in such a way that $\xi_{m-k+1}, \dots, \xi_m$ are algebraically independent over **C**. In view of the hypothesis of the theorem, Lemma 8 now implies that $\xi_i = 0$, $i = 1, \dots, m-k$, i.e., the forms $l_1, \dots, l_{m-k}$ are in the ideal $\mathcal{P}$. The ideal $(l_1, \dots, l_{m-k})$ is a prime ideal, and, since the forms $l_1, \dots, l_{m-k}$ are linearly independent over **C**, it has dimension $k = \dim \mathcal{P}$. Hence, $\mathcal{P} = (l_1, \dots, l_{m-k})$. The lemma is proved.

Lemma 10. *Let $\xi_i = l_i(\mathbf{h})$, $i = m-k+1, \dots, m$, and suppose that $\psi \in \mathcal{K}$ is such that ψ/ξ_m is algebraic over the field $\mathbf{C}(\xi_{m-k+1}/\xi_m, \dots, \xi_{m-1}/\xi_m)$. Then if $\delta\psi = \alpha_j \psi$ for some j, $1 \le j \le m-k$, ψ must be zero.*

Proof. The assumptions in the lemma imply that the elements $\psi, \xi_{m-k+1}, \dots, \xi_m$ are connected by a homogeneous algebraic equation with coefficients in **C**. The conclusion of the lemma now follows from Lemma 8, in view of the hypothesis of Theorem 7 and the fact that $\xi_{m-k+1}, \dots, \xi_m$ are algebraically independent over **C**, and so are nonzero.

§ 6. A lemma on linear approximation

The last part of the proof of the theorem is an inductive argument which shows that $g_1, \dots, g_m$ are linearly independent over a certain algebraic extension of $\mathbf{C}(z)$.

Lemma 11. *For every $r \in \mathbf{N}$ there exists a matrix*

$$B = B_r = \|b_{i,j}\|_{i,j=1,\dots,m}, \qquad b_{i,j} \in \mathbf{C}[z^{-1}],$$

such that

$$\deg b_{i,j} < \frac{r}{e}, \qquad v(B - E) \ge 1, \tag{24}$$

where E is the identity matrix,

$$V(\mathbf{g} - B\mathbf{h}) \ge r, \tag{25}$$

and for $r \ge 2$

$$v(B_r - B_{r-1}) \ge r - 1. \tag{26}$$

Remark. The inequality (25) means that the valuation V takes a value at least r at each component of the vector $\mathbf{g} - B\mathbf{h}$. One similarly interprets (26).

Proof. We use induction on r. If $r = 1$, the lemma holds with $B = E$, by (22). We suppose that the lemma is true for some r, $r \geq 1$, and we prove that it is then true for $r + 1$.

From (25) it follows that there exists a vector $\mathbf{c} \in \mathcal{K}^m$ such that

$$V(\mathbf{g} - B\mathbf{h} - t^r\mathbf{c}) \geq r + 1. \tag{27}$$

Since $v(B^{-1} - E) \geq 1$, from (27) we find that

$$V(\mathbf{h} - B^{-1}\mathbf{g} + t^r\mathbf{c}) \geq r + 1.$$

If we apply the operator G and use (19), we obtain:

$$V(G\mathbf{h} - GB^{-1}\mathbf{g} - B^{-1}G\mathbf{g} + t^r G\mathbf{c}) \geq r + 1.$$

Since $G\mathbf{g} = Q\mathbf{g}$, $v(GB^{-1}) \geq 1$, and $v(B^{-1}Q - A) \geq 1$, it follows from (27) that with $M = GB^{-1}B + B^{-1}QB$ we have

$$V(G\mathbf{h} - M\mathbf{h} + t^r(G\mathbf{c} - A\mathbf{c})) \geq r + 1. \tag{28}$$

Since for any $c \in \mathcal{K}$ we have

$$Gc - \delta c = \sum_{j=1}^{h} \frac{\partial c}{\partial g_j}(Gg_j - \delta g_j), \qquad \frac{\partial c}{\partial g_j} \in \mathcal{K}, \qquad j = 1, \ldots, k,$$

and for $j = 1, \ldots, k$ we have

$$Gg_j - \delta h_j = \sum_{i=1}^{m}(Q_{j,i}g_i - a_{j,i}h_i) = \sum_{i=1}^{m}((Q_{j,i} - a_{j,i})g_i + a_{j,i}(g_i - h_i)),$$

it follows that

$$V(Gc - \delta c) \geq 1, \tag{29}$$

and from (28) we obtain:

$$V(G\mathbf{h} - M\mathbf{h} + t^r(\delta\mathbf{c} - A\mathbf{c})) \geq r + 1. \tag{30}$$

If we apply G to the relations $l_i(\mathbf{h}) = 0$, $i = 1, \ldots, m - k$, we find that $l_i(G\mathbf{h}) = 0$, $i = 1, \ldots, m - k$, and from (30) we obtain

$$V(l_i(M\mathbf{h}) - t^r l_i(\delta\mathbf{c} - A\mathbf{c})) \geq r + 1, \qquad i = 1, \ldots, m - k. \tag{31}$$

We consider two cases.

1. e does not divide r. Then (31) means that $l_i(\delta\mathbf{c} - A\mathbf{c}) = 0$, $i = 1, \dots, m-k$, and we find that for some i, $1 \le i \le m-k$, $\psi_i = l_i(\mathbf{c})$ satisfies the relation

$$\delta\psi_i = l_i(\delta\mathbf{c}) = l_i(A\mathbf{c}) = Sl_i(\mathbf{c}) = \alpha_i\psi_i.$$

Since the elements g_j/g_1, h_j/g_1, $j = 1, \dots, m$, are algebraic over the field $\mathbf{C}(t, g_2/g_1, \dots, g_k/g_1)$, it follows that the elements

$$\frac{l_i(\mathbf{g} - B\mathbf{h})}{t^r l_m(\mathbf{h})}, \quad \frac{\xi_{m-k+1}}{\xi_m} = \frac{l_{m-k+1}(\mathbf{h})}{l_m(\mathbf{h})}, \quad \dots, \quad \frac{\xi_{m-1}}{\xi_m} = \frac{l_{m-1}(\mathbf{h})}{l_m(\mathbf{h})}$$

are algebraically dependent over $\mathbf{C}(t)$.

Because, by (27), we have

$$V\left(\frac{l_i(\mathbf{g} - B\mathbf{h})}{t^r\xi_m} - \frac{l_i(\mathbf{c})}{\xi_m}\right) \ge 1,$$

it follows just as in the proof of Lemma 6 that the elements

$$\frac{\psi_i}{\xi_m} = \frac{l_i(\mathbf{c})}{\xi_m}, \quad \frac{\xi_{m-k+1}}{\xi_m}, \quad \dots, \quad \frac{\xi_{m-1}}{\xi_m}$$

are algebraically dependent over $\mathbf{C}$, and the elements $\psi_i, \xi_{m-k+1}, \dots, \xi_m$ are connected by a homogeneous algebraic equation with coefficients in $\mathbf{C}$. Using Lemma 10, we find that $\psi_i = 0$. Thus,

$$l_i(\mathbf{c}) = 0, \qquad i = 1, \dots, m-k. \tag{32}$$

Since $g_i, h_i \in \mathcal{K}$, $i = 1, \dots, k$, it follows from (27) that

$$c_i = 0, \qquad i = 1, \dots, k. \tag{33}$$

If the linear forms

$$x_1, \dots, x_k, \; l_1(\mathbf{x}), \dots, l_{m-k}(\mathbf{x}) \tag{34}$$

are linearly dependent over $\mathbf{C}$ then, since $l_1, \dots, l_{m-k}$ are linearly independent, it follows that there exists a form $a_1x_1 + \cdots + a_kx_k \ne 0$ in the ideal $\mathcal{P}$. Then the elements $h_1, \dots, h_k$ are linearly dependent over $\mathbf{C}$. But this is impossible. Consequently, the forms (34) are linearly independent, and from (32) and (33) we then conclude that $c_i = 0$, $i = 1, \dots, m$. It now follows from (27) that the inequalities (24)-(26) hold for the matrix B with r replaced by $r+1$.

2. e divides r. Then $r = es$, $s \in \mathbf{N}$. From (31) we find that for each $i = 1, \dots, m-k$ there exists a linear form $m_i(\mathbf{h})$ with coefficients in $\mathbf{C}$ such that

$$m_i(\mathbf{h}) - l_i(\delta\mathbf{c} - A\mathbf{c}) = 0,$$

or

$$\delta\psi_i = \alpha_i\psi_i + m_i(\mathbf{h}), \qquad \psi_i = l_i(\mathbf{c}), \qquad i = 1, \ldots, m-k.$$

Let j be fixed, $1 \le j \le m-k$. The linear operator $\delta - \alpha_j E$, where E is the identity operator, maps to itself the space L of linear forms in $h_1, \ldots, h_m$ with coefficients in $\mathbf{C}$. Let $\psi \in L$ and $\delta\psi = \alpha_j\psi$. Then the elements ψ/ξ_m, $\xi_{m-k+1}/\xi_m, \ldots, \xi_{m-1}/\xi_m$ are algebraically dependent over the field $\mathbf{C}(g_2/g_1, \ldots, g_k/g_1)$. As in the first case above, we use Lemma 9 to find that $\psi = 0$. Thus, $\delta - \alpha_j E$ has kernel zero, and so there exists a linear form $n_j(\mathbf{h}) \in L$ such that

$$\delta n_j(\mathbf{h}) = \alpha_j n_j(\mathbf{h}) + m_j(\mathbf{h}), \qquad j = 1, \ldots, m-k,$$

or

$$\delta(\psi_j - n_j(\mathbf{h})) = \alpha_j(\psi_j - n_j(\mathbf{h})), \qquad j = 1, \ldots, m-k.$$

Again using Lemma 10, we find that

$$l_j(\mathbf{c}) = n_j(\mathbf{h}), \qquad j = 1, \ldots, m-k. \tag{35}$$

Since the forms (34) are linearly independent, we find from (33) and (35) that there exists a matrix C of complex numbers such that $\mathbf{c} = C\mathbf{h}$.

Now since $r = e \cdot s$ and $t^e = z^{-1}$, the inequality (27) can be rewritten in the form

$$V(\mathbf{g} - B^*\mathbf{h}) \ge r+1, \qquad B^* = B + z^{-s}C,$$

where the matrix entries in B^* are polynomials in z^{-1} of degree at most s, $s < (r+1)/e$. In addition, we have $v(B^* - E) = v(B - e) \ge 1$.

From this it follows that the lemma holds for $r+1$ if we take B^* for the matrix corresponding to B. By induction, the lemma holds for all $r \in \mathbf{N}$.

§ 7. End of the proof of Theorem 7

We prove part 1° of Theorem 7. In Lemma 11 we proved that for every $r \in \mathbf{N}$ there exists a matrix B with the properties (24), (25) and (26). If we let $B_0 = \lim_{r\to\infty} B_r$ (where this means that each entry in B_0 is the limit in $\mathcal{L}$ of the sequence of corresponding entries in B_r), then we have $V(\mathbf{g} - B_0\mathbf{h}) = \infty$, i.e., $\mathbf{g} = B_0\mathbf{h}$, where B_0 is a matrix whose entries are power series in z^{-1} with complex coefficients.

Since the forms (34) are linearly independent, we conclude from this that

$$g_i = \sum_{j=1}^{k} u_{i,j} g_j, \qquad i = k+1, \ldots, m,$$

where $u_{i,j}$ are power series in z^{-1}.

Let $u_{i,j,n}$ be the sum of the first n terms of the series for $u_{i,j}$ (where we formally regard the expansion of each $u_{i,j}$ as starting with the zero power of z^{-1}). Then

$$V\left(g_i - \sum_{j=1}^{k} u_{i,j,n} g_j\right) \geq en, \qquad i = k+1, \ldots, m.$$

Using (17), we hence obtain

$$V\left(\frac{\partial g_i}{\partial g_j} - u_{i,j,n}\right) \geq en - c_0,$$

and, letting n approach infinity, we find that

$$u_{i,j} = \frac{\partial g_i}{\partial g_j}, \qquad i = k+1, \ldots, m, \quad j = 1, \ldots, k. \tag{36}$$

The elements g_i are algebraic over the field $\mathbf{C}(z, g_1, \ldots, g_k)$; hence, the $u_{i,j}$ are also algebraic over this field. Since $z, g_1, \ldots, g_k$ are algebraically independent over $\mathbf{C}$ and

$$\frac{\partial u_{i,j}}{\partial g_l} = 0, \qquad l = 1, \ldots, k$$

(this is proved in the same way as (36)), we find that the $u_{i,j}$ are algebraic over the field $\mathbf{C}(z)$.

We let $\mathcal{F}$ denote the subfield of $\mathcal{L}$ obtained by adjoining to $\mathbf{C}(z)$ the elements $u_{i,j}$, $k+1 \leq i \leq m$, $1 \leq j \leq k$. Then $\mathcal{F} \supset \mathbf{C}(z)$ is an algebraic extension of finite degree. Further let

$$R_i(x_1, \ldots, x_m) = x_i - \sum_{j=1}^{k} u_{i,j} x_j, \qquad i = k+1, \ldots, m,$$

and let D be the differential operator in the proof of Lemma 1. From (13) and the vanishing of $R_i(g_1, \ldots, g_m)$ we obtain: $DR_i(g_1, \ldots, g_m) = G(R_i(g_1, \ldots, g_m)) = 0$. Since $g_1, \ldots, g_k$ are algebraically independent over $\mathbf{C}(z)$, these relations mean that the linear forms $DR_{k+1}, \ldots, DR_m$ can be expressed as a linear combination of the forms $R_{k+1}, \dots, R_m$ with coefficients in $\mathcal{F}$. That is, there exist $v_{ij} \in \mathcal{F}$, $i, j = k+1, \dots, m$, for which

$$DR_i(x_1, \dots, x_m) = \sum_{j=k+1}^{m} v_{ij} R_j(x_1, \dots, x_m), \qquad i = k+1, \dots, m.$$

From these equations and (9.15) it follows that, if the m-tuple $f_1(z), \dots, f_m(z)$ is any solution of the system of differential equations (9), then the functions $R_{k+1}(f_1, \dots, f_m), \dots, R_m(f_1, \dots, f_m)$ form a solution of the system of differential equations

$$y_i' = \sum_{j=k+1}^{m} v_{ij} y_j, \qquad i = k+1, \dots, m. \tag{37}$$

Thus, we obtain a linear map from the m-dimensional space of solutions of (9) to the $(m-k)$-dimensional space of solutions of (37). The kernel of this map has dimension at least $m - (m-k) = k$. This means that there exist k $\mathbf{C}$-linearly independent solutions

$$\varphi_l = (\varphi_{1,l}(z), \dots, \varphi_{m,l}(z)), \qquad l = 1, \dots, k,$$

of the system of differential equations (9) such that

$$R_i(\varphi_{1,l}, \dots, \varphi_{m,l}) = \varphi_{i,l} - \sum_{j=1}^{k} u_{i,j} \varphi_{j,l} = 0, \tag{38}$$

$$i = k+1, \dots, m, \qquad l = 1, \dots, k.$$

Setting $\varphi_j(z) = \varphi_{j,1}(z)$, $1 \le j \le m$, we obtain a solution $(\varphi_1(z), \dots, \varphi_m(z))$ of (9) whose components are linearly dependent over $\mathcal{F}$. This proves part 1° of Theorem 7.

We now prove part 2°. For fixed i, the equations in (38) may be regarded as a system of k linear equations in the k variables $u_{i,1}, \dots, u_{i,k}$. Since the solutions $\varphi_1, \dots, \varphi_k$ are linearly independent over $\mathbf{C}$, it follows that the matrix

$$\|\varphi_{j,l}\|, \qquad j = 1, \dots, m, \quad l = 1, \dots, k,$$

has rank k. From (38) it now follows that

$$|\varphi_{j,l}|_{j,l=1,\dots,k} \neq 0;$$

hence, each of the functions $u_{i,j}$ can be written as a ratio of two determinants in which the entries are the functions $\varphi_{j,l}$, $j = 1, \dots, m$, $l = 1, \dots, k$.

If the coefficients $Q_{i,j}(z)$ of the system of differential equations (9) have no poles in the region $z \neq 0$, then the functions $\varphi_{j,l}$ are analytic in a neighborhood of every finite point except perhaps $z = 0$. The above representation for the functions $u_{i,j}$ now implies that they do not have branch points anywhere in the finite plane, with the possible exception of the point $z = 0$. In the proof of part 1° we showed that the functions $u_{i,j}$ are single-valued in a neighborhood of $z = \infty$. But an algebraic function cannot have only one branch point in the extended complex plane. Consequently, the point $z = 0$ cannot be a branch point of $u_{i,j}$ either; thus, all of the $u_{i,j}$ are rational functions. This proves part 2°.

We note the following corollary of the proof of Theorem 7: *Under the conditions in Theorem 7, if among the functions $f_1(z), \dots, f_m(z)$ which form a solution of (9) there are k which are algebraically independent over $\mathbf{C}(z)$, then the solution $\varphi_1(z), \dots, \varphi_m(z)$ includes at most k functions which are linearly independent over $\mathcal{F}$ in part 1° and linearly independent over $\mathbf{C}(z)$ in part 2°.*

We now prove Theorem 2. We set $y_k = y^{(k-1)}$, $k = 1, \dots, p$, and we write the differential equation (4) as an equivalent system of differential equations:

$$\begin{aligned} y_k' &= y_{k+1}, \qquad k = 1, \dots, p-1, \\ y_p' &= -r_0 y_1 - r_1 y_2 - \cdots - r_{p-1} y_p. \end{aligned}$$

We now apply Theorem 7 to this system. We have

$$\mathrm{A} = \begin{pmatrix} 0 & 1 & 0 & \cdots & 0 & 0 \\ 0 & 0 & 1 & \cdots & 0 & 0 \\ \cdot & \cdot & \cdot & \cdots & \cdot & \cdot \\ 0 & 0 & 0 & \cdots & 0 & 1 \\ -\beta & 0 & 0 & \cdots & 0 & 0 \end{pmatrix}, \quad \mathrm{A} - \lambda \mathrm{E} = \begin{pmatrix} -\lambda & 1 & 0 & \cdots & 0 & 0 \\ 0 & -\lambda & 1 & \cdots & 0 & 0 \\ \cdot & \cdot & \cdot & \cdots & \cdot & \cdot \\ 0 & 0 & 0 & \cdots & -\lambda & 1 \\ -\beta & 0 & 0 & \cdots & 0 & -\lambda \end{pmatrix},$$

where β is the value of the function r_0 at $z = \infty$, $\beta \in \mathbf{C}$, $\beta \neq 0$. Since $|A - \lambda E| = (-1)^p(\beta + \lambda^p)$, as we easily see by expanding the determinant of $A - \lambda E$ along the first column, it follows that the roots of the characteristic polynomial of A are

$$\alpha_k = \gamma \zeta^k, \qquad \zeta = \exp\left(\frac{2\pi i}{p}\right), \qquad k = 1, \dots, p,$$

where γ is a fixed p-th root of $-\beta$. Since p is a prime number, the polynomial $x^{p-1} + \cdots + x + 1$ is irreducible. This implies that the matrix A satisfies the condition in Theorem 7. Hence, Theorem 2 follows from Theorem 7.

§ 8. Linear reducibility

We consider the hypergeometric function

$$\varphi = \varphi(z) = \sum_{n=0}^{\infty} \frac{1}{[\lambda_1 + 1, n] \cdots [\lambda_m + 1, n]} \left(\frac{z}{m}\right)^{mn}, \tag{39}$$

$$\lambda_i \in \mathbf{C}, \qquad -\lambda_i \notin \mathbf{N}, \qquad i = 1, \ldots, m, \qquad m \geq 2.$$

According to (5.5), the function (39) satisfies the linear differential equation

$$\left(\prod_{i=1}^{m} (\delta + m\lambda_i) - z^m\right) y = \prod_{i=1}^{m} (m\lambda_i), \qquad \delta = z\frac{d}{dz}. \tag{40}$$

We denote

$$G(\delta) = \prod_{i=1}^{m} (\delta + m\lambda_i). \tag{41}$$

Then the homogeneous linear differential equation corresponding to (40) has the form

$$Ly = (G(\delta) - z^m)y = 0. \tag{42}$$

In this section we find conditions for the differential equation (42) to be homogeneously linearly reducible over $\mathbf{C}(z)$. Here in order to simplify the argument it is convenient to reduce the proof to the case when the numbers $\lambda_1, \ldots, \lambda_m$ are in the strip $0 \leq \operatorname{Re} \lambda_i < 1$, $i = 1, \ldots, m$.

If two m-tuples of functions $F_1, \ldots, F_m$ and $\Phi_1, \ldots, \Phi_m$ analytic in some region are *algebraically equivalent* over $\mathbf{C}(z)$ (i.e., each function in one m-tuple is algebraically dependent over $\mathbf{C}(z)$ on the functions of the other m-tuple), then we shall write

$$(F_1, \ldots, F_m) \sim (\Phi_1, \ldots, \Phi_m).$$

Lemma 12. *Suppose that $\lambda = \lambda_j \in \{\lambda_1, \ldots, \lambda_m\}$, $\lambda \neq 0$, and the function φ_* is obtained from the function φ in (39) by replacing λ by $\lambda - 1$. Then*

$$(\varphi, \varphi', \ldots, \varphi^{(m-1)}) \sim (\varphi_*, \varphi_*', \ldots, \varphi_*^{(m-1)}). \tag{43}$$

Proof. We obviously have

$$\frac{1}{m\lambda}\delta\varphi + \varphi = \varphi_*, \tag{44}$$

from which we obtain

$$\frac{1}{m\lambda}\delta^{i+1}\varphi + \delta^i\varphi = \delta^i\varphi_*, \qquad i = 0, 1, \ldots, m-2. \tag{45}$$

Since the differential equation (40) and the relation (44) give

$$m\lambda \prod_{\substack{i=1\\ i\neq j}}^{m} (\delta + m\lambda_i)\varphi_* = \prod_{i=1}^{m}(\delta + m\lambda_i)\varphi = z^m\varphi + \prod_{i=1}^{m}(m\lambda_i),$$

it follows from this and from (45) that the following relation holds for some $q, q_0, \ldots, q_{m-1} \in \mathbf{C}$:

$$\delta^{m-1}\varphi_* = \frac{z^m}{m\lambda}\varphi + \sum_{i=0}^{m-1} q_i\delta^i\varphi + q. \tag{46}$$

From (45) and (46) we find that

$$\begin{pmatrix} \varphi_* \\ \delta\varphi_* \\ \vdots \\ \delta^{m-1}\varphi_* \end{pmatrix} =$$

$$= \begin{pmatrix} 1 & \frac{1}{m\lambda} & 0 & \cdots & \cdot & 0 \\ 0 & 1 & \frac{1}{m\lambda} & \cdots & \cdot & 0 \\ \cdot & \cdot & \cdot & \cdots & \cdot & \cdot \\ 0 & 0 & \cdot & \cdots & 1 & \frac{1}{m\lambda} \\ q_0 + \frac{z^m}{m\lambda} & q_1 & \cdot & \cdots & \cdot & q_{m-1} \end{pmatrix} \begin{pmatrix} \varphi \\ \delta\varphi \\ \vdots \\ \delta^{m-2}\varphi \\ \delta^{m-1}\varphi \end{pmatrix} + \begin{pmatrix} 0 \\ 0 \\ \vdots \\ 0 \\ q \end{pmatrix}. \tag{47}$$

The determinant of the square matrix in (47) is equal to $(-1)^{m-1}z^m/(m\lambda)^m + c$, where $c \in \mathbf{C}$. Hence, by Lemma 3 of Chapter 7, (47) implies that

$$(\varphi, \delta\varphi, \ldots, \delta^{m-1}\varphi) \sim (\varphi_*, \delta\varphi_*, \ldots, \delta^{m-1}\varphi_*). \tag{48}$$

Since $\delta^i\varphi$ can be expressed as a linear combination of $\varphi, \varphi', \ldots, \varphi^{(i)}$ with coefficients in $\mathbf{C}(z)$, $i = 0, 1, \ldots, m-1$, and since the linear transformation has nonzero determinant, it follows from Lemma 3 of Chapter 7 that

$$(\varphi, \varphi', \ldots, \varphi^{(m-1)}) \sim (\varphi, \delta\varphi, \ldots, \delta^{m-1}\varphi),$$

and similarly

$$(\varphi_*, \varphi'_*, \dots, \varphi_*^{(m-1)}) \sim (\varphi_*, \delta\varphi_*, \dots, \delta^{m-1}\varphi_*).$$

Hence (43) follows from (48).

In what follows we shall be considering the following types of differential operators: 1) $Q(\delta)$, $Q(x) \in \mathbf{C}[x]$; 2) $Q(z,\delta)$, $Q(z,x) \in \mathbf{C}[z][x]$, i.e.,

$$Q(z,\delta) = \sum_{i=1}^{m} P_i(z)\delta^i, \qquad P_i(z) \in \mathbf{C}[z], \qquad i = 1, \dots, m; \tag{49}$$

and 3) $R = Q(z,\delta)P(z)$, where $Q(z,\delta)$ is the operator (49) and $P(z) \in \mathbf{C}[z]$. When R is applied to a function $y = f(z)$, one must apply the operator $Q(z,\delta)$ to the product $P(z)y$.

Lemma 13. *Let* $\delta = z\frac{d}{dz}$, $s \in \mathbf{Z}^+$, *and* $Q(x) \in \mathbf{C}[x]$. *Then*

$$Q(\delta)z^s = z^s Q(\delta + s). \tag{50}$$

Proof. We must prove the equality

$$Q(\delta)z^s y = z^s Q(\delta + s)y, \tag{51}$$

where it is sufficient to prove this for the operators $Q(\delta) = \delta^k$, $k = 0, 1, 2, \dots$.

We use induction on k. The relation obviously holds if $k = 0$. Supposing now that $\delta^k z^s y = z^s(\delta + s)^k y$, we obtain

$$\begin{aligned}\delta^{k+1}z^s y &= \delta(\delta^k z^s y) = \delta(z^s(\delta+s)^k y) = \\ &= sz^s(\delta+s)^k y + z^s\delta(\delta+s)^k y = z^s(\delta+s)^{k+1}y.\end{aligned}$$

By induction, (51) holds for any $Q(x) \in \mathbf{C}[x]$.

Using (51), we can write the operator (49) in the following form:

$$\begin{aligned}Q(z,\delta) &= \sum_{i=1}^{m} P_i(z)\delta^i = \sum_{j=0}^{n} z^j Q_j(\delta) = \sum_{j=0}^{n} Q_j(\delta - j)z^j = \\ &= \sum_{j=0}^{n} Q_j^*(\delta)z^j, \qquad Q_j(x), Q_j^*(x) \in \mathbf{C}[x].\end{aligned} \tag{52}$$

Lemma 14. *If* $\lambda_i \in \mathbf{C}$, $0 \le \operatorname{Re}\lambda_i < 1$, $i = 1, \dots, m$, *then:*

1°. *The differential equation (42) is homogeneously linearly reducible over* $\mathbf{C}(z)$ *if and only if there exists a divisor* $l > 1$ *of* m *for which (see the notation defined before Theorem 5):*

$$\left\{\lambda + \frac{1}{l}\right\} \equiv \{\lambda\} \ (mod\ \mathbf{Z}), \qquad \{\lambda\} = \{\lambda_1, \dots, \lambda_m\}. \tag{53}$$

2°. *If $m = p$, where p is a prime number, then the differential equation (42) is homogeneously linearly reducible over $\mathbf{C}(z)$ if and only if there exists $c \in \mathbf{C}$ such that $0 \le \operatorname{Re} c < 1/p$ and*

$$\{\lambda\} = \left\{c, c + \frac{1}{p}, \dots, c + \frac{p-1}{p}\right\}. \tag{54}$$

Remark. Lemma 14 holds without the condition that $0 \le \operatorname{Re} \lambda_i < 1$ if we replace (54) by the condition

$$\{\lambda\} \equiv \left\{c, c + \frac{1}{p}, \dots, c + \frac{p-1}{p}\right\} \ (mod\ \mathbf{Z}).$$

Proof. 1. We first show that parts 1° and 2° are equivalent in the case $m = p$. After that it will remain for us only to prove part 1°.

Suppose that $m = p$, and 1° holds. Since p is a prime number, there is only one possible value for the divisor l: $l = p$. Let i_1 be such that $\operatorname{Re} \lambda_{i_1} = \min(\operatorname{Re} \lambda_1, \dots, \operatorname{Re} \lambda_m)$. We set $\lambda_{i_1} = c$, $0 \le \operatorname{Re} c < 1$. From (53) we find that there exists $i_2 \in \{1, \dots, p\}$ such that $c + 1/p \equiv \lambda_{i_2}$ $(mod\ \mathbf{Z})$. Repeating this argument, we obtain $i_1, \dots, i_p \in \{1, \dots, p\}$ such that $c + (j-1)/p \equiv \lambda_{i_j}$ $(mod\ \mathbf{Z})$, $j = 1, \dots, p$. However, $k_1/p \not\equiv k_2/p$ $(mod\ \mathbf{Z})$ if $0 \le k_1 < k_2 \le p-1$. Hence, the p-tuple $i_1, \dots, i_p$ is a permutation of $1, \dots, p$. Then

$$\{\lambda\} \equiv \left\{c, c + \frac{1}{p}, \dots, c + \frac{p-1}{p}\right\} \ (mod\ \mathbf{Z}). \tag{55}$$

From (55) it follows that there is only one of the numbers $\operatorname{Re} \lambda_i$, $i = 1, \dots, p$, which satisfies the condition $0 \le \operatorname{Re} \lambda_i < 1/p$. Because of the way we chose i_1, we have $i = i_1$ and $0 \le \operatorname{Re} c < 1/p$; we then obtain (54) from (55) and the inequalities $0 \le \operatorname{Re} \lambda_i < 1$, $i = 1, \dots, p$.

Now suppose that (54) holds. From that relation it follows that

$$\left\{\lambda + \frac{1}{p}\right\} = \left\{c + \frac{1}{p}, c + \frac{2}{p}, \dots, c + 1\right\},$$

from which (53) follows with $l = p$.

2. We now prove that, if (53) holds, then the operator L in (42) is reducible (i.e., it can be written as a product of two operators of the same type which contain

δ). It will then follow that the differential equation (42) is homogeneously linearly reducible over $\mathbf{C}(z)$.

It is not hard to verify that (54) and the condition $0 \le \operatorname{Re} \lambda_i < 1$ imply the relation

$$\{\lambda\} = \left\{\mu_1, \dots, \mu_k, \mu_1 + \frac{1}{l}, \dots, \mu_k + \frac{1}{l}, \dots, \mu_1 + \frac{l-1}{l}, \dots, \mu_k + \frac{l-1}{l}\right\},$$

$$\mu_i \in \mathbf{C}, \qquad 0 \le \operatorname{Re} \mu_i < \frac{1}{l}, \qquad i = 1, \dots, k; \qquad k = \frac{m}{l} \in \mathbf{N}, \qquad k < m.$$

We let ω denote a primitive l-th root of 1, and set

$$T_j = T_j(\delta) = \prod_{i=1}^{k} (\delta + m\mu_i + kj), \qquad j = 0, 1, \dots, l-1. \tag{56}$$

We shall show that

$$L = (T_0 - z^k)(T_1 - \omega z^k) \cdots (T_{l-1} - \omega^{l-1} z^k). \tag{57}$$

For every $r \in \mathbf{N}$, $1 \le r \le l-1$, we consider the following set of r-tuples $\mathbf{i} = (i_1, \dots, i_r)$:

$$J_r = \{(i_1, \dots, i_r) \mid 0 \le i_1 < i_2 < \dots < i_r \le l-1\}.$$

Since

$$x^l - 1 = \prod_{i=0}^{l-1} (x - \omega^i),$$

it follows that

$$\sum_{\mathbf{i}=(i_1,\dots,i_r) \in J_r} \omega^{i_1 + \dots + i_r} = 0, \qquad r = 1, \dots, l-1. \tag{58}$$

We have

$$\begin{aligned}(T_0 - z^k)(T_1 - \omega z^k) \cdots (T_{l-1} - \omega^{l-1} z^k) &= \\ = T_0 T_1 \cdots T_{l-1} - z^m + \sum_{r=1}^{l-1} \sum_{\mathbf{i} \in J_r} (-1)^r \omega^{i_1 + \dots + i_r} \prod_{j=0}^{l-1} D_{j,\mathbf{i}},\end{aligned} \tag{59}$$

where $D_{j,\mathbf{i}} = T_j$ if $j \notin \{i_1, \dots, i_r\}$ and $D_{j,\mathbf{i}} = z^k$ if $j \in \{i_1, \dots, i_r\}$.

From (56) and (50) it follows that

$$T_j(\delta)z^k = z^k T_j(\delta + k) = z^k T_{j+1}, \qquad 0 \le j \le l-2.$$

Hence,

$$\begin{aligned}\prod_{j=0}^{l-1} D_{j,\mathbf{i}} &= \left(\prod_{j=0}^{i_1-1} T_j\right) z^k \left(\prod_{j=i_1+1}^{i_2-1} T_j\right) z^k \cdots \left(\prod_{j=i_r+1}^{l-1} T_j\right) = \\ &= z^k \left(\prod_{j=1}^{i_2-1} T_j\right) z^k \left(\prod_{j=i_2+1}^{i_3-1} T_j\right) z^k \cdots \left(\prod_{j=i_r+1}^{l-1} T_j\right) = \\ &= z^{2k} \left(\prod_{j=2}^{i_3-1} T_j\right) z^k \cdots \left(\prod_{j=i_r+1}^{l-1} T_j\right) = \\ &= z^{rk} \prod_{j=r}^{l-1} T_j.\end{aligned} \tag{60}$$

Using (58) and (60), we obtain

$$\sum_{\mathbf{i}\in J_r} (-1)^r \omega^{i_1+\cdots+i_r} \prod_{j=0}^{l-1} D_{j,\mathbf{i}} = (-1)^r z^{rk} \prod_{j=r}^{l-1} T_j \sum_{\mathbf{i}\in J_r} \omega^{i_1+\cdots+i_r} = 0. \tag{61}$$

Furthermore,

$$T_0 T_1 \cdots T_{l-1} = \prod_{j=0}^{l-1} \prod_{i=1}^{k} \left(\delta + m\left(\mu_i + \frac{j}{l}\right)\right) = \prod_{i=1}^{m} (\delta + m\lambda_i). \tag{62}$$

From (59), (61) and (62) it follows that (57) holds, and this proves that the operator L is reducible.

3. We now prove that homogeneous linear reducibility over $\mathbf{C}(z)$ of the differential equation (42) implies the relation (53).

From the definition of homogeneous linear reducibility of a differential equation we know that there exists a solution y of (42), $y \not\equiv 0$, such that

$$L_1 y = \sum_{i=0}^{m-1} P_i(z)\delta^i y = 0, \qquad P_i(z) \in \mathbf{C}[z], \qquad i = 0, 1, \ldots, m-1, \tag{63}$$

where $P_i(z) \not\equiv 0$ for at least one index i.

We prove that the following relation holds:

$$L_1 = \sum_{i=1}^{m-1} P_i(z)\delta^i = Q(z^m - G(\delta)) + R, \tag{64}$$

where $Q, R \in \mathbf{C}[z][\delta]$, $\deg_z R \leq m-1$, and $G(\delta)$ is the operator defined in (41).

We first use induction to show that for all $i \geq m$ there exist $R_i(z,\delta)$, $Q_i(z,\delta) \in \mathbf{C}[z][\delta]$, $\deg_z R_i \leq m-1$, such that

$$z^i = Q_i(z,\delta)(z^m - G(\delta)) + R_i(z,\delta). \tag{65}$$

For $i = m$ we have

$$z^m = (z^m - G(\delta)) + G(\delta).$$

Suppose that (65) holds for $i = n \geq m$. We prove that it then holds for $i = n+1$ as well. The relation (52) allows us to write $R_n(z,\delta)$ in the form

$$R_n(z,\delta) = \sum_{j=0}^{m-1} R_{n,j}(\delta)z^j.$$

Using (65) with $i = n$ and also (50), we obtain

$$\begin{aligned} z^{n+1} &= zQ_n(z,\delta)(z^m - G(\delta)) + R_{n,m-1}(\delta-1)z^m + \sum_{j=1}^{m-1} R_{n,j-1}(\delta-1)z^j = \\ &= (zQ_n(z,\delta) + R_{n,m-1}(\delta-1))(z^m - G(\delta)) + \\ &\quad + \sum_{j=1}^{m-1} R_{n,j-1}(\delta-1)z^j + R_{n,m-1}(\delta-1)G(\delta), \end{aligned}$$

which proves that (65) holds for $i = n+1$, and hence, by induction, for all $i \geq m$.

If we use (52) to write the operator L_1 in the form

$$L_1 = \sum_{i=0}^{N} u_i(\delta)z^i$$

and take (65) into account, we obtain

$$L_1 = \sum_{i=0}^{m-1} u_i(\delta)z^i + \sum_{i=m}^{N} u_i(\delta)Q_i(z,\delta)(z^m - G(\delta)) + \sum_{i=m}^{N} u_i(\delta)R_i(z,\delta),$$

which implies (64).

We note that $R \neq 0$ in (64), since otherwise the degree of the operator L_1 in δ would be at least m. The relation (52) enables us to write the operator R in the form

$$R = \sum_{j=0}^{m-1} R_j(\delta) z^j, \qquad R_j \in \mathbf{C}[\delta],$$

where $R_j \neq 0$ for at least one value of j. From (63) and (64) we find that

$$\sum_{j=0}^{m-1} R_j(\delta) z^j y = 0. \tag{66}$$

We show that at least two of the operators $R_j(\delta)$ are nonzero in (66). In fact, otherwise, using (50), for some j we would have

$$R_j(\delta) z^j y = z^j R_j(\delta + j) y = 0, \qquad R_j \neq 0.$$

If we choose $r(\delta) \in \mathbf{C}[\delta]$ to be a polynomial of least degree such that $r(\delta)y = 0$, then by (42) and (57) we obtain:

$$0 = G(\delta) r(\delta) y = r(\delta) G(\delta) y = r(\delta) z^m y = z^m r(\delta + m) y,$$

i.e., $r(\delta + m)y = 0$. Then $r_0(\delta) = r(\delta + m) - r(\delta)$ is a polynomial of lower degree than $r(\delta)$ for which $r_0(\delta)y = 0$. But this contradicts our choice of $r(\delta)$.

Thus, any relation of the form (66) contains at least two nonzero operators $R_j(\delta)$. We now suppose that (66) is a relation with the minimal number of nonzero $R_j(\delta)$.

From (66) and (42), using (50) and (41), we find that

$$\begin{aligned}\prod_{k=0}^{m-1} G(\delta-k) \sum_{j=0}^{m-1} R_j(\delta) z^j y &= \sum_{j=0}^{m-1} \prod_{\substack{k=0\\k\neq j}}^{m-1} G(\delta-k) R_j(\delta) G(\delta-j) z^j y = \\ &= \sum_{j=0}^{m-1} \prod_{\substack{k=0\\k\neq j}}^{m-1} G(\delta-k) R_j(\delta) z^j G(\delta) y = \\ &= \sum_{j=0}^{m-1} \prod_{\substack{k=0\\k\neq j}}^{m-1} G(\delta-k) R_j(\delta) z^{j+m} y = \\ &= z^m \sum_{j=0}^{m-1} \prod_{\substack{k=0\\k\neq j}}^{m-1} G(\delta-k+m) R_j(\delta+m) z^j y = \\ &= 0.\end{aligned} \tag{67}$$

Zero terms go to zero terms in this transformation. Hence, the resulting relation (67) is an equation of the same type as (66): it also contains a minimal number of nonzero terms. Thus, if $R_s(\delta) \neq 0$ and $R_t(\delta) \neq 0$, $0 \leq s < t \leq m-1$, then the coefficients of $z^s y$ and $z^t y$ in (66) and in (67) are proportional. That is,

$$\frac{R_s(x)}{R_t(x)} = \frac{\prod_{k=0,\ k\neq s}^{m-1} G(x-k+m) R_s(x+m)}{\prod_{k=0,\ k\neq t}^{m-1} G(x-k+m) R_t(x+m)} = \frac{G(x-t+m) R_s(x+m)}{G(x-s+m) R_t(x+m)},$$

or

$$\frac{R_s(x) R_t(x+m)}{R_s(x+m) R_t(x)} = \frac{G(x-t+m)}{G(x-s+m)}. \tag{68}$$

If we replace x by $x+(u-1)m$ in (68), where $u = 1, \ldots, N$, $N \in \mathbf{N}$, and multiply together the resulting equalities, and if we use the expression for $G(\delta)$ in (41), we obtain

$$\begin{aligned}\Phi_N(x) = \frac{R_s(x) R_t(x+N+m)}{R_t(x) R_s(x+N+m)} &= \frac{\prod_{u=1}^{N} G(x+um-t)}{\prod_{u=1}^{N} G(x+um-s)} = \\ &= \frac{\prod_{i=1}^{m} \prod_{u=1}^{N} (z+m(\lambda_i+u))}{\prod_{i=1}^{m} \prod_{u=1}^{N} (z+m(\lambda_i+u+r))},\end{aligned}$$

where $z = x-t$, $r = (t-s)/m$, $0 < r < 1$, or $r = k/l$, $(k,l) = 1$, $l > 1$, with l a divisor of m.

We prove that

$$\{\lambda + r\} \equiv \{\lambda\} \ (mod\ \mathbf{Z}).$$

Suppose that this does not hold. Then for some $i \in \{1, \dots, m\}$ the function $\Phi_N(x)$ has poles at the points $x = t - m(\lambda_i + u + r)$, $u = 1, \dots, N$, which is impossible if N is sufficiently large, since the denominator of $\Phi_N(x)$ has bounded degree (at most $\deg R_t(x) + \deg R_s(x)$). Thus,

$$\left\{\lambda + \frac{k}{l}\right\} \equiv \{\lambda\} \ (mod\ \mathbf{Z}).$$

But then for any $a \in \mathbf{N}$

$$\left\{\lambda + \frac{ak}{l}\right\} \equiv \{\lambda\} \ (mod\ \mathbf{Z}).$$

We take a, $1 \le a \le l - 1$, to be the solution of the congruence $ka \equiv 1 \ (mod\ l)$. Then

$$\left\{\lambda + \frac{1}{l}\right\} \equiv \{\lambda\} \ (mod\ \mathbf{Z}),$$

and the lemma is proved.

We set

$$\begin{aligned}
\psi_1 &= \psi_1(z) = \varphi(z), \\
\psi_i &= \psi_i(z) = \sum_{n=0}^{\infty} \frac{(n+\lambda_1)\cdots(n+\lambda_{i-1})}{[\lambda_1+1,n]\cdots[\lambda_m+1,n]} \left(\frac{z}{m}\right)^{mn-i+1}, \\
&\qquad i = 2, 3, \dots, m,
\end{aligned} \tag{69}$$

where $\varphi(z)$ is the function (39). The functions (69) satisfy the following system S of linear differential equations:

$$S: \quad \begin{aligned}
y_1' &= -\frac{\beta_1}{z} y_1 + y_2, \\
&\cdots\cdots\cdots \\
y_{m-1}' &= -\frac{\beta_{m-1}}{z} y_{m-1} + y_m, \\
y_m' &= \frac{-\beta_m}{z} y_m + y_1 + \lambda_1 \cdots \lambda_m \left(\frac{z}{m}\right)^{-m}, \\
\beta_i &= m\lambda_i + i - 1, \quad i = 1, \dots, m.
\end{aligned} \tag{70}$$

Let $f_1 = f_1(z), \ldots, f_m = f_m(z)$ be either an arbitrary solution of the system of differential equations S, or else an arbitrary solution of the system of homogeneous differential equations S_0 which corresponds to S. Then

$$f_i = \left(\frac{d}{dz} + \frac{\beta_{i-1}}{z}\right) \cdots \left(\frac{d}{dz} + \frac{\beta_1}{z}\right) f_1, \qquad i = 1, \ldots, m.$$

This implies that

$$\begin{pmatrix} f_1 \\ f_2 \\ \vdots \\ f_m \end{pmatrix} = A \begin{pmatrix} f_1 \\ f_1' \\ \vdots \\ f_1^{(m-1)} \end{pmatrix}, \tag{71}$$

where A is a triangular matrix with 1's on the main diagonal, 0's above the main diagonal, and elements of $\mathbf{C}[1/z]$ elsewhere.

The equality (71) means that the functions $f_1, f_2, \ldots, f_m$ can be obtained from $f_1, f_1', \ldots, f_1^{(m-1)}$ by a linear transformation with coefficients in $\mathbf{C}[1/z]$ and nonzero determinant.

Using Lemma 3 of Chapter 7, we then obtain the following result in the non-homogeneous case.

Lemma 15. $(\psi_1, \psi_2, \ldots, \psi_m) \sim (\varphi, \varphi', \ldots, \varphi^{(m-1)})$.

In the homogeneous case, (71) implies that the system S_0 is homogeneous linearly irreducible over $\mathbf{C}(z)$ if and only if the differential equation (42) is homogeneously linearly irreducible over $\mathbf{C}(z)$, i.e., we have the following analogue of Lemma 14.

Lemma 16. *Let* $\lambda_i \in \mathbf{C}$, $0 \leq \operatorname{Re} \lambda_i < 1$, $i = 1, \ldots, m$. *Then:*

1°. *The system* S_0 *of homogeneous linear differential equations is homogeneously linearly reducible over* $\mathbf{C}(z)$ *if and only if there exists a divisor* l *of* m *such that* $2 \leq l \leq m$ *and (53) holds.*

2°. *If* $m = p$, *where* p *is a prime number, then the system* S_0 *is homogeneously linearly reducible over* $\mathbf{C}(z)$ *if and only if there exists* $c \in \mathbf{C}$ *such that* $0 \leq \operatorname{Re} c < 1/p$ *and (54) holds.*

§ 9. Proof of Theorems 6 and 5

Proof of Theorem 6. In view of Lemma 15, it is sufficient to prove just part 1°. By applying Lemma 12 as many times as necessary, we may suppose that the numbers $\lambda_1, \dots, \lambda_p$ satisfy the condition $0 \le \operatorname{Re} \lambda_i < 1$, $i = 1, \dots, p$. We consider two cases.

1. The condition (54) does not hold. By Lemma 14, the differential equation (42) is homogeneously linearly irreducible over $\mathbf{C}(z)$. Then, by Theorem 2, this differential equation is irreducible over $\mathbf{C}(z)$. Hence, by Theorem 4 of Chapter 9, the functions $\varphi(z), \varphi'(z), \dots, \varphi^{(p-1)}(z)$ are algebraically independent over $\mathbf{C}(z)$. In the present situation the condition (8) is automatically fulfilled, and so the assertion in Theorem 6 is true.

2. The condition (54) holds. We have

$$p^{pn} \prod_{j=1}^{p} [\lambda_j + 1, n] = p^{pn} \prod_{j=0}^{p-1} \prod_{i=1}^{n} \left(c + i + \frac{j}{p} \right) =$$
$$= \prod_{k=p}^{pn+p-1} (pc + k) = [\Lambda + 1, pn], \qquad \Lambda = pc + p - 1.$$

Hence, the function $\varphi(z)$ in (40) can be written in the form

$$\varphi(z) = \sum_{n=0}^{\infty} \frac{z^{pn}}{[\Lambda + 1, pn]}. \tag{72}$$

If

$$\varphi_\lambda(z) = \sum_{n=0}^{\infty} \frac{z^n}{(\lambda + 1) \cdots (\lambda + n)}$$

is the function (5.17) which we studied in Chapter 5, and if ζ is a primitive p-th root of 1, then

$$\frac{1}{p} \sum_{k=0}^{p-1} \varphi_\Lambda(\zeta^k z) = 1 + \frac{1}{p} \sum_{n=1}^{\infty} \sum_{k=0}^{p-1} \frac{\zeta^{kn} z^n}{(\Lambda + 1) \cdots (\Lambda + n)} =$$
$$= 1 + \sum_{N=1}^{\infty} \frac{z^{pN}}{(\Lambda + 1) \cdots (\Lambda + pN)} = \tag{73}$$
$$= \sum_{n=0}^{\infty} \frac{z^{pn}}{[\Lambda + 1, pn]},$$

since

$$\sum_{k=0}^{p-1} \zeta^{kn} = \begin{cases} 0, & \text{if } p \nmid n, \\ p, & \text{if } p | n. \end{cases}$$

From (72) and (73) we obtain:

$$\varphi(z) = \frac{1}{p}\sum_{k=0}^{p-1} \varphi_\Lambda(\zeta^k z). \tag{74}$$

From (5.18) it follows that the function $\varphi_\Lambda(\zeta^k z)$ is a solution of the differential equation

$$y' = \left(\zeta^k - \frac{\Lambda}{z}\right) y + \frac{\Lambda}{z}. \tag{75}$$

We use induction to show that the following relation holds for any y which satisfies the differential equation (75):

$$y^{(s)} = (\zeta^{ks} + r_{s,k})y + q_{s,k}, \qquad s \in \mathbf{N}, \tag{76}$$

where

$$r_{s,k} = \sum_{\nu=1}^{s} r_{s,k,\nu} z^{-\nu}, \qquad q_{s,k} = \sum_{\nu=1}^{s} q_{s,k,\nu} z^{-\nu}, \qquad r_{s,k,\nu}, q_{s,k,\nu} \in \mathbf{C}.$$

If $s = 1$, then (76) follows from (75). Suppose that (76) holds for $s = l$, $l \in \mathbf{N}$. We then differentiate the relation (76) with $s = l$ and use (75); as a result we find that

$$\begin{aligned} y^{(l+1)} &= (\zeta^{kl} + r_{l,k})y' + r'_{l,k}y + q'_{l,k} = \\ &= (\zeta^{kl} + r_{l,k})\left(\zeta^k - \frac{\Lambda}{z}\right) y + (\zeta^{kl} + r_{l,k})\frac{\Lambda}{z} + r'_{l,k}y + q'_{l,k} = \\ &= (\zeta^{k(l+1)} + r_{l+1,k})y + q_{l+1,k}, \end{aligned}$$

in which the functions

$$\begin{aligned} r_{l+1,k} &= \left(\zeta^k - \frac{\Lambda}{z}\right) r_{l,k} - \frac{\Lambda}{z}\zeta^{kl} + r'_{l,k}, \\ q_{l+1,k} &= (\zeta^{kl} + r_{l,k})\frac{\Lambda}{z} + q'_{l,k} \end{aligned}$$

have the required form. Consequently, (76) holds for $s = l + 1$, and so it holds for all $s \in \mathbf{N}$ by induction.

From (74) and (76) we obtain

$$\varphi^{(s)}(z) = \frac{1}{p}\sum_{k=0}^{p-1}(\zeta^{ks} + r_{s,k})\varphi_\Lambda(\zeta^k z) + \frac{1}{p}\sum_{k=0}^{p-1} q_{s,k}, \qquad s \in \mathbf{N}.$$

We set

$$q_s = \frac{1}{p}\sum_{k=0}^{p-1} q_{s,k}, \qquad r_{0,k} = 0, \qquad k = 0, 1, \dots, p-1,$$

$$B = \left\| \frac{1}{p}(\zeta^{ks} + r_{s,k}) \right\|_{s,k=0,1,\dots,p-1}.$$

Then

$$\begin{pmatrix} \varphi(z) \\ \varphi'(z) \\ \vdots \\ \varphi^{(p-1)}(z) \end{pmatrix} = B \begin{pmatrix} \varphi_\Lambda(z) \\ \varphi_\Lambda(\zeta z) \\ \vdots \\ \varphi_\Lambda(\zeta^{p-1} z) \end{pmatrix} + \begin{pmatrix} 0 \\ q_1 \\ \vdots \\ q_{p-1} \end{pmatrix}, \tag{77}$$

$$\det B = b_0 + \sum_{i=1}^{M} b_i z^{-i}, \qquad M = \frac{p(p-1)}{2},$$

$$b_i \in \mathbf{C}, \qquad i = 1, \dots, M, \qquad b_0 = \frac{1}{p^p} |\zeta^{ks}|_{s,k=0,1,\dots,p-1},$$

and $b_0 \neq 0$ since it is a Vandermonde determinant. Hence, by Lemma 3 of Chapter 7, it follows from (77) that

$$(\varphi(z), \varphi'(z), \dots, \varphi^{(p-1)}(z)) \sim (\varphi_\Lambda(z), \varphi_\Lambda(\zeta z), \dots, \varphi_\Lambda(\zeta^{p-1} z)). \tag{78}$$

There are two possible subcases: a) $c \neq 0$. Then $\Lambda = (pc + p - 1) \notin \mathbf{Z}$, and, by Theorem 4 of Chapter 5, the functions $\varphi_\Lambda(\zeta^k z)$, $k = 0, 1, \dots, p-1$, are algebraically independent over $\mathbf{C}(z)$. Then from (78) we find that the functions $\varphi(z)$, $\varphi'(z)$, ..., $\varphi^{(p-1)}(z)$ are algebraically independent over $\mathbf{C}(z)$. Since the condition (54) with $c \neq 0$ implies (8), it follows that Theorem 6 holds in this case.

b) $c = 0$. Then $\Lambda = (p-1) \in \mathbf{N}$,

$$\varphi_\Lambda(z) = \left(e^z - \sum_{k=0}^{p-1} \frac{z^k}{k!} \right) \frac{(p-1)!}{z^{p-1}},$$

and, by Lemma 3 of Chapter 7,

$$(\varphi_\Lambda(z), \varphi_\Lambda(\zeta z), \dots, \varphi_\Lambda(\zeta^{p-1} z)) \sim (e^z, e^{\zeta z}, \dots, e^{\zeta^{p-1} z}). \tag{79}$$

But

$$\prod_{i=0}^{p-1} e^{\zeta^i z} = 1.$$

Hence, it follows from (78) and (79) that the functions $\varphi(z)$, $\varphi'(z)$, ..., $\varphi^{(p-1)}(z)$ are algebraically dependent over $\mathbf{C}(z)$. In the present case (8) is not fulfilled. Hence, Theorem 6 again holds. Theorem 6 is proved.

The form of the equation connecting the functions $\psi_1(z), \dots, \psi_p(z)$ over $\mathbf{C}(z)$ when (8) fails to hold was found by Oleinikov. In this case we may suppose that $\lambda_i = (p-i)/p$, $i = 1, \dots, p$, and the system of differential equations (70) satisfied by $\psi_1, \dots, \psi_p$ takes the form

$$\begin{aligned} y_1' &= -\frac{p-1}{z} y_1 + y_2, \\ &\cdots\cdots\cdots \\ y_{p-1}' &= -\frac{p-1}{z} y_{p-1} + y_p, \\ y_p' &= -\frac{p-1}{z} y_p + y_1. \end{aligned}$$

We consider the determinant

$$D = D(\psi_1, \dots, \psi_p) = \begin{vmatrix} \psi_1 & \psi_2 & \cdots & \psi_p \\ \psi_2 & \psi_3 & \cdots & \psi_1 \\ \cdot & \cdot & \cdots & \cdot \\ \psi_p & \psi_1 & \cdots & \psi_{p-1} \end{vmatrix}.$$

If we differentiate the polynomial D by the derivative rule for determinants and use the above system of differential equations, we easily find that

$$D' = -\frac{p(p-1)}{z} D.$$

This implies that $D = c_1 z^{-p(p-1)}$, $c_1 \in \mathbf{C}$. Using (69), we find the value $c_1 = -p^{p(p-1)}$. Thus, the desired equation has the form

$$D(\psi_1, \dots, \psi_p) = -\left(\frac{p}{z}\right)^{p(p-1)}.$$

Proof of Theorem 5. We have already shown that the functions $\psi_1, \dots, \psi_p$ can be obtained from the functions $\varphi, \varphi', \dots, \varphi^{(p-1)}$ by a linear transformation with coefficients in $\mathbf{Q}[1/z]$ (since $\lambda_i \in \mathbf{Q}$) and with determinant 1. By Lemma 3 of Chapter 7, this implies that parts 1° and 2° of Theorem 5 are equivalent. Hence, it suffices to prove part 1°.

Since the function $\varphi(z) = \varphi_{\lambda_1,\dots,\lambda_p}(z)$ satisfies the differential equation (5), part 1° follows from Theorem 1 of Chapter 3 and Theorem 6.

It is not possible to prove part 2° of Theorem 5 by applying the Second Fundamental Theorem of Chapter 3 and Theorem 6 to the functions $\psi_1, \dots, \psi_p$, because the meromorphic functions $\psi_1, \dots, \psi_p$ are not E-functions. However, one can use those theorems to prove part 2° if one applies them to the E-functions $\psi_1^* = \psi$ and $\psi_i^* = \psi_i - \left(\frac{p}{z}\right)^{pn-i+1}$, $i = 2, \dots, p$.

Because of Lemma 15, to prove Theorem 6 it suffices to prove only one of the two parts 1° and 2°. Earlier we proved part 1°. But it is also easy to prove Theorem 6 by showing that part 2° holds. To do this it is not necessary to prove Theorem 2; rather, one applies Theorem 7 directly to the system of differential equations S_0. Then the matrices A and $A - \lambda E$ have the form

$$A = \begin{pmatrix} 0 & 1 & 0 & \cdots & 0 \\ 0 & 0 & 1 & \cdots & 0 \\ \cdot & \cdot & \cdot & \cdots & \cdot \\ 0 & 0 & 0 & \cdots & 1 \\ 1 & 0 & 0 & \cdots & 0 \end{pmatrix}, \quad A - \lambda E = \begin{pmatrix} -\lambda & 1 & 0 & \cdots & 0 & 0 \\ 0 & -\lambda & 1 & \cdots & 0 & 0 \\ \cdot & \cdot & \cdot & \cdots & \cdot & \cdot \\ 0 & 0 & 0 & \cdots & -\lambda & 1 \\ 1 & 0 & 0 & \cdots & 0 & -\lambda \end{pmatrix}.$$

As in the proof of Theorem 2 in § 7, we easily obtain $|A-\lambda E| = (-1)^{m+1}(1-\lambda^m)$, and if $m = p$ is a prime number, then the matrix A satisfies the conditions of Theorem 7. The rest of the proof of part 2° of Theorem 6 is similar to the proof of part 1°, except that we use Lemma 16 instead of Lemma 14.

In the homogeneous case with $p = 2$, Theorem 5 implies a result of Siegel (Theorem 2 of Chapter 6), and in the general case with $p = 2$ it implies a result of Belogrivov [3:3] and Oleinikov [56:7] (Theorem 8 of Chapter 6).

The notion of linear reducibility of a differential equation goes back to Frobenius ([21:1], 1873). The abstract study of this notion was a motivating factor for the development of differential algebra (see E. R. Kolchin [32:1]). It is remarkable that, over a century later, the same concept turned out to be useful when applying general theorems of transcendental number theory.

Remarks

In [56:3–5], after presenting general theorems Oleinikov uses his method to investigate the functions

$$\sum_{n=0}^{\infty} \frac{1}{[\lambda_1+1, n][\lambda_2+1, n][\lambda_3+1, n]} \left(\frac{z}{3}\right)^{3n}; \tag{80}$$

$$\sum_{n=0}^{\infty} \frac{[\mu, n]}{[\lambda_1 + 1, n][\lambda_2 + 1, n][\lambda_3 + 1, n]} \left(\frac{z}{2}\right)^{2n}, \tag{81}$$

$$\sum_{n=0}^{\infty} \frac{[\mu_1, n][\mu_2, n]}{[\lambda_1 + 1, n][\lambda_2 + 1, n][\lambda_3 + 1, n]} z^n. \tag{82}$$

He first examines algebraic independence of the values of the functions (80)–(82) when $\lambda_3 = 0$, and he then reduces the general case to that special case. However, there is a gap in the proof of the theorems for the functions (80) and (82): an assertion is used concerning the solutions of linear differential equations which has not yet been proved (or disproved). Using properties of the specific differential equations satisfied by the functions (80) and (82), it is not hard to fix up Oleinikov's proof in those two cases. However, his proof of the theorem for the functions (81) is erroneous.

In 1980, Salikhov [66:4] generalized Theorem 2 to the case of a set of several linear differential equations of prime orders. This enabled him to generalize Theorem 4 to values of the functions (6) at distinct algebraic points $\xi_1, \ldots, \xi_m$ which are linearly independent over the field $\mathbf{Q}(\zeta)$, where ζ is a primitive p-th root of 1. The proof of Theorem 5 made it possible for Salikhov to refine the latter result. Then in 1984, Salikhov [66:5] generalized Theorem 3 to a class of $(p+1)$-st order differential equations, and applied the result to prove a theorem on algebraic independence of the values at algebraic points of the function

$$\sum_{n=0}^{\infty} \frac{[\mu, n]}{[\lambda_1 + 1, n] \cdots [\lambda_{p+1} + 1, n]} \left(\frac{z}{p}\right)^{pn}$$

and its sequence of successive derivatives, where p is a prime number.

This last result with $p = 2$ implies the previously unproved claim of Oleinikov concerning the functions (81).

In [56:6], Oleinikov proved Theorem 5 with $p = 3$ and $\lambda_3 = 0$. This proof is based on a result in [56:5] concerning the functions (80). In light of the above remark, the proof should be viewed as being correct.

Chapter 11
The algebraic independence measure of values of IE-functions

§ 1. Definition of the measures

In transcendental number theory one looks for two types of results: qualitative and quantitative. In the first type of research one determines whether certain numbers or classes of numbers belong to some set. In particular, one proves irrationality, linear independence, transcendence, or algebraic independence. In the second type of research the qualitative results are amplified by quantitative information in the form of estimates for linear forms or polynomials in the given numbers with coefficients in **Z** or $\mathbf{Z_K}$.

The basic methods of transcendental number theory can be used to obtain both types of results: qualitative and quantitative. It should be noted that the quantitative information obtained for certain classes of numbers has found application in different questions of number theory and other branches of mathematics.

In Chapters 2–10, we examined qualitative problems of transcendental number theory. Chapter 1 contained the solutions to certain quantitative questions as well. It is to such quantitative aspects of the field that Chapters 11–13 are devoted.

In order to study quantitative problems of transcendental number theory, it is useful to introduce the notion of measures of irrationality, linear independence, transcendence, and algebraic independence.

Let $\alpha_1, \dots, \alpha_m$, $m \geq 2$, and α be numbers in **R** or **C**. By the *measure of linear independence of* $\alpha_1, \dots, \alpha_m$, we mean the following function of H:

$$L = L(\alpha_1, \dots, \alpha_m; H) = \min |a_1\alpha_1 + \cdots + a_m\alpha_m|, \tag{1}$$

$$H \in \mathbf{N}, \qquad a_k \in \mathbf{Z}, \ \ |a_k| \leq H, \ \ k = 1, \dots, m, \ \ |a_1| + \cdots + |a_m| > 0,$$

where the minimum is taken over all a_k satisfying the conditions indicated.

Clearly, L is a non-increasing function as H increases. If the numbers $\alpha_1, \dots$ $\dots, \alpha_m$ are linearly dependent over **Q**, then $L = 0$ for all H sufficiently large. Otherwise, always $L > 0$. Thus, $\alpha_1, \dots, \alpha_m$ *are linearly independent if and only if* $L(\alpha_1, \dots, \alpha_m; H) > 0$ *for all* $H \in \mathbf{N}$.

The *measure of irrationality of a single number* α is the function $L(1, \alpha; H)$.

The *measure of algebraic independence of the numbers* $\alpha_1, \dots, \alpha_m$, $m \geq 1$, is the following function of s and H:

$$\Phi = \Phi(\alpha_1, \dots, \alpha_m;\ s;\ H) = \min |P(\alpha_1, \dots, \alpha_m)|, \tag{2}$$

$$s, H \in \mathbf{N}, \quad P = P(x_1, \dots, x_m) \in \mathbf{Z}[x_1, \dots, x_m], \quad P \not\equiv 0, \quad \deg_{\mathbf{x}} P \leq s, \quad H_P \leq H,$$

where the minimum is taken over all polynomials P which satisfy the conditions indicated.

Φ is also a non-increasing function both with respect to s and H. If $\alpha_1, \dots, \alpha_m$ are algebraically dependent, then $\Phi = 0$ for all s and H sufficiently large. Otherwise, always $\Phi > 0$. Thus, *$\alpha_1, \dots, \alpha_m$ are algebraically independent if and only if $\Phi(\alpha_1, \dots, \alpha_m;\ s;\ H) > 0$ for all $s \in \mathbf{N}$ and $H \in \mathbf{N}$.*

The algebraic independence measure (2) is also called the *relative transcendence measure of the numbers $\alpha_1, \dots, \alpha_m$.*

If $m = 1$ and $\alpha_1 = \alpha$, we call (2) the *transcendence measure* of the number α. That is, the transcendence measure of α is the function

$$\Phi = \Phi(\alpha;\ s;\ H) = \min |P(\alpha)|, \qquad s, H \in \mathbf{N}, \tag{3}$$

$$P(x) \in \mathbf{Z}[x], \quad P(x) \not\equiv 0, \quad \deg P(x) \leq s, \quad H_P \leq H.$$

A number α is transcendental if and only if $\Phi(\alpha;\ s;\ H) > 0$ for all $s \in \mathbf{N}$ and $H \in \mathbf{N}$.

It is easy to see that

$$\Phi(\alpha;\ s;\ H) = L(1, \alpha, \dots, \alpha^s;\ H), \tag{4}$$

and that

$$\Phi(\alpha_1, \dots, \alpha_m;\ s;\ H) = L(1, \beta, \dots, \beta_\mu;\ H), \tag{5}$$

where the numbers

$$\beta_1, \dots, \beta_\mu, \qquad \mu = \mu_{s,m} = \frac{(s+m)!}{s!m!},$$

are the following monomials in the numbers $\alpha_1, \dots, \alpha_m$:

$$\alpha_1^{k_1} \cdots \alpha_m^{k_m}, \quad 0 \leq k_1 + \dots + k_m \leq s, \quad k_i \geq 0, \quad i = 1, \dots, m, \tag{6}$$

(numbered in any order).

If $m \geq 2$, we shall also consider the *homogeneous algebraic independence (relative transcendence) measure* $\Phi^0 = \Phi^0(\alpha_1, \dots, \alpha_m;\ s;\ H)$, which is defined in the same way as Φ, except with the condition that the $P(x_1, \dots, x_m)$ be homogeneous polynomials. *The numbers* $\alpha_1, \dots, \alpha_m$ *are homogeneously algebraically independent if and only if* $\Phi^0(\alpha_1, \dots, \alpha_m;\ s;\ H) > 0$ *for all* $s \in \mathbf{N}$ *and* $H \in \mathbf{N}$.

We have the analogous equality

$$\Phi^0(\alpha_1, \dots, \alpha_m;\ s;\ H) = L(\beta_1, \dots, \beta_{\mu^0};\ H), \tag{7}$$

where

$$\beta_1, \dots, \beta_{\mu^0}, \qquad \mu^0 = \mu^0_{s,m} = \frac{(s+m-1)!}{s!(m-1)!},$$

is the set of homogeneous monomials (6) with $k_1 + \cdots + k_m = s$.

The relations (4), (5) and (7) show that one can reduce the study of transcendence measures, algebraic independence measures and homogeneous algebraic independence measures to the study of linear independence measures.

We shall also consider the *algebraic independence measure of r groups of numbers* $\alpha_{1,1}, \dots, \alpha_{m_1,1};\ \dots;\ \alpha_{1,r}, \dots, \alpha_{m_r,r}$:

$$\Phi = \Phi(\alpha_{1,1}, \dots, \alpha_{m_1,1};\ \dots;\ \alpha_{1,r}, \dots, \alpha_{m_r,r};\ s_1, \dots, s_r;\ H), \tag{8}$$

which is defined as in (2), except that the polynomial $P = P(x_{1,1}, \dots, x_{m_1,1};\ \dots$ $\dots; x_{1,r}, \dots, x_{m_r,r})$ has total degree no greater than $s_1, \dots, s_r \in \mathbf{N}$ in each respective set of variables $x_{1,1}, \dots, x_{m_1,1};\ \dots;\ x_{1,r}, \dots, x_{m_r,r}$.

Finally, if $m_k \geq 2$ for $k = 1, \dots, r$, we shall consider the *homogeneous algebraic independence measure of r groups of numbers*:

$$\Phi^0 = \Phi^0(\alpha_{1,1}, \dots, \alpha_{m_1,1};\ \dots;\ \alpha_{1,r}, \dots, \alpha_{m_r,r};\ s_1, \dots, s_r;\ H), \tag{9}$$

which is defined in the same way, except with P homogeneous in each group of variables $x_{1,k}, \dots, x_{m_k,k}$, $k = 1, \dots, r$. The measures (8) and (9) can also be expressed in terms of the linear independence measures of certain sets of numbers.

Using the Dirichlet pigeon-hole principle, one can easily find upper bounds for the measures which hold for any m-tuple $\alpha_1, \dots, \alpha_m$ or for any α.

Theorem 5 of Chapter 1 immediately implies that

$$L(\alpha_1, \dots, \alpha_m;\ H) \leq cH^{1-\tau m}, \qquad c = c(\alpha_1, \dots, \alpha_m) = \sqrt{2}\sum_{k=1}^{m} |\alpha_k|, \tag{10}$$

where $\tau = 1$ if $\alpha_1, \dots, \alpha_m$ are all real numbers, and $\tau = 1/2$ if any of them belongs to $\mathbf{C} \setminus \mathbf{R}$.

From (4) and (10) we obtain:

$$\Phi(\alpha;\ s;\ H) < e^{\lambda s} H^{1-\tau(s+1)}, \qquad \lambda = \lambda(\alpha) > 0, \tag{11}$$

since in this case

$$c = \sqrt{2}\sum_{k=0}^{s} |\alpha|^k \le \sqrt{2}(1+|\alpha|)^s < e^{\lambda s}.$$

From (5) and (10) we find that

$$\Phi(\alpha_1, \dots, \alpha_m;\ s;\ H) < e^{\lambda s} H^{1-\tau\frac{(s+m)!}{s!m!}}, \tag{12}$$

$$\lambda = \lambda(\alpha_1, \dots, \alpha_m) > 0,$$

since

$$c = \sqrt{2} \sum_{k_1+\dots+k_m \le s} |\alpha_1|^{k_1} \cdots |\alpha_m|^{k_m} \le$$
$$\le \sqrt{2}(1+|\alpha_1|+\dots+|\alpha_m|)^s < e^{\lambda s}.$$

Finally, from (7) and (10) we have

$$\Phi^0(\alpha_1, \dots, \alpha_m;\ s;\ H) < e^{\lambda s} H^{1-\tau\frac{(s+m-1)!}{s!(m-1)!}}, \tag{13}$$

$$\lambda = \lambda(\alpha_1, \dots, \alpha_m) > 0.$$

In the same way, one can easily use (10) to obtain upper bounds for the measures (8) and (9).

The above notions of measures can be generalized, by allowing in the definitions linear forms and polynomials with algebraic integer coefficients. If $\mathbf{K}$ is an algebraic number field, then by the *size of the polynomial* $P(x_1, \dots, x_m) \in \mathbf{K}[x_1, \dots, x_m]$, denoted $\overline{|P|}$, we mean the maximum size of the coefficients of P, i.e., the maximum of the moduli of all of the coefficients and their conjugates in $\mathbf{K}$.

We shall consider *linear independence measures, transcendence measures, algebraic independence measures and homogeneous algebraic independence measures relative to the field* $\mathbf{K}$, which are defined just as in (1)–(3) and (7)–(9), except that the coefficients of the linear form or polynomial are in $\mathbf{Z_K}$ and it is the size of the linear form or polynomial which is $\le H$. These measures will be denoted by the same symbols as in (1)–(3) and (7)–(9), but with the subscript $\mathbf{K}$, i.e., $L_{\mathbf{K}}$, $\Phi_{\mathbf{K}}$ and $\Phi^0_{\mathbf{K}}$.

One can easily obtain upper bounds for the measures $L_{\mathbf{K}}$, $\Phi_{\mathbf{K}}$ and $\Phi^0_{\mathbf{K}}$, if one generalizes Lemma 5 of Chapter 1 to the case when the linear forms have coefficients in $\mathbf{Z_K}$. Here one must first prove the following fact: *If* $H \in \mathbf{R}$, $H \ge 1$, *and* N *is the number of* α *in* $\mathbf{Z_K}$ *for which* $\overline{|\alpha|} \le H$, *then there exist constants*

$\gamma_1 > 0$ *and* $\gamma_2 > 0$, *depending only on the field* $\mathbf{K}$, *such that* $\gamma_1 H^h \le N \le \gamma_2 H^h$, *where* $h = [\mathbf{K} : \mathbf{Q}]$.

In this way one proves estimates similar to those in (10)–(13):

$$L_{\mathbf{K}} = L_{\mathbf{K}}(\alpha_1, \dots, \alpha_m;\ H) < c_1 H^{1-\tau h m}, \tag{14}$$

$$\Phi_{\mathbf{K}} = \Phi_{\mathbf{K}}(\alpha;\ s;\ H) < c_2 H^{1-\tau h(m+1)}, \tag{15}$$

$$\Phi_{\mathbf{K}} = \Phi_{\mathbf{K}}(\alpha_1, \dots, \alpha_m;\ s;\ H) < c_3 H^{1-\tau h \frac{(s+m)!}{s!m!}}, \tag{16}$$

$$\Phi_{\mathbf{K}}^0 = \Phi_{\mathbf{K}}^0(\alpha_1, \dots, \alpha_m;\ s;\ H) < c_4 H^{1-\tau h \frac{(s+m-1)}{s!(m-1)!}}, \tag{17}$$

where τ is equal to 1 or $1/2$ depending on whether or not the field $\mathcal{K} = \mathbf{K}(\alpha_1, \dots$ $\dots, \alpha_m)$ (the field $\mathcal{K} = \mathbf{K}(\alpha)$) is real; all of the constants c_1, c_2, c_3 and c_4 depend on the field $\mathbf{K}$ and on the numbers $\alpha_1, \dots, \alpha_m$ (on the number α); and c_2, c_3 and c_4 also depend on s. There are similar bounds for the measures (7) and (8).

These upper bounds for the measures hold for any numbers under consideration. Hence, in transcendental number theory the point of interest is now to establish lower bounds for the measures which hold for various numbers or classes of numbers.

Any bounds for Φ^0 or $\Phi_{\mathbf{K}}^0$ imply corresponding bounds for Φ or $\Phi_{\mathbf{K}}$, if we introduce another number $\alpha_{m+1} = 1$; such bounds also give us bounds for Φ or $\Phi_{\mathbf{K}}$ applied to the numbers $\alpha_1/\alpha_m, \dots, \alpha_{m-1}/\alpha_m$.

If we have a lower bound for some measure, for example $\Phi_{\mathbf{K}} = \Phi_{\mathbf{K}}(\alpha_1, \dots$ $\dots, \alpha_m;\ s;\ H)$, then the same bound holds for any specific $|P(\alpha_1, \dots, \alpha_m)|$ with

$$P = P(x_1, \dots, x_m) \in \mathbf{Z}_{\mathbf{K}}[x_1, \dots, x_m], \qquad P \not\equiv 0, \ \deg P \le s, \ \overline{|P|} \le H,$$

since $\Phi_{\mathbf{K}}$ is the minimum modulus of the finite set of values of all such polynomials at the point $(\alpha_1, \dots, \alpha_m)$.

Since $\mathbf{Z} \subset \mathbf{Z}_{\mathbf{K}}$ for any $\mathbf{K}$, it follows that lower bounds for $\Phi_{\mathbf{K}}$ and $\Phi_{\mathbf{K}}^0$ imply the same bounds for Φ and Φ^0.

The numbers s and H in the above definitions will be called the *degree* and the *height* of the measure, respectively.

We note that there are various classifications of transcendental numbers based on the behavior of the transcendence measure, for example, the classification given by Mahler in 1932 in [42:1] (see also [31:1] and [73:1]).

We shall explain the idea behind one method of obtaining lower bounds for linear forms in a set of numbers.

Let $\alpha_1, \dots, \alpha_m$, $m \ge 2$, be complex numbers not all zero. To be definite, we shall assume that $\alpha_1 \ne 0$. Suppose that we have in some way constructed m

linearly independent forms in these numbers:

$$L_k = a_{k,1}\alpha_1 + \cdots + a_{k,m}\alpha_m, \qquad k = 1, \ldots, m,$$

$$a_{k,l} \in \mathbf{Z}, \qquad |a_{k,l}| \le H, \qquad k = 1, \ldots, m, \qquad H \in \mathbf{N}, \tag{18}$$

$$\max_{1 \le k \le m} |L_k| = \lambda.$$

Such forms exist for any H, and λ depends on H.

Let

$$L = a_1\alpha_1 + \cdots + a_m\alpha_m,$$

$$a_l \in \mathbf{Z}, \qquad |a_l| \le h_0, \qquad l = 1, \ldots, m, \qquad h_0 \in \mathbf{N},$$

be an arbitrary fixed linear form in $\alpha_1, \ldots, \alpha_m$. Since the linear forms in (18) are linearly independent, it is possible to choose $m-1$ of them, say $L_1, \ldots, L_{m-1}$, such that the forms $L_1, \ldots, L_{m-1}, L$ are linearly independent.

We consider the linear forms

$$\begin{aligned} L_1 &= a_{1,1}\alpha_1 + \cdots + a_{1,m}\alpha_m, \\ &\cdots\cdots\cdots \\ L_{m-1} &= a_{m-1,1}\alpha_1 + \cdots + a_{m-1,m}\alpha_m, \\ L &= a_1\alpha_1 + \cdots + a_m\alpha_m. \end{aligned} \tag{19}$$

Let Δ be the determinant of the matrix of coefficients of the forms (19). Since the forms are linearly independent, we have $\Delta \ne 0$. Since $\Delta \in \mathbf{Z}$, it follows that

$$|\Delta| \ge 1. \tag{20}$$

We let $\Delta_{k,i}$ denote the cofactor of the entry in the k-th row and i-th column of Δ. Then from (19) we have

$$\Delta\alpha_1 = \sum_{k=1}^{m-1} \Delta_{k,1} L_k + \Delta_{m,1} L,$$

and

$$|\Delta_{m,1}||L| \ge |\alpha_1||\Delta| - (m-1) \max_{1 \le k \le m-1} |\Delta_{k,1}| \max_{1 \le k \le m-1} |L_k|. \tag{21}$$

We have the obvious bounds

$$\begin{aligned} |\Delta_{k,1}| &\le (m-1)!h_0 H^{m-2}, \qquad k = 1, \dots, m-1, \\ |\Delta_{m,1}| &\le (m-1)! H^{m-1}. \end{aligned} \tag{22}$$

Substituting (20) and (22) in (21), we obtain

$$|L| \ge \frac{|\alpha_1|}{(m-1)!H^{m-1}} - \frac{(m-1)h_0\lambda}{H}. \tag{23}$$

Suppose that as H increases, the number λ, which depends on H, approaches zero faster than H^{2-m}. Then for H sufficiently large the inequality (23) gives a positive lower bound for $|L|$.

Linear forms (18) which satisfy these conditions are called numerical linear approximating forms. In Chapter 3 we constructed linear approximating forms in a set of values of E-functions. In the next section these forms will be employed to estimate measures of the values of E-functions using the above scheme.

This method can also be applied to find bounds for linear forms with algebraic integer coefficients, provided that we have a good enough lower bound for $|\Delta|$ which can be used in (21) to obtain a lower bound for $|L|$.

We note that the first bounds for transcendence measures were obtained in 1899 for the number e by E. Borel [6:1]. He proved that if H increases and s is bounded, we have

$$\Phi(e;\ s;\ H) > H^{-c \ln\ln H}, \qquad c = c(s) > 0.$$

This bound was improved significantly in 1928 by J. Popken [61:1], who showed that, under the same conditions on s and H,

$$\Phi(e;\ s;\ H) > H^{-s-\frac{\lambda}{\ln\ln H}}, \qquad \lambda = \lambda(s) > 0. \tag{24}$$

In 1932, K. Mahler [42:1] obtained the following more precise result under the same conditions:

$$\Phi(e;\ s;\ H) > H^{-s-\frac{cs^2\ln(s+1)}{\ln\ln H}}, \qquad c > 0, \qquad H \ge H_0(s), \tag{25}$$

where c is an absolute constant.

If we compare (24) and (25) with the upper bound in (11), we see that the lower bounds are very close to the natural upper limits. In particular, the principal term in the exponent is best possible in (24) and (25).

In 1929 in [72:3], Siegel used a method (a generalization of which was presented in Chapters 3 and 4) to prove that for any $\xi \in \mathbf{A} \setminus \{0\}$,

$$\Phi(J_0(\xi), J_0'(\xi);\ s;\ H) > \sigma(\xi, s) H^{-123h^3s^2}, \qquad \sigma(\xi, s) > 0, \tag{26}$$

where $J_0(z)$ is the Bessel function and $h = \deg \xi$. If we compare with (12), we see that the exponent in (26) has best possible order with respect to s.

Modern analytic methods of transcendental number theory enable one to derive estimates very close to the natural bounds for the measures of certain classes of numbers obtained as the values of analytic functions.

In the present chapter we shall only consider IE-functions and their values at points in the field **I**, in particular, QE-functions and their values at rational points. Thanks to the arithmetic properties of the field **I**, one can obtain bounds for the measures of the values of E-functions which are significantly more precise than in the case of an arbitrary algebraic number field **K**.

§ 2. The linear independence measure of values of IE-functions

Suppose that the m-tuple of KE-functions

$$f_1(z), \ldots, f_m(z) \tag{27}$$

satisfies the system of homogeneous linear differential equations

$$y'_k = \sum_{i=1}^{m} Q_{k,i} y_i, \qquad k = 1, \ldots, m, \qquad Q_{k,i} \in \mathbf{C}(z), \tag{28}$$

or the system of linear differential equations

$$y'_k = Q_{k,0} + \sum_{i=1}^{m} Q_{k,i} y_i, \qquad k = 1, \ldots, m, \qquad Q_{k,i} \in \mathbf{C}(z). \tag{29}$$

As before, we let $T(z)$ denote a least common denominator of all of the rational functions $Q_{k,i}$ in (28) or (29).

When one considers the measures of the values of the KE-functions (27) at a point $\xi \in \mathbf{A} \setminus \{0\}$, the number ξ may be chosen so that $\xi \notin \mathbf{K}$; in addition, in the definition of the measures the linear form or the polynomial P may be taken to have coefficients in an algebraic number field other than **K**. Thus, in Chapters 11–13 we shall suppose that two algebraic number fields **K** and $\mathbf{K}^*$ over **Q** are fixed (they may be the same field), and that $\xi \in \mathbf{A}$ is a fixed number with $\xi T(\xi) \neq 0$. We let $\mathbf{K}_0$ denote the smallest extension field of **Q** which contains **K**, $\mathbf{K}^*$ and ξ, and we set $h = [\mathbf{K}_0 : \mathbf{Q}]$. Here **K** will be a field which contains the power series coefficients of our KE-functions (27), and $\mathbf{K}^*$ will be the field with respect to which we are taking the measures of the values at ξ of the functions (27). We let $\xi_1 = \xi, \ldots, \xi_h$ denote the conjugates of ξ in $\mathbf{K}_0$.

Let $\mathbf{K}_{0,i}$, $i = 1, \ldots, h$, denote the fields conjugate to $\mathbf{K}_0$, $\mathbf{K}_{0,1} = \mathbf{K}_0$. For each $i = 1, \ldots, h$, the functions

$$f_{1,i}(z), \ldots, f_{m,i}(z) \tag{30}$$

are obtained from the functions (27) by replacing all of the power series coefficients by their conjugates in $\mathbf{K}_{0,i}$. We shall call the functions in (30) the *conjugates of the functions (27) relative to the field* $\mathbf{K}_{0,i}$.

In what follows we shall use analogous notation for certain other functions (including rational functions).

As in Chapter 3, we may suppose that the system of differential equations (28) (or (29)) has the property that all $Q_{k,i} \in \mathbf{K}(z)$. Then it is obvious that for each i the m-tuple (30) is a solution of the system of differential equations of the form (28) (or (29)), but with each of the rational functions $Q_{k,i}$ replaced by its conjugate function in $\mathbf{K}_{0,i}$.

Let $l_{\mathbf{K}^*} = l_{\mathbf{K}^*}(z_1, \ldots, z_m) \not\equiv 0$ denote an arbitrary linear form with coefficients in $\mathbf{Z}_{\mathbf{K}^*}$ and size $\overline{|l_{\mathbf{K}^*}|} = H$, and for $i = 1, \ldots, h$ let $l_i = l_i(z_1, \ldots, z_m)$ denote the forms obtained from $l_1 = l_{\mathbf{K}^*}$ by replacing all of the coefficients by their conjugates in $\mathbf{K}_{0,i}$.

Throughout Chapter 11, ϵ will denote an arbitrary real number with $0 < \epsilon < 1/2$. The positive constants b and c with or without subscripts will depend on the m-tuple of KE-functions (27), the integer m, and the numbers ξ and ϵ, and c will also depend on the numbers s or $s_1, \ldots, s_n, t_1, \ldots, t_\nu$ which appear in the algebraic independence measures which we are estimating. Here b and c denote different numbers in different places.

Theorem 1. *Let the m-tuple of KE-functions (27), $m \geq 2$, be a solution of the system of differential equations (28) and be linearly independent over $\mathbf{C}(z)$; and let $\xi \in \mathbf{A}$, $\xi T(\xi) \neq 0$. Then there exists a constant b such that*

$$\max_{1 \leq i \leq h} |l_i(f_{1,i}(\xi_i), \ldots, f_{m,i}(\xi_i))| > bH^{1-m-\epsilon}, \tag{31}$$

and if $\mathbf{K}_0 = \mathbf{I}$, then there exists b such that

$$L_{\mathbf{I}}(f_1(\xi), \ldots, f_m(\xi); H) > bH^{1-m-\epsilon}. \tag{32}$$

Proof. Let

$$L_0 = l_{\mathbf{K}^*}(f_1(\xi), \ldots, f_m(\xi)) = \sum_{l=1}^{m} a_l^0 f_l(\xi), \tag{33}$$

$$a_l^0 \in \mathbf{Z}_{\mathbf{K}^*}, \quad \overline{|a_l^0|} \leq H, \quad l = 1, \ldots, m, \quad |a_1^0| + \cdots + |a_m^0| > 0,$$

$$L_{0,i} = l_i(f_{1,i}(\xi_i), \dots, f_{m,i}(\xi_i)) = \sum_{l=1}^{m} a_{l,i}^0 f_{l,i}(\xi_i), \quad i = 1, \dots, h. \tag{34}$$

Under the conditions of Theorem 1 the hypothesis of Lemma 16 of Chapter 3 (with $\mathbf{K}$ replaced by $\mathbf{K}_0$) is satisfied. According to that lemma, for any natural number $n \geq n_0$, where n_0 is determined in (3.83), there exists an m-tuple of linearly independent forms in the numbers $f_1(\xi), \dots, f_m(\xi)$

$$L_k = \sum_{l=1}^{m} a_{k,l} f_l(\xi), \qquad k = 1, \dots, m, \qquad a_{k,l} \in \mathbf{Z}_{\mathbf{K}_0}, \tag{35}$$

such that

$$L_k = O(n^{-(m-1-\epsilon)n}), \qquad k = 1, \dots, m, \tag{36}$$

$$\overline{|a_{k,l}|} = O(n^{(1+\epsilon)n}), \qquad k, l = 1, \dots, m. \tag{37}$$

It is possible to choose $m - 1$ of the linear forms in (35) in such a way that they are linearly independent along with the form L_0. We may suppose that the first $m - 1$ forms have this property. For each $i = 1, \dots, h$, let

$$L_{k,i} = \sum_{l=1}^{m} a_{k,l,i} f_{l,i}(\xi_i), \qquad k = 1, \dots, m - 1, \tag{38}$$

be the forms obtained from L_k by passing to the conjugate field $\mathbf{K}_{0,i}$, i.e., by replacing the numbers $a_{k,l}$, the power series coefficients of $f_l(z)$ and the number ξ by their conjugates in $\mathbf{K}_{0,i}$.

We let Δ, $\Delta \not\equiv 0$, denote the determinant of the forms $L_1, \dots, L_{m-1}, L_0$, and we let Δ_i, $i = 1, \dots, h$, denote the determinant obtained from Δ by replacing its entries by their conjugates in $\mathbf{K}_{0,i}$. We further let $\Delta_{k,l,i}$ denote the cofactor of the entry in the k-th row and l-th column of Δ_i. Since $\Delta \in \mathbf{Z}_{\mathbf{K}_0}$ and $\Delta \not\equiv 0$, it follows that $|\mathcal{N}(\Delta)| = |\Delta_1| \cdots |\Delta_h| \geq 1$. Hence, there is some i for which

$$|\Delta_i| \geq 1, \qquad 1 \leq i \leq h. \tag{39}$$

From (34) and (38) we have for each i, $1 \leq i \leq h$:

$$\Delta_i f_{l,i}(\xi_i) = \Delta_{1,l,i} L_{1,i} + \cdots + \Delta_{m-1,l,i} L_{m-1,i} + \Delta_{m,l,i} L_{0,i}, \quad 1 \leq l \leq m, \tag{40}$$

We choose l so that $f_{l,i}(\xi_i) \neq 0$, which can be done because $T(\xi_i) \neq 0$. Then from (40) we find that

$$\overline{|\Delta_{m,l}|} |L_{0,i}| \geq |f_{l,i}(\xi_i)| |\Delta_i| - (m - 1) \max_{1 \leq k \leq m-1} \overline{|\Delta_{k,l}|} \max_{1 \leq k \leq m-1} |L_{k,i}|. \tag{41}$$

If we replace ϵ by $\epsilon/2(m-1)$ in (36) and (37) and note that (36) holds for all of the forms $L_{k,i}$, then from those inequalities we conclude that

$$\begin{aligned} \overline{|\Delta_{m,l}|} &\le b_1 n^{(m-1+\epsilon)n}, \\ \max_{1\le k\le m-1} \overline{|\Delta_{k,l}|} &\le b_2 H n^{(m-2+\epsilon/2)n}, \\ \max_{1\le k\le m-1} |L_{k,i}| &\le b_3 n^{-(m-1-\epsilon/2)n}. \end{aligned} \tag{42}$$

From (39), (41) and (42) we obtain:

$$|L_{0,i}| \ge b_4 n^{-(m-1+\epsilon)n}(1 - b_5 H n^{-(1-\epsilon)n}).$$

We choose the smallest n for which both $n \ge n_0$ and $n^{(1-\epsilon)n} > 2b_5 H$. Then

$$|L_{0,i}| > \frac{b_4}{2} n^{-(m-1+\epsilon)n}. \tag{43}$$

Under these assumptions, the following inequality holds for H sufficiently large:

$$(n-1)^{(1-\epsilon)(n-1)} \le 2b_5 H,$$

and hence

$$n^{(1-\epsilon)n} < b_6 H \ln H. \tag{44}$$

From (43) and (44), since $0 < \epsilon < 1/2$, we have

$$|L_{0,i}| > b_7 H^{1-m-2m\epsilon},$$

which implies that (31) holds for H sufficiently large. Decreasing the constant b_7 if necessary, we find that (31) holds for any H.

If $\mathbf{K}_0 = \mathbf{I}$, then (32) follows from (31), since complex conjugates have the same modulus.

The following result is a consequence of Theorem 1.

Theorem 2. *Suppose that the m-tuple of IE-functions (27), $m \ge 1$, is linearly independent over $\mathbf{C}(z)$ along with 1 and forms a solution of the system of linear differential equations (29); and let $\xi \in \mathbf{I}$, $\xi T(\xi) \neq 0$. Then there exists a constant b such that*

$$L_{\mathbf{I}}(1, f_1(\xi), \ldots, f_m(\xi); H) > bH^{-m-\epsilon}. \tag{45}$$

Theorems 1 and 2 can easily be modified for a function which satisfies a differential equation of the type

$$P_m y^{(m)} + \cdots + P_1 y' + P_0 y = 0, \tag{46}$$

$$P_i = P_i(z) \in \mathbf{C}[z], \qquad i = 0, 1, \ldots, m,$$

or of the type

$$P_m y^{(m)} + \cdots + P_1 y' + P_0 y = Q, \tag{47}$$

$$P_i = P_i(z) \in \mathbf{C}[z], \qquad i = 0, 1, \ldots, m, \qquad Q \in \mathbf{C}[z],$$

and its sequence of successive derivatives. We shall only give the theorem that corresponds to Theorem 2.

Theorem 3. *Let $f(z)$ be an IE-function which satisfies the linear differential equation (47) of order $m \geq 1$ and does not satisfy any linear differential equation of order less than m (is not a polynomial, if $m = 1$); and let $\xi \in \mathbf{I}$, $\xi T(\xi) \neq 0$. Then there exists a constant b such that*

$$L_{\mathbf{I}}(1, f(\xi), f'(\xi), \ldots, f^{(m-1)}(\xi);\, H) > bH^{-m-\epsilon}. \tag{48}$$

If we compare with (10) and (14), we see that in Theorems 1–3 the principal term in the exponent is best possible in the estimates (31), (32), (45) and (48).

§ 3. The algebraic independence measure of values of IE-functions which are not connected by algebraic equations over C(z)

From Theorem 1 one easily obtains several theorems bounding the algebraic independence measure of the values of E-functions. We shall prove these theorems in this section. As in Chapters 3 and 4, such theorems come in pairs — one for the homogeneous case and one for the non-homogeneous case. We shall give the proofs only in the homogeneous case, since the non-homogeneous case follows from the homogeneous case in the same way as in Chapter 4, where we proved the corresponding qualitative theorems.

Theorem 4. *Suppose that the m-tuple of IE-functions $f_1(z), \dots, f_m(z)$, $m \geq 2$ ($m \geq 1$), is a solution of the system of homogeneous linear differential equations (28) (the system of linear differential equations (29)) and is homogeneously algebraically independent (is algebraically independent) over $\mathbf{C}(z)$; and let $\xi \in \mathbf{I}$, $\xi T(\xi) \neq 0$. Then there exist constants c_0 and c_1 (a constant c) such that*

$$\Phi_{\mathbf{I}}^0(f_1(\xi), \dots, f_m(\xi); s; H) > c_0 H^{1-\frac{(s+m-1)!}{s!(m-1)!}-\epsilon}, \tag{49}$$

$$\Phi_{\mathbf{I}}(f_1(\xi)/f_m(\xi), \dots, f_{m-1}(\xi)/f_m(\xi); s; H) > c_1 H^{1-\frac{(s+m-1)!}{s!(m-1)!}-\epsilon}, \tag{50}$$

and in the non-homogeneous case

$$\Phi_{\mathbf{I}}(f_1(\xi), \dots, f_m(\xi); s; H) > cH^{1-\frac{(s+m)!}{s!m!}-\epsilon}. \tag{51}$$

Proof. We consider the homogeneous case, and look at the $\mu^0 = \mu^0_{s,m} = (s+m-1)!/s!(m-1)!$ functions

$$f_1^{k_1}(z)\cdots f_m^{k_m}(z), \qquad k_1 + \cdots + k_m = s, \qquad k_i \in \mathbf{Z}^+, \qquad i = 1, \dots, m.$$

We number them arbitrarily and denote them $\varphi_1(z), \dots, \varphi_{\mu^0}(z)$. By Lemma 18 of Chapter 3, this set of IE-functions forms a solution of a system of homogeneous linear differential equations of the form (28), in which m is replaced by μ^0 and which has no singular points other than those of the original system (28). Thus, if we use (7) and apply Theorem 1, we find that for some constant c_0 we have

$$\Phi_{\mathbf{I}}^0(f_1(\xi), \dots, f_m(\xi); s; H) = L(\varphi_1(\xi), \dots, \varphi_{\mu^0}(\xi); H) > c_0 H^{1-\mu^0-\epsilon}.$$

This proves (49). The inequality (50) is a consequence of (49). The non-homogeneous case (51) follows from the homogeneous case if we replace m by $m+1$ and add the function $f_0(z) \equiv 1$ to our m-tuple.

From Theorem 4 we obtain the following result.

Theorem 5. *Suppose that the IE-function $f(z)$ is a solution of the m-th order homogeneous linear differential equation (46) (the linear differential equation (47)), where $m \geq 2$ ($m \geq 1$), and is not a solution of any homogeneous algebraic differential equation (any algebraic differential equation) of order less than m (is not a polynomial, if $m = 1$); and let $\xi \in \mathbf{I}$, $\xi P_m(\xi) \neq 0$. Then there exist constants c_0 and c_1 (a constant c) such that*

$$\Phi_{\mathbf{I}}^0(f(\xi), f'(\xi), \dots, f^{(m-1)}(\xi); s; H) > c_0 H^{1-\frac{(s+m-1)!}{s!(m-1)!}-\epsilon}, \tag{52}$$

$$\Phi_{\mathbf{I}}(f'(\xi)/f(\xi), \ldots, f^{(m-1)}(\xi)/f(\xi);\ s;\ H) > c_1 H^{1-\frac{(s+m-1)!}{s!(m-1)!}-\epsilon}, \tag{53}$$

and in the non-homogeneous case

$$\Phi_{\mathbf{I}}(f(\xi), f'(\xi), \ldots, f^{(m-1)}(\xi);\ s;\ H) > cH^{1-\frac{(s+m)!}{s!m!}-\epsilon}. \tag{54}$$

We note that if $m = 2$, then (53) gives a bound for the transcendence measure of the logarithmic derivative of $f(z)$:

$$\Phi_{\mathbf{I}}(f'(\xi)/f(\xi);\ s;\ H) > cH^{-s-\epsilon}. \tag{55}$$

It is also worthwhile to take special note of the case $m = 1$ of Theorem 5.

Theorem 6. *If $f(z)$ is a transcendental IE-function which satisfies the first order linear differential equation*

$$P_1(z)y' + P_0(z)y = Q(z), \qquad Q(z), P(z), P_0(z) \in \mathbf{C}[z],$$

and if $\xi \in \mathbf{I}$, $\xi P_1(\xi) \neq 0$, then there exists a constant c such that

$$\Phi_{\mathbf{I}}(f(\xi);\ s;\ H) > cH^{-s-\epsilon}. \tag{56}$$

If $\alpha_{k,l}$, $k = 1, \ldots, m_l$, $l = 1, \ldots, n$, are complex numbers in the argument of the measures, then for brevity we shall write $\{\alpha_{k,l}\}$ instead of $\alpha_{1,1}, \ldots, \alpha_{m_1,1}; \ldots; \alpha_{1,n}, \ldots, \alpha_{m_n,n}$.

Theorem 7. *Suppose that $n, \nu \in \mathbf{Z}^+$, $n + \nu > 0$; each of the n sets of IE-functions*

$$\varphi_{1,l}(z), \ldots, \varphi_{m_l,l}(z), \qquad m_l \geq 1, \qquad l = 1, \ldots, n, \tag{57}$$

forms a solution of a corresponding system of linear differential equations

$$y'_{k,l} = Q_{k,l,0} + \sum_{i=1}^{m_l} Q_{k,l,i} y_{i,l}, \qquad Q_{k,l,i} \in \mathbf{C}(z), \tag{58}$$

$$k = 1, \ldots, m_l, \qquad l = 1, \ldots, n,$$

and each of the ν sets of IE-functions

$$\psi_{1,l}(z), \ldots, \psi_{\mu_l,l}(z), \qquad l = 1, \ldots, \nu, \tag{59}$$

forms a solution of a corresponding system of homogeneous linear differential equations

$$y'_{k,l} = \sum_{i=1}^{\mu_l} Q^*_{k,l,i} y_{i,l}, \qquad Q^*_{k,l,i} \in \mathbf{C}(z), \tag{60}$$

$$k = 1, \ldots, \mu_l, \qquad l = 1, \ldots, \nu.$$

Further suppose that the $m_1 + \cdots + m_n + \mu_1 + \cdots + \mu_\nu$ *functions in (57) and (59) are not connected by an algebraic equation with coefficients in* $\mathbf{C}[z]$ *which is homogeneous in each of the* ν *sets of functions (59). Finally, let* $\xi \in \mathbf{I}$ *be nonzero and not equal to a singular point of any of the systems in (58) or (60). Then there exist constants* c *and* c_1 *such that*

$$\Phi^*_{\mathbf{I}}(\{\varphi_{k,l}(\xi)\};\ \{\psi_{k,l}(\xi)\};\ s_1, \ldots, s_n;\ t_1, \ldots, t_\nu;\ H) > cH^{1-N-\epsilon}, \tag{61}$$

$$\begin{aligned}\Phi_{\mathbf{I}}(\{\varphi_{k,l}(\xi)\};\ &\psi_{1,1}(\xi)/\psi_{\mu_1,1}(\xi), \ldots, \psi_{\mu_1-1,1}(\xi)/\psi_{\mu_1,1}(\xi);\ \ldots \\ \ldots;\ &\psi_{1,\nu}(\xi)/\psi_{\mu_\nu,\nu}(\xi), \ldots, \psi_{\mu_\nu-1,\nu}(\xi)/\psi_{\mu_\nu,\nu}(\xi);\ s_1, \ldots, s_n;\ t_1, \ldots, t_\nu;\ H) > \\ &> c_1 H^{1-N-\epsilon},\end{aligned} \tag{62}$$

where

$$N = \left(\prod_{l=1}^{n} \frac{(s_l + m_l)!}{s_l! m_l!}\right) \prod_{k=1}^{\nu} \frac{(t_k + \mu_k - 1)!}{t_k!(\mu_k - 1)!},$$

and Φ^* *denotes the measure which is homogeneous only in the values of each of the sets of functions in (59).*

Proof. We consider the set of N IE-functions

$$\left(\prod_{j=1}^{n} \prod_{i=1}^{m_j} \varphi_{i,j}^{k_{i,j}}(z)\right) \prod_{q=1}^{\nu} \prod_{p=1}^{\mu_q} \psi_{p,q}^{\chi_{p,q}}(z), \tag{63}$$

$$0 \le \sum_{i=1}^{m_j} k_{i,j} \le s_j, \quad j = 1, \ldots, n; \qquad \sum_{p=1}^{\mu_q} \chi_{p,q} = t_q, \quad q = 1, \ldots, \nu.$$

Arguing just as in the proof of Lemma 18 of Chapter 3, we find that the set of N IE-functions in (63) forms a solution of a system of N first order homogeneous linear differential equations with coefficients in $\mathbf{C}[z]$ which has no singular points other than the singular points of all of the systems (58) and (60). Since, by assumption, the functions in (61) are linearly independent over $\mathbf{C}(z)$, Theorem 1 implies the inequality (61), from which (62) follows.

Theorem 7 contains Theorems 4–6. We now give some corollaries of Theorem 7.

Corollary 1. *If $n = 0$, then the set of functions in Theorem 7 consists only of the ν sets (59), and the measure is homogeneous in each of these sets of functions. Hence, the bounds (61) and (62) take the form*

$$\Phi_{\mathbf{I}}^0(\{\psi_{k,l}(\xi)\};\, t_1,\dots,t_\nu;\, H) > cH^{1-N_0-\epsilon}, \tag{64}$$

and

$$\begin{gathered}\Phi_{\mathbf{I}}(\psi_{1,1}(\xi)/\psi_{\mu_1,1}(\xi),\dots,\psi_{\mu_1-1,1}(\xi)/\psi_{\mu_1,1}(\xi);\,\cdots\\ \cdots;\,\psi_{1,\nu}(\xi)/\psi_{\mu_\nu,\nu}(\xi),\dots,\psi_{\mu_\nu-1,\nu}(\xi)/\psi_{\mu_\nu,\nu}(\xi);\, t_1,\dots,t_\nu;\, H) >\\ > c_0H^{1-N_0-\epsilon}, \qquad N_0 = \prod_{k=1}^{\nu}\frac{(t_k+\mu_k-1)!}{t_k!(\mu_k-1)!}.\end{gathered} \tag{65}$$

Corollary 2. *If $\nu = 0$, the set of functions in Theorem 7 consists only of the n sets (57). Hence, the bound (62) takes the form*

$$\Phi(\{\varphi_{k,l}(\xi)\};\, s_1,\dots,s_n;\, H) > cH^{1-N-\epsilon}, \qquad N = \prod_{k=1}^{n}\frac{(s_k+m_k)!}{s_k!m_k!}. \tag{66}$$

We note that the system (58) could possibly be homogeneous.

In Theorem 7 one could take the functions in (57) and (59) to be

$$\varphi_l(z), \varphi_l'(z), \dots, \varphi_l^{(m_l-1)}(z), \qquad l = 1,\dots,n,$$

and

$$\psi_l(z), \psi_l'(z), \dots, \psi_l^{(\mu_l-1)}(z), \qquad l = 1,\dots,\nu,$$

where the functions $\varphi_l(z)$ and $\psi_l(z)$ satisfy the corresponding differential equations

$$P_{m_l,l}\, y^{(m_l)} + \cdots + P_{1,l}\, y' + P_{0,l}\, y = Q_l, \qquad l = 1,\dots,n,$$

and

$$P_{\mu_l,l}^*\, y^{(\mu_l)} + \cdots + P_{1,l}^*\, y' + P_{0,l}^*\, y = 0, \qquad l = 1,\dots,\nu,$$

all of whose coefficients are in $\mathbf{C}[z]$.

We give a corollary of Theorem 1 which relates to this situation.

Corollary 3. *Suppose that each of the IE-functions $\psi_l(z)$, $l = 1, \ldots, \nu$, satisfies a second order homogeneous linear differential equation*

$$P_{2,l}(z)y'' + P_{1,l}(z)y' + P_{0,l}(z)y = 0, \quad l = 1, \ldots, \nu, \quad P_{k,l}(z) \in \mathbf{C}[z],$$

and does not satisfy any first order homogeneous algebraic differential equation with coefficients in $\mathbf{C}[z]$*; and let*

$$\xi \in \mathbf{A}, \qquad \xi P_{2,1}(\xi) \cdots P_{2,\nu}(\xi) \neq 0.$$

Then:

$$\Phi_{\mathbf{I}}^{0}(\psi_1(\xi), \psi_1'(\xi); \ldots; \psi_\nu(\xi), \psi_\nu'(\xi); t_1, \ldots, t_\nu; H) > cH^{1-(t_1+1)\cdots(t_\nu+1)-\epsilon}, \quad (67)$$

and

$$\Phi_{\mathbf{I}}(\psi_1'(\xi)/\psi_1(\xi), \ldots, \psi_\nu'(\xi)/\psi_\nu(\xi); t_1, \ldots, t_\nu; H) > c_1 H^{1-(t_1+1)\cdots(t_\nu+1)-\epsilon}. \quad (68)$$

Comparing with (10)–(17), we see that the principal term in the exponent is best possible in Theorems 4–7 and estimates (49)–(56), (61), (62), (64)–(68).

We make the following observation. Under the conditions of Theorem 4, the inequalities (49)–(51) hold for any s (here the constants c_0, c_1 and c depend on s). From the proof of the theorem it is clear that for fixed s these inequalities will still hold under weaker assumptions: it is enough to require that the functions $f_1(z), \ldots, f_m(z)$ not be connected by any homogeneous algebraic equation of degree s with coefficients in $\mathbf{C}[z]$. A similar remark applies to Theorems 5–7.

§ 4. Auxiliary results

We shall later establish bounds for the algebraic independence measures of values of sets of E-functions in the case when the functions are connected by algebraic equations with coefficients in $\mathbf{C}(z)$. This will require some preliminary results, which we shall prove in this section.

Auxiliary Theorem. *Suppose that an m-tuple of IE-functions $f_1(z), \ldots, f_m(z)$, $m \geq 2$ ($m \geq 1$), forms a solution of the system of homogeneous linear differential equations (28) (of linear differential equations (29)). Let l be the homogeneous transcendence degree (the transcendence degree) of these functions over* $\mathbf{C}(z)$, $2 \leq l \leq m$ ($1 \leq l \leq m$)*; and let $\xi \in \mathbf{I}$, $\xi T(\xi) \neq 0$. If the numbers*

$$f_1(\xi), \ldots, f_l(\xi) \quad (69)$$

are homogeneously algebraically independent (are algebraically independent), then there exist constants c_0 *and* c_1 *(a constant* c*) such that:*

$$\Phi^0_{\mathbf{I}}(f_1(\xi), \dots, f_l(\xi);\ s;\ H) > c_0 H^{-\gamma_0 s^{l-1}}, \tag{70}$$

$$\Phi_{\mathbf{I}}(f_1(\xi)/f_l(\xi), \dots, f_{l-1}(\xi)/f_l(\xi);\ s;\ H) > c_1 H^{-\gamma_0 s^{l-1}}, \tag{71}$$

and in the non-homogeneous case

$$\Phi_{\mathbf{I}}(f_1(\xi), \dots, f_l(\xi);\ s;\ H) > c H^{-\gamma_0 s^{l}}, \tag{72}$$

where γ_0 *is determined from the condition* $\dim \mathbf{L}^0_s \le \gamma_0 s^{l-1}$, $s = 1, 2, \dots$, *in which* $\mathbf{L}^0_s$ *is the vector space over* $\mathbf{C}(z)$ *spanned by the monomials*

$$f_1^{k_1}(z) \cdots f_m^{k_m}(z), \qquad k_1 + \dots + k_m = s, \qquad k_i \in \mathbf{Z}^+, \tag{73}$$

and γ *is determined from the condition* $\dim \mathbf{L}_s \le \gamma s^l$, $s = 1, 2, \dots$, *in which* $\mathbf{L}_s$ *is the vector space spanned by the monomials of the form (73) with* $k_1 + \dots + k_m \le s$, $k_i \in \mathbf{Z}^+$.

Proof. If $l = m$, then the theorem follows from Theorem 4, since the algebraic independence of the numbers (69) implies by Lemma 19 of Chapter 3 that the corresponding functions are algebraically independent over $\mathbf{C}(z)$. Here we can choose $\gamma_0 = m$ ($\gamma = m + 1$), since

$$\dim \mathbf{L}^0_s = \frac{(s+m-1)!}{s!(m-1)!} = m \frac{s+1}{2} \frac{s+2}{3} \cdots \frac{s+m-1}{m} \le m s^{l-1}.$$

Now suppose that $m > 2$ and $l < m$. Since the numbers (69) are algebraically independent, it follows that the $\mu^0_{s,l}$ numbers

$$f_1^{k_1}(\xi) \cdots f_l^{k_l}(\xi), \qquad k_1 + \dots + k_l = s, \qquad k_i \in \mathbf{Z}^+, \tag{74}$$

are linearly independent over $\mathbf{I}$. We number them arbitrarily, and denote them $\varphi_1(\xi), \dots, \varphi_{\mu^0_{s,l}}(\xi)$.

Suppose that the $\mu^0_{s,m}$ numbers

$$f_1^{k_1}(\xi) \cdots f_m^{k_m}(\xi), \qquad k_1 + \dots + k_m = s, \qquad k_i \in \mathbf{Z}^+, \tag{75}$$

have rank $\rho^0 = \rho^0_{s,l}$ relative to $\mathbf{I}$. We choose $\rho^0 - \mu^0_{s,l}$ of them which are different from the numbers in (74) and which, along with the numbers in (74), are linearly independent over $\mathbf{I}$; we denote these numbers $\varphi_{\mu^0_{s,l}+1}(\xi), \dots, \varphi_{\rho^0}(\xi)$.

If $\rho^0 < \mu^0_{s,m}$, then we further let $\varphi_{\rho^0+1}(\xi), \dots, \varphi_{\mu^0_{s,m}}(\xi)$ denote the numbers in (75) which do not occur among

$$\varphi_1(\xi), \dots, \varphi_{\rho^0}(\xi). \tag{76}$$

Let $\varphi_i(z)$ be the function in (73) corresponding to $\varphi_i(\xi)$, $i = 1, \dots, \mu^0_{s,m}$.

According to Lemma 18 of Chapter 3, the set of IE-functions in (73) forms a solution of a system of homogeneous linear differential equations of the form (28) with m replaced by $\mu^0_{s,m}$ and with no singular points other than the singular points of the system (28). Hence, Lemma 3 of Chapter 4 can be applied to this set of IE-functions, as a result of which we obtain $\rho^0 = \dim \mathbf{L}^0_s$. Since the numbers (76) are linearly independent over $\mathbf{I}$, it follows from Lemma 19 of Chapter 3 that the corresponding functions

$$\varphi_1(z), \dots, \varphi_{\rho^0}(z) \tag{77}$$

are linearly independent over $\mathbf{C}(z)$, and so form a basis of the vector space $\mathbf{L}^0_s$. Thus, when $\rho^0 < \mu^0_{s,m}$ we have

$$\varphi_j(z) = \sum_{i=1}^{\rho^0} D_{i,j}\varphi_i(z), \qquad j = \rho^0 + 1, \dots, \mu^0_{s,m}, \tag{78}$$

where all of the functions $D_{i,j}$ belong to $\mathbf{I}(z)$, by Lemma 4 of Chapter 4.

None of the functions $D_{i,j}$ can have a pole at the point $z = \xi$. In fact, otherwise, if we multiply both sides of the corresponding equation in (78) by a least common denominator of $D_{1,j}, \dots, D_{\rho^0,j}$ and set $z = \xi$, the resulting equality will contradict the linear independence of the numbers (75) over $\mathbf{I}$.

If $\rho^0 < \mu^0_{s,m}$, then in the system of homogeneous linear differential equations satisfied by the functions (73) we choose those equations whose left sides are the derivatives of the functions (77), and in the right sides of those equations we replace each function $\varphi_{\rho^0}(z), \dots, \varphi_{\mu^0_{s,m}}(z)$ by the corresponding sum in (78). Then when $\rho^0 \le \mu^0_{s,m}$ the set of functions (77) forms a solution of a system of homogeneous linear differential equations of the form (28) in which m is replaced by ρ^0 and $z = \xi$ is not a singular point.

Consequently, the set of functions (77) and the number ξ satisfy all of the conditions of Theorem 1 with m replaced by ρ^0. Theorem 1 then implies that

$$L_{\mathbf{I}}(\varphi_1(\xi), \dots, \varphi_{\rho^0}(\xi); H) > bH^{1-\rho^0-\epsilon}.$$

Hence, *a fortiori*

$$L_{\mathbf{I}}(\varphi_1(\xi), \dots, \varphi_{\mu^0_{s,l}}(\xi); H) > bH^{1-\gamma_0 s^{l-1}-\epsilon}. \tag{79}$$

But

$$\Phi_{\mathbf{I}}^{0}(f_1(\xi), \ldots, f_l(\xi);\ s;\ H) = L_{\mathbf{I}}(\varphi_1(\xi), \ldots, \varphi_{\mu_{s,l}^{0}}(\xi);\ H). \tag{80}$$

For fixed ϵ, $0 < \epsilon < 1/2$, we obtain (70) from (79) and (80). This completes the proof of the theorem.

Under the conditions of the Auxiliary Theorem, if $l = m$, then the First Fundamental Theorem (the Second Fundamental Theorem) of Chapter 3 tells us that the numbers (69) are homogeneously algebraically independent (algebraically independent) for any $\xi \in \mathbf{I}$, $\xi T(\xi) \neq 0$. Hence, (70) and (71) (respectively, (72)) hold for any such ξ.

In the case $l < m$, the analogous statements may fail to be true. In this case there may be additional exceptional points where (70) and (71) (respectively, (72)) do not hold.

As in Chapter 4, we shall say that an assertion holds for almost all numbers in a set M if it holds for all numbers in M except for those in a finite subset.

Under the conditions of the Auxiliary Theorem, it follows from Theorem 5 of Chapter 4 that for almost all $\xi \in \mathbf{I}$ the numbers (69) are homogeneously algebraically independent (algebraically independent). Hence for almost all $\xi \in \mathbf{I}$ we have (70) and (71), in which, by Lemma 6 of Chapter 4, we may take $\gamma_0 = (l+1)(kl)^{m-1}$ ($\gamma = (l+2)(k(l+1))^{m-1}$), where k is defined in that lemma.

We now prove two lemmas. Let $\mathbf{V}$ be a field, let $\mathbf{W}$ be an extension field, and let $u_1, \ldots, u_m \in \mathbf{W}$, $m \geq 2$. We consider the vector space $\mathbf{L}_N^0$ spanned over $\mathbf{V}$ by the monomials

$$u_m^{k_m} \cdots u_1^{k_1}, \qquad k_1 + \cdots + k_m = N, \qquad k_i \in \mathbf{Z}^+, \qquad N \in \mathbf{N}, \tag{81}$$

and the vector space $\mathbf{L}_N$ spanned by the monomials of the form (81) with $k_1 + \cdots + k_m \leq N$.

Lemma 1. *If the homogeneous transcendence degree (the transcendence degree) over* $\mathbf{V}$ *of the* m*-tuple* $u_1, \ldots, u_m$ *is equal to* $m-1$, *and if these elements are connected by the algebraic equation*

$$P = P(u_1, \ldots, u_m) = 0, \qquad P \in \mathbf{V}[u_1, \ldots, u_m], \qquad \deg_{\mathbf{u}} P = k, \qquad k \geq 1,$$

where P *is a homogeneous irreducible polynomial (an irreducible polynomial), then*

$$\dim \mathbf{L}_N^0 \leq k(m-1)N^{m-2} \qquad (\dim \mathbf{L}_N \leq kmN^{m-1}).$$

Proof. Let

$$u_m^{\kappa_m} \cdots u_1^{\kappa_1}, \qquad \kappa_1 + \cdots + \kappa_m = k,$$

be the leading term of P. We let B_N denote the subset of the set (81) consisting of $u_m^{k_m} \cdots u_1^{k_1}$ such that $k_i < \kappa_i$ for some index i, $1 \leq i \leq m$. The subset B_N forms a basis of $\mathbf{L}_N^0$, since obviously its elements are linearly independent over $\mathbf{V}$, and, using the equation $P = 0$, any element in (81) not in B_N can be represented as a linear combination of elements of B_N with coefficients in $\mathbf{V}$. The latter fact also follows from Lemma 17 of Chapter 4.

We let $B_{N,i}$, $i = 1, \dots, m$, denote the (perhaps empty) subset of the set (81) consisting of $u_m^{k_m} \cdots u_1^{k_1}$ for which $k_i < \kappa_i$. Then $B_N = \bigcup_{i=1}^m B_{N,i}$. The number of elements in $B_{N,i}$ is:

$$\sum_{k_i=0}^{\min(\kappa_i-1,N)} \mu_{N-k_i,m-1}^0 \leq \kappa_i \mu_{N,m-1}^0.$$

Hence, we have the inequality

$$\begin{aligned} \dim \mathbf{L}_N^0 &\leq \sum_{i=1}^m \kappa_i \mu_{N,m-1}^0 = k\frac{(N+m-2)!}{N!(m-2)!} = \\ &= k(m-1)\frac{N+1}{2}\frac{N+2}{3}\cdots\frac{N+m-2}{m-1} \leq k(m-1)N^{m-2}. \end{aligned}$$

Lemma 2. *If the homogeneous transcendence degree (the transcendence degree) over* $\mathbf{V}$ *of the* m*-tuple* $u_1, \dots, u_m$ *is equal to* l, $1 \leq l \leq m$, *and if these elements are connected by the algebraic equations*

$$P_\nu = P_\nu(u_1, \dots, u_\nu) = 0, \qquad P_\nu \in \mathbf{V}[u_1, \dots, u_\nu],$$

$$\deg_{\mathbf{u}} P_\nu = \kappa_\nu, \qquad \kappa_\nu \geq 1, \qquad \nu = l+1, \dots, m,$$

where P_ν *is a homogeneous irreducible polynomial (an irreducible polynomial) in which* u_ν *occurs to the power* κ_ν*, then:*

$$\dim \mathbf{L}_N^0 \leq klN^{l-1} \qquad (\dim \mathbf{L}_N \leq k(l+1)N^l), \qquad k = \prod_{\nu=l+1}^m \kappa_\nu.$$

Proof. The subset of the set (81) consisting of

$$u_m^{k_m} \cdots u_1^{k_1}, \qquad k_1 + \cdots + k_m = N, \quad k_i < \kappa_i, \quad i = l+1, \dots, m,$$

obviously contains a basis of $\mathbf{L}_N^0$. But this subset contains no greater a number of elements than the following set:

$$u_m^{k_m} \cdots u_1^{k_1}, \qquad k_1 + \cdots + k_l = N, \quad k_i < \kappa_i, \quad i = l+1, \ldots, m,$$

and the latter set has $k\mu_{N,l}^0$ elements. Hence,

$$\dim \mathbf{L}_N^0 \leq k\mu_{N,l}^0 = kl\frac{N+1}{2}\frac{N+2}{3}\cdots\frac{N+l-1}{l} \leq klN^{l-1}.$$

§ 5. The algebraic independence measure of values of IE-functions which are connected by algebraic equations over $\mathbf{C}(z)$

We now prove the quantitative analogue of Theorem 11 of Chapter 4.

Theorem 8. *Suppose that the m-tuple of IE-functions*

$$f_1(z), \ldots, f_m(z), \qquad m \geq 3 \ \ (m \geq 2), \tag{82}$$

forms a solution of the system of homogeneous linear differential equations (28) (of linear differential equations (29)); that the functions $f_1(z), \ldots, f_{m-1}(z)$ are homogeneously algebraically independent (are algebraically independent) over $\mathbf{C}(z)$; and that the functions (82) are connected by an algebraic equation

$$P = P(f_1(z), \ldots, f_m(z)) = 0,$$

where P is a homogeneous (an arbitrary) irreducible and primitive polynomial in the variables (82) with coefficients in $\mathbf{C}[z]$, $\deg_{\mathbf{f}} P = k$, $k \geq 1$. Further let Ω be the set of common zeros of all of the coefficients of the terms in P which contain $f_m(z)$, and let $\xi \in \mathbf{I}$, $\xi T(\xi) \neq 0$, $\xi \neq \Omega$. Then there exist constants c_0 and c_1 (a constant c) such that:

$$\Phi_{\mathbf{I}}^0(f_1(\xi), \ldots, f_{m-1}(\xi);\ s;\ H) > c_0 H^{-k(m-1)s^{m-2}}, \tag{83}$$

$$\Phi_{\mathbf{I}}(f_1(\xi)/f_{m-1}(\xi), \ldots, f_{m-2}(\xi)/f_{m-1}(\xi);\ s;\ H) > c_1 H^{-k(m-1)s^{m-2}}, \tag{84}$$

and in the non-homogeneous case

$$\Phi_{\mathbf{I}}(f_1(\xi), \ldots, f_{m-1}(\xi);\ s;\ H) > cH^{-kms^{m-1}}. \tag{85}$$

Proof. By Theorem 11 of Chapter 4, the numbers $f_1(\xi), \dots, f_{m-1}(\xi)$ are algebraically independent. Hence, applying the Auxiliary Theorem with $l = m - 1$ and Lemma 1, we obtain (83)–(85).

Remark. In (83) and (84) the exponent of H can be replaced by $-M$, where

$$M = \begin{cases} \frac{(s+m-1)!}{s!(m-1)!} - \frac{(s-k+m-1)!}{(s-k)!(m-1)!}, & \text{if } s \geq k, \\ \frac{(s+m-1)!}{s!(m-1)!}, & \text{if } s < k, \end{cases}$$

and in (85) the exponent can be replaced by the analogous expression with m in place of $m - 1$.

Using the Auxiliary Theorem, it is also easy to prove quantitative analogues of Theorems 6, 8 and 14 of Chapter 4 in the case $\mathbf{K} = \mathbf{I}$.

Theorem 9. *Suppose that the m-tuple of IE-functions (82) forms a solution of the system of homogeneous linear differential equations (28) (of linear differential equations (29)); that $f_1(z), \dots, f_l(z)$, $2 \leq l < m - 1$ ($1 \leq l \leq m - 1$), are homogeneously algebraically independent (are algebraically independent) over $\mathbf{C}(z)$; and that the functions (82) are connected by homogeneous (by arbitrary) algebraic equations*

$$P_\nu(z, f_1(z), \dots, f_\nu(z)) = 0,$$

$$P_\nu = P_\nu(z, z_1, \dots, z_\nu) \in \mathbf{C}[z, z_1, \dots, z_\nu], \qquad \nu = l + 1, \dots, m,$$

where P_ν is a homogeneous (an arbitrary) irreducible polynomial of degree κ_ν in $z_1, \dots, z_\nu$, $\kappa_\nu \geq 1$, which contains a term of the form

$$A_\nu(z) z_\nu^{\kappa_\nu}, \qquad A_\nu(z) \in \mathbf{C}[z], \qquad A_\nu(z) \not\equiv 0, \qquad \nu = l + 1, \dots, m.$$

Further let $\xi \in \mathbf{I}$, $\xi T(\xi) A_{l+1}(\xi) \cdots A_m(\xi) \neq 0$. Then there exist constants c_0 and c_1 (a constant c) such that:

$$\Phi_{\mathbf{I}}^0(f_1(\xi), \dots, f_l(\xi); s; H) > c_0 H^{-\gamma_0 s^{l-1}}, \tag{86}$$

$$\Phi_{\mathbf{I}}(f_1(\xi)/f_l(\xi), \dots, f_{l-1}(\xi)/f_l(\xi); s; H) > c_1 H^{-\gamma_0 s^{l-1}}, \tag{87}$$

where $\gamma_0 = l \kappa_{l+1} \cdots \kappa_m$, and in the non-homogeneous case

$$\Phi_{\mathbf{I}}(f_1(\xi), \dots, f_l(\xi); s; H) > c H^{-\gamma s^l}, \tag{88}$$

where $\gamma = (l + 1) \kappa_{l+1} \cdots \kappa_m$.

Proof. By Theorem 6 of Chapter 4, the numbers $f_1(\xi),\dots,f_l(\xi)$ are homogeneously algebraically independent. Hence, using the Auxiliary Theorem and Lemma 2, we obtain (86), (87) and (88).

Theorem 10. *Suppose that the m-tuple of IE-functions (82) forms a solution of the system of homogeneous linear differential equations (28) (of linear differential equations (29)); that*

$$\deg \operatorname{tr}^0_{\mathbf{C}(z)}\{f_1(z),\dots,f_m(z)\} = \deg \operatorname{tr}^0_{\mathbf{C}}\{f_1(z),\dots,f_m(z)\} = l, \quad 2 \le l \le m-1,$$

$$(\deg \operatorname{tr}_{\mathbf{C}(z)}\{f_1(z),\dots,f_m(z)\} = \deg \operatorname{tr}_{\mathbf{C}}\{f_1(z),\dots,f_m(z)\} = l, \quad 1 \le l \le m-1);$$

and that the functions $f_1(z),\dots,f_l(z)$ are homogeneously algebraically independent (are algebraically independent) over $\mathbf{C}(z)$. Further let $\xi \in \mathbf{A}$, $\xi T(\xi) \neq 0$. Then there exist constants c_0 and c_1 (a constant c) such that (86) and (87) hold, where $\gamma_0 = (l+1)(lk)^{m-l}$, and (88) holds in the non-homogeneous case, where $\gamma = (l+2)((l+1)k)^{m-l}$ and k is defined in Lemma 6 of Chapter 4.

Proof. By Theorem 8 of Chapter 4, the numbers $f_1(\xi),\dots,f_l(\xi)$ are homogeneously algebraically independent. Hence, applying the Auxiliary Theorem and Lemma 12 of Chapter 4, we obtain (86)–(88).

Theorem 11. *Suppose that the m-tuple of IE-functions (82) forms a solution of the system of homogeneous linear differential equations (28) (of linear differential equations (29)); that the homogeneous transcendence degree (the transcendence degree) of these functions over $\mathbf{C}(z)$ is equal to l, $2 \le l \le m-1$ $(1 \le l \le m-1)$; and that the functions $f_1(z),\dots,f_l(z)$ are homogeneously algebraically independent (are algebraically independent) over $\mathbf{C}(z)$. Further let*

$$A_1(z) f_m^{\kappa_{m,1}}(z)\cdots f_1^{\kappa_{1,1}}(z),\ \dots,\ A_\nu(z) f_m^{\kappa_{m,\nu}}(z)\cdots f_1^{\kappa_{1,\nu}}(z),$$

$$A_i(z) \in \mathbf{C}[z], \qquad i = 1,\dots,\nu,$$

be the set of leading terms of the minimal homogeneous (minimal arbitrary) equations of the functions (82) over $\mathbf{C}(z)$; and let $\xi \in \mathbf{A}$, $\xi T(\xi) A_1(\xi)\cdots A_\nu(\xi) \neq 0$. Then there exist constants c_0 and c_1 (a constant c) such that (86) and (87) hold with $\gamma_0 = l(m-l+1)(l\kappa)^{m-l}$ and (88) holds in the non-homogeneous case with $\gamma = (l+1)(m-l+1)((l+1)\kappa)^{m-1}$, where κ is defined in Lemma 18 of Chapter 4.

Proof. According to Theorem 14 of Chapter 4, the numbers $f_1(\xi),\dots,f_l(\xi)$ are homogeneously algebraically independent. Hence, if we apply the Auxiliary Theorem and Lemma 18 of Chapter 4, we obtain (86)–(88).

Comparing the lower bounds in Theorems 8–11 with the upper bounds (12), (13), (16) and (17), which were obtained using the Dirichlet pigeon-hole principle, we see that the exponents of H in (83)–(88) have best possible order in s.

Theorems 8–11 can be restated so as to apply to an IE-function satisfying a linear differential equation and its sequence of successive derivatives.

We note that in the theorems giving bounds for measures in this and the next two chapters the system of differential equations (29) may turn out to be homogeneous (i.e., $Q_{k,0} = 0$, $k = 1, \dots, m$). In this case the non-homogeneous cases of the theorems concern non-homogeneous measures of the values of E-functions which satisfy a homogeneous system.

Using Theorem 1, it is possible to prove analogues of Theorem 7 and its corollaries in the case when the functions (57) and (59) are algebraically dependent.

We note that, if we somewhat modify the definition of an E-function and refine the bounds in the earlier lemmas, then in Theorems 1–7 we can improve the remainder term in the exponent, replacing ϵ by a certain decreasing function of H. Such a refinement will be given in Chapter 13.

§ 6. Some applications of the general theorems

In Chapters 5–10 we proved theorems about the algebraic independence over $\mathbf{C}(z)$ of many classes of E-functions and obtained corresponding arithmetic results about the values of these functions at algebraic points. In the case $\mathbf{K} = \mathbf{I}$, the theorems of Chapter 11 enable us, without any additional work, to give bounds for the algebraic independence measures of these values. Thus, all of the qualitative arithmetic theorems of Chapters 5–10 obtain a quantitative dimension in the form of bounds for the corresponding measures. We give some simple examples.

The transcendental E-function

$$\varphi_\lambda(z) = \sum_{n=0}^{\infty} \frac{z^n}{(\lambda+1)\cdots(\lambda+n)}, \qquad \lambda \in \mathbf{Q}, \qquad \lambda \neq -1, -2, \dots,$$

satisfies the first order linear differential equation (5.18). Hence, by Theorem 6,

$$\Phi_{\mathbf{I}}(\varphi_\lambda(\xi);\, s;\, H) > cH^{-s-\epsilon}, \qquad \xi \in \mathbf{I}, \quad \xi \neq 0.$$

In particular, for $\lambda = 0$,

$$\Phi_{\mathbf{I}}(e^{\xi};\, s;\, H) > cH^{-s-\epsilon}, \qquad \xi \in \mathbf{I}, \quad \xi \neq 0.$$

If $\xi_1, \dots, \xi_m$ are distinct numbers in $\mathbf{I}$, then, by Lemma 20 of Chapter 3, the functions

$$y_k = e^{\xi_k z}, \qquad y_k' = \xi_k y_k, \qquad k = 1, \dots, m,$$

are linearly independent over $\mathbf{C}(z)$. By Theorem 1,

$$L_{\mathbf{I}}(e^{\xi_1}, \dots, e^{\xi_m}; H) > bH^{1-m-\epsilon}.$$

The E-function

$$K_\lambda(z) = \sum_{n=0}^{\infty} \frac{1}{n![\lambda+1, n]} \left(\frac{z}{2}\right)^{2n}, \tag{89}$$

$$\lambda \in \mathbf{Q}, \qquad \lambda \neq -k_1, \qquad k_1 \in \mathbf{N}, \qquad \lambda \neq \frac{2k_2 - 1}{2}, \qquad k_2 \in \mathbf{Z}^+,$$

satisfies the second order homogeneous linear differential equation (6.9), and, by Lemma 3 of Chapter 6, the functions $K_\lambda(z)$ and $K_\lambda'(z)$ are algebraically independent over $\mathbf{C}(z)$. Hence, by Theorem 5, for nonzero $\xi \in \mathbf{I}$ we have

$$\begin{aligned} &\Phi_{\mathbf{I}}^0(K_\lambda(\xi), K_\lambda'(\xi); s; H) > c_0 H^{-s-\epsilon}, \\ &\Phi_{\mathbf{I}}(K_\lambda'(\xi)/K_\lambda(\xi); s; H) > c_1 H^{-s-\epsilon}, \\ &\Phi_{\mathbf{I}}(K_\lambda(\xi), K_\lambda'(\xi); s; H) > cH^{-s(s+3)/2-\epsilon}. \end{aligned} \tag{90}$$

In particular, in view of (6.48) and (6.49), it follows from (90) that

$$\left| \frac{K_\lambda'(r)}{K_\lambda(r)} - \frac{p}{q} \right| > bq^{-2-\epsilon}, \qquad p \in \mathbf{Z}, \;\; q \in \mathbf{N}, \;\; r \in \mathbf{Q}, \;\; r \neq 0,$$

$$\Phi_{\mathbf{I}}(J_\lambda'(\xi)/J_\lambda(\xi); s; H) > cH^{-s-\epsilon}, \qquad \xi \in \mathbf{I}, \;\; \xi \neq 0,$$

and

$$\Phi(\alpha; s; H) > cH^{-2-\epsilon},$$

where α is the continued fraction

$$\alpha = 1 + \cfrac{1}{2 + \cfrac{1}{3 + \cfrac{1}{4 + \cdots}}}.$$

Suppose that $\lambda_i \in \mathbf{Q}$, $i = 1, \dots, m$, and $\xi_j \in \mathbf{I}$, $j = 1, \dots, n$, satisfy the conditions of Siegel's Theorem in Chapter 9. We number the functions $\{K_{\lambda_i}(\xi_j z)\}$ in an arbitrary way and denote them $\varphi_1(z), \dots, \varphi_{mn}(z)$. According to Corollaries 2 and 3 of Theorem 7 and Lemma 1 of Chapter 9, we have the lower bounds

$$\Phi_{\mathbf{I}}^0(\varphi_1(1), \varphi_1'(1); \dots; \varphi_{mn}(1), \varphi_{mn}'(1); s_1, \dots, s_{mn}; H) > \\ > c_0 H^{1-(s_1+1)\cdots(s_{mn}+1)-\epsilon},$$

$$\Phi_{\mathbf{I}}(\varphi_1'(1)/\varphi_1(1); \dots; \varphi_{mn}'(1)/\varphi_{mn}(1); s_1, \dots, s_{mn}; H) > \\ > c_1 H^{1-(s_1+1)\cdots(s_{mn}+1)-\epsilon},$$

$$\Phi_{\mathbf{I}}(\varphi_1(1), \varphi_1'(1); \dots; \varphi_{mn}(1), \varphi_{mn}'(1); s_1, \dots, s_{mn}; H) > cH^{1-N-\epsilon},$$

$$\text{where} \quad N = 2^{-mn} \prod_{k=1}^{mn} (s_k + 1)(s_k + 2).$$

We consider the E-functions

$$\psi_k(z) = \sum_{n=0}^{\infty} \frac{1}{(n!)^k} \left(\frac{z}{k}\right)^{kn}, \qquad k = 1, \dots, m,$$

which satisfy homogeneous linear differential equations with coefficients in $\mathbf{C}(z)$ and with singular point $z = 0$ (see Chapter 8 § 1). If we use the proof of Theorem 6 of Chapter 8, then from Theorem 5 and Corollary 1 of Theorem 7 we obtain the following bounds for nonzero $\xi \in \mathbf{I}$:

$$\Phi_{\mathbf{I}}(\psi_1(\xi); \psi_2(\xi), \psi_2'(\xi); \dots; \psi_m(\xi), \psi_m'(\xi), \dots, \psi_m^{(m-1)}(\xi); s; H) > \\ > cH^{1-\frac{(s+t)!}{s!t!}-\epsilon},$$

where $t = m(m+1)/2$, and

$$\Phi_{\mathbf{I}}(\psi_1(\xi); \psi_2(\xi), \psi_2'(\xi); \dots; \psi_m(\xi), \psi_m'(\xi), \dots, \psi_m^{(m-1)}(\xi); s_1, \dots, s_m; H) > \\ > c_1 H^{1-N-\epsilon},$$

where

$$N = \prod_{k=1}^{m} \frac{(s_k + k)!}{s_k! k!}.$$

We consider the E-functions

$$\operatorname{si} z = \int_0^z \frac{\sin t}{t}\, dt, \qquad \operatorname{ci} z = \int_0^z \frac{1 - \cos t}{t}\, dt.$$

In the proof of Theorems 2 and 5 of Chapter 8, we showed that the functions $\operatorname{si} z$ and $\sin z$, and also $\operatorname{ci} z$ and $\cos z$, are algebraically independent over $\mathbf{C}(z)$. Consider the two triples of functions

$$\operatorname{si} z, \quad \sin z, \quad \cos z; \qquad \operatorname{ci} z, \quad \cos z, \quad \sin z,$$

in each of which we have the algebraic equation $\sin^2 z + \cos^2 z = 1$ and the following system of differential equations, respectively:

$$\begin{aligned} y_1' &= y_2/z, & y_2' &= y_3, & y_3' &= -y_2, \\ y_1' &= (1-y_2)/z, & y_2' &= -y_3, & y_3' &= y_2. \end{aligned}$$

By Theorem 8 and the remark following it, we obtain the following inequalities for nonzero $\xi \in \mathbf{I}$:

$$\Phi_{\mathbf{I}}^0(\operatorname{si}\xi, \sin\xi;\ s;\ H) > c_0 H^{-(2s+1)}, \qquad \Phi_{\mathbf{I}}^0(\operatorname{ci}\xi, \cos\xi;\ s;\ H) > c_0 H^{-(2s+1)},$$

$$\Phi_{\mathbf{I}}(\operatorname{si}\xi, \sin\xi;\ s;\ H) > c_1 H^{-(s+1)^2}, \qquad \Phi_{\mathbf{I}}(\operatorname{ci}\xi, \cos\xi;\ s;\ H) > c_1 H^{-(s+1)^2}.$$

Let

$$\varphi_k(z) = 1 + \sum_{n=1}^{\infty} \frac{(-1)^n z^{2n}}{(2n)!(2n)^k}, \qquad k = 0, 1, \dots, m,$$

where $\varphi_0(z) = \cos z$. In the proof of Theorem 4 of Chapter 8 we showed that these functions are algebraically independent over $\mathbf{C}(z)$, and that, when $\sin z$ is added, they are connected by the equation $\sin^2 z + \cos^2 z = 1$ and satisfy a system of $m+2$ first order linear differential equations with coefficients in $\mathbf{C}(z)$ and with singular point at $z = 0$. Hence, by Theorem 8 and the remark following it, for nonzero $\xi \in \mathbf{I}$ we have

$$\Phi_{\mathbf{I}}(\varphi_0(\xi), \varphi_1(\xi), \dots, \varphi_m(\xi);\ s;\ H) > cH^{-(s+1)^{m+1}}.$$

We now return to the functions $K_\lambda(z)$ in (89). If $\lambda \in \mathbf{Q}$, $\lambda \neq k_1$, $k_1 \in \mathbf{Z}$, $\lambda \neq (2k_2 - 1)/2$, $k_2 \in \mathbf{Z}$, then in § 8 of Chapter 9 we proved that the functions $y_1 = K_\lambda(z)$, $y_2 = K_\lambda'(z)$, $y_3 = K_{-\lambda}(z)$, $y_4 = K_{-\lambda}'(z)$ have transcendence degree 3 over $\mathbf{C}(z)$ and are connected by the equation

$$y_1 y_4 - y_2 y_3 + \frac{2\lambda}{z} y_1 y_3 + \frac{2\lambda}{z} = 0.$$

They satisfy a system of first order homogeneous linear differential equations with singular point $z = 0$. Let $\varphi_1(z)$, $\varphi_2(z)$, $\varphi_3(z)$ denote any three of these functions.

Then, by Theorem 8 and the remark following it, for nonzero $\xi \in \mathbf{I}$ we obtain:

$$\Phi_{\mathbf{I}}(\varphi_1(\xi), \varphi_2(\xi), \varphi_3(\xi);\ s;\ H) > cH^{-(s+1)(s+2)(2s+3)/6}.$$

Remarks

The method of proof of the bounds for a linear form that was presented in § 1 is contained in Siegel's article [72:3]. In the same article Siegel states without proof that his method will give the bound

$$|a_1 J_0(r) + a_2 J_0'(r) + a_3| > cH^{-2-\epsilon}, \qquad a_i \in \mathbf{Z}, \qquad |a_i| \le H,$$

where $J_0(z)$ is the Bessel function, $r \in \mathbf{Q}$, $r \neq 0$, ϵ is an arbitrary positive number, and c depends only on r and ϵ.

Theorems 1–6 and Corollary 2 of Theorem 7 were published in 1967 in [70:21], and Theorem 7 for the measures L, Φ and Φ^0 appeared in 1977 in [70:27]. Theorem 1 in its final version and Theorems 8–11 were published in 1980 in [70:32]. Theorems with bounds for the measures of IE-functions which improve upon the remainder term in the exponent were announced in 1967 in the note [70:20] and were published with proofs in 1979 in [70:29]. Theorem 1 is also contained in [70:34].

Using the well-known Khinchin transfer principle [30:1], from Theorem 1 one can obtain a bound for simultaneous approximations of the numbers $f_1(\xi), \ldots$ $\ldots, f_m(\xi)$ considered in Theorem 1. One argues in the same way as in Galochkin's article [22:8], where a similar problem was solved for the values of certain hypergeometric functions. Analogous results concerning simultaneous approximation can also be obtained from the bounds for the measures which will be proved in Chapters 12 and 13.

We call attention to Makarov's articles [45:1–3], which give estimates for linear forms and polynomials in the values of E-functions which depend either on the sizes of all of the coefficients of the form or else on the heights of the polynomial with respect to several subsets of its coefficients. The same type of bound for linear forms in the values of E-functions is contained in articles by K. Väänänen [79:6,7,10] and R. Wallisser [85:1].

Chapter 12
The algebraic independence measure of values of KE-functions

§ 1. The fundamental lemma

In this chapter we derive bounds for the algebraic independence measures of the values of KE-functions at algebraic points in various cases depending on the number of algebraic equations which connect these functions over $\mathbf{C}(z)$ and their structure.

The power series coefficients of these functions will belong to an algebraic number field $\mathbf{K}$, and the coefficients of the polynomials in the definition of the measure will belong to an algebraic number field $\mathbf{K}^*$. We let $\mathbf{K}_0$ be an algebraic number field which contains $\mathbf{K}$, $\mathbf{K}^*$, and a fixed number $\xi \in \mathbf{A}$, and we set $h = [\mathbf{K}_0 : \mathbf{Q}]$. The m-tuple of KE-functions

$$f_1(z), \dots, f_m(z) \tag{1}$$

will form a solution of the system of homogeneous linear differential equations

$$y_k' = \sum_{i=1}^{m} Q_{k,i} y_i, \qquad k = 1, \dots, m, \qquad Q_{k,i} \in \mathbf{C}(z), \tag{2}$$

or the system of linear differential equations

$$y_k' = Q_{k,0} + \sum_{i=1}^{m} Q_{k,i} y_i, \qquad k = 1, \dots, m, \qquad Q_{k,i} \in \mathbf{C}(z), \tag{3}$$

and the polynomial $T = T(z)$ will have the same meaning as always.

In order to avoid repeating essentially the same argument, we shall describe the general nature of the bounds in the form of a fundamental lemma given below.

In § 1 we consider the homogeneous case. Suppose that the functions (1) satisfy the system (2), and for each N the symbol L_N^0 denotes, as before, the set of $\mu_{N,m}^0$ products of powers

$$f_m^{k_m}(z) \cdots f_1^{k_1}(z), \qquad k_i \in \mathbf{Z}^+, \qquad k_1 + \dots + k_m = N. \tag{4}$$

Let B_N, $N = 1, 2, \ldots$, denote sets of monomials in the functions (1) which satisfy the following conditions:

1°. $B_N \subset L_N^0$.

2°. The number of elements in B_N is given by $M = \varphi(N)$, where $\varphi(N)$ is a positive increasing function of N.

3°. The elements of B_N are linearly independent over $\mathbf{C}(z)$.

4°. The M-tuple of functions in B_N forms a solution of a system of M homogeneous linear differential equations of the form (2) (with m replaced by M) having singular points only in a set Λ, where Λ is a finite subset of $\mathbf{A}$ which contains zero and the roots of the polynomial $T(z)$.

There always exists a sequence of sets B_N, $N = 1, 2, \ldots$, for the KE-functions (1) satisfying (2). For example, if the functions (1) are homogeneously algebraically independent over $\mathbf{C}(z)$, then, by Lemma 18 of Chapter 3, we can set $B_N = L_N^0$ and $\varphi(N) = \mu_{N,m}^0$. On the other hand, if the functions (1) are homogeneously algebraically dependent over $\mathbf{C}(z)$, then, arguing as in § 1 of Chapter 4, we see that B_N can be taken to be, for example, any basis of the vector space $\mathbf{L}_N^0$ spanned over $\mathbf{C}(z)$ by the elements (4) of L_N^0.

Let $\xi \in \mathbf{A}$. We let $L_{N,\xi}^0$ and $B_{N,\xi}$ be the sets of numbers which are obtained from L_N^0 and B_N by setting $z = \xi$ in the functions which make up those sets.

For given N, $N \geq s$, and $\xi \in \mathbf{A}$, $\xi \notin \Lambda$, we say that a homogeneous polynomial

$$P = P(z_1, \ldots, z_m) \in \mathbf{Z}_{\mathbf{K}^*}[z_1, \ldots, z_m], \tag{5}$$

$$P \not\equiv 0, \qquad \deg_{\mathbf{z}} P = s, \qquad s \geq 1, \qquad \overline{|P|} = H,$$

is a $B_{N,\xi}$-*polynomial* if the following conditions are fulfilled:

1°. For any $f_m^{k_m}(z) \cdots f_1^{k_1}(z) \in B_{N-s}$ the polynomial in the numbers $f_1(\xi), \ldots, f_m(\xi)$

$$f_m^{k_m}(\xi) \cdots f_1^{k_1}(\xi) P(f_1(\xi), \cdots, f_m(\xi)), \tag{6}$$

$$k_i \in \mathbf{Z}^+, \qquad k_1 + \cdots + k_m = N - s,$$

can be written as a linear combination of elements in $B_{N,\xi}$ with coefficients in $\mathbf{K}_0$.

2°. All of the $\mu_{N-s,m}^0$ linear forms in the elements of $B_{N,\xi}$ obtained from the different polynomials (6) are linearly independent over $\mathbf{K}_0$.

3°. There exists a constant σ_0, depending on the KE-functions (1) and the numbers m, s, ξ and N, such that all of these $\mu_{N-s,m}^0$ linear forms can be multiplied by natural numbers so as then to have coefficients in $\mathbf{Z}_{\mathbf{K}_0}$ of size no greater than $\sigma_0 H$.

There are two situations which will be of interest in the subsequent proofs of the theorems bounding the measures.

1. After multiplying out each expression (6) term-by-term, these expressions become linear forms in the elements of $B_{N,\xi}$. Each coefficient of the forms is either zero or else a coefficient of P. It is then obvious that conditions 1°–3° hold for these forms, and P is a $B_{N,\xi}$-polynomial for such N and ξ.

2. All of the elements in the set $L^0_{N,\xi} \setminus B_{N,\xi}$ can be expressed as a linear combination of the elements of $B_{N,\xi}$ with coefficients in $\mathbf{K}_0$. Then, if we use these linear relations and multiply out each expression in (6), those expressions become linear forms in the elements of $B_{N,\xi}$. If these forms are linearly independent over $\mathbf{K}_0$, then P is a $B_{N,\xi}$-polynomial for such N and ξ, since conditions 1°–3° again hold for the linear forms.

If the KE-functions (1) are homogeneously algebraically independent over $\mathbf{C}(z)$, and if $B_N = L^0_N$, $N = 1, 2, \ldots$, then any polynomial (6) is a $B_{N,\xi}$-polynomial for any $N \geq s$ and any $\xi \in \mathbf{A}$, $\xi T(\xi) \neq 0$.

The following notation will be used in the rest of the chapter. The positive constants σ, c and γ with or without subscripts will depend on the functions and parameters as follows:

1) the constants σ (in § 1) will depend on the KE-functions (1), the numbers m, s and ξ, and the numbers N and ϵ which appear in § 1;

2) the constants c (in § 1–6) will depend on the KE-functions (1) and the numbers m, s and ξ;

3) the constants γ (in § 3–6) will depend on the KE-functions (1) and on the transcendence degree l of the functions (1) over $\mathbf{C}(z)$.

Fundamental Lemma. *Suppose we are given sets B_N, $N = 1, 2, \ldots$. Suppose that the polynomial (5) is a $B_{N,\xi}$-polynomial for certain N, $N \geq s$, and $\xi \in \mathbf{A}$, $\xi \notin \Lambda$. Let $v = \varphi(N - s)$ and $w = M - v$. If N has the property that*

$$M - (2n - 1)w > 0, \tag{7}$$

then there exists a constant c such that:

$$|P(\xi)| = |P(f_1(\xi), \ldots, f_m(\xi))| > cH^{-2hM}. \tag{8}$$

Proof. We number the KE-functions in B_N in an arbitrary way, and denote them

$$F_1(z), \ldots, F_M(z), \qquad M = \varphi(N). \tag{9}$$

We similarly list the functions in B_{N-s}

$$G_1(z), \ldots, G_v(z), \qquad v = \varphi(N - s). \tag{10}$$

We consider the following set of v polynomials in the numbers $f_1(\xi), \dots, f_m(\xi)$:

$$\Psi_i(\xi) = r_i G_i(\xi) P(f_1(\xi), \dots, f_m(\xi)), \qquad r_i \in \mathbf{N}, \qquad i = 1, \dots, v. \tag{11}$$

Then, by the definition of a $B_{N,\xi}$-polynomial, for a suitable choice of the numbers r_i the polynomials $\Psi_i(\xi)$ can be represented in the form

$$\Psi_i(\xi) = \sum_{j=1}^{M} b_{i,j} F_j(\xi), \qquad i = 1, \dots, v, \tag{12}$$

$$b_{i,j} \in \mathbf{Z}_{\mathbf{K}_0}, \qquad \overline{|b_{i,j}|} \le \sigma_1 H, \qquad i = 1, \dots, v, \qquad j = 1, \dots, M,$$

where the right side of (12) is a linear form in the numbers

$$F_1(\xi), \dots, F_m(\xi). \tag{13}$$

The defining properties of the set B_N imply that the M-tuple of KE-functions (9) satisfies all of the conditions of Lemma 16 of Chapter 3 with m replaced by M and the functions (1) replaced by the functions (9). According to that lemma, for any $\epsilon > 0$ and any $n \in \mathbf{N}$, $n \ge n_0$, where n_0 is given by (3.83), there exist M linearly independent linear forms in the numbers (13)

$$R_k(\xi) = \sum_{i=1}^{M} a_{k,i} F_i(\xi), \qquad k = 1, \dots, M, \qquad a_{k,i} \in \mathbf{Z}_{\mathbf{K}_0}, \tag{14}$$

such that

$$R_k(\xi) = O(n^{-(M-1-\epsilon)n}), \quad \overline{|a_{k,i}|} = O(n^{(1+\epsilon)n}), \quad k, i = 1, \dots, M. \tag{15}$$

We consider the numerical linear forms (12) in the numbers (13). From the definition of a $B_{N,\xi}$-polynomial it follows that these linear forms are linearly independent over $\mathbf{K}_0$. Hence, we can choose $w = M - v$ of the M linear forms (14)

$$R_{k_j}(\xi) = \sum_{i=1}^{M} a_{k_j,i} F_i(\xi), \qquad j = v+1, \dots, M, \tag{16}$$

which together with the v forms (12) are linearly independent over $\mathbf{K}_0$.

Let $\Delta(\xi) \in \mathbf{Z}_{\mathbf{K}_0}$, $\Delta(\xi) \ne 0$, denote the determinant of the matrix of coefficients of the M linear forms (12) and (16).

Since $\xi \notin \Lambda$, at least one of the numbers (13) is nonzero. Since we can change the numbering of the functions in (9), we may suppose that $F_1(\xi) \ne 0$.

From (12) and (16) we have

$$F_1(\xi)\Delta(\xi) = \sum_{j=1}^{v} \Delta_j \psi_j(\xi) + \sum_{j=v+1}^{M} \Delta_j R_{k_j}(\xi), \tag{17}$$

where the Δ_j, $j = 1, \ldots, M$, are the cofactors of the entries in the first column of the determinant $\Delta(\xi)$.

Using the bounds for the coefficients in (12) and (15), we find that

$$\begin{aligned} \Delta_j &= O(H^{v-1} n^{(1+\epsilon)wn}), \qquad & j &= 1, \ldots, v, \\ \Delta_j &= O(H^{v} n^{(1+\epsilon)(w-1)n}), \qquad & j &= v+1, \ldots, M. \end{aligned} \tag{18}$$

From (12) and (17), if we use the bounds in (15) and (18), we obtain:

$$\Delta(\xi) = O\left(H^v n^{(1+\epsilon)wn} \left(\frac{|P(\xi)|}{H} + n^{-Mn} \right) \right). \tag{19}$$

If $\mathcal{N}(\xi)$ is the norm of the element ξ of the algebraic number field $\mathbf{K}_0$, then, since $\Delta(\xi) \in \mathbf{Z}_{\mathbf{K}_0}$ and $\Delta(\xi) \neq 0$, we have

$$|\mathcal{N}(\Delta(\xi))| \geq 1. \tag{20}$$

Using (12) and (15) again, we find that

$$\overline{|\Delta(\xi)|} = O(H^v n^{(1+\epsilon)wn}). \tag{21}$$

The bounds in (19)–(21) enable us to conclude that

$$\sigma_2 H^{hv} n^{(1+\epsilon)hwn} \left(\frac{|P(\xi)|}{H} + n^{-Mn} \right) \geq 1. \tag{22}$$

We choose n to be the smallest integer for which both

$$n \geq n_0, \qquad n^{n/2} > (2\sigma_2)^{1/v} H^h. \tag{23}$$

Under these assumptions, for H sufficiently large we have $(n-1)^{(n-1)/2} \leq (2\sigma_2)^{1/v} H^h$, and hence

$$n^{n/2} < \sigma_3 H^h \ln H < \sigma_4 H^{h+\epsilon}. \tag{24}$$

Let N be chosen so that (7) holds. Then for ϵ sufficiently small we have

$$2M - 2(1+\epsilon)hw \geq v. \tag{25}$$

From (22), (23) and (25) it follows that

$$\sigma_2 H^{hv} n^{(1+\epsilon)hwn-Mn} < \frac{1}{2} n^{(1+\epsilon)hwn-Mn+vn/2} < \frac{1}{2}$$

and

$$2\sigma_2 H^{hv-1} n^{(1+\epsilon)hwn} |P(\xi)| \geq 1. \tag{26}$$

The inequalities (26) and (24) together give us the inequality

$$|P(\xi)| > \sigma H^{1-hv-2(1+\epsilon)h(h+\epsilon)w},$$

which for ϵ sufficiently small implies (8), in view of (25). The lemma is proved.

We use the Fundamental Lemma to prove Theorems 1–4 and 6–9 below. In those theorems and also in Theorem 5 the non-homogeneous case is a consequence of the homogeneous case, as can be seen in the same way as in Chapter 4, when we considered the corresponding qualitative theorems. We shall give the proofs of Theorems 1–9 only in the homogeneous case.

§ 2. Bounds for the measures of the values of E-functions which are not connected by algebraic equations over $\mathbf{C}(z)$

Theorem 1. *Suppose that the m-tuple of KE-functions $f_1(z), \dots f_m(z)$, $m \geq 2$ ($m \geq 1$), forms a solution of the system of homogeneous linear differential equations (2) (of linear differential equations (3)) and is homogeneously algebraically independent (is algebraically independent) over $\mathbf{C}(z)$, and let $\xi \in \mathbf{A}$, $\xi T(\xi) \neq 0$. Then there exist constants c_0 and c_1 (a constant c) such that:*

$$\Phi^0_{\mathbf{K}^*}(f_1(\xi), \dots, f_m(\xi);\ s;\ H) > c_0 H^{-2^m \frac{(m-1)^{m-1}}{(m-1)!} h^m s^{m-1}}, \tag{27}$$

$$\Phi_{\mathbf{K}^*}(f_1(\xi)/f_m(\xi), \dots, f_{m-1}(\xi)/f_m(\xi);\ s;\ H) > c_1 H^{-2^m \frac{(m-1)^{(m-1)}}{(m-1)!} h^m s^{m-1}}; \tag{28}$$

and in the nonhomogeneous case

$$\Phi_{\mathbf{K}^*}(f_1(\xi), \dots, f_m(\xi);\ s;\ H) > c H^{-2^{m+1} \frac{m^m}{m!} h^{m+1} s^m}. \tag{29}$$

Remark. Since $m! > 2(m/e)^m$, it follows that the exponent on the right in (27) and (28) can be replaced by $-(2e)^{m-1} h^m s^{m-1}$, and the exponent in (29) can be replaced by $-(2e)^m h^{m+1} s^m$.

Lemma 1. *Suppose that*

$$\varphi(N) = a_l N^l + \cdots + a_1 N + a_0, \qquad \varphi(N) \in \mathbf{R}[N], \qquad a_l > 0, \qquad l \geq 1,$$

and $s \in \mathbf{N}$. *Then:*

1) *there exists* $N_1 \in \mathbf{Z}^+$, *depending only on* $\varphi(N)$, *such that the following inequality holds for any* $N \in \mathbf{N}$, $N \geq N_1 + s$*:*

$$\varphi(N) - \varphi(N-s) < \frac{ls}{N - N_1}\varphi(N);$$

2) *if, on the other hand,*

$$a_l > 0, \qquad a_k \geq 0, \qquad k = 0, 1, \ldots, l-1, \tag{30}$$

then the following inequality holds for any $N \in \mathbf{N}$, $N \geq s$*:*

$$\varphi(N) - \varphi(N-s) < \frac{ls}{N}\varphi(N).$$

Proof. In part 2), the condition (30) implies that $\varphi'(N)$ is a monotonic increasing function. Hence, by Lagrange's Theorem, we have

$$\varphi(N) - \varphi(N-s) < \varphi'(N)s = \frac{s}{N} N\varphi'(N) \leq \frac{ls}{N}\varphi(N).$$

Part 1) reduces to part 2), because the condition $a_l > 0$ implies that there exists $N_1 \in \mathbf{Z}^+$ such that $\varphi(N_1) > 0$, $\varphi^{(k)}(N_1) \geq 0$, $k = 1, \ldots, l$. We set

$$\psi(x) = \varphi(x + N_1) = \sum_{k=0}^{l} \frac{\varphi^{(k)}(N_1)}{k!} x^k.$$

Then, by what was proved in part 2), for $x \geq s$ we have

$$\psi(x) - \psi(x-s) < \frac{ls}{x}\psi(x).$$

Setting $x = N - N_1 \geq 0$, we find that for $N \geq N_1 + s$

$$\varphi(N) - \varphi(N-s) < \frac{ls}{N - N_1}\varphi(N).$$

Proof of Theorem 1. In the Fundamental Lemma we set $B_N = L_N^0$, $N = 1, 2, \ldots$. Then we can take Λ to be the set consisting of zero and the roots of the polynomial $T(z)$.

Let P be an arbitrary homogeneous polynomial (5), $\overline{|P|} = H$. Then for any N, $N \geq s$, and for any ξ as in the theorem, P is a $B_{N,\xi}$-polynomial. Hence, by the Fundamental Lemma, if N is chosen so that (7) holds, then we have (8).

We have

$$M = \varphi(N) = \frac{(N+m-1)!}{N!(m-1)!}, \qquad v = \varphi(N-s), \qquad w = M - v.$$

Hence, by Lemma 1, for $N \geq s$ we have the inequality

$$\frac{w}{M} = \frac{\varphi(N) - \varphi(N-s)}{\varphi(N)} < \frac{(m-1)s}{N} = \frac{1}{2h-1},$$

if we set

$$N = (m-1)(2h-1)s. \tag{31}$$

This proves that, if N is chosen as in (31), then (7) is fulfilled.

In view of (31) we obtain:

$$M = \frac{1}{(m-1)!}(N+1)\cdots(N+m-1) \leq \frac{1}{(m-1)!}(N+m-1)^{m-1} =$$
$$= \frac{1}{(m-1)!}((m-1)(2h-1)s+m-1)^{m-1} \leq \frac{(2(m-1)hs)^{m-1}}{(m-1)!},$$

which, together with (8), gives us (27).

The inequality (28) is an obvious consequence of (27).

Theorem 1 was published in 1981 in [70:34]. Part 2) of the theorem (the bound (29)) in the case $\mathbf{K}^* = \mathbf{Q}$ is contained in a more general theorem that was proved by Galochkin in 1968 in [22:1]; this case was published in a less precise form in 1962 by S. Lang [37:2].

We now state Galochkin's theorem.

Theorem 2. *Suppose that each of the r sets of KE-functions*

$$f_{1,l}(z), \ldots, f_{m_l,l}(z), \qquad m_l \geq 1, \qquad l = 1, \ldots, r, \qquad r \geq 1,$$

satisfies a corresponding system of linear differential equations

$$y'_{k,l} = Q_{k,l,0} + \sum_{i=1}^{m_l} Q_{k,l,i} y_{i,l}, \qquad Q_{k,l,i} \in \mathbf{C}(z), \tag{32}$$

$$k = 1, \ldots, m_l, \qquad l = 1, \ldots, r,$$

and that all of these functions form an algebraically independent set over $\mathbf{C}(z)$. *Further suppose that* $\xi \in \mathbf{A}$ *is nonzero and is not a singular point of any of the systems (32), and* h *is the degree of the algebraic number field obtained by adjoining* ξ *to* $\mathbf{K}$. *Then:*

$$\Phi(f_{1,1}(\xi), \ldots, f_{m_1,1}(\xi); \ldots; f_{1,r}(\xi), \ldots, f_{m_r,r}(\xi); s_1, \ldots, s_r; H) >$$
$$> cH^{-\frac{2(2r)^m m_1^{m_1} \cdots m_r^{m_r}}{m_1! \cdots m_r!} h^{m+1} s_1^{m_1} \cdots s_r^{m_r}},$$

where $c > 0$ *is a constant depending only on the functions and the numbers* ξ, $m_1, \ldots, m_r$, *and* $s_1, \ldots, s_r$; *and* $m = m_1 + \cdots + m_r$.

It is easy to prove a homogeneous version of Theorem 2 giving a bound for Φ^0, and also to generalize Theorem 2 to give bounds for $\Phi^0_{\mathbf{K}^*}$ and $\Phi_{\mathbf{K}^*}$.

§ 3. Bounds for the measures of the values of E-functions which are connected by a single algebraic equation over $\mathbf{C}(z)$

Theorem 3. *Suppose that the* m*-tuple of KE-functions* $f_1(z), \ldots, f_m(z)$, $m \geq 3$ $(m \geq 2)$, *forms a solution of the system of homogeneous linear differential equations (2) (of linear differential equations (3)), that the functions* $f_1(z), \ldots, f_{m-1}(z)$ *are homogeneously algebraically independent (are algebraically independent) over* $\mathbf{C}(z)$, *and that they are related to* $f_m(z)$ *by an algebraic equation*

$$B = B(f_1(z), \ldots, f_m(z)) = 0, \tag{33}$$

where B *is a homogeneous (an arbitrary) irreducible and primitive polynomial of degree* k *in* m *variables with coefficients in* $\mathbf{C}[z]$. *The leading term of* B *has the form*

$$A(z) f_m^{\kappa_m}(z) \cdots f_{t+1}^{\kappa_{t+1}}(z), \qquad \kappa_i \in \mathbf{Z}^+, \qquad \kappa_{t+1} + \cdots + \kappa_m = k,$$

$$A(z) \in \mathbf{C}[z], \qquad A(z) \not\equiv 0, \qquad \kappa_m \geq 1, \qquad \kappa_{t+1} \geq 1,$$

where $2 \leq t \leq m-1$ $(1 \leq t \leq m-1)$. *Further let* $\xi \in \mathbf{A}$, $\xi T(\xi) A(\xi) \neq 0$. *Then there exist constants* c_0 *and* c_1 *(a constant* c*) such that:*

$$\Phi^0_{\mathbf{K}^*}(f_1(\xi), \ldots, f_t(\xi); s; H) > c_0 H^{-\frac{k(2(m-1)h)^{m-1}}{(m-1)!} s^{m-2}}, \tag{34}$$

$$\Phi_{\mathbf{K}^*}(f_1(\xi)/f_t(\xi), \ldots, f_{t-1}(\xi)/f_t(\xi);\ s;\ H) > c_1 H^{-\frac{k(2(m-1)h)^{m-1}}{(m-1)!} s^{m-2}}, \qquad (35)$$

and in the non-homogeneous case

$$\Phi_{\mathbf{K}^*}(f_1(\xi), \ldots, f_t(\xi);\ s;\ H) > cH^{-\frac{k(2mh)^m}{m!} s^{m-1}}. \qquad (36)$$

Remark. As in the case of Theorem 1, the exponent in (34) and (35) can be replaced by $-k2^{m-2}(eh)^{m-1}s^{m-2}$, and the exponent in (36) can be replaced by $-k2^{m-1}(eh)^m s^{m-1}$.

Proof. According to Lemma 5 of Chapter 4, the numerical coefficients of B, and hence also of $A(z)$, can be chosen in $\mathbf{Z_K}$. We consider the set L_N^0 in (4) for any $N \geq 1$. We choose $B_N \subset L_N^0$ as follows: an element $f_m^{k_m}(z) \cdots f_1^{k_1}(z)$ is in B_N if $k_i < \kappa_i$ for at least one i, $t+1 \leq i \leq m$. Using the equation (33) and Lemma 19 of Chapter 4, we easily see that B_N satisfies all of the conditions in the definition of a system of sets B_N in § 1 for $N = 1, 2, \ldots$; here Λ consists of zero and the roots of the polynomials $T(z)$ and $A(z)$, and

$$M = \varphi(N) = \begin{cases} \frac{(N+m-1)!}{N!(m-1)!}, & \text{if } N < k, \\ \frac{(N+m-1)!}{N!(m-1)!} - \frac{(N-k+m-1)!}{(N-k)!(m-1)!}, & \text{if } N \geq k. \end{cases} \qquad (37)$$

We consider an arbitrary homogeneous polynomial

$$P(f_1(z), \ldots, f_t(z)), \qquad P = P(z_1, \ldots, z_t) \in \mathbf{Z}_{\mathbf{K}^*}[z_1, \ldots, z_t], \qquad (38)$$

$$P \not\equiv 0, \qquad \overline{|P|} = H, \qquad \deg_{\mathbf{z}} P = s, \qquad s \geq 1.$$

Then for any $N \geq s$ and any ξ as in the theorem, P is obviously a $B_{N,\xi}$-polynomial. Hence, by the Fundamental Lemma, the inequality (8) holds if N is chosen in such a way that (7) is fulfilled.

We have $M = \varphi(N)$, $v = \varphi(N-s)$, $w = M - v$. We determine N from the equation (31). We consider three possible cases.

1. $k \leq N - s$. Then $N \geq k + s$. We set $\psi(N) = (N+1) \cdots (N+m-1)$, and, using Lemma 1, because of (37) we obtain

$$\begin{aligned} \frac{w}{M} &= 1 - \frac{\varphi(N-s)}{\varphi(N)} = 1 - \frac{\psi(N-s)}{\psi(N)} \frac{1 - \psi(N-k-s)/\psi(N-s)}{1 - \psi(N-k)/\psi(N)} < \\ &< \frac{\psi(N) - \psi(N-s)}{\psi(N)} < \frac{(m-1)s}{N} = \frac{1}{2h-1}, \end{aligned} \qquad (39)$$

since

$$\frac{1-\psi(N-k-s)/\psi(N-s)}{1-\psi(N-k)/\psi(N)} = \frac{1-\left(1-\frac{k}{N-s+1}\right)\cdots\left(1-\frac{k}{N-s+m-1}\right)}{1-\left(1-\frac{k}{N+1}\right)\cdots\left(1-\frac{k}{N+m-1}\right)} > 1.$$

The inequality (39) shows that (7) holds in this case.

We set $\psi_1(N)=\psi(N-1)$. Since $\psi_1(N)$ satisfies the conditions of part 2) of Lemma 1, it follows from Lemma 1 that, using (37) and (31), we have

$$\begin{aligned} M &= \frac{1}{(m-1)!}(\psi_1(N+1)-\psi_1(N+1-k)) < \frac{\psi_1(N+1)(m-1)k}{(m-1)!(N+1)} = \\ &= \frac{\psi(N)k}{(m-2)!(N+1)} \le \frac{k}{(m-2)!}(N+m-1)^{m-2} = \\ &= \frac{k}{(m-2)!}((m-1)(2h-1)s+m-1)^{m-2} \le \frac{k}{(m-2)!}(2(m-1)hs)^{m-2}. \end{aligned} \tag{40}$$

Now (34) follows from (40) and (8) in this case.

2. $N-s<k<N$. In this case $k<N<k+s$. Then from (37) we have

$$\begin{aligned} M = \varphi(N) &= \frac{(N+m-1)!}{N!(m-1)!} - \frac{(N-k+m-1)!}{(N-k)!(m-1)!}, \\ v = \varphi(N-s) &= \frac{(N-s+m-1)!}{(N-s)!(m-1)!}, \qquad w = M-v. \end{aligned} \tag{41}$$

Using (31) and (41) and arguing as in the first case, we obtain

$$\begin{aligned} \frac{w}{M} &= 1-\frac{v}{M} = 1-\frac{\psi(N-s)}{\psi(N)-\psi(N-k)} = \\ &= 1-\frac{\psi(N-s)}{\psi(N)}\frac{1}{1-\psi(N-k)/\psi(N)} < \\ &< \frac{\psi(N)-\psi(N-s)}{\psi(N)} < \frac{(m-1)s}{N} - \frac{1}{2h-1}. \end{aligned} \tag{42}$$

The inequality (42) shows that the condition (7) again holds. Since (40) is also fulfilled, by the first equality in (41), it follows that (34) holds in this case as well.

3. $k>N$. For such N, the elements of L_N^0 are linearly independent over $\mathbf{C}(z)$, $B_N=L_N^0$, $\varphi(N)=(N+m-1)!/N!(m-1)!$, and the inequality (39) is obtained in exactly the same way as in the proof of Theorem 1.

Furthermore, because of (31) and the inequality $N+1 \leq k$, we find that

$$M = \frac{1}{(m-1)!}(N+1)\cdots(N+m-1) \leq$$
$$\leq \frac{k(N+m-1)^{m-2}}{(m-1)!} < \frac{k}{(m-2)!}(2(m-1)hs)^{m-2}.$$

This means that in case 3 we also have (40), and so (34) as well, by the Fundamental Lemma. The inequality (35) is a consequence of (34).

The most interesting case of Theorem 3 is the case $t = m-1$, when B has leading term of the form $A(z)f_m^k(z)$.

§ 4. Bounds for the measures of the values of E-functions which are connected by arbitrary algebraic equations over $\mathbf{C}(z)$

Theorem 4. *Suppose that the m-tuple of KE-functions $f_1(z), \dots, f_m(z)$, $m \geq 3$ ($m \geq 2$), forms a solution of the system of homogeneous linear differential equations (2) (of linear differential equations (3)); that this m-tuple has homogeneous transcendence degree (has transcendence degree) over $\mathbf{C}(z)$ equal to l, $2 \leq l \leq m-1$ ($1 \leq l \leq m-1$); and that the functions*

$$f_1(z), \dots, f_l(z) \tag{43}$$

are homogeneously algebraically independent (are algebraically independent) over $\mathbf{C}(z)$. Further let

$$A_1(z)f_m^{\kappa_{m,1}}(z)\cdots f_{t+1}^{\kappa_{t+1,1}}(z),\ \dots,\ A_\nu(z)f_m^{\kappa_{m,\nu}}(z)\cdots f_{t+1}^{\kappa_{t+1,\nu}}(z), \tag{44}$$

$$1 \leq t \leq l, \quad \max_{1\leq j\leq \nu} \kappa_{t+1,j} \geq 1, \quad A_i(z) \in \mathbf{C}[z], \quad i = 1, \dots, \nu,$$

be the set of leading terms of the minimal homogeneous equations (the minimal equations) of the functions $f_1(z), \dots, f_m(z)$ over $\mathbf{C}(z)$, and let $\xi \in \mathbf{A}$, $\xi T(\xi)A_1(\xi) \cdots A_\nu(\xi) \neq 0$. Then there exist constants c_0, c_1 and γ_0 (constants c and γ) such that

$$\Phi^0_{\mathbf{K}^*}(f_1(\xi), \dots, f_t(\xi); s; H) > c_0 H^{-\gamma_0 h^l s^{l-1}}, \tag{45}$$

$$\Phi_{\mathbf{K}^*}(f_1(\xi)/f_t(\xi), \dots, f_{t-1}(\xi)/f_t(\xi); s; H) > c_1 H^{-\gamma_0 h^l s^{l-1}}, \tag{46}$$

and in the non-homogeneous case

$$\Phi_{\mathbf{K}^*}(f_1(\xi), \ldots, f_t(\xi); s; H) > cH^{-\gamma h^{l+1} s^l}. \tag{47}$$

Proof. We consider the set L_N^0 in (4) and the set of minimal homogeneous equations satisfied by $f_1(z), \ldots, f_m(z)$ over $\mathbf{C}(z)$. Let B_N be a basis of $\mathbf{L}_N^0$ obtained using the minimal equations (Lemma 17 of Chapter 4), and let $M = \varphi(N) = r_N^0$, where r_N^0 is the number of elements in the basis B_N.

For any $N \in \mathbf{N}$ the set B_N is a B_N-set in the sense of the definition in § 1, as we see using Theorem 13 and Lemma 19 of Chapter 4. Here Λ is the set consisting of zero and the roots of all of the polynomials $T(z), A_1(z), \ldots, A_\nu(z)$.

We consider an arbitrary homogeneous polynomial (38). By the hypothesis of the theorem, P is a $B_{N,\xi}$-polynomial for any $N \geq s$ and any $\xi \in \mathbf{A}$, $\xi \notin \Lambda$. Hence, by the Fundamental Lemma, if N is chosen so that the condition (7) is fulfilled, then (8) holds.

By Theorem 13 of Chapter 4, there exists $\gamma_1 \in \mathbf{N}$ such that for $N \geq \gamma_1$ we have $r_n^0 = \alpha_{l-1}N^{l-1} + \cdots + \alpha_1 N + \alpha_0$, $\alpha_{l-1} > 0$, and hence

$$M = \varphi(N) = r_N^0 \leq l \max_{0 \leq i \leq l-1} |\alpha_i| N^{l-1} = \gamma_2 N^{l-1}. \tag{48}$$

We have $v = \varphi(N-s) = r_{N-s}^0$, $w = M - v$. By Lemma 1, there exists $\gamma_3 \in \mathbf{N}$ such that the following inequality holds for $N \geq \max(\gamma_1, \gamma_3 + s)$:

$$\begin{aligned} \frac{w}{M} = \frac{\varphi(N) - \varphi(N-s)}{\varphi(N)} < \frac{(l-1)s}{N - \gamma_3} = \frac{l-1}{N}\frac{Ns}{N-\gamma_3} \leq \\ \leq \frac{(l-1)(\gamma_3+1)s}{N} \leq \frac{\gamma_4 s}{N}, \end{aligned} \tag{49}$$

where

$$\gamma_4 = \max((l-1)(\gamma_3+1), \gamma_1), \qquad \gamma_4 \in \mathbf{N}. \tag{50}$$

We set

$$N = \gamma_4(2h-1)s. \tag{51}$$

From (50) and (51) it follows that $N \geq \gamma_4 s \geq \gamma_1$, $N \geq \gamma_4 s \geq (\gamma_3+1)s \geq \gamma_3 + s$. Hence, $N \geq \max(\gamma_1, \gamma_3 + s)$, and so from (49) and (51) we conclude that (7) is fulfilled, and consequently (8) holds.

From (48) and (51) we find that

$$M \leq \gamma_2(\gamma_4(2h-1)s)^{l-1} < \gamma_5 h^{l-1} s^{l-1}.$$

From (8) and this last inequality we obtain the inequality (45), and (46) is a consequence of (45).

§ 5. Algebraic independence of the values of E-functions in conjugate fields

If we take the leading terms of the minimal equations to be arbitrary in Theorem 4 (while keeping the other assumptions), we cannot obtain bounds for the measures of $f_1(\xi), \ldots, f_l(\xi)$. Similarly, in the same situation in Theorem 15 of Chapter 4 we could not prove algebraic independence of $f_1(\xi), \ldots, f_l(\xi)$. Such a result was only obtained in the case $\mathbf{K}_0 = \mathbf{I}$ in Theorem 11 of Chapter 11 and Theorem 14 of Chapter 4.

Using Theorem 1 of Chapter 11, in the $\mathbf{K}_0$ case one can only prove that in at least one of the conjugate fields $\mathbf{K}_{0,i}$ the numbers $f_{1,i}(\xi_i), \ldots, f_{l,i}(\xi_i)$ are algebraically independent.

Theorem 5. *Suppose that the KE-functions $f_1(z), \ldots, f_m(z)$, $m \geq 3$ ($m \geq 2$), satisfy the system of homogeneous linear differential equations (2) (of linear differential equations (3)); this set of functions has homogeneous transcendence degree (has transcendence degree) over $\mathbf{C}(z)$ equal to l, $2 \leq l \leq m-1$ ($1 \leq l \leq m-1$); and the functions $f_1(z), \ldots, f_l(z)$ are homogeneously algebraically independent (are algebraically independent) over $\mathbf{C}(z)$. Further let*

$$A_1(z) f_m^{\kappa_{m,1}}(z) \cdots f_1^{\kappa_{1,1}}(z), \ \ldots, \ A_\nu(z) f_m^{\kappa_{m,\nu}}(z) \cdots f_1^{\kappa_{1,\nu}}(z),$$

$$A_i(z) \in \mathbf{C}[z], \qquad i = 1, \ldots, \nu,$$

be the set of leading terms of the homogeneous minimal equations (the minimal equations) satisfied by $f_1(z), \ldots, f_m(z)$ over $\mathbf{C}(z)$; and let $\xi \in \mathbf{A}$, $\xi T(\xi) A_1(\xi) \cdots \cdots A_\nu(\xi) \neq 0$. Then there exists i, $1 \leq i \leq h$, such that the numbers $f_{1,i}(\xi_i), \ldots \ldots, f_{l,i}(\xi_i)$ are homogeneously algebraically independent (are algebraically independent).

Proof. Suppose the contrary, i.e., that for any i, $i = 1, \ldots, h$, the numbers $f_{1,i}(\xi_i), \ldots, f_{l,i}(\xi_i)$ are homogeneously algebraically dependent. Then, by Lemma 1 of Chapter 3, there exist polynomials

$$P_i = P_i(z_1, \ldots, z_l) \in \mathbf{Q}[z_1, \ldots, z_l], \qquad P_i \not\equiv 0, \qquad i = 1, \ldots, h,$$

such that $P_i(f_{1,i}(\xi_i), \ldots, f_{l,i}(\xi_i)) = 0$, $i = 1, \ldots, h$. We set $P = P(z_1, \ldots, z_l) = P_1 \cdots P_h$. Then

$$P(f_{1,i}(\xi_i), \ldots, f_{l,i}(\xi_i)) = 0, \qquad i = 1, \ldots, m. \tag{52}$$

Let $\deg_{\mathbf{z}} P = k$. We consider the set of products of powers

$$f_m^{k_m}(z) \cdots f_1^{k_1}(z), \qquad k_i \in \mathbf{Z}^+, \qquad k_1 + \cdots + k_m = k,$$

and the basis B_k of the vector space $\mathbf{L}_k^0$ spanned over $\mathbf{C}(z)$ by these functions which is constructed using the minimal homogeneous equations satisfied by $f_1(z), \ldots, f_m(z)$ (Lemma 17 of Chapter 4). By Lemma 19 of Chapter 4, the functions in the basis B_k form a solution of a system of homogeneous linear differential equations of the form (2) with m replaced by $r_k^0 = \mu_{k,m}^0$ and with no singular points other than the singular points of the system (2) and the roots of the polynomials $A_1(z), \ldots, A_\nu(z)$. Since the functions in B_k are linearly independent over $\mathbf{C}(z)$, and since the functions $f_l^{k_l}(z) \cdots f_1^{k_1}(z)$, $k_1 + \cdots + k_l = k$, occur in B_k, it follows by Theorem 1 of Chapter 11 that

$$\max_{1 \le i \le h} |P(f_{1,i}(\xi_i), \ldots, f_{l,i}(\xi_i))| > 0,$$

and this contradicts (52). This contradiction proves the theorem.

§ 6. An auxiliary theorem

The Fundamental Lemma, which was used to prove Theorems 1–4, fails to give us bounds for the measures in certain cases when the KE-functions

$$f_1(z), \ldots, f_m(z) \tag{53}$$

are connected by other types of algebraic equations over $\mathbf{C}(z)$. But it turns out that using the Fundamental Lemma in conjunction with Theorem 1 of Chapter 11 enables us to obtain bounds for the measures in some additional cases. We shall use the Fundamental Lemma and Theorem 1 of Chapter 11 to prove an auxiliary theorem, from which several theorems with bounds for measures will follow.

Let $\mathbf{K}_{0,i}$, $i = 1, \ldots, h$, be the algebraic number fields which are conjugate to $\mathbf{K}_0$. As in § 2 of Chapter 11, along with the functions (53) we shall consider the conjugate functions $f_{1,i}(z), \ldots, f_{m,i}(z)$ relative to $\mathbf{K}_{0,i}$, and along with ξ we shall consider its conjugates $\xi_1, \ldots, \xi_h$. We shall also use the subscript i with various other functions to denote the conjugate relative to $\mathbf{K}_{0,i}$.

In what follows we shall let

$$g_{\mathbf{K}^*} = g_{\mathbf{K}^*}(z_1, \ldots, z_m) \not\equiv 0, \qquad \overline{|g_{\mathbf{K}^*}|} = H, \qquad \deg_{\mathbf{z}} g_{\mathbf{K}^*} = s,$$

denote an arbitrary polynomial with coefficients in $\mathbf{Z}_{\mathbf{K}^*}$, and we shall let $g_i = g_i(z_1, \ldots, z_m)$, $i = 1, \ldots, h$, denote the polynomials obtained from $g_{\mathbf{K}^*}$ by replacing all of the coefficients by their conjugates in $\mathbf{K}_{0,i}$.

The notation $g^0_{\mathbf{K}^*} = g^0_{\mathbf{K}^*}(z_1, \ldots, z_m)$ and $g^0_i = g^0_i(z_1, \ldots, z_m)$ will have the same meaning as $g_{\mathbf{K}^*}$ and g_i in the case when the polynomial under consideration is homogeneous.

Theorem 6. *Suppose that the m-tuple of KE-functions (53), $m \geq 2$ ($m \geq 1$) forms a solution of the system of homogeneous linear differential equations (2) (of linear differential equations (3)) and has homogeneous transcendence degree (has transcendence degree) over $\mathbf{C}(z)$ equal to l, $1 \leq l \leq m$. Further suppose that $\xi \in \mathbf{A}$, $\xi T(\xi) \neq 0$, and the polynomial $g^0_{\mathbf{K}^*}$ (the polynomial $g_{\mathbf{K}^*}$) and the number ξ are such that*

$$g^0_i(f_{1,i}(\xi_i), \ldots, f_{m,i}(\xi_i)) \neq 0, \qquad i = 1, \ldots, h, \tag{54}$$

$$(g_i(f_{1,i}(\xi_i), \ldots, f_{m,i}(\xi_i)) \neq 0, \qquad i = 1, \ldots, h). \tag{55}$$

Then there exist constants c_0, c_1 and γ_0 (constants c and γ) such that

$$|g^0_i(f_{1,i}(\xi_i), \ldots, f_{m,i}(\xi_i))| > c_0 H^{-\gamma_0 h^l s^{l-1}}, \qquad i = 1, \ldots, h,$$

and in the non-homogeneous case

$$|g_i(f_{1,i}(\xi_i), \ldots, f_{m,i}(\xi_i))| > c_1 H^{-\gamma h^{l+1} s^l}, \qquad i = 1, \ldots, h;$$

in particular,

$$|g^0_{\mathbf{K}^*}(f_1(\xi), \ldots, f_m(\xi))| > c H^{-\gamma_0 h^l s^{l-1}}, \tag{56}$$

and in the non-homogeneous case

$$|g_{\mathbf{K}^*}(f_1(\xi), \ldots, f_m(\xi))| > c H^{-\gamma h^{l+1} s^l}. \tag{57}$$

Proof. First suppose that $l < m$. The proof begins in the same way as the proof of Theorem 7 of Chapter 4; however, because of the more general situation, it becomes more complicated to pass to the conjugate fields.

For each $N = 1, 2, \ldots$, we consider the set L^0_N in (4) and the vector space $\mathbf{L}^0_N$ it spans over $\mathbf{C}(z)$. If $\xi \in \mathbf{A}$ with $\xi T(\xi) \neq 0$, then, by Lemmas 1 and 4 of

Chapter 4 applied to the functions in (4), there exists a basis B_N of the vector space $\mathbf{L}_N^0$ with r_N^0 elements

$$F_{N,\nu}(z), \qquad \nu = 1, \dots, r_N^0, \tag{58}$$

where this basis is made up of functions in (4) and has the following properties. If

$$F_{N,\nu}(z), \qquad \nu = r_N^0 + 1, \dots, \mu_{N,m}^0, \tag{59}$$

are the elements of L_N^0 not in B_N, and if we write the equations for them in terms of the functions (58) in the form

$$F_{N,\nu}(z) = \sum_{k=1}^{r_N^0} R_{N,\nu,k}(z) F_{N,k}(z), \qquad R_{N,\nu,k}(z) \in \mathbf{K}(z), \tag{60}$$

$$\nu = r_N^0 + 1, \dots, \mu_{N,m}^0,$$

then none of the functions $R_{N,\nu,k}(z)$ has a pole at the point $z = \xi$.

If we use Lemma 19 of Chapter 4, Lemma 2, and Theorem 13 of Chapter 4, we see that B_N satisfies all of the conditions for B_N-sets in § 1, where Λ is the set consisting of zero, the roots of the polynomial $T(z)$, and the poles of all of the functions $R_{N,\nu,k}(z)$ in (60) for $N = 1, 2, \dots$; thus, Λ does not contain ξ.

We prove that $g_{\mathbf{K}^*}^0$ is a $B_{N,\xi}$-polynomial for any N, $N \geq s$, and for any $\xi \in \mathbf{A}$ as in the theorem. We consider the expressions

$$F_{N-s,\nu}(\xi) g_{\mathbf{K}^*}^0(f_1(\xi), \dots, f_m(\xi)) = \sum_{k=1}^{\mu_{N,m}^0} \beta_{N,\nu,k} F_{N,k}(\xi), \tag{61}$$

$$\beta_{N,\nu,k} \in \mathbf{Z}_{\mathbf{K}^*}, \qquad \nu = 1, \dots, r_{N-s}^0,$$

where all of the nonzero $\beta_{N,\nu,k}$ are coefficients of the polynomial $g_{\mathbf{K}^*}^0$. From (60) with $z = \xi$ we have

$$F_{N,\nu}(\xi) - \sum_{k=1}^{r_N^0} \lambda_{N,\nu,k} F_{N,k}(\xi) = 0, \tag{62}$$

$$\lambda_{N,\nu,k} = R_{N,\nu,k}(\xi) \in \mathbf{K}_0, \qquad \nu = r_N + 1, \dots, \mu_{N,m}^0.$$

We shall see that the linear forms (61) and (62) in the numbers

$$F_{N,1}(\xi), \dots, F_{N,\mu^0_{N,m}}(\xi) \tag{63}$$

with coefficients in $\mathbf{K}_0$ are linearly independent over $\mathbf{K}_0$ as linear forms in $\mu^0_{N,m}$ independent variables.

Suppose that this is not the case. Then the following relation holds identically in the numbers (63):

$$\sum_{\nu=1}^{r^0_{N-s}} a_\nu F_{N-s,\nu}(\xi) g^0_{\mathbf{K}^*}(f_1(\xi), \dots, f_m(\xi)) = \sum_{\nu=1}^{r^0_{N-s}} a_\nu \sum_{k=1}^{\mu^0_{N,m}} \beta_{N,\nu,k} F_{N,k}(\xi) =$$
$$= \sum_{\nu=r^0_N+1}^{\mu^0_{N,m}} a_\nu \left(F_{N,\nu}(\xi) - \sum_{k=1}^{r^0_N} \lambda_{N,\nu,k} F_{N,k}(\xi) \right), \tag{64}$$

$$a_\nu \in \mathbf{K}_0, \qquad \nu = 1, \dots, r^0_{N-s}, r^0_N + 1, \dots, \mu^0_{N,m},$$

where at least one of the numbers a_ν is nonzero. Since the expressions on the left in (62) are linearly independent over $\mathbf{K}_0$ as linear forms in the numbers (63), it follows that at least one of the numbers $a_1, \dots, a_{r^0_{N-s}}$ is nonzero. In view of (62), from this and (64) we obtain:

$$\sum_{\nu=1}^{r^0_{N-s}} a_\nu F_{N-s,\nu}(\xi) g^0_{\mathbf{K}^*}(f_1(\xi), \dots, f_m(\xi)) = 0. \tag{65}$$

We now pass to the conjugate fields $\mathbf{K}_{0,i}$. Let $F_{N,k,i}(z)$ and $R_{N,\nu,k,i}(z)$, $i = 1, \dots, h$, be the functions conjugate to $F_{N,k}(z)$ and $R_{N,\nu,k}(z)$, and let $\beta_{N,\nu,k,i}$, $\lambda_{N,\nu,k,i} = R_{N,\nu,k,i}(\xi_i)$ and $a_{\nu,i}$, $i = 1, \dots, h$, be the conjugates of the numbers $\beta_{N,\nu,k}$, $\lambda_{N,\nu,k}$ and a_ν. Then (61) takes the form

$$F_{N-s,\nu,i}(\xi_i) g^0_i(f_{1,i}(\xi_i), \dots, f_{m,i}(\xi_i)) = \sum_{k=1}^{\mu^0_{N,m}} \beta_{N,\nu,k,i} F_{N,k,i}(\xi_i), \tag{66}$$

$$\beta_{N,\nu,k,i} \in \mathbf{Z}_{\mathbf{K}_{0,i}}, \qquad \nu = 1, \dots, r^0_{N-s}, \qquad i = 1, \dots, h.$$

The equations (60) are preserved when we pass to conjugate fields, and for $z = \xi_i$ they give us

$$F_{N,\nu,i}(\xi_i) - \sum_{k=1}^{r_N^0} \lambda_{N,\nu,k,i} F_{N,k,i}(\xi_i) = 0, \tag{67}$$

$$\lambda_{N,\nu,k,i} \in \mathbf{K}_{0,i}, \qquad \nu = r_N^0 + 1, \ldots, \mu_{N,m}^0, \qquad i = 1, \ldots, h;$$

and the equations (64) become

$$\begin{aligned}\sum_{\nu=1}^{r_{N-s}^0} a_{\nu,i} F_{N-s,\nu,i}(\xi_i) g_i^0(f_{1,i}(\xi_i), \ldots, f_{m,i}(\xi_i)) = \\ = \sum_{\nu=1}^{r_{N-s}^0} a_{\nu,i} \sum_{k=1}^{\mu_{N,m}^0} \beta_{N,\nu,k,i} F_{N,k,i}(\xi_i) = \\ = \sum_{\nu=r_N^0+1}^{\mu_{N,m}^0} a_{\nu,i} \left(F_{N,\nu,i}(\xi_i) - \sum_{k=1}^{r_N^0} \lambda_{N,\mu,k,i} F_{N,k,i}(\xi_i) \right),\end{aligned} \tag{68}$$

$$a_{\nu,i} \in \mathbf{K}_{0,i}, \qquad \nu = 1, \ldots, r_{N-s}^0, r_N^0 + 1, \ldots, \mu_{N,m}^0, \qquad i = 1, \ldots, h.$$

By (67), from this it follows that

$$\sum_{\nu=1}^{r_{N-s}^0} a_{\nu,i} F_{N-s,\nu,i}(\xi_i) g_i^0(f_{1,i}(\xi_i), \ldots, f_{m,i}(\xi_i)) = 0, \quad i = 1, \ldots, h. \tag{69}$$

By Lemma 18 of Chapter 3 and Lemma 2 of Chapter 4, the r_{N-s}^0-tuple of functions in the basis B_{N-s} and the number ξ satisfy all of the conditions in Theorem 1. Theorem 1 then says that

$$\max_{1 \le i \le h} \left| \sum_{\nu=1}^{r_{N-s}^0} a_{\nu,i} F_{N-s,\nu,i}(\xi_i) \right| > 0, \tag{70}$$

since at least one of the numbers $a_1, \ldots, a_{r_{N-s}^0}$ is nonzero.

From (70) and (54) it follows that (69) cannot hold, and so our supposition is false, i.e., the linear forms (61) and (62) are linearly independent over $\mathbf{K}_0$, as are the linear forms which are obtained from them by passing to the conjugate fields $\mathbf{K}_{0,i}$.

From the above we conclude that, if we use (67), then the equations (66) take the form

$$F_{N-s,\nu,i}(\xi_i)g_i^0(f_{1,i}(\xi_i),\ldots,f_{m,i}(\xi_i)) = \sum_{k=1}^{r_N^0} \delta_{N,\nu,k,i}F_{N,k,i}(\xi_i), \tag{71}$$

$$\delta_{N-s,\nu,k,i} \in \mathbf{K}_{0,i}, \qquad \nu = 1,\ldots,r_{N-s}^0, \qquad i = 1,\ldots,h.$$

In the case $l < m$, (71) shows that g_i^0 is a B_{N,ξ_i}-polynomial for any ξ as in the theorem, any N, $N \geq s$, and any $i = 1,\ldots,h$. This is also obviously true if $l = m$.

The rest of the proof of the theorem in the general case $1 \leq l \leq m$ proceeds using the Fundamental Lemma, in exactly the same way as in the case of Theorem 3.

Corollary. *Under the conditions of Theorem 4, suppose that for each* $i = 1,\ldots,h$ *the* l *numbers*

$$f_{1,i}(\xi_i),\ldots,f_{l,i}(\xi) \tag{72}$$

are homogeneously algebraically independent (are algebraically independent). Then there exist constants c_0, c_1 *and* γ_0 *(constants* c *and* γ*) such that*

$$\Phi_{\mathbf{K}^*}^0(f_1(\xi),\ldots,f_l(\xi);\ s;\ H) > c_0H^{-\gamma_0h^ls^{l-1}}, \tag{73}$$

$$\Phi_{\mathbf{K}^*}(f_1(\xi)/f_l(\xi),\ldots,f_{l-1}(\xi)/f_l(\xi);\ s;\ H) > c_1H^{-\gamma_0h^ls^{l-1}}, \tag{74}$$

and in the non-homogeneous case

$$\Phi_{\mathbf{K}^*}(f_1(\xi),\ldots,f_l(\xi);\ s;\ H) > cH^{-\gamma h^{l+1}s^l}. \tag{75}$$

In fact, under the conditions of the corollary, for any polynomial

$$P = P(x_1,\ldots,x_l) \in \mathbf{Z}_{\mathbf{K}^*}[x_1,\ldots,x_l], \qquad P \not\equiv 0,$$

we have

$$P(f_{1,i}(\xi_i),\ldots,f_{l,i}(\xi_i)) \neq 0, \qquad i = 1,\ldots,h,$$

and then the corollary follows from Theorem 4.

§ 7. Consequences of the auxiliary theorem

If we use the general theorems on algebraic independence of the values of E-functions that were proved in Chapters 3–10, then, based on the corollary to Theorem 5, we can obtain corresponding bounds for the algebraic independence measures of the values of the E-functions.

For example, if we use the First and Second Fundamental Theorems of Chapter 3 (which remain in force when we pass to conjugate fields), then from the corollary to Theorem 5 we obtain all of the inequalities in Theorem 1, except with the constants γ_0 and γ on the right.

Similarly, if we use Theorems 12 and 14 of Chapter 4, then the corollary to Theorem 5 gives us Theorems 3 and 4.

The following result follows from Theorem 5 of Chapter 4 and the corollary to Theorem 6.

Theorem 7. *Suppose that the KE-functions $f_1(z), \dots, f_m(z)$, $m \geq 2$ ($m \geq 1$), satisfy the system of homogeneous linear differential equations (2) (of linear differential equations (3)) and have homogeneous transcendence degree (have transcendence degree) over $\mathbf{C}(z)$ equal to l, $1 \leq l \leq m$ ($1 \leq l \leq m$), and suppose that the functions $f_1(z), \dots, f_l(z)$ are algebraically independent over $\mathbf{C}(z)$. Then there exist constants γ_0, c_0 and c_1 (constants γ and c) such that the inequalities (73) and (74) (the inequality (75) in the non-homogeneous case) hold for almost all $\xi \in \mathbf{A}$.*

Proof. By Theorem 5 of Chapter 4, under the conditions of the theorem the numbers $f_1(\xi), \dots, f_l(\xi)$ are algebraically independent for almost all $\xi \in \mathbf{A}$. This result remains in force when we pass to the conjugate fields $\mathbf{K}_{0,i}$. Consequently, the l numbers (72) are algebraically independent for almost all $\xi \in \mathbf{A}$ and for all i, $i = 1, \dots, h$. Then the corollary to Theorem 5 implies that (73) and (74) hold (and (75) in the non-homogeneous case).

We similarly obtain a quantitative theorem from Theorem 6 of Chapter 4 and the corollary to Theorem 6.

Theorem 8. *Suppose that the KE-functions $f_1(z), \dots, f_m(z)$, $m \geq 3$ ($m \geq 2$), satisfy the system of homogeneous linear differential equations (2) (of linear differential equations (3)) and have homogeneous transcendence degree (have transcendence degree) over $\mathbf{C}(z)$ equal to l, $2 \leq l \leq m-1$ ($1 \leq l \leq m-1$). Further suppose that the functions $f_1(z), \dots, f_l(z)$ are homogeneously algebraically independent (are algebraically independent) over $\mathbf{C}(z)$, and the functions $f_1(z), \dots, f_m(z)$ are connected by algebraic equations*

$$P_\nu = P_\nu(f_1(z), \dots, f_\nu(z)) = 0, \qquad \nu = l+1, \dots, m,$$

where P_ν is a homogeneous (an arbitrary) polynomial in ν variables with coefficients in $\mathbf{A}[z]$ whose leading term has the form

$$A_\nu(z) f_\nu^{s_\nu}(z), \qquad A_\nu \in \mathbf{A}[z], \qquad A_\nu(z) \not\equiv 0, \qquad s_\nu \geq 1.$$

Further let $\xi \in \mathbf{A}$, $\xi T(\xi) A_{l+1}(\xi) \cdots A_m(\xi) \neq 0$. Then (73) and (74) hold (and (75) in the non-homogeneous case).

Finally, we obtain the following results from Theorems 7 and 8 of Chapter 4 and the corollary to Theorem 6.

Theorem 9. *Suppose that the KE-functions $f_1(z), \ldots, f_m(z)$, $m \geq 2$ ($m \geq 1$), satisfy the system of homogeneous linear differential equations (2) (of linear differential equations (3)) and have homogeneous transcendence degree (have transcendence degree) either over $\mathbf{C}(z)$ or over $\mathbf{C}$ equal to l, $2 \leq l \leq m$ ($1 \leq l \leq m$). Further suppose that the functions $f_1(z), \ldots, f_l(z)$ are homogeneously algebraically independent (are algebraically independent) over $\mathbf{C}(z)$ and that*

$$P = P(z_1, \ldots, z_m) \in \mathbf{A}[z_1, \ldots, z_m],$$

$$\deg_{\mathbf{z}} P = s, \qquad s \geq 1, \qquad \overline{|P|} = H,$$

is any homogeneous (any arbitrary) polynomial such that

$$P(f_1(z), \ldots, f_m(z)) \not\equiv 0$$

as a function of z. Let $\xi \in \mathbf{A}$, $\xi T(\xi) \neq 0$. Then there exist constants γ_0, c_0, c_1 and c_2 (constants γ, c and c_4) such that

$$|P(f_1(\xi), \ldots, f_m(\xi))| > c_2 H^{-\gamma_0 h^l s^{l-1}},$$

the inequalities (73) and (74) hold, and in the non-homogeneous case we have

$$|P(f_1(\xi), \ldots, f_m(\xi))| > c_4 H^{-\gamma h^{l+1} s^l}$$

along with (75).

It is not hard to give versions of Theorems 1–9 which apply to a function $f(z)$ and its sequence of successive derivatives, where $f(z)$ satisfies a linear differential equation with coefficients in $\mathbf{C}[z]$.

We note that in this chapter, as in Chapter 4, we have not obtained definitive results in all cases when our set of E-functions is algebraically dependent over $\mathbf{C}(z)$. We have not treated the case when the functions (1) are connected by arbitrary algebraic equations over $\mathbf{C}(z)$, or the case when the minimal polynomials satisfied by the functions (1) over $\mathbf{C}(z)$ have arbitrary leading terms.

§ 8. Some applications of the general theorems

If we use the results known about algebraic independence over $\mathbf{C}(z)$ of concrete sets of E-functions along with facts about the basic algebraic equations satisfied by the E-functions over $\mathbf{C}(z)$, then from Theorems 1–4, 6–9 we can obtain bounds for the algebraic independence measures of the values of many sets of E-functions. We shall give some simple examples.

Suppose that the KE-function $f(z)$ is a solution of the second order homogeneous linear differential equation

$$P_2(z)y'' + P_1(z)y' + P_0(z)y = 0$$

with coefficients in $\mathbf{C}[z]$. If the functions $f(z)$ and $f'(z)$ are homogeneously algebraically independent (are algebraically independent) over $\mathbf{C}(z)$, then by Theorem 1 we have the following inequalities for any $\xi \in \mathbf{A}$, $\xi P_2(\xi) \neq 0$:

$$\Phi^0_{\mathbf{K}^*}(f(\xi), f'(\xi);\ s;\ H) > c_0 H^{-4h^2 s}, \tag{76}$$

$$\Phi_{\mathbf{K}^*}(f'(\xi)/f(\xi);\ s;\ H) > c_1 H^{-4h^2 s}, \tag{77}$$

and in the non-homogeneous case

$$\Phi_{\mathbf{K}^*}(f(\xi), f'(\xi);\ s;\ H) > cH^{-16h^3 s^2}. \tag{78}$$

Using Lemmas 3, 4, 7 and 8 of Chapter 6, we find that the inequalities (76), (77) and (78) hold for the E-functions $f(z) = K_\lambda(z)$ in (6.8) and $f(z) = A_{\lambda,\nu}(z)$ in (6.60) under the restrictions on λ and ν in those lemmas, for any nonzero $\xi \in \mathbf{A}$. The constant 16 in (78) for $K_\lambda(z)$ is better than the value 123 in Siegel's bound (11.26).

Now suppose that the E-function $f(z)$ satisfies the second order linear differential equation

$$P_2(z)y'' + P_1(z)y' + P_0(z)y = Q(z) \tag{79}$$

with coefficients in $\mathbf{C}[z]$. If the functions $f(z)$ and $f'(z)$ are algebraically independent over $\mathbf{C}(z)$, then (78) holds for any $\xi \in \mathbf{A}$ with $\xi P_2(\xi) \neq 0$.

Using the proofs of Theorems 6 and 7 in Chapter 6, we find that (78) holds for the functions $f(z) = K_{\lambda,\mu}(z)$ in (6.78) and $f(z) = K_{\lambda,\mu,\nu}(z)$ in (6.80) under the restrictions on λ, μ and ν in those theorems, for any $\xi \in \mathbf{A}$ for which $\xi P_2(\xi) \neq 0$.

Now suppose that $f(z)$ is a transcendental E-function which satisfies the linear differential equation (79) (which may be homogeneous, i.e., $Q(z) \equiv 0$). If $f(z)$ and $f'(z)$ are algebraically dependent over $\mathbf{C}(z)$, then it follows from Lemma 1 of Chapter 6 that they are related by an algebraic equation of the form

$$P = P(z, f(z), f'(z)) = 0, \qquad P \in \mathbf{C}[z, f(z), f'(z)],$$

where P has total degree k in $f(z)$ and $f'(z)$, $k \geq 1$, and contains the term

$$A(z)(f'(z))^k, \qquad A(z) \in \mathbf{C}[z], \qquad A(z) \not\equiv 0.$$

Hence, by Theorem 3, the following bound holds for any $\xi \in \mathbf{A}$, $\xi P_2(\xi)A(\xi) \neq 0$:

$$\Phi(f(\xi);\ s;\ H) > cH^{-8kh^2 s}. \tag{80}$$

In the case of the transcendental functions $f(z) = \sin z$ ($f(z) = \cos z$), which satisfy the differential equation $y'' + y = 0$ and the algebraic relation $f(z)^2 + f'(z)^2 = 1$, the inequalities (76), (77) and (78) take the form

$$\Phi^0_{\mathbf{K}^*}(\sin\xi, \cos\xi;\ s;\ H) > c_0 H^{-4h^2 s}, \qquad \Phi_{\mathbf{K}^*}(\sin\xi;\ s;\ H) > cH^{-16h^2 s},$$

$$\Phi_{\mathbf{K}^*}(\tan\xi;\ s;\ H) > c_1 H^{-4h^2 s}, \qquad \Phi_{\mathbf{K}^*}(\cos\xi;\ s;\ H) > cH^{-16h^2 s},$$

for any nonzero $\xi \in \mathbf{A}$.

We consider the $2m$ E-functions

$$u_k = \sin\xi_k z, \qquad v_k = \cos\xi_k z, \qquad k = 1, \ldots, m, \qquad m \geq 1, \tag{81}$$

where $\xi_1, \ldots, \xi_m$ are algebraic numbers which are linearly independent over $\mathbf{Q}$.

Lemma 2. *The functions $u_1, \ldots, u_m$ ($v_1, \ldots, v_m$) are algebraically independent over $\mathbf{C}(z)$.*

Proof. By Euler's formula we have

$$2iw_k u_k = w_k^2 - 1, \qquad w_k = e^{i\xi_k z}, \qquad k = 1, \ldots, m.$$

These equations show that the functions $u_1, \ldots, u_m$ depend algebraically on $w_1, \ldots, w_m$ both over $\mathbf{C}(z)$ and over $\mathbf{C}$, and conversely. Hence, these m-tuples of functions have the same transcendence degree over $\mathbf{C}(z)$ or over $\mathbf{C}$. But by Lemma 21 of Chapter 3, the m-tuple $w_1, \ldots, w_m$ is an algebraically independent set over $\mathbf{C}(z)$; hence, the functions $u_1, \ldots, u_m$ are algebraically independent over $\mathbf{C}(z)$.

We similarly see that the functions $v_1, \ldots, v_m$ are algebraically independent over $\mathbf{C}(z)$. Because of the relations

$$u_k^2 + v_k^2 = 1, \qquad k = 1, \ldots, m, \tag{82}$$

we conclude that the $2m$ functions in (81) have transcendence degree m either over $\mathbf{C}(z)$ or over $\mathbf{C}$.

The functions (81) form a solution of the system of linear differential equations

$$v_k' = \xi_k v_k, \qquad v_k' = -\xi_k u_k, \qquad k = 1, \ldots, m,$$

and so, using either Theorem 9 or Theorem 8, one can obtain the bound

$$\Phi_{\mathbf{K}^*}(\sin \xi_1, \ldots, \sin \xi_m;\ s;\ H) > cH^{-\gamma h^{m+1} s^m}. \tag{83}$$

This bound also holds if $\sin \xi_k$ is replaced by $\cos \xi_k$ for $k = 1, \ldots, m$.

The functions (81) are an example of a set of functions for which the number of minimal equations they satisfy over $\mathbf{C}(z)$ is equal to the difference between the number of functions in the set and their transcendence degree over $\mathbf{C}(z)$.

We write the functions (81) in the following order: $v_m, \ldots, v_1, u_m, \ldots, u_1$. We show that the equations (82) are then the minimal equations for the $2m$-tuple of functions (81) over $\mathbf{C}(z)$. To do this it is obviously sufficient to show that for any $N \in \mathbf{N}$ the set of monomials

$$v_m^{k_m} \cdots v_1^{k_1} u_m^{\kappa_m} \cdots u_1^{\kappa_1}, \quad \sum_{i=1}^{m} (k_i + \kappa_i) \leq N, \quad 0 \leq \kappa_i \leq 1, \quad k_i \geq 0,$$

forms a basis of the vector space $\mathbf{L}_N$ over $\mathbf{C}(z)$ which is spanned by the monomials

$$v_m^{k_m} \cdots v_1^{k_1} u_m^{\kappa_m} \cdots u_1^{\kappa_1}, \quad \sum_{i=1}^{m} (k_i + \kappa_i) \leq N, \quad k_i \geq 0, \quad \kappa_i \geq 0.$$

We prove a lemma which implies this claim.

Lemma 3. *The functions (81) are not related by any algebraic equation over* $\mathbf{C}(z)$ *in which the functions* $v_m, \ldots, v_1$ *appear to powers no greater than 1.*

Proof. We use induction on the number l of functions $v_1, \ldots, v_l$ which appear in a relation between $u_1, \ldots, u_m, v_1, \ldots, v_l$. If $l = 0$, the result holds by Lemma 2. We suppose that it holds for $l - 1$, and we prove that it then holds for the l functions $v_1, \ldots, v_l$.

Suppose that we have an equation of the form

$$Av_l + B = 0, \qquad A \not\equiv 0, \qquad B \not\equiv 0, \tag{84}$$

$$A, B \in \mathbf{C}[z][u_1, \ldots, u_m, v_1, \ldots, v_{l-1}].$$

Let $\deg_{u_l} A = n$, $n \geq 0$, and $\deg_{u_l} B = \nu$, $\nu \geq 0$. We rewrite (84) as follows:

$$v_l(A_n u_l^n + \cdots + A_0) + B_\nu u_l^\nu + \cdots + B_0 = 0, \qquad A_n \not\equiv 0, \qquad B_\nu \not\equiv 0. \quad (85)$$

$$A_k, B_s \in \mathbf{C}[z][u_1, \ldots, u_{l-1}, u_{l+1}, \ldots, u_m, v_1, \ldots, v_{l-1}],$$

$$k = 0, 1, \ldots, n, \qquad s = 0, 1, \ldots, \nu.$$

On the left in (85) we set

$$u_k = \frac{w_k - w_k^{-1}}{2i}, \qquad v_k = \frac{w_k + w_k^{-1}}{2}, \qquad w_k = e^{i\xi_k z}, \qquad k = 1, \ldots, m.$$

Then, by Lemma 21 of Chapter 3, the left side of (85) is identically zero as a function of $w_1, \ldots, w_m$.

We consider four cases: 1) $n \geq \nu$; 2) $n = \nu - 1$, $A_n \not\equiv iB_n$; 3) $n = \nu - 1$, $A_n = iB_n$; 4) $n < \nu - 1$. In the first two cases, we compare the coefficients of the terms with the highest power of w_l on the left in (85); in the last two cases, we compare the coefficients of the terms with the lowest power of w_l. In all cases, the identity (85) cannot hold. This contradiction proves the lemma.

Lemma 3 enables us to apply Theorem 4 to prove (83).

We consider the E-function

$$\varphi(z) = \sum_{n=0}^{m} \frac{1}{[\lambda_1 + 1, n] \cdots [\lambda_p + 1, n]} \left(\frac{z}{p}\right)^{pn},$$
$$\lambda_i \in \mathbf{Q}, \quad \lambda_i \neq -k, \quad k \in \mathbf{N}, \quad i = 1, \ldots, p,$$

where p is a prime number. This function satisfies the p-th order linear differential equation (10.5) with coefficients in $\mathbf{C}(z)$ and with its only singular point at $z = 0$. If we use Theorem 6 of Chapter 10 and Theorem 1, restated so as to apply to an E-function which satisfies a linear differential equation and its sequence of successive derivatives, we obtain the following bound for $\lambda_1, \ldots, \lambda_p$ satisfying the conditions in Theorem 6 of Chapter 10 and for any nonzero $\xi \in \mathbf{A}$:

$$\Phi_{\mathbf{K}^*}(\varphi(\xi), \varphi'(\xi), \ldots, \varphi^{(p-1)}(\xi); s; H) > cH^{-2^{p+1} \frac{p^p}{p!} h^{p+1} s^p}.$$

Remarks

The first bound for the algebraic independence measure of the values of E-functions was established by Siegel in 1929 in [72:3] (see (11.26)), using the method in Chapters 11 and 12.

After the proof of the Second Fundamental Theorem of Chapter 3 in 1955 in [70:2,8], it became easy, using Siegel's method, to obtain a general bound for the algebraic independence measure of the values of a set of E-functions which satisfy the conditions of the Second Fundamental Theorem. This was pointed out in the report [70:15] and the article [70:8]. Such a bound was published by S. Lang in 1962 in [37:2] in the homogeneous case, although the non-homogeneous case is analogous. Lang showed that, under the conditions of the Second Fundamental Theorem (with a homogeneous system of differential equations), the following bound holds for any $\xi \in \mathbf{A}$ with $\xi T(\xi) \neq 0$:

$$\Phi(f_1(\xi), \ldots, f_m(\xi); s; H) > \sigma H^{-\gamma s^m}, \tag{86}$$

where $\sigma > 0$ is a constant depending only on the functions (1) and the numbers ξ, m and s, and γ is a positive constant depending only on the number m and the degree h of the algebraic number field obtained by adjoining ξ to the field $\mathbf{K}$.

In 1968 in [22:1], Galochkin proved Theorem 2, replacing the constant γ in (86) by a specific function of m and h, and he generalized Lang's result to a more general measure.

The Fundamental Lemma and Theorems 1, 3 and 8 were published in [70:30, 34,36], Theorem 4 appeared in [70:33,34], and Theorems 6, 7 and 9 appeared in [70:31,34]. Theorem 5 was the first to appear in print.

We note that the idea of considering $B_{N,\xi}$-polynomials, which is basic in Chapter 12, can be found in [70:11].

The articles [79:2,8,9] by Väänänen are devoted to bounds for the algebraic independence measure of the values of E-functions and to the approximation of certain numbers connected with the values of E-functions.

Chapter 13
Effective bounds for measures

§ 1. Definitions and notation

The bounds for the measures of the values of a set of analytic functions $f_1(z), \ldots$ $\ldots, f_m(z)$ which are obtained by the methods of transcendental number theory contain various constants. These constants depend on the class of functions to which $f_1(z), \ldots, f_m(z)$ belong, the number m, the values of z that are being considered, and the degree s of the measure. The constants which occur in the bounds may be either effective or non-effective. Without delving into the logical foundations of the notion of effectiveness, in this chapter we shall understand the term "effective" in the following sense.

We shall say that a constant in a bound for a measure is an *effective constant* if it can be computed using a finite number of operations (the four arithmetic operations, raising to powers, taking logarithms, finding the largest or smallest element in a finite set) involving the parameters used to define the class of functions to which $f_1(z), \ldots, f_m(z)$ belong, the number m, the parameters which determine the values of z under consideration, and the degree s of the measure. Otherwise the constant will be called a *non-effective constant.*

We say that a bound is *effective* if all of the constants which appear in it are effective constants. We shall sometimes speak of bounds which are effective with respect to certain but not all of their parameters.

The bounds for the measures that were proved in Chapters 11 and 12 were non-effective, since there was no indication in the proofs of how to compute the constants in the right side of the inequalities. However, if we carefully examine the proofs of the lemmas in the method presented in Chapter 3, we observe that all of the constants which arise in the computations are effective constants in the above sense, with the single exception of the number n_0 in Lemma 9 of Chapter 3.

In the proof of Siegel's general theorem [72:4] and its generalization in [70:9], the number n_0 in the lemmas corresponding to Lemma 9 of Chapter 3 is an effective constant. These lemmas are proved under much more stringent restrictions than simply linear independence of the functions over $\mathbf{C}(z)$. In Siegel's paper one needs the *normality condition* and in the other paper the *irreducibility condition* for the functions $f_1(z), \ldots, f_m(z)$. The irreducibility condition is less restrictive than the normality condition, and so it enables one to apply the method more widely to specific functions. After the First and Second Fundamental Theorems of Chapter 3 were proved, this condition lost its usefulness for proving algebraic independence theorems for the values of E-functions. However, the irreducibility condition turned out to be important in bounds for measures of the values of E-functions, because it gives us an effective constant n_0. Thus, in this chapter we shall give a variant of the method which uses the condition of irreducibility of the set of monomials in the functions under consideration.

If the functions $f_1(z), \ldots, f_m(z)$ are analytic in some region and form a solution of the system of homogeneous linear differential equations

$$y_k' = \sum_{i=1}^{m} Q_{k,i} y_i, \qquad k = 1, \ldots, m, \qquad Q_{k,i} \in \mathbf{C}(z), \tag{1}$$

then we say that this m-tuple is an *irreducible set of functions* if none of the functions is identically zero and if a relation of the form

$$\sum_{k=1}^{m} P_k(z) y_k = 0, \qquad P_k(z) \in \mathbf{C}[z], \qquad k = 1, \ldots, m,$$

where $y_1, \ldots, y_m$ is any solution of (1), is possible only if $P_k(z) y_k = 0$ identically in z for $k = 1, \ldots, m$.

In other words, this definition says that if we have an irreducible set of functions, then the nonzero components of any solution of the system (1) are linearly independent over $\mathbf{C}(z)$. It follows that, if $f_1(z), \ldots, f_m(z)$ form an irreducible set of functions, then they are linearly independent over $\mathbf{C}(z)$.

In the present chapter, after modifying somewhat the definition of an E-function, we shall prove theorems bounding the linear and algebraic independence measures of the values of IE-functions which are effective or at least are effective with respect to the degree of the measure. They give refinements of some of the results of Chapter 11 with better remainder terms in the exponent in the bounds, in which ϵ is replaced by a function of H. In addition, in the bounds for algebraic independence measures in these theorems the degree of the measure is allowed to grow with the height up to a certain point, unlike in the analogous theorems of Chapters 11 and 12.

In order to simplify the derivation of effective bounds, it is worthwhile to modify somewhat the definition of an E-function. We shall now call an entire function

$$f(z) = \sum_{n=0}^{\infty} c_n \frac{z^n}{n!} \tag{2}$$

an E-function if it satisfies the three conditions defining an E-function in § 1 of Chapter 3, but with the bounds for c_n and q_n replaced by the requirements:

$$\overline{|c_n|} = O(c^n), \qquad q_n = O(c^n),$$

where $c \geq 1$ is some fixed number.

It is easy to see that, as in the case of the original definition, the set of all E-functions is a ring of functions which is closed under differentiation, integration from 0 to z, and substitution of λz for z, where λ is an algebraic number.

We note that all of the known E-functions which satisfy linear differential equations with coefficients in $\mathbf{C}(z)$, are also E-functions in the sense of the modified definition. In particular, from Lemmas 1 and 2 of Chapter 5 it follows that this is true of all of the hypergeometric E-functions.

We let $\mathbf{E}^*$ denote *the class of all E-functions which satisfy the new definition and which are solutions of linear differential equations with coefficients in* $\mathbf{C}(z)$. Then obviously $\mathbf{E}^* \subset \mathbf{E}$, where $\mathbf{E}$ is *the class of E-functions in the sense of the old definition which are solutions of such differential equations.* We do not know if the converse is true. It would be very natural to conjecture that $\mathbf{E}^* = \mathbf{E}$. This conjecture is weaker than Siegel's conjecture that any E-function in the sense of the earlier definition can be written as a polynomial with algebraic coefficients in z and in a finite number of hypergeometric E-functions and functions obtained from them by replacing z by λz for $\lambda \in \mathbf{A}$ (see § 1 of Chapter 5).

We make a further refinement in our definition of an E-function which will enable us to characterize how the constants in the bounds depend on the E-functions and the numbers m, ξ and s. We shall say that the entire function (2) is an *E-function in the class* $KE(\lambda, c; \mu, g)$ if:

1) $c_n \in \mathbf{K}$, $n = 0, 1, 2, \ldots$;

2) there exist constants $\lambda \geq 1$ and $c \geq 1$ such that $\overline{|c_n|} \leq \lambda c^n$, $n = 0, 1, 2, \ldots$;

3) there exist a sequence $\{g_n\}$, $g_n \in \mathbf{N}$, and constants $\mu \geq 1$, $g \geq 1$, such that $g_n c_k \in \mathbf{Z_K}$, $k = 0, 1, \ldots, n$, for all $n = 0, 1, 2, \ldots$, and $g_n \leq \mu g^n$, $n = 0, 1, 2, \ldots$.

We consider a set of functions

$$f_k(z) = \sum_{n=0}^{\infty} \frac{c_{k,n}}{n!} z^n, \qquad k = 1, \ldots, m, \tag{3}$$

in the class $KE(\lambda, c; \mu, g)$. Then there exists a sequence

$$\{g_{0,n}\}, \qquad g_{0,n} \leq \mu_0 g_0^n, \qquad \mu_0 \geq \mu, \qquad g_0 \geq g, \tag{4}$$

having the property that

$$g_{0,n} c_{k,\nu} \in \mathbf{Z_K}, \qquad \nu = 0, 1, \ldots, n, \qquad k = 1, \ldots, m. \tag{5}$$

In fact, we can take such a sequence to be, for example,

$$g_{0,n} = g_{1,n} g_{2,n} \cdots g_{m,n}, \qquad n = 0, 1, 2, \ldots, \tag{6}$$

where $\{g_{k,n}\}$, $k = 1, \ldots, m$, are the sequences corresponding by the definition to each of the KE-functions in (3). In this situation we may set

$$\mu_0 = \mu^m, \qquad g_0 = g^m. \tag{7}$$

In certain cases there exist sequences of the type (4) which satisfy (5) and have the property that $\mu_0 < \mu^m$ and $g_0 < g^m$, for example, in Lemma 7 below.

In this chapter, whenever we consider a set of KE-functions (3), the numbers μ_0 and g_0 will always have the meaning described above.

In what follows we shall consider a set of E-functions (3) which satisfies the system of homogeneous linear differential equations (1) or the system of linear differential equations

$$y_k' = Q_{k,0} + \sum_{i=1}^{m} Q_{k,i} y_i, \qquad k = 1, \ldots, m, \qquad Q_{k,i} \in \mathbf{C}(z). \tag{8}$$

These functions will be linearly independent over $\mathbf{C}(z)$. Hence, by Lemma 3 of Chapter 3, all of the $Q_{k,i}$ belong to $\mathbf{K}(z)$. The polynomial $T = T(z)$ will be defined as before for the systems of differential equations (1) and (3); by the same lemma, we have $T \in \mathbf{Z_K}[z]$, and all $TQ_{k,i}$ belong to $\mathbf{Z_K}[z]$.

We let q denote the maximum degree of the m^2+1 polynomials T and $TQ_{k,i}$, $k, i = 1, \ldots, m$; we let G denote the maximum modulus of all of the coefficients of these polynomials and their conjugates; and we set $p = \min_{1 \le i \le m} \operatorname{ord} f_i(z)$.

If our set of E-functions splits into ν groups, each satisfying a subsystem of differential equations of the form (1) or (3), then the notation $T_l = T_l(z)$, $l = 1, \ldots, \nu$, will have the same meaning as the symbol T for the systems (1) and (3).

In what follows we agree always to let the symbol σ with or without subscripts denote positive constants which depend only on the class $KE(\lambda, c; \mu, g)$ to which all of our E-functions belong, i.e., on the field $\mathbf{K}$ and the numbers λ, c, μ and g. The symbol γ with or without subscripts will denote positive constants which depend only on $KE(\lambda, c; \mu, g)$, the numbers p, q and G, and the number $\xi \in \mathbf{A}$. We note that the numbers σ, γ, σ_k and γ_k do not depend on the number m of functions in our set or on the degrees of the measures.

The positive constants Ω, Ω_0 and Ω_1 will depend on the class $KE(\lambda, c; \mu, g)$, the system of differential equations satisfied by our functions, and the numbers p, q, G, ξ and m.

All of the constants σ, γ, σ_k and γ_k will be effective constants, i.e., for any specific set of E-functions they can be computed if we know the parameters. In general, Ω, Ω_0 and Ω_1 will be non-effective constants.

§ 2. Refinement of the fundamental lemmas

Let

$$f_1(z), \ldots, f_m(z), \qquad m \geq 2, \tag{9}$$

be a set of analytic functions which satisfy the system of homogeneous linear differential equations (1), let $y_1, \ldots, y_m$ be an arbitrary solution of that system, and let

$$R_1 = \sum_{i=1}^{m} P_{1,i} y_i, \qquad P_{1,i} \in \mathbf{C}[z], \tag{10}$$

be any linear form in $y_1, \ldots, y_m$. As in § 4 of Chapter 3, we define the linear forms

$$R_k = \sum_{i=1}^{m} P_{k,i} y_i, \qquad P_{k,i} \in \mathbf{C}[z], \qquad k = 1, 2, \ldots, \tag{11}$$

where $R_k = TDR_{k-1} = TR'_{k-1}$, $k = 2, 3, \ldots$, and we consider the determinant of the coefficient matrix of the first m forms in (11):

$$\Delta = \Delta(z) = |P_{k,i}|_{k,i=1,\ldots,m}. \tag{12}$$

We shall show that, if the functions in (9) form an irreducible set of functions, and if the linear form R_1 has a large order zero at $z = 0$ for $y_i = f_i(z)$, as in Lemma 9 of Chapter 3, then the number n_0 which gives us values of n for which $\Delta(z) \not\equiv 0$ can be effectively computed.

We suppose that at least one of the polynomials $P_{1,i}$ in R_1 is nonzero. It may turn out that some of the columns of the determinant Δ in (12) have all zero entries. Since we have the numbering of the functions in (9) at our disposal, we may assume that every column of Δ with index $1, \ldots, M$, $1 \leq M \leq m$, has at least one nonzero entry, and the columns with index $M + 1, \ldots, m$ (if $M < m$) consist entirely of zeros. Then the linear forms $R_1, \ldots, R_M$ take the form

$$R_k = \sum_{i=1}^{M} P_{k,i} y_i, \qquad k = 1, \ldots, M. \tag{13}$$

We consider the determinant Δ_0 of the linear forms $R_1, \ldots, R_M$ in (13). We may suppose that this determinant now has at least one nonzero entry in each column, since otherwise we could repeat the above argument until we arrive at such a system of linear forms (but with a smaller M, $M \geq 1$).

Lemma 1. *Suppose that the m-tuple of analytic functions in (9) forms a solution of the system of homogeneous linear differential equations (1) and is an irreducible set of functions. Suppose that the linear form R_1 in (10) has the property that at least one of the polynomials $P_{1,i}$, $1 \le i \le m$, is nonzero. Then the determinant $\Delta_0(z)$ of the system of linear forms $R_1, \ldots, R_M$ in (13) is not identically zero as a function of z.*

Proof. Suppose that $\Delta_0(z) \equiv 0$. Then the linear forms in (13) are linearly dependent. Suppose they have rank l. Then $1 \le l < M$. By Lemma 7 of Chapter 3, there exists a fundamental system of solutions of the system of differential equations (1)

$$y_{i,s}, \qquad i = 1, \ldots, m, \quad s = 1, \ldots, m, \tag{14}$$

such that all of the forms $R_1, R_2, \ldots$ vanish when the first $\mu = m - l$ solutions in (14) are substituted into them. Thus,

$$R_{k,s} = \sum_{i=1}^{M} P_{k,i} y_{i,s} = 0, \qquad k = 1, \ldots, M, \quad s = 1, \ldots, \mu. \tag{15}$$

In Δ_0 each column has at least one nonzero entry. Hence, for each i, $1 \le i \le M$, there is at least one of the linear forms $R_1, \ldots, R_M$, say the one with index k_i, $1 \le k_i \le M$, for which $P_{k_i,i} \not\equiv 0$. But the functions (9) form an irreducible set; hence it follows from (15) that $y_{i,s} = 0$, $s = 1, \ldots, \mu$. Since this holds for each $i = 1, \ldots, M$, we have

$$y_{i,s} = 0, \qquad i = 1, \ldots, M, \quad s = 1, \ldots, \mu. \tag{16}$$

The equalities in (16) mean that the upper-left corner of the matrix of fundamental solutions of (14) consists of an $M \times \mu$ block of zeros. Since we have $l < M \le m$, it follows that $M + \mu = M + m - l \ge m + 1$. Hence, by Laplace's Theorem, the determinant of the $m \times m$ matrix of solutions of (14) is identically zero. But the determinant of a fundamental system of solutions of a system of homogeneous linear differential equations cannot be identically zero. This contradiction proves the lemma.

We recall the notation:

$$p = \min_{1 \le i \le m} \operatorname{ord} f_i(z), \qquad q = \max(\deg T, \max_{\substack{1 \le i \le m \\ 1 \le k \le m}} \deg T Q_{k,i}), \tag{17}$$

$$n_0 = q(m-1)m + 2\tau_0, \tag{18}$$

where τ_0 was defined in Lemma 8 of Chapter 3,

$$s \in \mathbf{Z}^+, \qquad 0 \le s \le \left[\frac{n}{2}\right], \qquad n \in \mathbf{N}, \tag{19}$$

$$t = q\frac{(m-1)m}{2} + s + p. \tag{20}$$

Lemma 2. *Suppose that the functions (9) are analytic in a region containing the point $z = 0$, form a solution of the system of homogeneous linear differential equations (1), and are linearly independent over $\mathbf{C}(z)$. Further suppose that*

$$R_1 = \sum_{i=1}^{m} P_{1,i} f_i(z), \qquad R_1 \not\equiv 0, \tag{21}$$

$$P_{1,i} \in \mathbf{C}[z], \qquad \deg P_{1,i} \le n, \qquad n \in \mathbf{N}, \qquad i = 1, \dots, m,$$

is any linear form in the functions (9) for which

$$\operatorname{ord} R_1(z) \ge m(n+1) - s + 1. \tag{22}$$

Then:

1) *for any $n \ge n_0$ the linear forms $R_1, \dots, R_m$ in (11) obtained from the form R_1 in (21) are linearly independent, and the determinant $\Delta(z)$ in (12) has the form*

$$\Delta(z) = z^{mn-s-p}\Delta_1(z), \qquad \Delta_1(z) \not\equiv 0, \tag{23}$$

$$\Delta_1(z) \in \mathbf{C}[z], \qquad \deg \Delta_1(z) = \kappa, \qquad 0 \le \kappa \le t;$$

2) *if, in addition, the functions (9) form an irreducible set of functions, then*

$$n_0 = q(m-1)m + 2p. \tag{24}$$

Proof. Part 1) is a restatement of Lemma 9 of Chapter 3. We now prove part 2).

Let the numbers M and l for the form R_1 in (21) have the same meaning as in Lemma 1. We consider the linear forms $R_1, \dots, R_M$ in (13) with $y_i = f_i(z)$, $i = 1, \dots, M$, and the corresponding determinant $\Delta_0(z)$. By Lemma 1, we have $M = l$ and $\Delta_0(z) \not\equiv 0$. We denote $p_0 = \min_{1 \le i \le M} \operatorname{ord} f_i(z)$. The rest of the proof is a repetition of the proof of Lemma 9 of Chapter 3, with τ replaced by p_0. As a result we find that $l = m$ and $p_0 = p$, and then (23) and (24) are fulfilled.

Lemma 3. *Suppose that the functions (9) are analytic in a region containing the point $z = 0$, form a solution of the system of homogeneous linear differential equations (1), and are linearly independent over $\mathbf{C}(z)$ (are an irreducible set of functions). Let s and t be given by (17), let n_0 be given by (18) (by (24)), and let p and q be given by (19) and (20). Further suppose that the linear form R_1 in (21) has the property that $n \geq n_0$ and (22) holds. Let $\alpha \in \mathbf{C}$, $\alpha T(\alpha) \neq 0$. Then the coefficient matrix*

$$\|P_{k,i}(\alpha)\|, \qquad i = 1, \ldots, m, \quad k = 1, \ldots, m + t, \tag{25}$$

of the linear forms

$$R_k(\alpha) = \sum_{i=1}^{m} P_{k,i}(\alpha) f_i(\alpha), \qquad k = 1, \ldots, m + t, \tag{26}$$

has rank m, and so it is possible to choose m linearly independent forms from among the forms in (26).

Lemma 3 is a restatement of Lemma 10 of Chapter 3, which holds in each of the two cases of Lemma 2.

Lemma 4. *Suppose that $f_k(z) \in KE(\lambda, c; \mu, g)$, $k = 1, \ldots, m$, $m \geq 2$, $n \in \mathbf{N}$, and $\omega = \omega(n)$ is a non-decreasing function of n, $2 \leq \omega(n) \leq n$. Then there exist m polynomials*

$$P_k(z) = \sum_{l=0}^{n} b_{k,l} z^l, \qquad P_k(z) \not\equiv 0, \qquad k = 1, \ldots, m,$$

with the following properties:

1) $b_{k,l} \in \mathbf{Z_K}$, $l = 0, 1, \ldots, n$, $k = 1, \ldots, m$,

$$\overline{|b_{k,l}|} < e^{\sigma \ln(2cg_0) m^2 \omega n} n^n, \quad l = 0, 1, \ldots, n, \quad k = 1, \ldots, m; \tag{27}$$

2) *the linear form*

$$R = \sum_{k=1}^{m} P_k(z) f_k(z) = \sum_{\nu=0}^{\infty} a_\nu \frac{z^\nu}{\nu!} \tag{28}$$

has the property that

$$\operatorname{ord} R \geq \tau, \qquad \tau = m(n+1) - \left[\frac{n}{\omega}\right] - 1; \tag{29}$$

3) *the coefficients* a_ν *in (28) satisfy the condition*

$$\overline{|a_\nu|} < e^{\sigma_0 \ln(2cg_0) m^2 \omega n} n^n (2c)^\nu, \qquad \nu \geq \tau. \tag{30}$$

Proof. We denote

$$P_k(z) = n! \sum_{\nu=0}^{n} B_{k,\nu} \frac{z^\nu}{\nu!}, \qquad k = 1, \ldots, m, \tag{31}$$

where the $B_{k,\nu}$ are chosen to be in $\mathbf{Z_K}$. Then $P_k(z) \in \mathbf{Z_K}[z]$.

If we write our KE-functions in the form (3), we obtain

$$P_k(z) f_k(z) = n! \sum_{\nu=0}^{\infty} d_{k,\nu} \frac{z^\nu}{\nu!}, \qquad k = 1, \ldots, m, \tag{32}$$

where

$$d_{k,\nu} = \sum_{\rho=0}^{\min(n,\nu)} \binom{\nu}{\rho} B_{k,\rho} c_{k,\nu-\rho}, \tag{33}$$

since $B_{k,\rho} = 0$ for $\rho > n$. Then from (28) and (32) we find that

$$a_\nu = n! \sum_{k=1}^{m} d_{k,\nu}, \qquad \nu = 0, 1, 2, \ldots. \tag{34}$$

If we multiply both sides of (34) by $g_{0,\nu}/n!$, where $\{g_{0,\nu}\}$ is the sequence in (4) which satisfies (5), then because of (29) and (33) we obtain τ homogeneous linear equations in the $m(n+1)$ unknowns $B_{k,\rho}$:

$$\sum_{k=1}^{m} \sum_{\rho=0}^{n} g_{0,\nu} \binom{\nu}{\rho} c_{k,\nu-\rho} B_{k,\rho} = 0, \qquad \nu = 0, 1, \ldots, \tau - 1. \tag{35}$$

Since $\nu \leq m(n+1)$ in (35), it follows from (7) that

$$g_{0,\nu} \leq \mu_0 g_0^{m(n+1)} \leq \mu^m g_0^{m(n+1)}, \qquad \binom{\nu}{\rho} \leq 2^{m(n+1)}, \qquad \overline{|c_{k,\nu}|} \leq \lambda c^{m(n+1)},$$

$$\overline{\left| g_{0,\nu} \binom{\nu}{\rho} c_{k,\nu-\rho} \right|} \leq \lambda \mu^m (2cg_0)^{m(n+1)}.$$

We apply Lemma 13 of Chapter 3 to the system of equations (35) with

$$p = \lambda = m(n+1) - \left[\frac{n}{\omega}\right] - 1, \quad q = m(n+1), \quad A = \lambda\mu^m(2cg_0)^{m(n+1)}.$$

Then, in view of the inequality

$$\frac{p}{q-p} = \left(m(n+1) - \left[\frac{n}{\omega}\right] - 1\right) \Big/ \left(\left[\frac{n}{\omega}\right] + 1\right) < m(n+1)/(n/\omega) \leq 2m\omega,$$

we find that the $B_{k,\rho}$ belong to $\mathbf{Z_K}$, are not all zero, and, by (3.109), have the property that

$$\overline{|B_{k,\rho}|} < \sigma_1(\sigma_1 mn\lambda\mu^m(2cg_0)^{m(n+1)})^{2m\omega} < e^{\sigma_2 \ln(2cg_0)m^2\omega n}. \tag{36}$$

From (31) and (36) we obtain (27), and from (33), (34) and (36) we obtain (30). The lemma is proved.

In Lemma 2 we set $s = \left[\frac{n}{\omega}\right]$, where $n \geq n_0$, and $R_1 = R$, where R is the approximating form constructed in Lemma 4. Since $T(z) \in \mathbf{Z_K}[z]$, it follows that all of the linear forms $R_1, R_2, \ldots$ have coefficients in $\mathbf{Z_K}[z]$. By Lemma 3, the numerical linear approximating forms $R_1(\xi), \ldots, R_{m+t}(\xi)$ have rank m for any $\xi \in \mathbf{A}$ with $\xi T(\xi) \neq 0$. The next lemma, whose proof is similar to the proof of Lemma 15 of Chapter 3, enables us to estimate $\overline{|P_{k,i}(\xi)|}$ and $|R_k(\xi)|$.

Lemma 5. *Suppose that the functions $f_i(z) \in KE(\lambda, c; \mu, g)$, $i = 1, \ldots, m$, $m \geq 2$, form a solution of the system of homogeneous linear differential equations (1) and are linearly independent over $\mathbf{C}(z)$, and let $\xi \in \mathbf{A}$. Further suppose that the linear form $R_1 = R_1(z)$ is constructed from Lemma 4 for a value of n satisfying*

$$n \geq \max(n_0, \exp(4m^2 \ln(2cg_0)), \tag{37}$$

and for a value of ω to be chosen later, and let t be given by (20). Then there exist constants γ_1 and γ_2 such that the following inequalities hold for any k satisfying $k \leq m+t$:

$$\overline{|P_{k,i}(\xi)|} < e^{\gamma_1\sqrt{\ln(2cg_0)}mn\sqrt{\ln n}} n^n, \qquad i = 1, \ldots, m, \tag{38}$$

$$|R_k(\xi)| < e^{\gamma_2\sqrt{\ln(2cg_0)}mn\sqrt{\ln n}} n^{-(m-1)n}, \tag{39}$$

and (39) remains in force if all of the power series coefficients of $R_k(z)$ and the number ξ are replaced by their conjugates in the field $\mathbf{K}_{0,i}$ conjugate to $\mathbf{K}$.

Proof. Using the notation in Lemma 15 of Chapter 3, we have

$$T(z) \ll G(1+z)^q, \qquad T(z)Q_{k,i} \ll G(1+z)^q,$$

where q and G were defined in § 1.

If we repeat the argument from Lemma 15 of Chapter 3 and use (3.37), by induction we obtain the following results:

$$P_{k+1,i}(z) \ll G^k(1+z)^{kq+n} e^{\sigma \ln(2cg_0)m^2\omega n} n^n \prod_{\nu=0}^{k-1} (\nu q + m + n), \qquad i = 1, \ldots, m, \quad k = 0, 1, 2, \ldots; \tag{40}$$

$$R_{k+1} \ll G^k(1+z)^{kq} \prod_{\nu=0}^{k-1} \left(\nu q + \frac{d}{dz}\right) \widehat{R}_1(z), \quad k = 0, 1, 2, \ldots, \tag{41}$$

where (40) and (41) remain in force if the coefficients of the polynomial $P_{k+1,i}(z)$ and the power series R_{k+1} are replaced by their conjugates in any of the fields $\mathbf{K}_{0,i}$.

By assumption, (37) holds. We suppose that n and ω are such that

$$\frac{n}{\omega} > 2m^2 \ln 2. \tag{42}$$

Then, using (20), we find that

$$k \le m + t = m + \left[\frac{n}{\omega}\right] + p + q\frac{(m-1)m}{2} < \gamma_3 \frac{n}{\omega}. \tag{43}$$

It now follows from (40) and (43) that

$$\overline{|P_{k,i}(\xi)|} < e^{\sigma \ln(2cg_0)m^2\omega n + \gamma_4 \frac{n}{\omega} \ln n} n^n, \qquad k \le m + t. \tag{44}$$

Furthermore, from (43) and (30) we have

$$\prod_{\nu=0}^{k-1}\left(\nu q+\frac{d}{dz}\right)\widehat{R}_1(z) \ll e^{\gamma_5\frac{n}{\omega}\ln n}\left(1+\frac{d}{dz}\right)^k \widehat{R}_1(z) \ll$$

$$\ll e^{\sigma_0 \ln(2cg_0)m^2\omega n+\gamma_5\frac{n}{\omega}\ln n}n^n\left(1+\frac{d}{dz}\right)^k\sum_{\nu=\tau}^{\infty}(2c)^\nu\frac{z^\nu}{\nu!} \ll$$

$$\ll e^{\sigma_0 \ln(2cg_0)m^2\omega n+\gamma_5\frac{n}{\omega}\ln n}n^n\sum_{\rho=0}^{k}\sum_{\nu=\tau}^{\infty}\binom{k}{\rho}(2c)^\nu\frac{z^{\nu-\rho}}{(\nu-\rho)!} \ll$$

$$\ll e^{\sigma_0 \ln(2cg_0)m^2\omega n+\gamma_5\frac{n}{\omega}\ln n}n^n 2^k\sum_{\nu=\tau-k}^{\infty}(2c)^{\nu+k}\frac{z^\nu}{\nu!},$$

where τ is defined in (29). We hence obtain:

$$\left|\prod_{\nu=0}^{k-1}\left(\nu q+\frac{d}{dz}\right)\widehat{R}_1(z)\right|_{z=\xi} < e^{\sigma_0 \ln(2cg_0)m^2\omega n+\gamma_6\frac{n}{\omega}\ln n}n^{-(m-1)n}. \tag{45}$$

From (41) and (43) we find that

$$|R_k(\xi)| < e^{\sigma_0 \ln(2cg_0)m^2\omega n+\gamma_7\frac{n}{\omega}\ln n}n^{-(m-1)n}, \qquad k \le m+t. \tag{46}$$

Equating the terms in the exponent on the right of (44), we set

$$\omega = \frac{1}{m}\sqrt{\frac{\ln n}{\ln(2cg_0)}}. \tag{47}$$

Since $\omega \ge 2$ by assumption, it follows that n must satisfy the condition $\ln n \ge 4m^2\ln(2cg_0)$, which holds because of (37). Then

$$\frac{n}{\omega} = \frac{\sqrt{\ln(2cg_0)}\,mn}{\sqrt{\ln n}} > \sqrt{\ln 2}\,m\sqrt{n} > 2m^2\ln 2,$$

since $cg_0 > 1$ and $\sqrt{n} > \sqrt{\ln n} > 2m\sqrt{\ln 2}$. This means that the condition (42) is fulfilled.

If we substitute the value of ω from (47) into the bounds (44) and (46), we obtain the inequalities (38) and (39) of the lemma.

Lemma 6. *Suppose that the KE-functions $f_1(z),\dots,f_m(z)$, $m \ge 2$, form a solution of the system of homogeneous linear differential equations (1) and are linearly independent over $\mathbf{C}(z)$ (form an irreducible set of functions), and let $\xi \in \mathbf{A}$, $\xi T(\xi) \ne 0$. Then there exist constants γ_8 and γ_9 such that for any $n \in \mathbf{N}$ with*

$n \geq \max(n_0, \exp(4m^2 \ln(2cg_0))$, where n_0 is defined by (18) (by (24)), there exists an m-tuple of linearly independent linear forms in the numbers $f_1(\xi), \dots, f_m(\xi)$

$$L_k = \sum_{l=1}^{m} a_{k,l} f_l(\xi), \qquad k = 1, \dots, m, \qquad a_{k,l} \in \mathbf{Z}_{\mathbf{K}_0}, \tag{48}$$

such that

$$\overline{|a_{k,l}|} \leq e^{\gamma_8 \sqrt{\ln(2cg_0)}\, mn\sqrt{\ln n}}\, n^n, \qquad k, l = 1, \dots, m, \tag{49}$$

$$|L_k| \leq e^{\gamma_9 \sqrt{\ln(2cg_0)}\, mn\sqrt{\ln n}}\, n^{-(m-1)n}, \qquad k = 1, \dots, m, \tag{50}$$

where (50) remains in force if all of the $a_{k,l}$ in (48), the power series coefficients of all of the functions $f_1(z), \dots, f_m(z)$, and the number ξ are replaced by their conjugates in any of the conjugate fields $\mathbf{K}_{0,i}$ of $\mathbf{K}$.

Proof. The proof uses Lemma 5 (the bounds (38) and (39)) in the same way as Lemma 15 of Chapter 3 was used to prove Lemma 16 of Chapter 3. Here we need only observe that the relations (17), (19) and (20) imply the bounds

$$m + t = m + q\frac{(m-1)m}{2} + s + p \leq \gamma_{10} n, \qquad n + q(m + t - 1) \leq \gamma_{11} n,$$

and, if we choose $a \in \mathbf{N}$ so that $a\xi \in \mathbf{Z}_{\mathbf{K}_0}$, then

$$a^{\gamma_{11} n} P_{k,l}(\xi) \in \mathbf{Z}_{\mathbf{K}_0}, \qquad k = 1, \dots, m + t, \quad l = 1, \dots, m.$$

§ 3. Bounds for linear independence measures

The algebraic number fields $\mathbf{K}$, $\mathbf{K}^*$, $\mathbf{K}_0$ and the conjugate fields $\mathbf{K}_{0,i}$, $i = 1, \dots, h$, have the same meaning in this section as in Chapters 11 and 12.

As in § 2 of Chapter 11, $l_{\mathbf{K}^*} = l_{\mathbf{K}^*}(z_1, \dots, z_m) \not\equiv 0$ denotes an arbitrary linear form with coefficients in $\mathbf{Z}_{\mathbf{K}^*}$ and size $\overline{|l_{\mathbf{K}^*}|} = H$, and $l_i = l_i(z_1, \dots, z_m)$, $i = 1, \dots, h$, denote the linear forms obtained from $l_{\mathbf{K}^*} = l_1$ by replacing its coefficients by their conjugates in $\mathbf{K}_{0,i}$.

We generalize Theorem 1 of Chapter 11.

Theorem 1. *Suppose that the KE-functions $f_1(z), \dots, f_m(z)$, $m \geq 2$, form a solution of the system of homogeneous linear differential equations (1) and are* *linearly independent over* $\mathbf{C}(z)$, *and let* $\xi \in \mathbf{Z}$, $\xi T(\xi) \neq 0$. *Then there exist constants* γ *and* Ω *such that for* $\ln\ln H > \Omega$ *one has*

$$\max_{1 \le i \le h} |l_i(f_{1,i}(\xi_i), \ldots, f_{m,i}(\xi_i))| > H^{1-m-\frac{\gamma m^3\sqrt{m}}{\sqrt{\ln\ln H}}}, \tag{51}$$

and if $\mathbf{K}_0 = \mathbf{I}$, *then one has*

$$L_{\mathbf{I}}(f_1(\xi), \ldots, f_m(\xi); H) > H^{1-m-\frac{\gamma m^3\sqrt{m}}{\sqrt{\ln\ln H}}}. \tag{52}$$

Corollary. *Under the conditions of Theorem 1, there exist constants* γ *and* Ω_1 *such that for any* H *one has*

$$L_{\mathbf{I}}(f_1(\xi), \ldots, f_m(\xi); H) > \Omega_1 H^{1-m-\frac{\gamma m^3\sqrt{m}}{\sqrt{\ln\ln(H+2)}}}.$$

Theorem 1′. *Under the conditions of Theorem 1, if the* m*-tuple* $f_1(z), \ldots, f_m(z)$ *is an irreducible set of functions, then in Theorem 1 and its corollary one can take*

$$\Omega = \gamma^2 m^5,$$

$$\Omega_1 = \begin{cases} 1, & \text{if } \rho < 1, \\ \exp(-(2m-1)\exp(\gamma^2 m^5)), & \text{if } \rho \ge 1, \end{cases}$$

where

$$\rho = \frac{\gamma m^2\sqrt{m}}{\sqrt{\ln\ln(H+2)}}.$$

Theorem 1′ says that all of the constants in Theorem 1 and its corollary are effective constants if our m-tuple of functions satisfies the irreducibility condition.

The proof of Theorem 1 is analogous to that of Theorem 1 of Chapter 11, except that Lemma 6 is used instead of Lemma 16 of Chapter 3.

We consider an arbitrary linear form L_0 in $f_1(\xi), \ldots, f_m(\xi)$, as in (11.33), along with the conjugate linear forms $L_{0,1}, \ldots, L_{0,h}$ in (11.34), which are obtained from L_0 by replacing its coefficients, the power series coefficients of $f_1(z), \ldots, f_m(z)$, and the number ξ by their conjugates in the fields $\mathbf{K}_{0,i}$ which are conjugate to $\mathbf{K}_0$.

By Lemma 6, for $n \ge \max(n_0, \exp(4m^2\ln(2cg_0)))$ there exist m linearly independent linear forms (48) in $f_1(\xi), \ldots, f_m(\xi)$ with coefficients in $\mathbf{K}_0$ which satisfy (49) and (50). One can choose $m-1$ of these forms in such a way that they together with L_0 will be linearly independent. To be definite, suppose that these $m-1$ forms are $L_1, \dots, L_{m-1}$. We also consider the conjugate forms $L_{1,i}, \dots, L_{m-1,i}$ in (11.38).

We let Δ denote the determinant of the linear forms $L_1, \dots, L_{m-1}, L_0$, $\Delta \in \mathbf{Z}_{\mathbf{K}_0}$, $\Delta \not\equiv 0$, and we let Δ_i, $i = 1, \dots, h$, denote the determinant obtained from Δ by replacing its entries by their conjugates in $\mathbf{K}_{0,i}$. We further let $\Delta_{k,l,i}$ denote the cofactor of the entry in the k-th row and l-th column of Δ_i. Since $\Delta \in \mathbf{Z}_{\mathbf{K}_0}$ and $\Delta \not\equiv 0$, it follows that $|\mathcal{N}(\Delta)| = |\Delta_1 \cdots \Delta_h| \geq 1$. Hence, we have (11.39) for some i. Applying (11.34) and (11.38) for each l, $1 \leq l \leq m$, we find that (11.40) is fulfilled. We choose l so that $f_{l,i}(\xi_i) \neq 0$, which can be done, since $T(\xi) \neq 0$. Then from (11.40) we obtain the inequality (11.41).

If we use (49) and (50) and observe that (50) holds for all of the forms $L_{k,i}$, we obtain:

$$\overline{|\Delta_{m,l}|} < e^{\gamma_{12}\sqrt{\ln(2cg_0)}m^2 n\sqrt{\ln n}} n^{(m-1)n}, \tag{53}$$

$$\max_{1 \leq k \leq m-1} \overline{|\Delta_{k,l}|} < e^{\gamma_{13}\sqrt{\ln(2cg_0)}m^2 n\sqrt{\ln n}} n^{(m-2)n} H, \tag{54}$$

$$\max_{1 \leq k \leq m-1} |L_{k,i}| < e^{\gamma_9\sqrt{\ln(2cg_0)}\, mn\sqrt{\ln n}} n^{-(m-1)n}. \tag{55}$$

From (11.39), (11.41) and (53)–(55) we find that

$$|L_{0,i}| > e^{-\gamma_{14}\sqrt{\ln(2cg_0)}\, m^2 n\sqrt{\ln n}} n^{-(m-1)n} \left(1 - H e^{\gamma_{15}\sqrt{\ln(2cg_0)}\, m^2 n\sqrt{\ln n}} n^{-n}\right). \tag{56}$$

We choose n to be the smallest integer for which the following condition holds:

$$n^n e^{-\gamma_{15}\sqrt{\ln(2cg_0)}\, m^2 n\sqrt{\ln n}} > 2H. \tag{57}$$

From (57) we have $2H < n^n$. Taking the logarithm twice, we obtain

$$\gamma_{16} \ln\ln H < \ln n.$$

In § 1 it was indicated that $g_0 \leq g^m$. Hence there exists a constant γ_{17} such that for

$$\ln\ln H > \gamma_{17} \max(m^5, \ln n_0)$$

one has the inequalities

$$n \geq n_0, \qquad \sqrt{\ln n} > 2\gamma_{18}\sqrt{\ln(2cg_0)}\, m^2, \qquad \gamma_{18} = \max(1, \gamma_{15}). \tag{58}$$

It then follows from (58) that the conditions on n in Lemma 6 are fulfilled, and from (56) and (57) we have

$$|L_{0,i}| > \frac{1}{2} e^{-\gamma_{14}\sqrt{\ln(2cg_0)}\, m^2 n\sqrt{\ln n}}\, n^{-(m-1)n}. \tag{59}$$

Because of our choice of n, we have the inequality

$$(n-1)^{(n-1)} e^{-\gamma_{15}\sqrt{\ln(2cg_0)}\, m^2 (n-1)\sqrt{\ln(n-1)}} \leq 2H,$$

from which

$$\begin{aligned} & n^n e^{-\gamma_{15}\sqrt{\ln(2cg_0)}\, m^2 n\sqrt{\ln n}} \leq \\ & \leq 2H\left(1+\frac{1}{n-1}\right)^{n-1} e^{-\gamma_{15}\sqrt{\ln(2cg_0)}\, m^2 (n\sqrt{\ln n}-(n-1)\sqrt{\ln(n-1)})}\, n < \\ & < 2enH. \end{aligned} \tag{60}$$

Taking the logarithm, we obtain

$$n\ln n - \gamma_{15}\sqrt{\ln(2cg_0)}\, m^2 n\sqrt{\ln n} < \ln(2e) + \ln n + \ln H,$$

and then, using (58), we have

$$n \ln n < \gamma_{19} \ln H, \tag{61}$$

from which, by (57), (60) and (61), we find that

$$2H < n^n e^{-\gamma_{15}\sqrt{\ln(2cg_0)}\, m^2 n\sqrt{\ln n}} < \gamma_{20} H \frac{\ln H}{\ln\ln H}, \tag{62}$$

and

$$\ln\ln H > \gamma_{21} \ln n. \tag{63}$$

From (59), (61) and (62) we obtain:

$$|L_{0,i}| > H^{1-m-\frac{\gamma_{22}\sqrt{\ln(2cg_0)}\, m^2}{\sqrt{\ln\ln H}}}, \tag{64}$$

and since $g_0 \leq g^m$ and the inequalities (58) and (63) are fulfilled, it follows that

$$|L_{0,i}| > H^{1-m-\frac{\gamma_{23} m^3\sqrt{m}}{\sqrt{\ln\ln H}}}. \tag{65}$$

We set

$$\gamma = \max(\gamma_{23}, \sqrt{\gamma_{17}}), \qquad \Omega = \gamma^2 \max(m^5, \ln n_0).$$

Then for $\ln\ln H > \Omega$ the inequality (65) implies (51).

If $\mathbf{K}_0 = \mathbf{I}$, then (52) follows from (51), since complex conjugate numbers have the same absolute value. The corollary and Theorem 1$'$ follow from Theorem 1 and the relations (24) and (65), if we observe that for $\ln\ln H \le \gamma^2 m^5$ one has

$$L_{\mathbf{I}}(f_{1,i}(\xi_i), \ldots, f_{m,i}(\xi_i);\ H) \ge L_{\mathbf{I}}(f_{1,i}(\xi_i), \ldots, f_{m,i}(\xi_i);\ \exp\exp\gamma^2 m^5).$$

Theorem 1 also implies the following theorem, which is useful in applications to specific functions.

Theorem 2. *Suppose that the KE-functions $f_1(z), \ldots, f_m(z)$, $m \ge 1$, form a solution of the system of linear differential equations (8) and are linearly independent together with 1 over $\mathbf{C}(z)$, and let $\xi \in \mathbf{A}$, $\xi T(\xi) \ne 0$. Then there exist constants γ and Ω such that for* $\ln\ln H > \Omega$ *one has*

$$\max_{1\le i\le h} |l_i(1, f_{1,i}(\xi_i), \ldots, f_{m,i}(\xi_i))| > H^{-m-\frac{\gamma m^3\sqrt{m}}{\sqrt{\ln\ln H}}},$$

and if $\mathbf{K}_0 = \mathbf{I}$, then one has

$$L_{\mathbf{I}}(1, f_1(\xi), \ldots, f_m(\xi);\ H) > H^{-m-\frac{\gamma m^3\sqrt{m}}{\sqrt{\ln\ln H}}}. \tag{66}$$

We have a corollary of Theorem 2 and a companion Theorem 2$'$ which are analogous to the corollary of Theorem 1 and Theorem 1$'$.

§ 4. Bounds for algebraic independence measures

From the proof of Theorem 1 and Lemmas 7, 8 and 9 below we obtain a theorem bounding the algebraic independence measure of the values of IE-functions.

Theorem 3. *Suppose that the IE-functions $f_1(z), \ldots, f_m(z)$, $m \ge 2$ ($m \ge 1$), form a solution of the system of homogeneous linear differential equations (1) (of linear differential equations (8)) and are homogeneously algebraically independent (are algebraically independent) over $\mathbf{C}(z)$, and let $\xi \in \mathbf{I}$, $\xi T(\xi) \ne 0$. Then there exist constants γ_0 and Γ_0 (constants γ and Ω) such that*

$$\Phi^0_{\mathbf{I}}(f_1(\xi), \ldots, f_m(\xi);\ s;\ H) > c_0 H^{1-M-\delta};$$

$$\Phi_{\mathbf{I}}(f_1(\xi)/f_m(\xi), \ldots, f_{m-1}(\xi)/f_m(\xi);\ s;\ H) > c_1 H^{1-M-\delta},$$

where

$$M = \frac{(s+m-1)!}{s!(m-1)!}, \quad \delta = \frac{\gamma_0 (s+1)^{3(m-1)} \sqrt{m \ln(s+1)}}{\sqrt{\ln\ln(H+2)}}, \quad c_1 = |f_m^{-s}(\xi)| c_0;$$

$$c_0 = \begin{cases} 1, & \text{if } \rho < 1, \\ \exp\Big\{-\Big((s+1)^{m-1}\Big(1+\frac{\gamma_0}{\Omega_0}\Big)-1\Big) \times \\ \qquad \times \exp(\Omega_0^2 m (s+1)^{4(m-1)} \ln(s+1))\Big\}, & \text{if } \rho \geq 1; \end{cases}$$

$$\rho = \frac{\Omega_0 (s+1)^{2(m-1)} \sqrt{m \ln(s+1)}}{\sqrt{\ln\ln(H+2)}},$$

and in the non-homogeneous case

$$\Phi_{\mathbf{I}}(f_1(\xi), \ldots, f_m(\xi); s; H) > cH^{-M-\delta},$$

where

$$M = \frac{(s+m)!}{s!m!}, \qquad \delta = \frac{\gamma(s+1)^{3m} \sqrt{m \ln(s+1)}}{\sqrt{\ln\ln(H+2)}};$$

$$c = \begin{cases} 1, & \text{if } \rho < 1, \\ \exp\Big\{-\big((s+1)^m\big(1+\frac{\gamma}{\Omega}\big)-1\big) \times \\ \qquad \times \exp(\Omega^2 m (s+1)^{4m} \ln(s+1))\Big\}, & \text{if } \rho \geq 1; \end{cases}$$

$$\rho = \frac{\Omega (s+1)^{2m} \sqrt{m \ln(s+1)}}{\sqrt{\ln\ln(H+2)}}.$$

Theorem 3′. *Under the conditions of Theorem 3, if the set of products of powers*

$$f_m^{k_m}(z) \cdots f_1^{k_1}(z), \qquad k_1 + \cdots + k_m = N \quad (k_1 + \cdots + k_m \leq N), \tag{67}$$

is an irreducible set of functions for any $N = 1, 2, \ldots$, then $\Omega_0 = \gamma_0$ ($\Omega = \gamma$) in Theorem 3.

Lemma 7. *If the functions $f_k(z)$ belong to $KE(\lambda, c; \mu, g)$, $k = 1, \dots, m$, and if $N \in \mathbf{N}$, then the $M = (N+m-1)!/N!(m-1)!$ (the $M = (N+m)!/N!m!$) products of powers in (67) belong to the class*

$$KE(\lambda, Nc;\, \mu^{mN},\, g^{m(\ln N+1)}). \tag{68}$$

Here the sequence $\{g_n\}$ in the definition of a function in the class (68) can be chosen to be the same for all of the M functions in (67), i.e.,

$$\mu_0 = \mu^{mN}, \qquad g_0 = g^{m(\ln N+1)}. \tag{69}$$

Proof. We consider any of the products of powers in (67) for which $k_1 + \dots + k_m = s$, $1 \le s \le N$. This product may be regarded as a product of s of the functions $f_1(z), \dots, f_m(z)$, perhaps with some of these functions occurring more than once. We number these functions in any order, and we let $\gamma_{l,n}/n!$, $n = 0, 1, 2, \dots$, $l = 1, \dots, s$, be their power series coefficients. Then

$$f_m^{k_m}(z) \cdots f_1^{k_1}(z) = \sum_{n=0}^{\infty} \frac{\gamma_n}{n!} z^n,$$

where

$$\gamma_n = \sum_{n_1+\dots+n_s=n} \frac{n!}{n_1! \cdots n_s!} \gamma_{1,n_1} \cdots \gamma_{s,n_s}.$$

From the definition of the class $KE(\lambda, c; \mu, g)$ we have

$$\overline{|\gamma_n|} = \sum_{n_1+\dots+n_s=n} \frac{n!}{n_1! \cdots n_s!} \overline{|\gamma_{1,n_1}|} \cdots \overline{|\gamma_{s,n_s}|} \le$$
$$\le \lambda^s c^n \sum_{n_1+\dots+n_s=n} \frac{n!}{n_1! \cdots n_s!} = \lambda^s c^n s^n \le \lambda^N (cN)^n.$$

For each n we write the factors γ_{l,n_l} in order of non-increasing second subscript, i.e.,

$$\gamma_{1,n_1} \cdots \gamma_{s,n_s} = \gamma_{l_1,n_{l_1}} \cdots \gamma_{l_s,n_{l_s}}, \qquad n_{l_1} \ge n_{l_2} \ge \dots \ge n_{l_s}.$$

Then obviously $n_{l_1} \le n$, $n_{l_2} \le \frac{n}{2}$, ..., $n_{l_s} \le \frac{n}{s}$. Hence, if we set

$$g_n = g_{0,n} g_{0,[n/2]} \cdots g_{0,[n/N]},$$

where $\{g_{0,n}\}$ is the sequence (6), then we have $g_n\gamma_l \in \mathbf{Z_K}$, $l = 0, 1, \ldots, n$, for $n = 1, 2, \ldots$, and from (4) and (7) we find that

$$g_n \le \mu^{ms} g^{m\left(1+\frac{1}{2}+\cdots+\frac{1}{N}\right)^n} \le \mu^{mN} g^{m(\ln(N+1))n}.$$

The lemma is proved.

Lemma 8 (Yu. V. Nesterenko). *Suppose that*

$$P \in \mathbf{C}[z, x_1, \ldots, x_m], \qquad P \not\equiv 0, \qquad \deg_z P \le n, \qquad \deg_{\mathrm{x}} P \le N;$$

the functions $f_1(z), \ldots, f_m(z)$ are analytic at $z = 0$, form a solution of the system of homogeneous linear differential equations (1), and are homogeneously algebraically independent over $\mathbf{C}(z)$. Further set $R(z) = P(z, f_1(z), \ldots, f_m(z))$, and suppose that the $\mathbf{C}(z)$-vector space spanned by the functions $R^{(i)}(z)$, $i = 0, 1, 2, \ldots$, is l-dimensional. Then

$$\operatorname{ord} R(z) \le ln + \eta N^\tau, \qquad \tau = m^m + m,$$

where η depends on the system of differential equations (1) satisfied by $f_1(z), \ldots$ $\ldots, f_m(z)$ and on the number m.

We shall not give a proof of Lemma 8. The reader can find a proof in Nesterenko's paper [52:5] (see Theorem 3).

Suppose that $f_1(z), \ldots, f_m(z)$ are KE-functions which satisfy the system of homogeneous linear differential equations (1). We consider the set of $M = (N + m - 1)!/N!(m - 1)!$ products of powers of these functions in (67). According to Lemma 18 of Chapter 3, the functions in (67) form a solution of a system of homogeneous linear differential equations of the form (1), in which m is replaced by M and the coefficients of the system have no poles other than the poles of the original system (1).

We list the M functions in (67) in an arbitrary order, and denote them

$$\varphi_1(z), \ldots, \varphi_M(z). \tag{70}$$

Using Lemma 7, we easily see that Lemma 4 with m replaced by M applies to the functions (70). If we suppose that the KE-functions $f_1(z), \ldots, f_m(z)$ are homogeneously algebraically independent over $\mathbf{C}(z)$ (that the functions in (70) form an irreducible set of functions), then Lemmas 2, 3, 5 and 6, again with m replaced by M, hold for the functions (70). Taking Lemma 8 into account, we can now give a refinement of Lemma 2.

Lemma 9. *Under the conditions of Lemma 2, if m is replaced by M and the KE-functions $f_1(z), \ldots, f_m(z)$ are replaced by the KE-functions (70), where the* *functions $f_1(z), \dots, f_m(z)$ are assumed to be homogeneously algebraically independent over $\mathbf{C}(z)$, then in Lemma 2 one has*

$$n_0 = \eta_1 N^\tau, \qquad \tau = m^m + m,$$

where η_1 depends on the system of differential equations (1) which is satisfied by our functions and on the number m.

Proof. Let $R_1(z)$ be the linear form in the functions (70) which was constructed in Lemma 4. According to that lemma, we have $\operatorname{ord} R_1(z) \geq Mn - \left[\frac{n}{\omega}\right] - 1$. But then, by Lemma 8,

$$Mn - \left[\frac{n}{\omega}\right] - 1 \leq ln + \eta N^\tau, \qquad \tau = m^m + m.$$

From this we obtain $\left(M - l - \frac{1}{\omega}\right) n < \eta N^\tau + 2$. If we suppose that $l < M$, then $M - l \geq 1$, and, by the conditions in Lemma 4, we have $\frac{1}{\omega} < \frac{1}{2}$. Hence, $n < 2\eta N^\tau + 4$. Thus, $l = M$ for $n \geq 2\eta N^\tau + 4$, and this implies the lemma.

Proof of Theorem 3. Because of the equality

$$\Phi^0_{\mathbf{I}}(f_1(\xi), \dots, f_m(\xi);\ s;\ H) = L_{\mathbf{I}}(\varphi_1(\xi), \dots, \varphi_M(\xi);\ H),$$

where $\varphi_1(z), \dots, \varphi_M(z)$ are the functions in (70) with $N = s$, we can follow the argument in the proof of Theorem 1, using Lemmas 1–6 with m replaced by M and $f_1(z), \dots, f_m(z)$ replaced by the functions (70), and also using Lemmas 7 and 9. Here we note that the parameter G in the proof of Lemma 5 must be replaced by a parameter G_s which depends on s. Using the proof of Lemma 18 of Chapter 3, we see that we can set $G_s = Gs$. But it easily follows from the proof of Lemma 5 that the conclusion remains valid (except with new constants γ_1 and γ_2 in (38) and (39)) if G is replaced by Gs.

From (64) and the analogue of (66) obtained using Lemma 7, we find that if

$$\begin{aligned}\ln\ln(H+2) &> \max(\gamma^2 m M^4 \ln(s+1), \gamma_{19} \ln n_0) = \\ &= \max(\gamma^2 m M^4 \ln(s+1), \gamma_{19}(\tau \ln(s+1) + \ln \eta)),\end{aligned}$$

in which case

$$\ln\ln(H+2) > \Omega^2 m (s+1)^{4(s-1)} \ln(s+1),$$

we have Theorem 3 with $c_0 = 1$. The proof of the theorem in the case

$$\ln\ln(H+2) \leq \Omega^2 m (s+1)^{4(s-1)} \ln(s+1)$$

is just as in Theorem 1.

Theorem 3′ follows from the proof of Theorem 3, if we use the relations (24) for the functions (70).

Theorem 4. *Suppose that $m, \nu \in \mathbf{Z}^+$, $n+\nu > 0$; each of the m sets of IE-functions*

$$\varphi_{i,l}(z), \ldots, \varphi_{m_l,l}(z), \qquad m_l \geq 1, \qquad l = 1, \ldots, n, \tag{71}$$

forms a solution of a corresponding system of linear differential equations

$$S_l: \ y'_{k,l} = Q_{k,l,0} + \sum_{i=1}^{m_l} Q_{k,l,i} y_{i,l}, \qquad Q_{k,l,i} \in \mathbf{C}(z),$$

$$k = 1, \ldots, m_l, \qquad l = 1, \ldots, n;$$

and each of the ν sets of IE-functions

$$\psi_{1,l}(z), \ldots, \psi_{\mu_l,l}, \qquad \mu_l \geq 1, \qquad l = 1, \ldots, \nu, \tag{72}$$

forms a solution of a corresponding system of homogeneous linear differential equations

$$S_l^0: \ y'_{k,l} = \sum_{i=1}^{\mu_l} Q^0_{k,l,i} y_{i,l}, \quad Q^0_{k,l,i} \in \mathbf{C}(z), \quad k = 1, \ldots, \mu_l, \ \ l = 1, \ldots, \nu.$$

Further suppose that the $m = m_1 + \cdots + m_n + \mu_1 + \cdots + \mu_\nu$ functions in (71) and (72) do not satisfy any algebraic equation with coefficients in $\mathbf{C}(z)$ which is homogeneous in each of the ν sets of functions in (72). Let ξ be a nonzero element of $\mathbf{I}$ which is not a singular point of any of the systems S_l or S_l^0. Then there exist constants γ and Ω such that the following inequalities hold:

$$\Phi^*_{\mathbf{I}}(\{\varphi_{k,l}(\xi)\}; \{\psi_{k,l}(\xi)\}; s_1, \ldots, s_n; t_1, \ldots, t_\nu; H) > cH^{1-M-\delta},$$

$$\Phi_{\mathbf{I}}(\{\varphi_{k,l}(\xi)\}; \psi_{1,1}(\xi)/\psi_{\mu_1,1}(\xi), \ldots, \psi_{\mu_1-1,1}(\xi)/\psi_{\mu_1,1}(\xi); \ldots; \psi_{1,\nu}(\xi)/\psi_{\mu_\nu,\nu}(\xi),$$
$$\ldots, \psi_{\mu_\nu-1,\nu}(\xi)/\psi_{\mu_\nu,\nu}(\xi); s_1, \ldots, s_n; t_1, \ldots, t_\nu; H) > c_1 H^{1-M-\delta},$$

$$M = \left(\prod_{l=1}^{n} \frac{(s_l + m_l)!}{s_l! m_l!} \right) \prod_{k=1}^{\nu} \frac{(t_k + \mu_k - 1)!}{t_k! (\mu_k - 1)!},$$

$$c = \begin{cases} 1, & \text{if } \rho < 1, \\ \exp(-A \exp B^2), & \text{if } \rho \geq 1, \end{cases}$$

$$A = \left(\prod_{l=1}^{n} (s_l + 1)^{m_l}\right) \left(\prod_{k=1}^{\nu} (t_k + 1)^{\mu_k - 1}\right) \left(1 + \frac{\gamma}{\Omega}\right) - 1,$$

$$B = \Omega \left(\prod_{l=1}^{n} (s_l + 1)^{2m_l}\right) \left(\prod_{k=1}^{\nu} (t_k + 1)^{2(\mu_k - 1)}\right) \times$$
$$\times \sqrt{m \ln(s_1 + \cdots + s_n + t_1 + \cdots + t_\nu + 1)},$$

$$\rho = \frac{B}{\sqrt{\ln \ln (H + 2)}}, \qquad c_1 = \frac{c}{|\psi_{\mu_1,1}(\xi)|^{t_1} \cdots |\psi_{\mu_\nu,\nu}(\xi)|^{t_\nu}}.$$

The proof of Theorem 4 is analogous to the proof of Theorem 3, if we consider the set of M products of powers

$$\varphi_{m_n,n}^{k_{m_n,n}} \cdots \varphi_{1,n}^{k_{1,n}} \cdots \varphi_{m_1,1}^{k_{m_1,1}} \cdots \varphi_{1,1}^{k_{1,1}} \psi_{\mu_\nu,\nu}^{\kappa_{\mu_\nu,\nu}} \cdots \psi_{1,\nu}^{\kappa_{1,\nu}} \cdots \psi_{\mu_1,1}^{\kappa_{\mu_1,1}} \cdots \psi_{1,1}^{\kappa_{1,1}}, \tag{73}$$

$$k_{1,l} + \cdots + k_{m_l,l} \leq s_l, \quad l = 1, \ldots, n; \quad \kappa_{1,k} + \cdots + \kappa_{\mu_k,k} = t_k, \quad k = 1, \ldots, \nu.$$

Theorem 4′. *Under the conditions of Theorem 4, if the products of powers in (73) form an irreducible set of functions for any $s_l \in \mathbf{N}$, $l = 1, \ldots, n$, $t_k \in \mathbf{N}$, $k = 1, \ldots, \nu$, then $\Omega = \gamma$ in Theorem 4.*

The special cases of Theorem 4 when $n > 0$, $\nu = 0$ and when $n = 0$, $\nu > 0$ are of particular interest.

Theorems 1–3 can be restated so as to apply to an E-function which satisfies a linear differential equation and its sequence of successive derivatives, and Theorem 4 can be restated for a set of such functions.

In all of the Theorems 1–4, γ is an effective constant, and Ω_0, Ω, η and η_1 are non-effective. Nesterenko's result in Lemma 8 enables one to make the constants in the bounds effective as functions of the degrees of the measures, but it does not make the constants fully effective.

In Theorems 1′–4′, all of the constants in the bounds for the measures are effective constants. However, it should be noted that it is usually difficult to prove that a concrete set of functions is irreducible, and that this can be done only in cases when the functions are solutions of differential equations of not very high order.

The bounds in these theorems hold for any number m of functions, any degrees of the measures and any height H. These bounds become coarse when m and the degrees become sufficiently large. But the bounds for the measures remain rather accurate if m and the degrees of the measures increase along with H up to a certain limit.

It is clear from the proofs of Theorems 1 and 2 that m may increase as a function of increasing H. However, the limits of this increase are not given effectively. In Theorems 1′ and 2′, on the other hand, effective limits are established. For example, in Theorem 1′ we have a rather accurate bound for $L_{\mathbf{I}}$ when $m < \gamma^{-2}(\ln\ln H)^{1/5}$. But if $m \geq \gamma^{-2}(\ln\ln H)^{1/5}$, then it becomes a coarse bound, and becomes worse as m increases.

In the same way, in Theorem 3 a rather accurate bound for $\Phi_{\mathbf{I}}^{0}$ holds for $(s+1)^{4(m-1)}\ln(s+1) < \frac{1}{m\Omega_0}\ln\ln H$, but with Ω_0 a non-effective constant. This bound becomes coarse if $(s+1)^{4(m-1)}\ln(s+1) \geq \frac{1}{m\Omega_0}\ln\ln H$. In Theorem 3′, the corresponding result is effective: we have an accurate bound for $(s+1)^{4(m-1)}\ln(s+1) < \frac{1}{m\gamma_0}\ln\ln H$.

Similar results hold in the case of Theorems 4 and 4′.

§ 5. Some applications of the general theorems

The general theorems proved above can be applied to many concrete E-functions. We shall give only some simple examples.

Let $\xi_1, \ldots, \xi_m$ be distinct number in $\mathbf{I}$. By Lemma 20 of Chapter 3, the functions $e^{\xi_1 z}, \ldots, e^{\xi_m z}$ are then linearly independent over $\mathbf{C}(z)$. They satisfy the system of differential equations $y_k' = \xi_k y_k$, $k = 1, \ldots, m$, and they obviously form an irreducible set of functions. Hence, by Theorem 1, we have:

$$L_{\mathbf{I}}\left(e^{\xi_1}, \ldots, e^{\xi_m}; H\right) > cH^{1-m-\frac{\gamma m^3\sqrt{m}}{\sqrt{\ln\ln(H+2)}}},$$

where

$$c = \begin{cases} 1, & \text{if } \rho < 1, \\ \exp(-(2m-1)\exp(\gamma^2 m^5)), & \text{if } \rho \geq 1, \end{cases}$$

$$\rho = \frac{\gamma m^2\sqrt{m}}{\sqrt{\ln\ln(H+2)}}.$$

Now suppose that $\xi_1, \ldots, \xi_m \in \mathbf{I}$ are linearly independent over $\mathbf{Q}$. By Lemma 21 of Chapter 3, the functions $u_1 = e^{\xi_1 z}, \ldots, u_m = e^{\xi_m z}$ are algebraically independent over $\mathbf{C}(z)$. Then it is clear that for any $N = 1, 2, \ldots$ the monomials

$$u_m^{k_m} \cdots u_1^{k_1} = e^{(k_1\xi_1 + \cdots + k_m\xi_m)z}, \qquad k_i \in \mathbf{Z}^+,$$

$$k_1 + \cdots + k_m = N \qquad (k_1 + \cdots + k_m \leq N)$$

form an irreducible set of functions. Hence, by Theorem 3′, we have

$$\Phi_{\mathbf{I}}^0(e^{\xi_1}, \ldots, e^{\xi_m};\, s;\, H) > c_0 H^{1-M-\delta},$$

where

$$M = \frac{(s+m-1)!}{s!(m-1)!}, \qquad \delta = \frac{\gamma_0 (s+1)^{3(m-1)} \sqrt{m \ln(s+1)}}{\sqrt{\ln\ln(H+2)}}, \tag{74}$$

$$c_0 = \begin{cases} 1, & \text{if } \rho < 1, \\ \exp\left(-(s+1)^{m-1} \exp(\gamma_0^2 m (s+1)^{4(m-1)} \ln(s+1))\right), & \text{if } \rho \geq 1, \end{cases} \tag{75}$$

$$\rho = \frac{\gamma_0 (s+1)^{2(m-1)} \sqrt{m \ln(s+1)}}{\sqrt{\ln\ln(H+2)}};$$

and

$$\Phi_{\mathbf{I}}(e^{\xi_1}, \ldots, e^{\xi_m};\, s;\, H) > c H^{1-M-\delta},$$

where M, δ and c are given by (74) and (75) with $m-1$ replaced by m in the exponent of $s+1$ and γ_0 and c_0 replaced by γ and c.

We consider the E-function

$$K_\lambda(z) = \sum_{n=0}^{\infty} \frac{(-1)^n}{n!(\lambda+1)\cdots(\lambda+n)} \left(\frac{z}{2}\right)^{2n}, \qquad \lambda \neq -1, -2, \ldots,$$

which satisfies the linear differential equation

$$y'' + \frac{2\lambda+1}{z} y' + y = 0. \tag{76}$$

Let $\lambda \in \mathbf{Q}$. If $\lambda \neq \frac{2k-1}{2}$, $k \in \mathbf{Z}$, and if y is an arbitrary nontrivial solution of (76), then, by Lemma 3 of Chapter 6, the functions y and y' are linearly independent together with 1 over $\mathbf{C}(z)$. It is not hard to see that the same is true even if

$\lambda = \frac{2k-1}{2}$. The functions $y_1 = y$, $y_2 = y'$ and $y_3 = 1$ form a solution of the system of differential equations

$$y_1' = y_2, \qquad y_2' = -\frac{2\lambda+1}{z} y_2 - y_1, \qquad y_3' = 0.$$

Consequently, they are an irreducible set of functions. Then for any nonzero $\xi \in \mathbf{I}$, Theorem $2'$ tells us that there exists a constant γ such that

$$L_{\mathbf{I}}(1, K_\lambda(\xi), K_\lambda'(\xi);\, H) > cH^{-2-\frac{\gamma}{\sqrt{\ln\ln(H+2)}}},$$

where

$$c = \begin{cases} 1, & \text{if } \rho < 1, \\ \exp(-5\exp\gamma^2), & \text{if } \rho \geq 1, \end{cases} \qquad \rho = \frac{\gamma}{\sqrt{\ln\ln(H+2)}}.$$

Lemma 10. *Let $\lambda \in \mathbf{C}$, $\lambda \neq \frac{2k-1}{2}$, $k \in \mathbf{Z}$, $\lambda \neq -1, -2, \ldots$, and $f_1(z) = K_\lambda(z)$, $f_2(z) = K_\lambda'(z)$. Then for any $N \in \mathbf{N}$ the products of powers*

$$F_{k_2,k_1}(z) = f_2^{k_2}(z) f_1^{k_1}(z), \tag{77}$$

$$k_1, k_2 \in \mathbf{Z}^+, \qquad k_1 + k_2 = N \quad (k_1 + k_2 \leq N)$$

form an irreducible set of functions.

Proof. We consider the non-homogeneous case. The homogeneous case is simpler to prove, and it follows from the non-homogeneous case. Let N be fixed. If we differentiate the functions (77) and use the differential equation (76), we find that the set of functions (77), none of which is identically zero, is a solution of the system of homogeneous linear differential equations

$$U_{k_2,k_1}' = k_1 U_{k_2+1,k_1-1} - k_2 \frac{2\lambda+1}{z} U_{k_2,k_1} - k_2 U_{k_2-1,k_1+1}, \quad k_1 + k_2 \leq N, \tag{78}$$

where $U_{-1,k_1} = U_{k_2,-1} = 0$.

In order to prove that the functions in (77) form an irreducible set, it suffices to show that an equation of the form

$$\sum_{k_1+k_2 \leq N} P_{k_2,k_1} U_{k_2,k_1} = 0, \tag{79}$$

where $P_{k_2,k_1} \in \mathbf{C}[z]$ and $\{U_{k_2,k_1}\}_{k_2+k_1 \leq N}$ is an arbitrary solution of the system (78), is impossible if any of the products $P_{k_2,k_1} U_{k_2,u_1}$ is nonzero. To do this, we use the general solution of the differential equation (76) to construct a fundamental system of solutions of the system of differential equations (78).

Let y be an arbitrary solution of the differential equation (76). We set $v = z^{\lambda} y$. Then v is an arbitrary solution of the Bessel differential equation (9.42). If v is a nontrivial solution, then from Lemma 3 of Chapter 6 it obviously follows that the functions v and v' are algebraically independent over $\mathbf{C}(z)$. And if v_1 and v_2 are two linearly independent solutions of the Bessel differential equation, then, by Lemma 4 of Chapter 9, the three functions v_1, v_1' and v_2 are algebraically independent over $\mathbf{C}(z)$.

The function $v = c_1 v_1 + c_2 v_2$ is the general solution of the Bessel differential equation, and the function

$$y = c_1 y_1 + c_2 y_2 = z^{-\lambda}(c_1 v_1 + c_2 v_2), \qquad c_1, c_2 \in \mathbf{C}, \tag{80}$$

is the general solution of the differential equation (76). Using Liouville's formula for the Bessel differential equation: $v_1 v_2' - v_1' v_2 = \frac{a}{z}$, $a \in \mathbf{C}$, $a \neq 0$, we obtain:

$$y' = z^{-\lambda}\left(\left(\frac{v_1'}{v_1} - \frac{\lambda}{z}\right)(c_1 v_1 + c_2 v_2) + \frac{a c_2}{z v_1}\right). \tag{81}$$

We let the matrix of a fundamental set of solutions of (78) be denoted

$$\|U_{k_2,k_1;\rho_2,\rho_1}\|, \qquad k_2 + k_1 \leq N, \quad \rho_2 + \rho_1 \leq N. \tag{82}$$

Given any solution y of the differential equation (76), the functions

$$U_{k_2,k_1} = (y')^{k_2} y^{k_1} = (c_1 y_1' + c_2 y_2')^{k_2}(c_1 y_1 + c_2 y_2)^{k_1}, \qquad k_2 + k_1 \leq N. \tag{83}$$

obviously form a solution of the system (78). For $k_2 + k_1 \leq N$ and for all ρ_2 and ρ_1 satisfying

$$\rho_2 + \rho_1 = k_2 + k_1, \tag{84}$$

we define the functions $U_{k_2,k_1;\rho_2,\rho_1}$ by the following relations, which hold as identities in c_1 and c_2:

$$U_{k_2,k_1} = \sum_{\rho_2+\rho_1=k_2+k_1} c_2^{\rho_2} c_1^{\rho_1} U_{k_2,k_1;\rho_2,\rho_1}, \qquad k_2 + k_1 \leq N, \tag{85}$$

and for all other values of ρ_2 and ρ_1 we set

$$U_{k_2,k_1;\rho_2,\rho_1} = 0. \tag{86}$$

From (80), (81), (83) and (85) it follows that

$$\sum_{\rho_2+\rho_1=k_2+k_1} c_2^{\rho_2} c_1^{\rho_1} U_{k_2,k_1;\rho_2,\rho_1} =$$

$$= z^{-\lambda(k_2+k_1)} \left(\left(\frac{v_1'}{v_1} - \frac{\lambda}{z} \right) (c_1 v_1 + c_2 v_2) + \frac{a c_2}{z v_1} \right)^{k_2} (c_1 v_1 + c_2 v_2)^{k_1}, \tag{87}$$

$$k_2 + k_1 \leq N.$$

To see that the set of functions $U_{k_2,k_1;\rho_2,\rho_1}$ defined in this way is a solution of (78) for any fixed ρ_2 and ρ_1 with $\rho_2 + \rho_1 \leq N$, it suffices to differentiate the functions U_{k_2,k_1} with respect to c_i ρ_i times, $i = 1, 2$, and then set $c_1 = c_2 = 0$, at which point, by the linearity of the system (78), one obtains a set of functions which satisfy the system of differential equations.

We show that the $\mu_{N,2}$ solutions

$$U_{k_2,k_1;\rho_2,\rho_1}, \qquad k_2 + k_1 \leq N, \qquad \rho_2 + \rho_1 \leq N,$$

of the system (78) defined by (84), (85) and (86), form a fundamental set of solutions. In fact, (78) is a homogeneous system which splits into $N+1$ subsystems corresponding to the products of powers of $f_2(z)$ and $f_1(z)$ of fixed total degree:

$$F_{k_2,k_1}(z) = f_2^{k_2}(z) f_1^{k_1}(z), \qquad k_2 + k_1 = k, \qquad k = 0, 1, \ldots, N;$$

and the $U_{k_2,k_1;\rho_2,\rho_1}$ are nonzero only when $\rho_2 + \rho_1 = k_2 + k_1$, $k_2 + k_1 \leq N$, i.e., we have

$$U_{k_2,k_1;\rho_2,\rho_1} = 0, \qquad \rho_2 + \rho_1 \neq k_2 + k_1. \tag{88}$$

Because of (87), our claim follows from the fact that we cannot have a linear equation with constant coefficients between the functions

$$U_{0,k;\rho_2,\rho_1} = \frac{k! z^{-\lambda k}}{\rho_1!(k - \rho_1)!} v_2^{k-\rho_1} v_1'^{\rho_1}, \qquad 0 \leq \rho_1 \leq k,$$

for any $k = 0, 1, \ldots, N$, since such an equation would lead to an algebraic equation for v_1/v_2, which would contradict the linear independence of v_1 and v_2.

The relations (87) show that the $U_{k_2,k_1;\rho_2,\rho_1}$ are polynomials in v_1, v_1^{-1}, v_1' and v_2. We shall later need to know $U^0_{k_2,k_1;\rho_2,\rho_1}$, i.e., the leading terms with respect to v_1, v_1' and v_2 which occur in each of the functions $U_{k_2,k_1;\rho_2,\rho_1}$. From (87) and

(88) we find that

$$U^0_{k_2,k_1;\rho_2,\rho_1} = \frac{(k_2+k_1)!}{\rho_2!\rho_1!} z^{-\lambda(k_2+k_1)} \left(\frac{v_1'}{v_1} - \frac{\lambda}{z}\right)^{k_2} v_1^{\rho_1} v_2^{\rho_2}, \quad \rho_2+\rho_1 = k_2+k_1. \tag{89}$$

We now suppose the contrary of what we want to prove, i.e., that (79) holds for some solution $\{U_{k_2,k_1}\}_{k_2+k_1\le N}$ of the system (78), where at least one of the products $P_{k_2,k_1}U_{k_2,k_1}$ is nonzero. On the left side of (79) we replace the functions U_{k_2,k_1} by their expressions as linear combinations of the matrix entries in (82) given by (85), (86) and (87) with constant coefficients c_{ρ_2,ρ_1}.

Let r be the least nonnegative integer such that $r-\lambda>0$, and let the natural number q be the denominator of λ. Then the function $z^{N(r+1)}v_1^N U_{k_2,k_1;\rho_2,\rho_1}$ is a polynomial in v_1, v_1', v_2 and $z^{1/q}$. If we multiply the left side of (79) by $z^{(r+1)N}v_1^N$, we obtain an algebraic relation between v_1, v_1', v_2 and $z^{1/q}$. By Lemma 8 of Chapter 5, this algebraic equation is an identity in these variables, since the functions v_1, v_1' and v_2 are algebraically independent over $\mathbf{C}(z)$. Consequently, the left side of (79) is identically zero as a function of v_1, v_1', v_2 and $z^{1/q}$.

After making our substitution, from (85) and (88) we see that (79) takes the form

$$\sum_{k_2+k_1\le N} P_{k_2,k_1} \sum_{\rho_2+\rho_1=k_2+k_1} c_{\rho_2,\rho_1} U_{k_2,k_1;\rho_2,\rho_1} = 0. \tag{90}$$

In the identity (90) at least one of the products $P_{k_2,k_1}c_{\rho_2,\rho_1}$ is nonzero, since otherwise all of the products $P_{k_2,k_1}U_{k_2,k_1}$ in (79) would be zero. Suppose that all of the products $P_{k_2,k_1}c_{\rho_2,\rho_1}$ are zero whenever $k_2+k_1>k$, where $1\le k\le N$, but that there is at least one such product with $k_2+k_1=k$ which is nonzero. If we collect terms of maximal degree k in v_1, v_1' and v_2 in (90), we obtain the following identity in v_1, v_1', v_2 and $z^{1/q}$:

$$\sum_{k_2+k_1=k} P_{k_2,k_1} \sum_{\rho_2+\rho_1=k_2+k_1} c_{\rho_2,\rho_1} U^0_{k_2,k_1;\rho_2,\rho_1} = 0.$$

From this relation, using (89), we find that

$$\sum_{k_2+k_1=k} P_{k_2,k_1} \sum_{\rho_2+\rho_1=k} \frac{c_{\rho_2,\rho_1}}{\rho_2!\rho_1!} \left(\frac{v_1'}{v_1} - \frac{\lambda}{z}\right)^{k_2} \left(\frac{v_2}{v_1}\right)^{\rho_2} = 0.$$

Since at least one of the products $P_{k_2,k_1}c_{\rho_2,\rho_1}$ is nonzero, this last identity contradicts the algebraic independence of the functions v_1, v_1' and v_2 over $\mathbf{C}(z)$. The lemma is proved.

Now suppose that $\lambda \in \mathbf{C}$, $\lambda \neq \frac{2k-1}{2}$, $k \in \mathbf{Z}$, $\lambda \neq -1, -2, \ldots$, and $\xi \in \mathbf{I}$, $\xi \neq 0$. By Theorem 3′ there exist constants γ_0 and γ such that:

$$\Phi_{\mathbf{I}}^0(K_\lambda(\xi), K_\lambda'(\xi);\, s;\, H) > c_0 H^{-s-\frac{\gamma_0(s+1)^3\sqrt{\ln(s+1)}}{\sqrt{\ln\ln(H+2)}}},$$

$$\Phi_{\mathbf{I}}(K_\lambda'(\xi)/K_\lambda(\xi);\, s;\, H) > c_1 H^{-s-\frac{\gamma_0(s+1)^3\sqrt{\ln(s+1)}}{\sqrt{\ln\ln(H+2)}}}, \tag{91}$$

$$c_0 = \begin{cases} 1, & \text{if } \rho < 1, \\ \exp\left(-(s+1)\exp(2\gamma_0^2(s+1)^4\ln(s+1))\right), & \text{if } \rho \geq 1, \end{cases}$$

$$\rho = \frac{\gamma_0(s+1)^2\sqrt{2\ln(s+1)}}{\sqrt{\ln\ln(H+2)}}, \qquad c_1 = |K^{(s)}(\xi)|c_0,$$

and

$$\Phi_{\mathbf{I}}(K_\lambda(\xi), K_\lambda'(\xi);\, s;\, H) > cH^{-\frac{s(s+3)}{2}-\frac{\gamma(s+1)^3\sqrt{\ln(s+1)}}{\sqrt{\ln\ln(H+2)}}},$$

where

$$c = \begin{cases} 1, & \text{if } \rho < 1, \\ \exp\left(-(s+1)^2\exp(\gamma^2 2(s+1)^8\ln(s+1))\right), & \text{if } \rho \geq 1, \end{cases}$$

$$\rho = \frac{\gamma(s+1)^4\sqrt{\ln(s+1)}}{\sqrt{\ln\ln(H+2)}}.$$

Because of the relation (6.48), the bound (91) remains in force if $K_\lambda'(\xi)/K_\lambda(\xi)$ is replaced by $J_\lambda'(\xi)/J_\lambda(\xi)$ or by α, where

$$\alpha = 1 + \cfrac{1}{2 + \cfrac{1}{3 + \cfrac{1}{4 + \cdots}}},$$

since $\alpha = i\,K_0'(2i)/K_0(2i)$.

The irreducibility of some other sets of E-functions was proved in 1959 in [70:9].

Remarks

All of the results in this chapter were proved in 1979 in [70:29]. The note [70:20], published in 1967, contains the statements of theorems with bounds for the measures of the values of IE-functions which have remainder terms of the same type as in this chapter and which allow the degree of the measure to increase along with the height up to a non-effective limit. A theorem with a bound of the same type for KE-functions which is effective in the degree of the measure was proved in Nesterenko's papers [52:2–5], published in 1972–77. This theorem uses Nesterenko's remarkable result bounding the order of zero of a polynomial in several analytic functions (Lemma 8).

We note that the size of the constant τ in Lemma 8 has been lowered in papers by Brownawell [8:1], Bertrand and Beukers [4:1], Nesterenko [52:11], and others. In this book we used Nesterenko's first result, because it is more accessible to the Soviet reader. In 1981–83 in [53:1], Nguyen Ten Tai generalized Lemma 8 to the case of functions which are algebraically dependent over $\mathbf{C}(z)$.

Effective bounds for the algebraic independence measures of the values of KE-functions are proved in still unpublished work by Yu. N. Makarov and V. A. Gorelov. We also note Salikhov's article [66:3], which contains effective bounds for the measures of the values of certain concrete E-functions.

The algorithmic effectiveness of the method in Chapter 3 is established in Osgood's 1981 paper [57:3].

Concluding remarks

The material in this book shows that the applications of these techniques to the arithmetic properties of the values of E-functions have been largely carried to their natural limits. In order to apply the general theorems proved above to new classes of concrete E-functions, it would be necessary to develop methods for proving that functions are algebraically independent over $\mathbf{C}(z)$.

It would be of interest to generalize these results to the values of E-functions at certain transcendental numbers. Some theorems have been proved in this direction. In 1970 in [22:3], Galochkin proved an algebraic independence theorem for the values of a set of E-functions at values of the argument which have a sufficiently "good" approximation by algebraic numbers.

Theorem. *Suppose that the E-functions*

$$f_1(z), \ldots, f_m(z), \qquad m \geq 1, \tag{1}$$

form a solution of the system of linear differential equations (13.8) and are algebraically independent over $\mathbf{C}(z)$. *Let* $\xi \in \mathbf{C}$ *be a root of the equation*

$$P = P(z, f_1(z), \ldots, f_m(z)) = 0,$$

$$P \in \mathbf{A}[z, f_1(z), \ldots, f_m(z)], \qquad P \not\equiv 0.$$

Then for any fixed positive numbers ϵ *and* κ, *the inequality*

$$|\xi - \theta| < \exp(-h_\theta^\epsilon) \tag{2}$$

has finitely many solutions in numbers $\theta \in \mathbf{A}$ *of height* h_θ *and of degree at most* κ.

If, in addition, $P(\xi, x_1, \ldots, x_m) \not\equiv 0$ *and* $\xi T(\xi) \neq 0$, *then the following inequality holds for any* $\epsilon > 0$ *and any polynomial* $g(z) \in \mathbf{Z}[z]$, $g(z) \not\equiv 0$, *of height* H *and of degree at most* s:

$$|g(\xi)| > c\exp(-H^\epsilon), \tag{3}$$

where $c > 0$ *is a constant which depends only on* ϵ, ξ, s, *and the functions* $f_1(z), \ldots, f_m(z)$.

Corollary. *Suppose that the E-functions (1) form a solution of the system of linear differential equations (13.8) and are algebraically independent over* $\mathbf{C}(z)$. *Let* $\xi \in \mathbf{C}$ *be such that for some positive* ϵ *and* κ *the inequality (2) has infinitely many solutions in numbers* $\theta \in \mathbf{A}$ *of height* h_θ *and of degree at most* κ. *Then the numbers*

$$\xi, f_1(\xi), \ldots, f_m(\xi) \tag{4}$$

are algebraically independent.

This theorem and its corollary can be applied to many of the E-functions which were considered in Chapters 5–10. In particular, the inequality (3) holds for algebraic A-points of the functions $K_\lambda(z)$ and their derivatives, which we studied in Chapter 6, for suitable values of λ, and so it holds for the zeros of the Bessel functions $J_\lambda(z)$ and their derivatives.

In 1979 in [70:28] some conjectures were made. Suppose that the E-functions (1) satisfy the system of differential equations (13.8) and are algebraically independent over $\mathbf{C}(z)$. It is easy to see that the numbers (4) are algebraically independent for almost all $\xi \in \mathbf{C}$ in the sense of Lebesgue measure. However, when one attempts to prove that these numbers are algebraically independent for a particular transcendental ξ, one usually encounters insurmountable obstacles.

It is natural to make the following conjecture: *if the functions (1) satisfy the system (13.8) and are algebraically independent over* $\mathbf{C}(z)$, *and if* $\xi \in \mathbf{C}$, $\xi T(\xi) \neq 0$, *then at least* m *of the* $m+1$ *numbers in (4) are algebraically independent.* This conjecture would have several interesting consequences:

1°. $m = 2$, $f_1(z) = e^{iz}$, $f_2(z) = e^z$, $\xi = \pi$. *Then the numbers* π *and* e^π *are algebraically independent.*

2°. $m = k+1$, $f_1(z) = e^{\beta_1 z}, \ldots, f_k(z) = e^{\beta_k z}$, $f_{k+1}(z) = e^z$, *where* $\beta_1, \ldots, \beta_k$ *are algebraic numbers which, along with 1, are linearly independent over* $\mathbf{Q}$, $\xi = \ln\alpha$, *where* $\alpha \in \mathbf{A}$, $\alpha \neq 0, 1$. *Then the numbers* $\ln\alpha$, $\alpha^{\beta_1}, \ldots, \alpha^{\beta_k}$ *are algebraically independent.*

3°. *If* ξ *is a zero of one of the functions in (1), say* $f_m(z)$, *or a point at which that function takes an algebraic value, then the numbers* $\xi, f_1(\xi), \ldots, f_{m-1}(\xi)$ *are algebraically independent.*

4°. *Let* ξ *be a root of the equation* $P = P(z, f_1(z), \ldots, f_m(z)) = 0$, *where* P *is an irreducible polynomial in* $m+1$ *variables with coefficients in* $\mathbf{A}$. *Then any* m *of the numbers in (4), where the excluded value is the value at* ξ *of one of the functions (1) which occurs in* P, *are algebraically independent.*

We note that consequences 1° and 2° are contained in a well-known conjecture of Schanuel, to the effect that, *if* $\alpha_1, \ldots, \alpha_m$ *are complex numbers which are linearly independent over* $\mathbf{Q}$, *then the transcendence degree of the set* $\alpha_1, \ldots, \alpha_m$, $e^{\alpha_1}, \ldots, e^{\alpha_m}$ *is at least* m. At the present time one cannot see any approaches to proving either Schanuel's conjecture or the general form of the other conjecture stated above. Some important special cases of the latter conjecture (the Second Fundamental Theorem of Chapter 3 and the corollary of the theorem of Galochkin mentioned above) have been proved. It is natural to expect that the addition of some new ideas to the method will make it possible to prove it in its general form.

Another natural conjecture is the following: *if the set of functions (1) has transcendence degree* l *over* $\mathbf{C}(z)$, *where* $1 \leq l \leq m$, *and if* $\xi \in \mathbf{C}, \xi T(\xi) \neq 0$, *then the set of numbers (4) has transcendence degree at least* l. From this conjecture one can also obtain a series of consequences.

In 1929 in [72:3], Siegel indicated that his method can be applied to the study of certain arithmetic properties of the values of another class of analytic functions. Unlike E-functions, these functions have power series with a finite radius of convergence. He called them G-functions.

A function

$$f(z) = \sum_{n=0}^{\infty} c_n z^n \tag{5}$$

is said to be a G-function if the coefficients c_n *satisfy the same three conditions as the numbers* c_n *in the second definition of an E-function.*

Like E-functions, G-functions form a ring of functions which is closed under differentiation and integration from 0 to z and under the substitution of λz in place of z, where $\lambda \in \mathbf{A}$.

A simple example of a G-function is a geometric progression. By a well-known theorem of Eisenstein, an algebraic function is a G-function if it is regular in a neighborhood of $z = 0$ and satisfies an algebraic equation with algebraic coefficients. Hence, the integral from 0 to z of such a function is also a G-function. Another example of a G-function is the Gauss hypergeometric series

$$\sum_{n=0}^{\infty} \frac{[\alpha, n][\beta, n]}{n![\gamma, n]} z^n \tag{6}$$

with rational values of the parameters α, β and γ.

In studying abelian integrals of the form $\int_0^z f(t)dt$, where $y = f(t)$ is an algebraic function which is analytic in a neighborhood of $z = 0$ and has algebraic Taylor series coefficients, Siegel recommends that one consider more general approximating forms in the functions under consideration, having coefficients which are polynomials in z and y.

Without giving proofs, Siegel made a series of assertions about the values at algebraic points of certain classes of G-functions (in particular, the values of abelian integrals and the hypergeometric series (5)) being irrational or not belonging to an algebraic number field of bounded degree.

When applied to G-functions, Siegel's method and the generalization considered in this book enable one to obtain arithmetic results only at points of sufficiently small modulus. This is connected with the fact that the small size of the numerical linear approximating forms in the values of G-functions that are constructed by this method is achieved not from a factor of $1/n!$ in the general term of the series, as in the case of E-functions, but rather from the factor $|\xi|^n$, where ξ is the value of the argument in our functions. The same consideration explains why transcendence and algebraic independence of the values of G-functions cannot be proved by the method in this book.

In 1971 in the papers [55:1,2] by M. S. Nurmagomedov, the method described in this book was first applied to study the arithmetic properties of the values of G-functions at algebraic points. In the first of these articles the G-functions have power series coefficients in **I**, and in the second article in **K**. These papers prove several interesting theorems with bounds for linear forms, and polynomials in the values of arbitrary G-functions. However, they have the drawback that the point ξ at which the bounds are given has a magnitude depending on the height of the linear form or polynomial. Some of the theorems are effective.

Nurmagomedov applied his general theorems to the functions

$$\varphi_{\lambda,k}(z) = \sum_{n=0}^{\infty} \frac{z^n}{(n+\lambda)^k}, \qquad \lambda \in \mathbf{Q}, \quad \lambda \neq -1, -2, \dots, \quad k = 1, \dots, m,$$

and

$$\ln(1 - \beta_k z) = -\sum_{n=1}^{\infty} \frac{\beta_k^n}{n} z^n, \qquad \beta_k \in \mathbf{I}, \qquad k = 1, \dots, m, \tag{7}$$

and, in joint work with V. G. Chirskii [13:1], to the functions

$$\psi(z) = \sum_{n=0}^{\infty} \frac{[\mu, n]}{[\lambda, n]} z^n \tag{8}$$

and the hypergeometric series (6).

We shall give the statement of only one of Nurmagomedov's results, which is an application of one of his general theorems.

Theorem. *Suppose that β_k are distinct numbers in $\mathbf{I}$; $H, b, N \in \mathbf{N}$; $a, u_1, \dots, u_m \in \mathbf{Z_I}$, $a \neq 0$, $|u_i| \leq H$, $i = 1, \dots, m$, $|u_1| + \dots + |u_m| > 0$; and σ_1, σ_2 and n_1 are given by the relations*

$$\sigma_1 = (m+1)(10m+1)\sqrt{m(\beta e^2)}, \qquad \beta = \max_{1 \leq i \leq m} |\beta_i|,$$

$$\sigma_2 = m_1(10m_1 - 9)\sqrt{\ln(2\beta e^{2m(1+\ln H)})}, \qquad m_1 = \binom{N+m}{m}.$$

$$n_1 = \left[\max\left(\exp((m+1)^2(2m^2+2m+3)\ln(\beta e^2)), \frac{4\ln 4(m+1)|d|^{\frac{3}{2}}}{(m+1)\ln(\beta e^2)}\right)\right] + 1,$$

where d is the discriminant of the field **I**. *Then:*

1) *for*

$$b > \exp\left(\sigma_1\sqrt{\ln n_1} + 2(m+n)\ln|a| + \frac{1}{n_1}\ln H\right),$$

the following inequality holds:

$$\left|u_0 + u_1\ln\left(1-\beta_1\frac{a}{b}\right) + \cdots + u_m\ln\left(1-\beta_m\frac{a}{b}\right)\right| \geq$$
$$\geq \exp\left(-\sigma_1 n_1\sqrt{\ln n_1}\right) b^{-(2m+1)n_1};$$

2) *if* $P = P(z_1,\ldots,z_m) \in \mathbf{Z_I}[z_1,\ldots,z_m]$, $P \not\equiv 0$, $\deg_z P = N$, *then there exists* $n_2 \in \mathbf{N}$, *depending only on* $\beta_1,\ldots,\beta_m$, m, N *and* **I**, *such that for*

$$b \geq \exp\left(\sigma_2\sqrt{\ln n_2} + 2m_1\ln|a| + \frac{\ln H}{n_2}\right)$$

the following inequality holds:

$$\left|P\left(\ln\left(1-\beta_1\frac{a}{b}\right),\ldots,\ln\left(1-\beta_m\frac{a}{b}\right)\right)\right| \geq \exp\left(-\sigma_2 n_2\sqrt{\ln n_2}\right) b^{-(2m_1-1)n_2}.$$

In part 2) of this theorem for the first time we have a lower bound for a polynomial in the values of a set of logarithms at algebraic points.

In 1973 in [12:1], Chirskii proved two theorems analogous to those of Nurmagomedov in the case when the G-functions are connected by algebraic equations over $\mathbf{C}(z)$, and in [12:2] he applied the method to elliptic integrals.

In 1973, Galochkin continued Nurmagomedov's line of study in [22:5,6]. Using Siegel's suggestion in [72:3] of reducing the coefficients of the linear approximating forms, he obtained bounds for the modulus of polynomials in the values of a subclass of G-functions at points whose magnitude does not depend on the height of the polynomial. In particular, these results were proved for the functions (7). He also stated a theorem giving an effective bound for a linear form in G-functions of a certain class; the proof of this theorem was published in [22:7].

In 1977, Chirskii [12:4,8] returned to the study of elliptic integrals, and proved that their values at sufficiently small points are irrational; in [12:5], he obtained the same result for the values of the hypergeometric functions (8). The same year, N. D. Nagaev [51:1] investigated the arithmetic properties of elliptic integrals.

Interest in the study of G-functions has grown noticeably in recent years. The articles [48:1,2] by E. M. Matveev are devoted to bounds for linear forms in the values of G-functions and applications to Diophantine equations. We also call attention to papers by Nurmagomedov [55:3], Philippon [59:1], Reyssat [62:1], and Matala-Aho and Väänänen [47:1].

Y. Z. Flicker [19:1] and Väänänen [79:11] have studied p-adic G-functions. In 1981 Bombieri's deep work [5:1] on the properties of G-functions appeared. In particular, Bombieri proved the above-mentioned statements of Siegel on the values of G-functions. In 1979, E. M. Nikishin [54:1,2] used methods different from those in this book in order to obtain bounds for linear forms in certain G-functions.

In 1966 in [70:18] the problem was posed of proving irrationality of the values at algebraic points of hypergeometric functions with algebraic values of the parameters, and some results of this type were announced for certain concrete functions. V. G. Sprindzhuk [73:2] examined a similar problem. He studied entire functions with algebraic Taylor series coefficients which satisfy first order linear differential equations with coefficients in $\mathbf{C}(z)$ and which are in a larger class than that of E-functions, and he obtained an irrationality result for the values of such functions. In connection with Sprindzhuk's work we also mention the article [34:1] by S. V. Kotov. In [50:1], S. M. Molchanov obtained a certain quantitative generalization of Sprindzhuk's result for simultaneous approximation of the values of the functions.

In [12:3,5–7,10], Chirskii considers a class of functions with algebraic Taylor series coefficients which contains the class of E-functions and also a class of G-functions whose Taylor series coefficients have least common denominator which may grow faster than in the case of E-functions. He finds bounds for polynomials with integer coefficients in the values of such functions at sufficiently small points. As an example, he gives a set of functions of the form e^z and $\ln(1+z)$ at different algebraic points; this set contains both E-functions and G-functions.

In Chapter 3 we used the Dirichlet pigeon-hole principle to prove the existence of functional approximating forms in a set of E-functions which have the properties we need. In Chapter 11 we used them for the field $\mathbf{I}$ to obtain bounds for linear forms and polynomials in the values of our E-functions. These bounds have a best possible principal term in the exponent of H, and they have a remainder term of the form $\lambda/\sqrt{\ln\ln H}$, where λ does not depend on H. The same type of bound has been obtained in several papers on G-functions.

In the Hermite-Lindemann method (see Chapter 2), the functional approximating forms in the set of exponential functions are constructed in an exact fashion: the polynomial coefficients of these forms are given by certain formulas. Hence, one obtains more precise bounds in the quantitative applications of this method. Their remainder term in the exponent has the form $\lambda/\ln\ln H$. Examples of this type of bound were given in Chapter 11 (see Popken's bound (11.24) and Mahler's bound (11.25)). N. I. Fel'dman [17:4] obtained a bound of the same kind for the functions $\varphi_\lambda(z)$ in (5.17) with various values of the parameter λ.

Bounds for linear forms in the values of hypergeometric functions at algebraic points have been obtained by several authors using other methods. In 1966–67,

Osgood [57:1,2] used the ideas of Hermite–Lindemann to prove several theorems on Diophantine approximation of the values at algebraic points of certain hypergeometric functions and their derivatives with rational and algebraic parameters.

Generalizing the same method, Galochkin [22:2,3,8–12] obtained a series of bounds for linear forms in the values of hypergeometric functions. In [22:8], for certain of these functions he reduced the remainder term in the bounds for linear forms to $\lambda \ln\ln H/\ln H$. We also mention Galochkin's paper [22:4] on approximation of values of the exponential function and values of the solutions of certain transcendental equations.

In 1983, A. N. Korobov [33:1] proved best possible bounds for linear forms in the values of certain hypergeometric functions. In 1984, Galochkin [22:13] obtained similar bounds in a more general case.

It would be interesting to supplement the method in this book with some new ideas, so that it could be applied to larger classes of functions, and in the case of G-functions and their generalizations to larger sets of values of the argument for which one can prove irrationality and obtain other arithmetic results.

Many deep results related to bounds for the measures of the values of certain classes of analytic functions were obtained by using and developing the method of Hermite–Lindemann and the methods which arose from the solution of Hilbert's Seventh Problem. We call attention to the articles by Popken [61:1], Mahler [42:1], Gel'fond [24:5–7], Fel'dman [17:1–5], and Baker [2:1,2,4,5]. The reader can become familiar with many of these results from the books by Gel'fond [24:8], Fel'dman [17:8] and Cijsow [14:1].

We note that at present transcendental number theory still has a small collection of facts and methods to obtain them. The reason is not for want of important and interesting problems. To the contrary, there are many good problems, but they are usually too difficult and cannot be solved by any of the known methods.

For example, no one has been able to generalize Lindemann's classical theorem on linear independence of the values of the exponential function at distinct algebraic points to the case when an algebraic number is taken to be the base of the exponential function.

We know that e and π are transcendental. But we do not know whether or not they are algebraically independent. It is not even known whether the numbers $e \pm \pi$ and $e\pi$ are irrational.

We know that π and e^{π} are transcendental numbers. But we do not know whether they are algebraically independent.

Still nothing is known about the arithmetic nature of Euler's constant $c = \lim_{n\to\infty}\left(1+\frac{1}{2}+\cdots+\frac{1}{n}-\ln n\right)$ or any of the values of the Riemann zeta-function $\zeta(s) = \sum_{n=1}^{\infty}\frac{1}{n^s}$ at odd integers $s \geq 5$. We cannot even say whether these values are irrational. In 1978, Apéry's proof of the irrationality of $\zeta(3)$ produced a big sensation at the International Congress of Mathematicians in Helsinki. For the proof of this result, see Apéry [1:1] or Van der Poorten [60:1].

In conclusion, we would like to express the hope that some new ideas will be introduced into the techniques presented in this book, so that they can be applied to larger classes of functions and to certain sets of transcendental values of the argument.

Supplementary remarks on recent work for the English edition

In the time since the preparation of the manuscript of the Russian edition, several papers have appeared which relate to the material covered in the book. For this reason the author decided to write this supplement to the English translation, in which we discuss the most important of these recent papers and briefly mention some others.

1. We first discuss results concerning effective bounds for the algebraic independence measure of the values at algebraic points of a set of E-functions.

Suppose that the E-functions

$$f_1(z), \dots, f_m(z) \tag{1}$$

form a solution of the system of linear differential equations

$$y_k' = Q_{k,0} + \sum_{i=1}^{m} Q_{k,i} y_i, \qquad k = 1, \dots, m, \qquad Q_{k,i} \in \mathbf{C}(z). \tag{2}$$

In 1985, D. Brownawell [8:1] showed that the estimate of the measure

$$\Phi(f_1(\xi), \dots, f_m(\xi); s; H), \qquad \xi \in \mathbf{A}, \qquad \xi T(\xi) \neq 0, \tag{3}$$

using the method in the book, can be made effective if the components of an arbitrary solution $y_1, \dots, y_m$ of (2) are algebraically independent over $\mathbf{C}(z)$.

In 1988 in [104:1], Nesterenko gave a less restrictive condition for the estimate of the measure (3) to be effective. His condition is that the components of any solution $y_1, \dots, y_m$ of (2) consisting of nonzero functions be algebraically independent over $\mathbf{C}(z)$. This result carries over to the case of a set consisting of several subsystems of E-functions, each of which is a solution of a corresponding system of differential equations of the form (2).

If either Brownawell's or Nesterenko's condition is fulfilled, then estimates of the type established in Chapter 13 will hold for the corresponding algebraic independence measure of the values of IE-functions. In the case of KE-functions, the measures in papers by Lang [37:2], Galochkin [22:1], Shidlovskii [70:33,36] and others (see Chapters 11 and 12) become effective. Nesterenko's estimate in [52:5] becomes fully effective, and the effective bounds of Makarov [103:1] and Gorelov [98:1,2] are also valid.

2. Let

$$\varphi(z) = \sum_{n=0}^{\infty} \frac{[\mu_1, n] \cdots [\mu_l, n]}{[\lambda_1, n] \cdots [\lambda_m, n]} \left(\frac{z}{t}\right)^{tn}, \qquad (4)$$

$$\lambda_i \neq 0, -1, -2, \ldots, \qquad m > l \geq 0, \qquad m - l = t, \qquad \lambda_i, \mu_j \in \mathbf{C},$$

be a hypergeometric function satisfying the linear differential equation

$$\left(\prod_{i=1}^{m}(\delta + t(\lambda_i - 1)) - z^t \prod_{j=1}^{l}(\delta + t\mu_j)\right) y = t^m \prod_{i=1}^{m}(\lambda_i - 1), \qquad \delta = z\frac{d}{dz}. \quad (5)$$

Beginning in 1977, Salikhov (see [66:2,4,5] and [107:1,2,3]) studied the analytic properties of the functions (4) and the arithmetic properties of their values in the case $l = 0$ and $m = p$, where p is a prime number. He developed an analytic method, which he used to give sufficient conditions on the parameters λ_i which ensure irreducibility of the differential equation (5) and the corresponding homogeneous differential equation. In the case when the parameters have rational values — in which case the function in (4) is an E-function — this enabled him to prove a result on algebraic independence of the values at algebraic points of the functions (4) and their successive derivatives (see Theorem 5 of Chapter 10); here he used the general Theorem 3 in § 13 of Chapter 3 and Nesterenko's theorem (Theorem 2 in § 6 of Chapter 9). An important consideration in Salikhov's method is that to prove that a homogeneous linear differential equation is irreducible it suffices to show that it is linearly irreducible, and this is a simpler problem.

We note that proofs of Salikhov's Theorems 5 and 6 in § 1 of Chapter 10 appeared in 1988 in [107:3], and the results of Salikhov and Kazakov on linear irreducibility of the differential equation (5) and the corresponding homogeneous equation (referred to on p. 322) were published in 1986 in [107:2] and in 1987 in [101:1].

In some preprints of Salikhov that will soon appear in print, he further develops his method, applying it to the general hypergeometric function (4) with some conditions on the parameters λ_i and μ_j. We shall state his results below.

A completely unexpected and deep result was published in 1988 by Beukers, Brownawell and Heckman [91:1], who showed that the Siegel normality condition holds for a product of powers of a hypergeometric function of the form (4) satisfying a homogeneous differential equation of the form (5) ($\lambda_m = 1$) and its successive derivatives (see p. 8–9 of the Introduction). On p. 9 we stated that the normality condition for a product of powers of several functions had been verified only for sets of functions each of which is a solution of a first or second order linear differential equation. Thus, the paper by Beukers, Brownawell and Heckman is of great interest.

Siegel's normality condition, the author's irreducibility condition for a set of functions, and algebraic independence of the functions over $\mathbf{C}(z)$ — all of these are sufficient conditions for a general algebraic independence theorem to hold for the values of E-functions at algebraic points. Of these, Siegel's condition is the most restrictive, and the most difficult to verify.

We consider an m-tuple of functions (1) which form a solution of a system (2) of homogeneous linear differential equations (i.e., all of the $Q_{k,0}$ are zero). In proving Siegel's theorem [72:4] and the fundamental theorems of Chapter 3, one constructs a linear form

$$R = P_1 f_1(z) + \cdots + P_m f_m(z), \qquad P_i \in \mathbf{C}[z], \qquad i = 1, \ldots, m,$$

which has a sufficiently large order of zero at $z = 0$. Let $y_1, \ldots, y_m$ be an arbitrary solution of the homogeneous system (2). We denote

$$R_1 = P_{1,1} y_1 + \cdots + P_{1,m} y_m, \qquad P_{1,i} = P_i, \qquad i = 1, \ldots, m,$$

and, as in § 4 of Chapter 3, we define $R_{k+1} = TDR_k = TR_k'$. Then

$$R_k = P_{k,1} y_1 + \cdots + P_{k,m} y_m, \qquad k = 1, \ldots, m, \qquad P_{k,i} \in \mathbf{C}[z].$$

We denote $\Delta = |P_{k,i}|_{k,i=1,\ldots,m}$. To prove the general theorems one must find a condition under which $\Delta \not\equiv 0$.

It may happen that the matrix

$$Q = \|Q_{k,i}\|_{k,i=1,\ldots,m}$$

of the homogeneous system (2) splits into r square submatrices

$$Q_t = \|Q_{k,i,t}\|_{k,i=1,\ldots,m_t}, \qquad t = 1, \ldots, r, \qquad m_1 + \cdots + m_r = m.$$

Here we mean that the submatrices Q_t are located along the main diagonal of Q, and that all of the entries outside these submatrices are zero:

$$Q : \left\| \begin{matrix} Q_1 & 0 & 0 & \cdots & 0 \\ 0 & Q_2 & 0 & \cdots & 0 \\ 0 & 0 & Q_3 & \cdots & 0 \\ & \cdots & & \cdots & \cdots \\ 0 & 0 & 0 & \cdots & Q_r \end{matrix} \right\| .$$

This decomposition is clearly unique, if we take the largest possible r.

Corresponding to the decomposition of Q, the homogeneous system (2) splits into r subsystems

$$y_{k,t}' = \sum_{i=1}^{m_t} Q_{k,i,t} y_{i,t}, \qquad k = 1, \ldots, m_t, \qquad t = 1, \ldots, r. \tag{6}$$

Let

$$\|y_{k,i,t}\|_{k,i=1,\dots,m_t}, \qquad t = 1,\dots,r,$$

be matrices of fundamental solutions of the systems (6).

We now give some definitions.

A solution of the system (6) is said to be *admissible* if it is formed from nontrivial solutions of all of the subsystems.

The submatrices in Q are said to be *independent* if any admissible solution of the system (6) is linearly independent over $\mathbf{C}(z)$.

We say that the set of functions (1) is a *normal set* if none of the functions is identically zero, and if the functions form a solution of a homogeneous system (2) whose matrix Q has independent submatrices.

To prove Siegel's theorem it is enough to know that for any $N = 1, 2, \dots$ the functions

$$f_1^{k_1}(z)\cdots f_m^{k_m}(z), \qquad 0 \le k_1 + \cdots + k_m \le N,$$

form a normal set of functions.

We note that the Siegel normality condition fails to hold in the case of a non-homogeneous system (2). This follows because in Siegel's condition the submatrices into which the solution matrix of (2) decomposes cannot have any entries identically zero. In the non-homogeneous case this fails even for $N = 1$. On the other hand, the author's irreducibility condition and the condition that the functions (1) be algebraically independent over $\mathbf{C}(z)$ do not rule out the possibility of zeros in these submatrices. These conditions make it possible to obtain arithmetic results in the non-homogeneous case as well.

After proving certain criteria for normality, the authors of [91:1] showed that the normality condition holds for products of powers of the function (4) and its successive derivatives (with $\lambda_m = 1$), and, using Siegel's theorem [72:4], they proved an algebraic independence result for the values of the function (4) and its derivatives. We shall state this result below.

Let $m \in \mathbf{N}$, $l \in \mathbf{Z}^+$, $m > l \ge 0$, $m \ge 2$. We let S denote the set of values of the parameters:

$$S = \{\mu_1, \dots, \mu_l; \lambda_1, \dots, \lambda_m\}, \qquad \lambda_i, \mu_j \in \mathbf{R}.$$

We say that the set S of values of the parameters is *admissible* if it satisfies one of the following conditions:

A) $(\lambda_i - \mu_j) \notin \mathbf{Z}$ for all pairs i, j, $1 \le i \le m$, $1 \le j \le l$; and all sums $\lambda_i + \lambda_j$, $1 \le i \le j \le m$, are distinct modulo $\mathbf{Z}$.

B) $l = 0$, $m = 2$ or m is odd, and the set $\{\lambda_1, \dots, \lambda_m\}$ modulo $\mathbf{Z}$ is not a union of arithmetic progressions $\{\lambda, \lambda + 1/d, \dots, \lambda + (d-1)/d\}$ of fixed length d, $d|m$, $d > 1$.

We make one further definition. Two sets of parameter values

$$S = \{\mu_1, \dots, \mu_l; \lambda_1, \dots, \lambda_m\} \qquad \text{and} \qquad S' = \{\mu'_1, \dots, \mu'_{l'}; \lambda'_1, \dots, \lambda'_{m'}\}$$

are said to be *similar* if:

a) $l' = l$, $m' = m$;

b) there exist $\mu_0, \lambda_0 \in \mathbf{R}$ and $r \in \{0, 1\}$ such that with a suitable ordering of the μ_i and λ_j we have

$$\mu'_i \equiv \mu_0 + (-1)^r \mu_i \ (mod\ \mathbf{Z}), \qquad i = 1, \dots, l,$$

$$\lambda'_j \equiv \lambda_0 + (-1)^r \lambda_j \ (mod\ \mathbf{Z}), \qquad j = 1, \dots, m.$$

We denote

$$\varphi_S(z) = \sum_{n=0}^{\infty} \frac{[\mu_1, n] \cdots [\mu_l, n]}{[\lambda_1, n] \cdots [\lambda_{m-1}, n] n!} \left(\frac{z}{t}\right)^{tn}, \qquad \lambda_i \neq 0, -1, -2, \dots.$$

Theorem. *Let $S_1, \dots, S_k$ be admissible sets (not necessarily all distinct) of parameter values with $\lambda_i, \mu_j \in \mathbf{Q}$, $\lambda_i \neq 0, -1, -2, \dots$, and $\lambda_m = 1$. Let $\xi_1, \dots, \xi_k$ be nonzero algebraic numbers such that, if S_i and S_j, $i \neq j$, are similar sets of parameter values of length l, m, then $\pm\xi_i/\xi_j$ is not a t-th root of unity. Further let $\eta_1, \dots, \eta_s$ be algebraic numbers which are linearly independent over $\mathbf{Q}$. Then the numbers*

$$e^{\eta_1}, \dots, e^{\eta_s},$$

$$\varphi_{S_1}(\xi_1),\ \varphi'_{S_1}(\xi_1), \dots,\ \varphi_{S_1}^{(m(1)-1)}(\xi_1), \qquad (7)$$

$$\dots \quad \dots \quad \dots \quad \dots$$

$$\varphi_{S_k}(\xi_k),\ \varphi'_{S_k}(\xi_k), \dots,\ \varphi_{S_k}^{(m(k)-1)}(\xi_k),$$

are algebraically independent.

In the proof of this theorem the normality condition is shown to hold for products of powers of the functions; hence, effective estimates can be written out for the algebraic independence measure of the numbers (7), as we mentioned above.

The method in [91:1] is different from Salikhov's method. It uses techniques of differential algebra, the theory of algebraic groups, and the theory of Lie groups.

The conditions on the parameters that are given in the theorem are only sufficient conditions for the theorem to hold.

We now state the theorems proved by Salikhov. First of all, he generalized Theorem 5 of Chapter 10 to the case when the prime number p is replaced by an odd number m, and he also extended it to the case of distinct values of the argument, with certain conditions needed on the parameters $\lambda_1, \dots, \lambda_m$.

Theorem 1. *Let $\varphi(z)$ be defined by (4), where $l = 0$, m is an odd number, $\lambda_i \neq 0, -1, -2, \dots$; and let $\xi_1, \dots, \xi_k$ be nonzero algebraic numbers such that $\xi_i/\xi_j \notin \mathbf{Q}(\zeta)$ for any distinct $i, j \in \{1, \dots, k\}$, where ζ is a primitive m-th root of unity. Then the set of km numbers $\varphi^{(i)}(\xi_j)$, $i = 0, 1, \dots, m-1$, $j = 1, \dots, k$, are algebraically independent if and only if the numbers $m\lambda_1, \dots, m\lambda_m$ are not integers which form a complete system of residues modulo m.*

Let $\overline{a} = (a_1, \dots, a_s)$, $\overline{b} = (b_1, \dots, b_s)$, $a_i, b_j \in \mathbf{C}$. We shall write $\overline{a} \sim \overline{b}$ if there exists a permutation $i_1, \dots, i_s$ of $1, \dots, s$ such that $(a_j - b_{i_j}) \in \mathbf{Z}$, $j = 1, \dots, s$. If $c \in \mathbf{C}$, then we denote $(\overline{a} + c) = (a_1 + c, \dots, a_s + c)$. We set $\overline{\lambda} = (\lambda_1, \dots, \lambda_m)$, $\overline{\mu} = (\mu_1, \dots, \mu_l)$.

We now state some other theorems about the function $\varphi(z)$ in (4). In all of these theorems we shall assume that $\lambda_i, \mu_j \in \mathbf{Q}$, $\lambda_i, \mu_j \neq 0, -1, -2, \dots$, $i = 1, \dots, m$, $j = 1, \dots, l$.

Theorem 2. *Suppose that t is an odd number, $l > 0$, and the following conditions hold:*

1) $(\lambda_i - \mu_j) \notin \mathbf{Z}$, $i = 1, \dots, m$, $j = 1, \dots, l$;

2) there does not exist an integer $d > 1$ which is a common divisor of t and l and has the property that

$$(\overline{\lambda} + \frac{1}{d}) \sim \overline{\lambda}, \qquad (\overline{\mu} + \frac{1}{d}) \sim \overline{\mu}.$$

Then the numbers

$$\varphi(\xi), \varphi'(\xi), \dots, \varphi^{(m-1)}(\xi), \qquad \xi \in \mathbf{A}, \qquad \xi \neq 0, \tag{8}$$

are algebraically independent.

Theorem 3. *Suppose that t is an even number, $t = 2\tau$, $l \geq 0$, $\tau + l \geq 2$; one has $l > 3$ if $t = 6$; and conditions 1) and 2) above are fulfilled, along with the conditions:*

a) if $l = 0$, then there do not exist $x_0, x_1, \dots, x_{\tau-1} \in \mathbf{Q}$ such that

$$(\overline{\lambda} + x_0) \sim (0, \frac{1}{2}, x_1, -x_1, \dots, x_{\tau-1}, -x_{\tau-1}); \tag{9}$$

b) if $l > 0$ is an even number, $l = 2s$, then there do not exist $x_0, x_1, \dots, x_{\tau-l-1} \in$ $\mathbf{Q}$ such that

$$(\overline{\lambda} + x_0) \sim (0, \frac{1}{2}, x_1, -x_1, \dots, x_{\tau-s-1}, -x_{\tau-s-1}),$$
$$(\overline{\mu} + x_0) \sim (x_{\tau+s}, 1 - x_{\tau+s}, \dots, x_{\tau+l-1}, 1 - x_{\tau+l-1}); \tag{10}$$

c) condition b) holds with the pair of relations (10) replaced by

$$(\overline{\lambda} + x_0) \sim (x_1, -x_1, x_2, -x_2, \dots, x_{\tau+s}, -x_{\tau+s}),$$
$$(\overline{\mu} + x_0) \sim (0, \frac{1}{2}, x_{\tau+s+1}, 1 - x_{\tau+s+1}, \dots, x_{\tau+l-1}, 1 - x_{\tau+l-1}); \tag{11}$$

d) if l is odd, $l = 2s + 1$, then there do not exist $x_0, x_1, \dots, x_{\tau+l-1} \in \mathbf{Q}$ such that

$$(\overline{\lambda} + x_0) \sim (0, x_1, -x_1, \dots, x_{\tau+s}, -x_{\tau+s}),$$
$$(\overline{\mu} + x_0) \sim (\frac{1}{2}, x_{\tau+s+1}, 1 - x_{\tau+s+1}, \dots, x_{\tau+l-1}, 1 - x_{\tau+l-1}). \tag{12}$$

Then the numbers (8) are algebraically independent.

Conditions a)–d) are close to being necessary conditions, as one sees from the next result.

Theorem 4. *Let t be an even number, $t = 2\tau$, $l \geq 0$, $\tau + l \geq 2$. Suppose that there exists i, $1 \leq i \leq m$, such that $\lambda_i \in \mathbf{Z}^+$ and one of the sets of relations (9)–(12) is fulfilled. Then the numbers (8) are algebraically dependent.*

In the case $t = 6$, $l \leq 3$, one can also write out sufficient conditions on the parameters λ_i and μ_j which ensure that the numbers (8) are algebraically independent.

Since in several important cases Salikhov establishes irreducibility of the differential equation (5) in the course of proving his theorems, it follows from what was said above that one has effective bounds for the algebraic independence measure of the numbers (8) in those cases.

It is worth noting that the method of [91:1] enables one simultaneously to consider different hypergeometric functions (4). However, this method can only be applied to functions which satisfy a homogeneous differential equation (5).

Salikhov's method makes it possible to consider the non-homogeneous case as well. His results for $l > 0$ are more general than the results of [91:1]. For $l = 0$ and odd m, Salikhov's statement gives not only sufficiency, but necessity as well.

3. In 1984 in [95:1], G. V. Chudnovsky proved some theorems on approximation of the values at points of $\mathbf{Q}$ of E-functions having rational power series coefficients. We describe one of his basic results.

We consider E-functions (1) with rational coefficients of powers of z, each of which is a solution of a linear differential equation with polynomial coefficients. Let $r \in \mathbf{Q}$, $r \neq 0$; $H_i \in \mathbf{Z}$, $H_i \neq 0$, $i = 1, \ldots, m$, and $H = \max(|H_1|, \ldots, |H_m|)$. Under certain assumptions concerning the functions (1), for any $\epsilon > 0$ there exists a constant $c_0 = c_0(\epsilon, r, f_1(z), \ldots, f_m(z)) > 0$ such that for $H \geq c_0$ whenever

$$H_1 f_1(r) + \cdots + H_m f_m(r) \neq 0,$$

one has

$$|H_1 f_1(r) + \cdots + H_m f_m(r)| > (H_1 \cdots H_m)^{-1} H^{1-\epsilon}.$$

4. We call attention to some other publications.

Fel'dman's article [96:1] gives an improvement of the constant in the bound for the transcendence measure of π, and Popov's article [106:1] deals with the approximation of certain powers of e by rational numbers.

In [102:1,2], V. A. Kulagin establishes algebraic independence of the values of certain E-functions. In particular, his articles include results on the exceptional cases in the theorems of Salikhov and Beukers-Brownawell-Heckman. In [98:1], Gorelov generalizes a result of Siegel from [72:3] concerning the arithmetic properties of the values of the functions (6.8) and the exponential function with an effective bound on the corresponding algebraic independence measure. In [92:1] Brownawell renders effective some bounds on algebraic independence measures arising out of the work of Tai [53:1].

In [100:1], P. L. Ivankov considers the problem of approximating the roots of certain transcendental equations connected with E-functions, and in [100:2] he obtains a bound for linear forms in the values of certain hypergeometric functions. In [97:1], Galochkin studies simultaneous approximation of the values of certain hypergeometric functions.

Articles by O. N. Vasilenko [108:1,2], G. V. Chudnovsky [95:2], D. V. Chudnovsky and G. V. Chudnovsky [94:1], Xu Guangchan [99:1] are devoted to the arithmetic properties of the values of G-functions. C. Osgood's paper [105:1] is concerned with approximation of the values of E-functions and G-functions.

Bibliography

[1] Apéry R.

1. Irrationalité de $\zeta(2)$ et $\zeta(3)$, *Astérisque*, 1979, Vol. 61, p. 11-13.

[2] Baker A.

1. On some Diophantine inequalities involving the exponential function, *Canad. J. Math.*, Vol. 17, No. 4, p. 616-626.
2. Linear forms in the logarithms of algebraic numbers, *Mathematika*, 1966, Vol. 13, p. 204-216.
3. Contributions to the theory of Diophantine equations, 1, 2, *Philos. Trans. Roy. Soc. London*, 1968, Vol. 263, No. 1139, p. 173-208.
4. Effective methods in the theory of numbers, *Proc. Intern. Congress Math. (Nice, 1970)*, Vol. 1, p. 19-26.
5. *Transcendental Number Theory*, Cambridge Univ. Press, 1975.
6. The theory of linear forms in logarithms, *Transcendence Theory: Advances and Applications*, Academic Press, 1977, p. 1-27.

[3] Belogrivov I. I.

1. On transcendence and algebraic independence of the values of certain E-functions, *Vestnik Moskov. Univ.*, Ser. 1, 1967, No. 2, p. 55-62.
2. On transcendence and algebraic independence of the values of certain hypergeometric E-functions, *Dokl. Akad. Nauk SSSR*, 1967, Vol. 174, No. 2, p. 267-270.
3. On transcendence and algebraic independence of the values of certain hypergeometric E-functions, *Mat. Sb.*, 1970, Vol. 82 (124), No. 3 (7), p. 387-408.
4. On transcendence and algebraic independence of the values of Kummer functions, *Sibirsk. Mat. Zh.*, 1971, Vol. 12, No. 5, p. 961-982.
5. On transcendence and algebraic independence of the values of a class of E-functions, *Sibirsk. Mat. Zh.*, 1973, Vol. 14, No. 1, p. 16-35.

[4] Bertrand D., Beukers F.

1. Equations differentielles linéaires et majorations de multiplicités, *Ann. Sci. Ecole Norm. Sup.*, 4 sér., 1985, Vol. 18, p. 181-192.

[5] Bombieri E.

1. On G-functions, *Recent Progress in Analytic Number Theory*, Vol. 2, Academic Press, 1981, p. 1-68.

[6] Borel E.

1. Sur la nature arithmétique du nombre *e*, *C. R. Acad. Sci. Paris*, 1899, Vol. 128, p. 596-599.

[7] Borevich Z. I., Shafarevich I. R.

1. *Number Theory*, Academic Press, 1966.

[8] Brownawell W. D.

1. Effectivity in independence measures for values of E-functions, *J. Austral. Math. Soc.*, Ser. A, 1985, Vol. 39, No. 2, p. 227-240.

[9] Cantor G.

1. Über eine Eigenschaft des Inbegriffs aller reellen algebraischen Zahlen, *J. Reine Angew. Math.*, 1874, Vol. 77, p. 258-262.

[10] Cassels J. W. S.

1. *An Introduction to Diophantine Approximation*, Cambridge Univ. Press, 1957.

[11] Chevalley C.

1. *Introduction to the Theory of Algebraic Functions of One Variable*, Amer. Math. Soc., 1951.

[12] Chirskii V. G.

1. On arithmetic properties of the values of analytic functions which are connected by algebraic equations over the field of rational functions, *Mat. Zametki*, 1973, Vol. 14, No. 1, p. 83-94.
2. On arithmetic properties of the values of elliptic integrals, *Vestnik Moskov. Univ.*, Ser. 1, 1973, No. 5, p. 57-64.
3. On arithmetic properties of the values of certain analytic functions, *Theoretical and Applied Questions in Algebra and Differential Equations*, Kiev, 1976, p. 116-117.
4. On arithmetic properties of the values of elliptic integrals, *Uspekhi Mat. Nauk*, 1977, Vol. 32, No. 1 (193), p. 211-212.
5. On arithmetic properties of the values of analytic functions with algebraic Taylor series coefficients, *Vestnik Moskov. Univ.*, Ser. 1, 1978, No. 2, p. 41-47.
6. On arithmetic properties of the values of analytic functions with irrational algebraic Taylor series coefficients, *Vestnik Moskov. Univ.*, Ser. 1, 1978, No. 3, p. 29-34.
7. On arithmetic properties of the values of hypergeometric functions with irrational parameters, *Vestnik Moskov. Univ.*, Ser. 1, 1978, No. 5, p. 3-8.
8. On arithmetic properties of the values of elliptic integrals, *Vestnik Moskov. Univ.*, Ser. 1, 1979, No. 1, p. 3-11.
9. An estimate for a linear form in the values of certain functions, *Vestnik Moskov. Univ.*, Ser. 1, 1982, No. 5, p. 14-17.
10. Estimates for linear forms with algebraic coefficients, *Vestnik Moskov. Univ.*, Ser. 1, 1984, No. 3, p. 32-35.

[13] Chirskii V. G., Nurmagomedov M. S.

1. On arithmetic properties of the values of certain functions, *Vestnik Moskov. Univ.*, Ser. 1, 1973, No. 1, p. 19-26, and 1973, No. 2, p. 38-45.

[14] Cijsow P. L.

1. *Transcendence Measures*, Academic Service (Amsterdam), 1972.

[15] Dyson F. J.

1. The approximation to algebraic numbers by rationals, *Acta Math.*, 1947, Vol. 79, p. 225-240.

[16] Euler L.

1. *Opera Omnia, Introductio in Analysis Infinitorum*, Lipsiae et Berolini, 1922, 1945.

[17] Fel'dman N. I.

1. Approximation of certain transcendental numbers, I, *Izv. Akad. Nauk SSSR*, Ser. Mat., 1951, Vol. 15, No. 1, p. 53-74.
2. Approximation of certain transcendental numbers, II, *Izv. Akad. Nauk SSSR*, Ser. Mat., 1951, Vol. 15, No. 2, p. 153-176.
3. On the transcendence measure of π, *Izv. Akad. Nauk SSSR*, Ser. Mat., 1960, Vol. 24, No. 3, p. 357-368.

4. Lower bounds for certain linear forms, *Vestnik Moskov. Univ.*, Ser. 1, 1967, No. 2, p. 63-72.
5. Estimates for linear forms in the logarithms of algebraic numbers, *Mat. Sb.*, 1968, Vol. 76 (118), No. 2, p. 304-319.
6. An effective power sharpening of a theorem of Liouville, *Izv. Akad. Nauk SSSR*, Ser. Mat., 1971, Vol. 35, No. 5, p. 973-990.
7. *Approximation of Algebraic Numbers*, Izd. Moskov. Univ., 1981.
8. *Hilbert's Seventh Problem*, Izd. Moskov. Univ., 1982.

[18] Fel'dman N. I., Shidlovskii A. B.
1. The history and current state of the theory of transcendental numbers, *Uspekhi Mat. Nauk*, 1967, Vol. 22, No. 3 (135), p. 3-81.

[19] Flicker Y. Z.
1. On p-adic G-functions, *J. London Math. Soc.*, 1977, Vol. 15, No. 3, p. 395-402.

[20] Fourier C.
1. *Mélanges d'analyse algébrique*, Stainville, 1815.

[21] Frobenius G.
1. Über den Begriff der Irreducibilität in der Theorie der linearen Differentialgleichungen, *J. Reine Angew. Math.*, 1873, Vol. 76, p. 236-270.

[22] Galochkin A. I.
1. Estimate for the relative transcendence measure of the values of E-functions, *Mat. Zametki*, 1968, Vol. 3, No. 4, p. 377-386.
2. Lower bound for linear forms in the values of certain hypergeometric functions, *Mat. Zametki*, 1970, Vol. 8, No. 1, p. 19-28.
3. On algebraic independence of the values of E-functions at certain transcendental points, *Uspekhi Mat. Nauk*, 1970, Vol. 25, No. 4, p. 168, and *Vestnik Moskov. Univ.*, Ser. 1, 1970, No. 5, p. 58-63.
4. On Diophantine approximation of values of the exponential function and solutions of certain transcendental equations, *Vestnik Moskov. Univ.*, Ser. 1, 1972, No. 3, p. 16-23.
5. Lower bounds for polynomials in several logarithms of algebraic numbers close to one, *Uspekhi Mat. Nauk*, 1973, Vol. 29, No. 2 (170), p. 235.
6. Lower bounds for polynomials in the values of a class of analytic functions, *Mat. Sb.*, 1974, Vol. 95 (137), No. 3 (11), p. 396-417.
7. Lower bounds for linear forms in the values of certain G-functions, *Mat. Zametki*, 1975, Vol. 18, No. 4, p. 541-552.
8. Refinement of estimates for certain linear forms, *Mat. Zametki*, 1976, Vol. 20, No. 1, p. 35-45.
9. Arithmetic properties of the values of certain hypergeometric entire functions, *Sibirsk. Mat. Zh.*, 1976, Vol. 17, No. 6, p. 1220-1235.
10. On Diophantine approximation of the values of certain entire functions with algebraic coefficients, I, *Vestnik Moskov. Univ.*, Ser. 1, 1978, No. 6, p. 25-32, and II, 1979, No. 1, p. 26-30.
11. A criterion for hypergeometric Siegel functions to belong to the class of E-functions, *Mat. Zametki*, 1981, Vol. 29, No. 1, p. 3-14.
12. Sur des estimations précises par rapport à la hauteur de certaines formes linéaires, *Diophantine Approximation and Transcendental Numbers (Luminy, 1982)*, Birkhäuser, 1983, p. 95-98.

13. On bounds for certain linear forms which are best possible with respect to height, *Mat. Sb.*, 1984, Vol. 124 (166), No. 3 (7), p. 416-430.

[23] Gantmacher F. R.

1. *The Theory of Matrices*, Chelsea, 1959.

[24] Gel'fond A. O.

1. Sur les propriétés arithmétiques des fonctions entières, *Tôhoku Math. J.*, 1929, Vol. 30, No. 3-4, p. 280-285.
2. Sur les nombres transcendantes, *C. R. Acad. Sci. Paris*, 1929, Vol. 189, p. 1224-1228.
3. On Hilbert's seventh problem, *Dokl. Akad. Nauk SSSR*, 1934, Vol. 2, No. 1, p. 1-6.
4. Sur le septième problème de Hilbert, *Izv. Akad. Nauk SSSR*, Ser. Mat., 1934, Vol. 7, p. 623-630.
5. On approximating transcendental numbers by algebraic numbers, *Dokl. Akad. Nauk SSSR*, 1935, Vol. 2, No. 3-4, p. 177-182.
6. Approximation of algebraic irrationalities and their logarithms, *Vestnik Moskov. Univ.*, 1948, No. 9, p. 3-25.
7. On algebraic independence of certain classes of transcendental numbers, *Uspekhi Mat. Nauk*, 1949, Vol. 4, No. 5 (33), p. 14-48.
8. *Transcendental and Algebraic Numbers*, Dover, 1960.
9. *Calcul des différences finies*, Dunod, 1963.
10. *Selected Works*, Moscow, 1973.

[25] Gorelov V. A.

1. On algebraic independence of the values of certain E-functions, *Vestnik Moskov. Univ.*, Ser. 1, 1981, No. 1, p. 47-51.

[26] Hecke E.

1. *Vorlesungen über die Theorie der algebraischen Zahlen*, Leipzig, 1923.

[27] Hermite C.

1. Sur la fonction exponentielle, *C. R. Acad. Sci. Paris*, 1873, Vol. 77, p. 18-24, 74-79, 221-233, 285-293.

[28] Hilbert D.

1. Über die Transzendenz der Zahlen e und π, *Nachr. Ges. Wiss. Göttingen*, 1893, p. 113-116, and *Math. Ann.*, 1893, Vol. 43, p. 216-220.
2. Mathematische Probleme, *Nachr. Ges. Wiss. Göttingen*, 1900, p. 253-297.
3. Über die Theorie der algebraischen Formen, *Math. Ann.*, 1890, Vol. 36, p. 473-534.

[29] Hurwitz A.

1. Über die Kettenbruchentwicklung der Zahl e, *Schriften der phys.-ökon. Ges. Königsberg*, 32. Jahrgang, 1891, Sitzungsber., p. 59-62.
2. Beweis der Transzendenz der Zahl e, *Nachr. Ges. Wiss. Göttingen*, 1893, p. 153-155.

[30] Khinchin A. Ya.

1. Zur metrischen Theorie der diophantischen Approximationen, *Math. Z.*, 1926, Vol. 24, No. 4, p. 706-714.
2. *Continued Fractions*, Univ. Chicago Press, 1964.

[31] Koksma J. F.

1. *Diophantische Approximationen*, Ergeb. Math. Grenzgeb., Vol. 4, Springer-Verlag, 1936.

[32] Kolchin E. R.

1. Algebraic matrix groups and the Picard-Vessiot theory of homogeneous linear ordinary differential equations, *Ann. of Math.*, 1948, Vol. 49, No. 1, p. 1-42.

2. Algebraic groups and algebraic dependence, *Amer. J. Math.*, 1968, Vol. 90, No. 4, p. 1151-1164.

[33] Korobov A. N.

1. Estimates for certain linear forms, *Vestnik Moskov. Univ.*, Ser. 1, 1983, No. 6, p. 36-40.

[34] Kotov S. V.

1. Irrationality of the values of certain transcendental functions, *Izv. Akad. Nauk Belorus. SSR*, 1971, No. 5, p. 32-38.

[35] Kuz'min P. O.

1. On a new class of transcendental numbers, *Izv. Akad. Nauk SSSR*, Ser. Mat., 1930, Vol. 3, p. 583-597.

[36] Lambert J. H.

1. Mémoire sur quelques propriétés remarquables des quantités transcendantes circulaires et logarithmiques, *Hist. de l'Acad. Roy. de Sci. et des Belles Lettres de Berlin*, Année 1761 (1768), p. 265-322.

[37] Lang S.

1. *Introduction to Transcendental Numbers*, Addison-Wesley, 1961.
2. A transcendence measure for E-functions, *Mathematika*, 1962, Vol. 9, p. 157-161.
3. *Algebra*, Addison-Wesley, 1965.
4. *Algebraic Numbers*, Addison-Wesley, 1964.
5. *Introduction to Diophantine Approximations*, Addison-Wesley, 1966.

[38] Legendre A. M.

1. *Éléments de géométrie*, Paris, 1855, Note 4.

[39] LeVeque W. J.

1. *Topics in Number Theory*, Vol. 2, Addison-Wesley, 1956.

[40] Lindemann F.

1. Über die Zahl π, *Math. Ann.*, 1882, Vol. 20, p. 213-225.
2. Über die Ludolph'sche Zahl, *Sitzungsber. Preuss. Akad. Wiss. Berlin*, 1882, p. 679-682.

[41] Liouville J.

1. Sur des classes très étendues de quantités dont la valeur n'est ni algébrique, ni même reductible à des irrationelles algébriques, *C. R. Acad. Sci. Paris*, 1844, Vol. 18, p. 883-885.
2. Nouvelle démonstration d'un théorème sur les irrationelles algébriques inséré dans le Compte Rendu de la dernière séance, *C. R. Acad. Sci. Paris*, 1844, Vol. 18, p. 910-911.
3. Sur l'irrationalité du nombre e, *J. Math. Pures Appl. (1)*, 1840, Vol. 5, p. 192.
4. Addition à la note sur l'irrationalité du nombre e, *J. Math. Pures Appl. (1)*, 1840, Vol. 5, p. 193.

[42] Mahler K.

1. Zur Approximation der Exponentialfunktion und des Logarithmus, 1, *J. Reine Angew. Math.*, 1932, Vol. 166, p. 118-136.
2. Ein Analogon zu einem Schneiderschen Satz, *Nederl. Akad. Wetensch. Proc. Ser. A*, 1936, Vol. 39, p. 633-640, 729-737.
3. Arithmetische Eigenschaften einer Klasse von Dezimalbrüchen, *Proc. Acad. Wet. Amst.*, 1937, Vol. 40, p. 421-428.
4. On a lemma of A. B. Shidlovskii, *Mat. Zametki*, 1967, Vol. 2, No. 1, p. 25-32.

5. Applications of a theorem by A. B. Shidlovskii, *Proc. Roy. Soc.*, Ser. A, 1968, Vol. 305, p. 149-173.
6. Lectures on transcendental numbers, *Proc. Sympos. Pure Math.*, 1971, Vol. 20, p. 248-274.
7. *Lectures on Transcendental Numbers*, Springer-Verlag, 1976.

[43] Maier W.
1. Potenzreihen irrationalen Grenzwertes, *J. Reine Angew. Math.*, 1927, Vol. 156, p. 93-148.

[44] Maillet E.
1. *Introduction à la théorie des nombres transcendants et les propriétés arithmétiques des fonctions*, Gauthier-Villars, 1906.

[45] Makarov Yu. N.
1. On estimates for linear independence measures for the values of a class of E-functions, *Theoretical and Applied Questions in Algebra and Differential Equations*, Kiev, 1976, p. 105-106.
2. On estimates for linear independence measures for the values of E-functions, *Vestnik Moskov. Univ.*, Ser. 1, 1978, No. 2, p. 3-12.
3. Toward estimates for linear forms in the values of E-functions, *Vestnik Moskov. Univ.*, Ser. 1, 1984, No. 2, p. 37-42.

[46] Markov A. A.
1. Proof of the transcendence of e and π (impossibility of squaring the circle): After the articles by Hermite and Lindemann, Saint Petersburg, 1899, *Selected Works*, Izd. AN SSSR, 1951.
2. On Sylvester's proof of the transcendence of π, *Protokoly Sankt-Peterburgskogo Mat. Obshchestva*, Saint Petersburg, 1899.

[47] Matala-Aho T., Väänänen K.
1. On the arithmetic properties of certain values of one Gauss hypergeometric function, *Acta Univ. Oulu.*, Ser. A, 1981, No. 4, p. 1-26.

[48] Matveev E. M.
1. Estimates for linear forms in the values of G-functions and their applications to the solution of Diophantine equations, *Uspekhi Mat. Nauk*, 1981, Vol. 36, No. 2, p. 191-192.
2. Linear forms in the values of G-functions and Diophantine equations, *Mat. Sb.*, 1982, Vol. 117 (159), No. 3, p. 379-396.

[49] Mertens F.
1. Über die Transzendenz der Zahlen e und π, *Sitzungsber. Akad. Wiss. Wien*, 1896, Vol. 105, p. 839-855.

[50] Molchanov S. M.
1. On simultaneous approximation of the values of certain entire functions, *Vestnik Moskov. Univ.*, Ser. 1, 1981, No. 4, p. 28-32.

[51] Nagaev N. D.
1. Incomplete elliptic integrals of the first kind as Siegel G-functions, *XXX Gertsenovskie Chteniya. Mat. Nauch. Dokl.*, Leningrad, 1977, p. 17-19.

[52] Nesterenko Yu. V.
1. On algebraic independence of the values of E-functions which satisfy non-homogeneous linear differential equations, *Mat. Zametki*, 1969, Vol. 5, No. 5, p. 587-589.

2. Estimate for the algebraic independence measure for the values of E-functions with increasing exponent and height, *Uspekhi Mat. Nauk*, 1972, Vol. 27, No. 4 (166), p. 218.
3. Estimates for the order of zero of a certain class of analytic functions and applications to the theory of transcendental numbers, *Dokl. Akad. Nauk SSSR*, 1972, Vol. 205, No. 2, p. 292-295.
4. On algebraic independence of the components of solutions of a system of linear differential equations, *Izv. Akad. Nauk SSSR*, Ser. Mat., 1974, Vol. 38, No. 3, p. 492-512.
5. Estimates for the order of zero of a certain class of functions and applications to the theory of transcendental numbers, *Izv. Akad. Nauk SSSR*, Ser. Mat., 1977, Vol. 41, No. 2, p. 253-284.
6. On the algebraic independence of algebraic numbers to algebraic powers, *Diophantine Approximation and Transcendental Numbers (Luminy, 1982)*, Birkhäuser, 1983, p. 199-220.
7. Estimates for the characteristic function of a prime ideal, *Mat. Sb.*, 1984, Vol. 123 (165), No. 1, p. 11-34.
8. On algebraic independence of algebraic powers of algebraic numbers, *Mat. Sb.*, 1984, Vol. 123 (165), No. 4, p. 435-459.
9. On the algebraic independence measure of the values of certain functions, *Mat. Sb.*, 1985, Vol. 128 (170), No. 4, p. 545-568.
10. On the algebraic independence measure of the values of the exponential function, *Dokl. Akad. Nauk SSSR*, 1986, Vol. 286, No. 4, p. 817-821.
11. On the algebraic independence measure of the values of an elliptic function at algebraic points, *Uspekhi Mat. Nauk*, 1985, Vol. 40, No. 4, p. 221-222.

[53] Nguyen Ten Tai
1. On bounds for the order of zero of polynomials in analytic functions and applications to estimates for the relative transcendence measure of the values of E-functions, *Dokl. Akad. Nauk SSSR*, 1981, Vol. 259, No. 6, p. 1302-1306, and *Mat. Sb.*, 1983, Vol. 120 (162), No. 1, p. 112-142.

[54] Nikishin E. M.
1. On irrationality of the values of the function $F(x, s)$, *Mat. Sb.*, 1979, Vol. 109 (151), No. 3, p. 410-417.
2. On the logarithms of natural numbers, *Izv. Akad. Nauk SSSR*, Ser. Mat., 1979, Vol. 43, No. 6, p. 1319-1327, and 1980, Vol. 44, No. 4, p. 972.

[55] Nurmagomedov M. S.
1. On arithmetic properties of the values of a class of analytic functions, *Mat. Sb.*, 1971, Vol. 85 (127), No. 3 (7), p. 339-365.
2. On arithmetic properties of the values of G-functions, *Vestnik Moskov. Univ.*, Ser. 1, 1971, No. 6, p. 79-85.
3. Lower bound for a polynomial in the values of a class of analytic functions which are connected by special types of algebraic equations, *Izv. Severo-Kavkaz. Nauchn. Tsentra Vyssh. Shkoly Estestv. Nauk*, 1982, No. 3, p. 35-38.

[56] Oleinikov V. A.
1. On transcendence and algebraic independence of the values of certain E-functions, *Vestnik Moskov. Univ.*, Ser. 1, 1962, No. 6, p. 34-38.

2. On certain properties of algebraically dependent quantities, *Vestnik Moskov. Univ.*, Ser. 1, 1962, No. 5, p. 11-17.
3. On transcendence and algebraic independence of the values of E-functions which satisfy a third order linear differential equation, *Dokl. Akad. Nauk SSSR*, 1966, Vol. 166, No. 3, p. 540-543.
4. On algebraic independence of the values of E-functions which satisfy a third order non-homogeneous linear differential equation, *Dokl. Akad. Nauk SSSR*, 1966, Vol. 169, No. 1, p. 32-34.
5. On transcendence and algebraic independence of the values of certain entire functions, *Izv. Akad. Nauk SSSR*, Ser. Mat., 1968, Vol. 32, No. 1, p. 63-92.
6. On algebraic independence of the values of E-functions, *Mat. Sb.*, 1969, Vol. 78 (120), No. 2, p. 301-306.
7. On differential irreducibility of a non-homogeneous linear equation, *Dokl. Akad. Nauk SSSR*, 1970, Vol. 194, No. 5, p. 1017-1020.
8. On conditions for relative transcendence of hypergeometric series, *Vestnik Moskov. Univ.*, Ser. 1, 1986, No. 1, p. 20-25.

[57] Osgood C. F.
1. Some theorems on Diophantine approximation, *Trans. Amer. Math. Soc.*, 1966, Vol. 123, No. 1, p. 64-87.
2. Theorems about the derivatives of certain entire functions at algebraic points, *Nederl. Akad. Wetensch. Indag. Math.*, 1967, Vol. 29, No. 4, p. 431-443, and No. 5, p. 447-555.
3. Nearly perfect systems and effective generalizations of Shidlovskii's theorem, *J. Number Theory*, 1981, Vol. 13, p. 515-540.

[58] Pershikova T. B.
1. On transcendence of the values of certain E-functions, *Vestnik Moskov. Univ.*, Ser. 1, 1966, No. 2, p. 55-61.

[59] Philippon P.
1. Indépendance algébrique par la méthode des G-fonctions, *Publ. Math. Univ. Pierre et Marie Curie*, 1979-1980, p. 1-10.
2. Critères pour l'indépendance algébrique de familles de nombres, *C. R. Math. Rep. Acad. Sci. Canada*, 1984, Vol. 6, No. 5, p. 285-290.

[60] Van der Poorten
1. A proof that Euler missed... Apéry's proof of the irrationality of $\zeta(3)$, *Math. Intelligencer*, 1979, Vol. 1, No. 4, p. 195-203.

[61] Popken J.
1. Zur Transzendenz von e, *Math. Z.*, 1929, Vol. 29, p. 525-541.
2. Zur Transzendenz von π, *Math. Z.*, 1929, Vol. 29, p. 542-548.

[62] Reyssat E.
1. Nature arithmétique de valeurs de G-fonctions, *Publ. Math. Univ. Pierre et Marie Curie*, 1979-1980, p. 1-13.

[63] Roth K. F.
1. Rational approximations to algebraic numbers, *Mathematika*, 1955, Vol. 2, p. 1-20, and corrigendum on p. 168.

[64] Rouché E.
1. Note sur l'impossibilité de la quadrature de cercle, *Nouv. Ann. Math.*, 1883, Vol. 2, p. 5-16.

[65] Rudio F.
1. *Archimedes, Huygens, Lambert, Legendre. Vier Abhandlungen über die Kreismessung*, Teubner (Leipzig), 1892.

[66] Salikhov V. Kh.
1. On algebraic independence of the values of E-functions which satisfy first order linear differential equations, *Mat. Zametki*, 1973, Vol. 13, No. 1, p. 29-40.
2. On differential irreducibility of a class of differential equations, *Dokl. Akad. Nauk SSSR*, 1977, Vol. 235, No. 1, p. 30-33, and *Izv. Akad. Nauk SSSR*, Ser. Mat., 1980, Vol. 44, No. 1, p. 176-202.
3. On bounds for linear independence and transcendence measures for the values of certain E-functions at rational points, *Vestnik Moskov. Univ.*, Ser. 1, 1978, No. 3, p. 9-18.
4. Algebraic irreducibility of a set of linear differential equations, *Dokl. Akad. Nauk SSSR*, 1980, Vol. 254, No. 4, p. 806-808, and *Izv. Akad. Nauk SSSR*, Ser. Mat., 1985, Vol. 49, No. 1, p. 194-210.
5. On algebraic independence of the values of hypergeometric E-functions, *Uspekhi Mat. Nauk*, 1984, Vol. 39, No. 2 (236), p. 185-186.

[67] Schmidt W. M.
1. Simultaneous approximation to algebraic numbers by rationals, *Acta Math.*, 1970, Vol. 125, No. 3-4, p. 189-201.
2. Linear forms with algebraic coefficients, *J. Number Theory*, 1971, Vol. 3, p. 253-277.
3. Linearformen mit algebraischen Koeffizienten, 2, *Math. Ann.*, 1971, Vol. 191, p. 1-20.
4. *Diophantine Approximation*, Springer-Verlag, 1980.

[68] Schneider T.
1. Transzendenzuntersuchungen periodischer Funktionen, 1, *J. Reine Angew. Math.*, 1934, Vol. 172, p. 65-69.
2. Transzendenzuntersuchungen periodischer Funktionen, 2, *J. Reine Angew. Math.*, 1934, Vol. 172, p. 70-74.
3. Über die Approximation algebraischer Zahlen, *J. Reine Angew. Math.*, 1936, Vol. 175, p. 182-192.
4. Arithmetische Untersuchungen elliptischer Integrale, *Math. Ann.*, 1937, Vol. 113, p. 1-13.
5. Zur Theorie der Abelschen Funktionen und Integrale, *J. Reine Angew. Math.*, 1941, Vol. 183, p. 110-128.
6. *Einführung in die transzendenten Zahlen*, Springer-Verlag, 1957.

[69] Schottky F.
1. Zu den Beweisen des Lindemannschen Satzes, *Math. Abh. Schwarz, 50 jähr. Doktorjubiläum*, Springer-Verlag, 1914, p. 384-389.

[70] Shidlovskii A. B.
1. On transcendence and algebraic independence of the values of certain classes of entire functions, *Dokl. Akad. Nauk SSSR*, 1954, Vol. 96, No. 4, p. 697-700.
2. A criterion for algebraic independence of the values of a class of entire functions, *Dokl. Akad. Nauk SSSR*, 1955, Vol. 100, No. 2, p. 221-224.
3. On certain classes of transcendental numbers, *Dokl. Akad. Nauk SSSR*, 1955, Vol. 103, No. 6, p. 977-980.

4. On transcendence of the values of a class of entire functions which satisfy linear differential equations, *Dokl. Akad. Nauk SSSR*, 1955, Vol. 105, No. 1, p. 35-37.
5. A new criterion for transcendence and algebraic independence of the values of a class of entire functions, *Dokl. Akad. Nauk SSSR*, 1956, Vol. 106, No. 3, p. 399-400.
6. On algebraic independence of a class of transcendental numbers, *Dokl. Akad. Nauk SSSR*, 1956, Vol. 108, No. 3, p. 400-403.
7. On a class of transcendental numbers, *Trudy III Vsesoyuznogo Mat. S'ezda, 1*, Izd. Akad. Nauk SSSR, 1956, p. 15-16.
8. A criterion for algebraic independence of the values of a class of entire functions, *Izv. Akad. Nauk SSSR*, Ser. Mat., 1959, Vol. 23, No. 1, p. 35-66.
9. On transcendence and algebraic independence of the values of certain classes of entire functions, *Uchen. Zap. Moskov. Univ.*, 1959, No. 186, p. 11-70.
10. On transcendence and algebraic independence of the values of certain functions, *Trudy Moskov. Mat. Obshch.*, 1959, Vol. 8, p. 283-320, and corrigendum in *Trudy Moskov. Mat. Obshch.*, 1960, Vol. 10, p. 438.
11. On transcendence and algebraic independence of the values of E-functions which are connected by an algebraic equation over the field of rational functions, *Vestnik Moskov. Univ.*, Ser. 1, 1960, No. 5, p. 19-28.
12. A generalization of a theorem of Lindemann, *Dokl. Akad. Nauk SSSR*, 1961, Vol. 138, No. 6, p. 1301-1304.
13. On transcendence and algebraic independence of the values of certain E-functions, *Vestnik Moskov. Univ.*, Ser. 1, 1961, No. 5, p. 44-59.
14. On transcendence and algebraic independence of the values of E-functions which are connected by an arbitrary number of algebraic equations over the field of rational functions, *Izv. Akad. Nauk SSSR*, Ser. Mat., 1962, Vol. 26, No. 6, p. 877-910.
15. Transcendence of the values of E-functions, *Trudy IV Vsesoyuznogo Mat. S'ezda*, 1961, Nauka, 1964, p. 147-158.
16. On transcendence and algebraic independence of the values of E-functions which satisfy second order non-homogeneous linear differential equations, *Dokl. Akad. Nauk SSSR*, 1966, Vol. 169, No. 1, p. 42-45.
17. Transcendence and algebraic independence of the values of E-functions, *Abstracts of Invited Talks at the Intern. Congress Math. (Moscow, 1966)*, p. 177-180.
18. Transcendence of the values of certain classes of functions, *Litovsk. Mat. Sb.*, 1966, Vol. 6, No. 1, p. 129-130.
19. Toward a general theorem on algebraic independence of the values of E-functions, *Dokl. Akad. Nauk SSSR*, 1966, Vol. 171, No. 4, p. 810-813.
20. On estimates for the transcendence measure of the values of E-functions, *Uspehki Mat. Nauk*, 1967, Vol. 22, No. 3 (135), p. 245-246.
21. On estimates for the transcendence measure of the values of E-functions, *Mat. Zametki*, 1967, Vol. 2, No. 1, p. 33-44.
22. Transcendence and algebraic independence of the values of E-functions, *Proc. Intern. Congress Math. (Moscow, 1966)*, p. 299-307.
23. On algebraic independence of the values of certain hypergeometric E-functions, *Trudy Moskov. Mat. Obshch.*, 1967, Vol. 18, p. 55-64.
24. On a theorem of Siegel, *Vestnik Moskov. Univ.*, Ser. 1, 1969, No. 6, p. 39-42.
25. On arithmetic properties of the values of analytic functions, *Proc. Intern. Number Theory Conf. (Moscow, 1971)*; *Trudy Mat. Inst. Steklov.*, 1971, Vol. 132, p. 169-202.

26. Some problems in the theory of transcendental numbers, Appendix to the collection of translations *Problems in the Theory of Diophantine Approximation*, Mir, 1974, p. 139-160.
27. On estimates for the transcendence measure of the values of E-functions, *Vestnik Moskov. Univ.*, Ser. 1, 1977, No. 6, p. 3-10.
28. On arithmetic properties of the values of certain classes of analytic functions, *Vestnik Moskov. Univ.*, Ser. 1, 1979, No. 6, p. 24-31.
29. On the estimates of the algebraic independence measures of the values of E-functions, *J. Austral. Math. Soc.*, Ser. A, 1979, Vol. 27, p. 385-407.
30. On arithmetic properties of the values of E-functions at algebraic points, *Dokl. Akad. Nauk SSSR*, 1979, Vol. 248, No. 1, p. 50-55.
31. On arithmetic properties of the values of E-functions which are connected by algebraic equations over the field of rational functions, *Dokl. Akad. Nauk SSSR*, 1980, Vol. 251, No. 4, p. 808-812.
32. On arithmetic properties of polynomials in the values of E-functions which are connected by algebraic equations over the field of rational functions, *Acta Arith.*, 1980, Vol. 37, p. 405-426.
33. On estimates for the relative transcendence measure of values of E-functions which are connected by arbitrary algebraic equations over $\mathbf{C}(z)$, *Uspekhi Mat. Nauk*, 1980, Vol. 35, No. 2 (212), p. 227-228.
34. On estimates for polynomials in the values of E-functions, *Mat. Sb.*, 1981, Vol. 115 (157), No. 1 (5), p. 3-39.
35. *Diophantine Approximation and Transcendental Numbers*, Izd. Moskov. Univ., 1982.
36. Estimates for the moduli of polynomials with algebraic coefficients at the values of E-functions, *Studies in Pure Math., To the Memory of Paul Turán*, Akadémiai Kiadó, Budapest, 1983, p. 635-657.
37. On a fundamental lemma used to prove algebraic independence of the values of E-functions, *Vestnik Moskov. Univ.*, Ser. 1, 1983, No. 6, p. 94-98.
38. On algebraic independence of the values of E-functions at algebraic points, *Vestnik Moskov. Univ.*, Ser. 1, 1987, No. 1, p. 30-33.

[71] Shmelev A. A.
1. On algebraic independence of the values of certain E-functions, *Izv. Vyssh. Uchebn. Zaved. Mat.*, 1969, No. 4 (83), p. 103-111.
2. Concerning algebraic independence of powers of algebraic numbers, *Mat. Zametki*, 1972, Vol. 11, No. 6, p. 635-644.

[72] Siegel C. L.
1. Approximation algebraischer Zahlen, *Math. Z.*, 1921, Vol. 10, p. 173-213.
2. Über Näherungswerte algebraischer Zahlen, *Math. Ann.*, 1921, Vol. 84, p. 80-99.
3. Über einige Anwendungen Diophantischer Approximationen, *Abh. Preuss. Akad. Wiss., Phys.-Math. Kl.*, 1929-1930, No. 1, p. 1-70.
4. *Transcendental Numbers*, Princeton Univ. Press, 1949.

[73] Sprindzhuk V. G.
1. *Mahler's Problem in Metric Number Theory*, Amer. Math. Soc., 1969.
2. Irrationality of the values of certain transcendental functions, *Izv. Akad. Nauk SSSR*, Ser. Mat., 1968, Vol. 32, No. 1, p. 93-107.
3. *The Metric Theory of Diophantine Approximations*, Wiley, 1979.

[74] Stieltjes T. J.
1. Sur une fonction exponentielle, *C. R. Acad. Sci. Paris*, 1890, Vol. 110, p. 267-270.

[75] Stolarsky K. B.
1. *Algebraic Numbers and Diophantine Approximation*, M. Dekker, 1974.

[76] Stridsberg E.
1. Sur quelques propriétés arithmétiques de certaines fonctions transcendantes, *Acta Math.*, 1910, Vol. 33, p. 233-292.

[77] Sylvester J. J.
1. Sur le rapport de la circonférence au diamètre, *C. R. Acad. Sci. Paris*, 1890, Vol. 111, p. 778-780.
2. Preuve que π ne peut pas être racine d'une équation algébrique à coefficients entiers, *C. R. Acad. Sci. Paris*, 1890, Vol. 111, p. 866-871.

[78] Thue A.
1. Bemerkungen über gewisse Näherungsbrüche algebraischer Zahlen, *Norske Vid. Selsk. Skr.*, 1908, No. 3, p. 1-34.
2. Über Annäherungswerte algebraischer Zahlen, *J. Reine Angew. Math.*, 1909, Vol. 135, p. 284-305.

[79] Väänänen K.
1. On a conjecture of Mahler concerning the algebraic independence of the values of some E-functions, *Ann. Acad. Sci. Fenn.*, Ser. A, 1972, Vol. 512, p. 3-46.
2. Estimates of the conjugate transcendence measure for the values of some E-functions, *Ann. Acad. Sci. Fenn.*, Ser. A, 1973, Vol. 536, p. 3-12.
3. On the transcendence and algebraic independence of the values of certain E-functions, *Ann. Acad. Sci. Fenn.*, Ser. A, 1973, Vol. 537, p. 3-15.
4. On the algebraic independence of the values of some E-functions related to Kummer's functions, *Ann. Acad. Sci. Fenn.*, Ser. A, 1975, Vol. 1, p. 183-194.
5. On the algebraic independence of the values of some E-functions, *Ann. Acad. Sci. Fenn.*, Ser. A, 1975, Vol. 1, p. 93-109.
6. On linear forms of the values of one class of E-functions, *Acta Univ. Oulu.*, Ser. A, 1976, Vol. 12, p. 5-19.
7. On lower estimates for linear forms involving certain transcendental numbers, *Bull. Austral. Math. Soc.*, 1976, Vol. 14, p. 161-179.
8. On lower bounds for polynomials in the values of E-functions, *Manuscripta Math.*, 1977, Vol. 21, p. 173-180.
9. On the simultaneous approximation of certain numbers, *J. Reine Angew. Math.*, 1977, Vol. 296, p. 205-211.
10. On linear forms of values of one class of E-functions, *Acta Univ. Oulu.*, Ser. A, 1979, Vol. 41, p. 403-404.
11. On linear forms of a certain class of G-functions and p-adic G-functions, *Acta Arith.*, 1980, Vol. 36, p. 273-295.

[80] Vahlen K. T.
1. Beweis des Lindemannschen Satzes über die Exponentialfunction, *Math. Ann.*, 1900, Vol. 53, p. 457-460.

[81] Valiron G.
1. *Fonctions Analytiques*, Presses Universitaires de France, 1954.

[82] Veblen O.
1. The transcendence of π and e, *Amer. Math. Monthly*, 1904, Vol. 11, p. 219-223.

[83] Van der Waerden B. L.
1. *Algebra*, Ungar, 1970.
[84] Waldschmidt M.
1. *Nombres Transcendants*, Springer-Verlag, 1974.
2. Indépendance algébrique par la méthode de G. V. Chudnovskij, *Sém. Delange-Pisot-Poitou*, 1974-1975.
[85] Wallisser R.
1. Application de la méthode de Siegel-Shidlovskii à une fonction hipergéométrique, *Publ. Math. Univ. Pierre et Marie Curie*, 1979-1980, p. 1-11.
[86] Wantzel P. L.
1. Recherches sur les moyens de reconnaître si un problème de géométrie peut se resoudre avec la règle et compos, *J. Math. Pures Appl. (1)*, 1837, Vol. 2, p. 366-372.
[87] Weierstrass K.
1. Zu Lindemann's Abhandlung. Über die Ludolphsche Zahl, *Sitzungsber. Preuss. Akad. Wiss. Berlin*, 1885, p. 1067-1085.
[88] Weyl H.
1. *Algebraic Theory of Numbers*, Princeton Univ. Press, 1940.
[89] Wirsing E.
1. On approximation of algebraic numbers by algebraic numbers of bounded degree, *Proc. Sympos. Pure Math.*, 1971, Vol. 20, p. 275-286.
[90] Yakabe I.
1. Note on the arithmetic properties of Kummer's functions, *Mem. Fac. Sci. Kyushu Univ.*, Ser. A, 1959, Vol. 13, No. 1, p. 49-52.

Supplementary Bibliography of Recent Publications

[91] Beukers F., Brownawell W. D., Heckman G.
1. Siegel normality, *Ann. of Math.*, 1988, Vol. 127, p. 279-308.
[92] Brownawell W. D.
1. The effectivity of certain measures of algebraic independence, in Séminaire de Théorie des Nombres, Paris 1984-85, Catherine Goldstein, ed., Birkhäuser; 1986, p. 41-50.
[93] Chirskii V. G.
1. Estimate for polynomials in the values of elliptic integrals, *Diophantine Approximation, Part I*, 1985, Izd. Moskov. Univ., p. 106-109.
[94] Chudnovsky D. V., Chudnovsky G. V.
1. Applications of Padé approximations to diophantine equalities in values of G-functions, *Springer Lecture Notes in Math.*, 1985, Vol. 1135, p. 9-51.
[95] Chudnovsky G. V.
1. On some applications of diophantine approximations, *Proc. Nat. Acad. Sci. USA. Math.*, 1984, Vol. 81, p. 1926-1930.
2. On applications of diophantine approximations, *Proc. Nat. Acad. Sci. USA. Math.*, 1984, Vol. 81, p. 7261-7265.

[96] Fel'dman N. I.

1. The constant in the bound for the transcendence measure of π, *Diophantine Approximation, Part I*, 1985, Izd. Moskov. Univ., p. 97-106.

[97] Galochkin A. I.

1. On simultaneous approximation of the values of certain entire functions, *Diophantine Approximation, Part I*, 1985, Izd. Moskov. Univ., p. 17-24.

[98] Gorelov V. A.

1. On a result of Siegel, *Diophantine Approximation, Part I*, 1985, Izd. Moskov. Univ., p. 25-36.
2. Effective bounds for the algebraic independence measure of the values of certain hypergeometric functions, *Diophantine Approximation, Part II*, 1986, Izd. Moskov. Univ., p. 50-63.

[99] Guangchan Xu

1. A note on linear forms in a class of E and G-functions, *J. Austral. Math. Soc.*, 1983, Vol. 35, p. 338-348.

[100] Ivankov P. L.

1. On approximation of the roots of certain transcendental equations by algebraic numbers, *Diophantine Approximation, Part I*, 1985, Izd. Moskov. Univ., p. 36-42.
2. Lower bounds for certain linear forms in the values of hypergeometric functions, *Diophantine Approximation, Part II*, 1986, Izd. Moskov. Univ., p. 34-41.

[101] Kazakov V. V.

1. On linear irreducibility of a type of linear differential equation, *Mat. Zametki*, 1987, Vol. 44, p. 48-56.

[102] Kulagin V. A.

1. On algebraic independence of the values of a subclass of E-functions, *Diophantine Approximation, Part II*, 1986, Izd. Moskov. Univ., p. 50-63.
2. On algebraic independence of the values of certain subclasses of E-functions, *Uspekhi Mat. Nauk*, 1987, Vol. 42, No. 2 (252), p. 233-234.

[103] Makarov Yu. N.

1. Effective bounds for the relative transcendence measure of the values of E-functions, *Diophantine Approximation, Part I*, 1985, Izd. Moskov. Univ., p. 47-56.

[104] Nesterenko Yu. V.

1. Effective bounds for the algebraic independence measure of the values of E-functions, *Vestnik Moskov. Univ.*, Ser. 1, 1988, No. 4, p. 85-88.

[105] Osgood C.

1. Product type bounds on the approximation of values of E and G functions, *Monatsh. Math.*, 1986, Vol. 102, p. 7-25.

[106] Popov A. Yu.

1. Approximation of certain powers of e, *Diophantine Approximation, Part I*, 1985, Izd. Moskov. Univ., p. 77-85.

[107] Salikhov V. Kh.

1. On algebraic independence of the values of a class of hypergeometric E-functions, *Diophantine Approximation, Part II*, 1986, Izd. Moskov. Univ., p. 95-114.
2. On linear irreducibility of differential equations and algebraic independence of the values of hypergeometric E-functions, *Uspekhi Mat. Nauk*, 1986, Vol. 41, No. 3 (249), p. 201-202.

3. Formal solutions of linear differential equations and their application in transcendental number theory, *Trudy Moskov. Mat. Obshch.*, 1988, Vol. 51.

[108] Vasilenko O. N.

1. On approximation of hypergeometric functions and their values, *Diophantine Approximation, Part I*, 1985, Izd. Moskov. Univ., p. 10-16.
2. On linear independence of the values of certain functions, *Diophantine Approximation, Part II*, 1986, Izd. Moskov. Univ., p. 3-16.

Bei Fragen zur Produktsicherheit wenden Sie sich bitte an:
If you have any questions regarding product safety,
please contact:

Walter de Gruyter GmbH
Genthiner Straße 13
10785 Berlin
productsafety@degruyterbrill.com